Emmanuel Gillain (Ed.)
Demystifying Artificial Intelligence

Also of Interest

Quantum Machine Learning
Siddhartha Bhattacharyya, Indrajit Pan, Ashish Mani, Sourav De, Elizabeth
Behrman, Susanta Chakraborti (Eds.), 2020
ISBN 978-3-11-067064-6, e-ISBN 978-3-11-067070-7

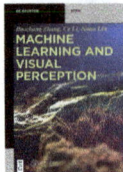

Machine Learning and Visual Perception
Baochang Zhang, Ce Li, Nana Li, 2020
ISBN 978-3-11-059553-6, e-ISBN 978-3-11-059556-7

Maschinelles Lernen
Ethem Alpaydin, 2022
ISBN 978-3-11-074014-1, e-ISBN 978-3-11-074019-6

Künstliche Intelligenz und menschliche Gesellschaft
László Kovács, 2023
ISBN 978-3-11-103449-2, e-ISBN 978-3-11-103470-6

Demystifying Artificial Intelligence

—

Symbolic, Data-Driven, Statistical and Ethical AI

Edited by
Emmanuel Gillain

DE GRUYTER

Editor
Emmanuel Gillain
Rue Sainte-Wivine, 4A
1315 Incourt
Belgium
emmanuel.gillain@outlook.be

Royalties from the sale of each physical copy are donated to a charitable cause. Please consider supporting this initiative by purchasing a physical copy.

The citation of registered names, trade names, trade marks, etc. in this work does not imply, even in the absence of a specific statement, that such names are exempt from laws and regulations protecting trade marks etc. and therefore free for general use.

ISBN 978-3-11-142567-2
e-ISBN (PDF) 978-3-11-142614-3
e-ISBN (EPUB) 978-3-11-142630-3
DOI https://doi.org/10.1515/9783111426143

Library of Congress Control Number: 2024934642

Bibliographic information published by the Deutsche Nationalbibliothek
The Deutsche Nationalbibliothek lists this publication in the Deutsche Nationalbibliografie; detailed bibliographic data are available on the Internet at http://dnb.dnb.de.

© 2024 the author(s), editing © 2024 Emmanuel Gillain, published by Walter de Gruyter GmbH, Berlin/Boston. The book is published open access at www.degruyter.com.
Cover image: ipopba / iStock / Getty Images Plus
Typesetting: VTeX UAB, Lithuania

www.degruyter.com

Acknowledgment

This book is a collective effort of volunteers who have the same passion for education as a public good: none of the contributors will receive any monetary benefits from this book. It took more time than I expected to finish this book. In March 2019, I began to gather a small group of professors and industry experts, persuading them to devote some of their spare time to impart some of their insights and expertise with artificial intelligence (AI). More than 50 people ended up contributing to my initiative, despite personal, family, or COVID related challenges.

First and foremost, I want to express my gratitude to my children, **Marjorie, Matthieu, and Quentin**, for their enduring patience throughout the countless hours that I dedicated to this book; my brother, **Luc**, and my parents, **Josette and Alain**, for their indefectible support during the difficult moments of my life. I am also grateful to my friends, whose encouragement helped me persevere through the myriad obstacles encountered on this 5-year journey.

None of this would have been possible without the active contribution of the professors, both as authors and advisors. I would like to thank **Professor Hendrik Blockeel, Professor Bart Bogaerts, Professor Walter Daelemans, Professor Yves Deville, Professor Isabelle Linden, Professor Erik Mannens, and Professor Aleksandra Pizurica**. Special thanks to

- **Professor Hendrik Blockeel**, not only for his authorship, but also for his academic review and advice all along;
- **Raphaele Moeremans**, our first reviewer;
- **Moritz Ritzl**, for his precious and tenacious help to go up to finishing lines;
- **Dr. Barak Chizi**, as an executive sponsor and reviewer; and
- **David Carmona**, as an executive sponsor opened the right doors when needed.

In addition, I'd like to thank the numerous authors from the industry, along with their current or former employers, for their valuable contributions that have helped illustrate the theoretical aspects in this book: **Agrimetrics, ArcelorMittal, ASML, EY, Icertis, Imandra, KBC Group, Maersk Container Industry A/S, Microsoft, ONTOFORCE, Pythagoria, Robovision, Veranneman Technical Textiles, and Viu More.**

Publishing an Open Access book requires a budget. I am therefore also extremely grateful to our financial sponsors: **KBC Group**, the **Flanders AI Research Program**, and the **University of Antwerp, Ghent, KU Leuven, UCLouvain, Vrije Universiteit Brussels.**

Lastly, to all those who have contributed, one way or another, big or small, to this journey, thank you all.

Contents

Yves Deville

Erik Mannens

Oussama Chelly and Hendrik Blockeel

List of chapter authors

Professor Yves Deville

Yves Deville is a Professor at the Louvain School of Engineering (UCLouvain, Belgium). He is also a member of the Institute for Information and Communication Technologies, Electronics, and Applied Mathematics. His research interests are in artificial intelligence, constraint programming, and optimization and constraint satisfaction. Since 2014, he has been an Advisor to the Rector for the Digital University and chaired the Information System Governance of UCLouvain.

Industry examples: Karen Veldeman, Sanjib Dutta, Dennis Conrad, Nikolaj Bjorner

Professor Bart Bogaerts

Bart Bogaerts obtained his PhD in Computer Science at the KU Leuven and UHasselt in June 2015. After PostDoc positions in Aalto University and KU Leuven, he joined the Artificial Intelligence Lab of the Vrije Universiteit Brussel, where he is now an Associate Professor. He has a strong focus on making combinatorial optimization algorithms trustworthy, on the one hand, by developing mechanisms for explaining their inferences in human-understandable terms, and on the other hand, by investigating how to get formal 100 % sure guarantees of correctness of their output.

Industry examples: Djordje Markovic, Grant Passmore, Pieter Van Hertum, Thomas Nagele

Professor Isabelle Linden

With a Master's degree in Mathematics and in Philosophy and a Doctorate in Computer Sciences, Isabelle Linden is a Professor in Information Management at the University of Namur (UNamur). Her main interest is on AI and knowledge models in management systems with a particular focus on modes of knowledge representation and reasoning, as well as the techniques to understand, formalize, and structure them. Isabelle Linden chairs the Research Group on Foundations of Computer Sciences (Focus) at UNamur, and from 2019 to 2023, she chaired the Euro Working Group on Decision Support Systems.

Industry examples: Erik Mannens, Filip Pattyn

Professor Aleksandra Pizurica

Aleksandra Pizurica is a full Professor in Statistical image modelling at Ghent University, Belgium, where she is leading the research group Artificial Intelligence and Sparse Modelling. Her research includes probabilistic graphical models and Bayesian inference, spatial context modelling with Markov Random Fields, sparse coding, deep learning, and image reconstruction, restoration, and analysis. Some of her most important results are in the domain of context-aware Bayesian inference.

Industry examples: Nicolas Vercheval, Matthew Smith, Pierre Stratonovitch, Richard Tiffin, Karishma Dixit, Emmanuel Gillain

Professor Hendrik Blockeel

Hendrik Blockeel is a full Professor of Computer Science at KU Leuven, Belgium. His research focuses on machine learning and data science and their role in artificial intelligence. He has made fundamental contributions to areas such as decision tree learning, inductive logic programming, clustering, experiment databases, and constrained machine learning. He is a Fellow of the European Association for Artificial Intelligence and is Editor-in-Chief of the Machine Learning journal.

Industry examples: Oussama Chelly, Nicolae Duta, Saheli Datta, Jonathan Kesteloot, Michaël Mariën, Abrao Aqueri, Emmanuel Gillain

Professor Walter Daelemans

Walter Daelemans is the Professor of Computational Linguistics at the University of Antwerp where he directs the CLiPS computational linguistics research group. His research interests are in applying machine learning techniques to natural language processing; computational psycholinguistics (modeling how people acquire and use language); computational stylometry, with a focus on authorship attribution and author profiling from text; and Language technology applications, for example, biomedical information extraction and conversational agents using large language models.

Industry examples: Anneleen Artois, Sunu Engineer, Pierre-Yves Thomas, Parag Agrawal, Achraf Chalabi

Professor Erik Mannens

Professor dr. ir. Erik Mannens is a Director at the imec UAntwerp IDLab and Professor at UAntwerp (Sustainable AI) and at Ghent University (Semantic Intelligence). He received his PhD degree in Computer Science Engineering (2011) at UGent, his Master's degree in Computer Science (1995) at K. U. Leuven University and his Master's degree in Electro-Mechanical Engineering (1992) at KAHO Ghent. He headed a 50+ researchers' team at UGent on Semantics and AI (2005–2022), and since 2022, he has been heading a 125+ researchers' team at UAntwerp on Wireless Communication and Sustainable AI. Being 50+, he tends to only take on new applied and fundamental projects who will have a positive and lasting impact on society and "Spaceship Earth" as a whole!

Industry examples: Joakim Åström, Yanyun Hu, Mario Schlener, Jason Tuo, Yara Elias

Emmanuel Gillain

1 Preface and introduction

Academic review by Professor Hendrik Blockeel

1.1 Fast progress in artificial intelligence (AI)

The artificial intelligence (AI) wave comes like a tsunami at a speed that our common senses have difficulties to fully grasp and the pace of progress is accelerating. The last decade has seen rapid progress in AI research, with advancements in machine learning, deep learning specifically, fast progress in computer vision and natural language processing, with an increased focus on ethical considerations.

Figure 1.1, from *"The 2023 AI Index Annual report,*[1]*"* illustrates the global growth in AI publications over the years. The period from 2010 to 2021 saw a more than twofold

Number of AI Publications in the World, 2010–21
Source: Center for Security and Emerging Technology, 2022 | Chart: 2023 AI Index Report

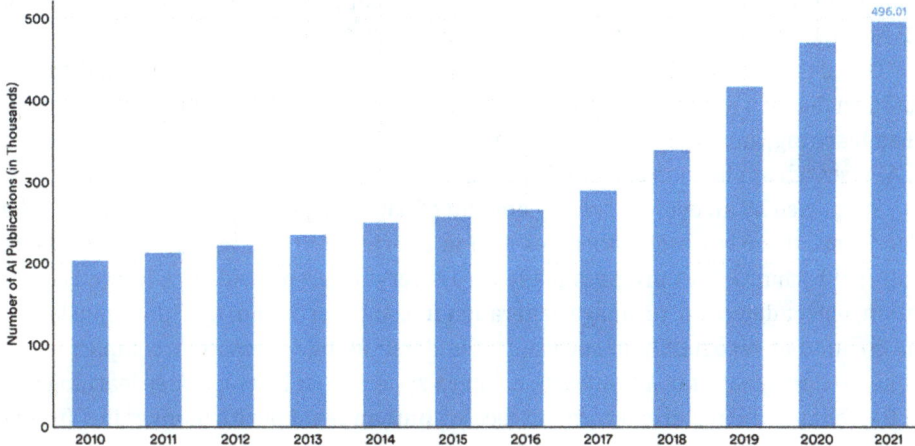

Figure 1.1: Number of AI publications worldwide, 2010–2021. Chart from public data, available at Charts – Google Drive, source: Center for Security and Emerging Technology, 2022 | Chart: 2023 AI Index Report.

---

1 Nestor Maslej, Loredana Fattorini, Erik Brynjolfsson, John Etchemendy, Katrina Ligett, Terah Lyons, James Manyika, Helen Ngo, Juan Carlos Niebles, Vanessa Parli, Yoav Shoham, Russell Wald, Jack Clark,

---

Disclaimer: Due to the rapidly evolving nature of AI developments and the 5-year book writing process, the information provided may not be the most up to date. Nevertheless, the knowledge acquired should empower the reader with concepts that facilitate an understanding of more recent advancements.

AI Patent Applications○ in 2017-2021

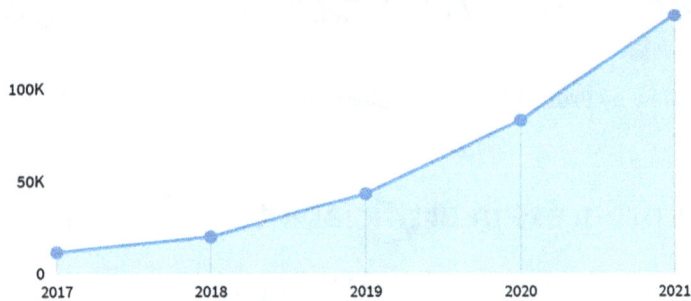

Figure 1.2: Number of AI patent applications from 2017 to 2021, according to the public Global AI Vibrancy tool.[2]

increase in the number of publications, with a noticeable acceleration over the past 4 to 5 years. *"The Global AI Vibrancy Tool,"* an interactive visualization tool featured on the same AI Index website, also reports a remarkable annual compounded growth rate of 235 % in the number of AI patent applications from 2017 to 2021 (Fig. 1.2).

Out of the different fields of AI, much of the attention over the last years has been focused on the field of **machine learning (ML)**, its subfield zrtificial neural network with **deep learning**, and **natural language processing (NLP)**, as illustrated by the evolution of AI research application areas in Figure 1.3.

Supported by an ever-increasing amount of data and processing power to train the algorithms, those research efforts resulted in a series of breakthroughs that can be illustrated by impressive progress made across very different fields like image classification, object detection, or image generation in computer vision, machine translation, the creation or recognition of speech, text understanding and writing, complex board games, etc. In a matter of a few years, complex systems that leverage deep learning and NLP techniques hit historic records, exceeding **human-level performance**[3] in different tasks:

– In the field of **computer vision**, for example, algorithms that could meet and sometimes exceed human-level performance on image classification tasks were already announced in December 2015 (see, e. g., ImageNet Computer Vision Challenge, De-

and Raymond Perrault, "The AI Index 2023 Annual Report," AI Index Steering Committee, Institute for Human-Centered AI, Stanford University, Stanford, CA, April 2023. The "AI Index Report 2023" is an independent initiative at the Stanford Institute for Human-Centered artificial intelligence. It is led by the AI Index Steering Committee, an interdisciplinary group of experts from across academia and industry.

2 https://aiindex.stanford.edu/vibrancy/

3 often called "human parity," which doesn't mean perfect.

AI application areas by country

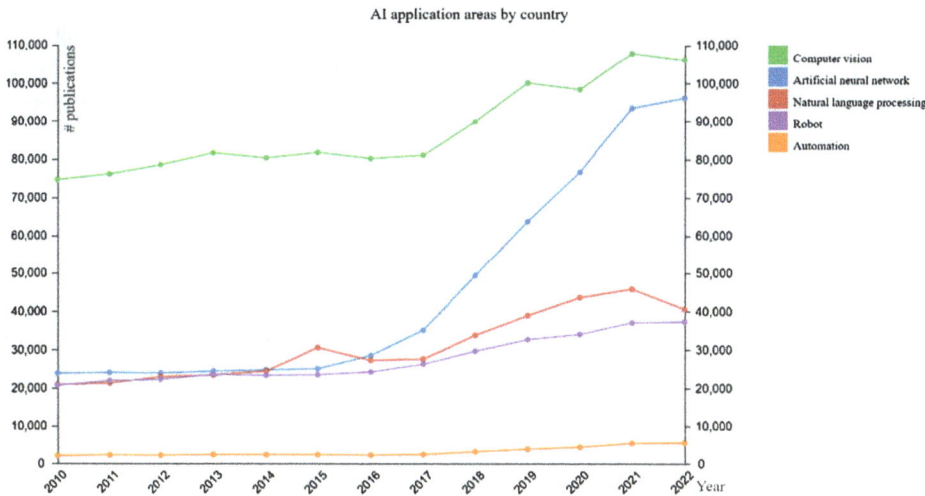

Source: OECD.AI (2023), visualisations powered by JSI using data from OpenAlex., accessed on 8/12/2023, www.oecd.ai

Figure 1.3: Source of data: OpenAlex. OECD.AI (2023), visualizations powered by JSI using data from OpenAlex, accessed on 8/12/2023, www.oecd.ai.

cember 2015[4]). Performance and accuracy consistently increased, surpassing 90 % by 2020–2021.[5]

– In the field of **speech recognition**, a team of researchers at Microsoft Corporation (further referred to as Microsoft) announced in October 2016 a speech recognition system achieving 5.9 % Word Error Rate claiming the same or fewer errors than professional transcribers, followed less than a year later by a new industry milestone reaching 5.1 % Word Error Rate.[6]

– In the field of **translation**, Microsoft AI & Research announced in 2018 a system that translates Chinese to English news, and performs on par with professional human translators

– In the field of **machine reading comprehension** of text,[7] the first models that scored higher than humans in the SQuAD[8] test were announced in January 2018 by Microsoft and Alibaba, at least for one of the two performance metrics. More recent models outperform human performance on both metrics.

---

4 Microsoft researchers win ImageNet computer vision challenge—The AI Blog.

5 top-1 accuracy, how well the algorithm can assign the correct label to an image.

6 a common metric defined as the sum of the number of substitutions, insertions, and omissions, divided by the number of words in the reference. The Microsoft 2017 Conversational Speech Recognition System.

7 useful to have search engines and virtual assistants to answer questions from text.

8 Stanford Question Answering Dataset, The Stanford Question Answering Dataset tests the ability of a system to answer reading comprehension questions.

- At the intersection of **computer vision and natural language processing**, Microsoft Research announced in October 2020[9] that it achieved human parity in Novel Object Captioning, which not only describe novel objects in an image but also the relation between those objects (describing an image with a caption such as "a man with glasses sitting on a chair").

As the field advances very quickly, more demanding benchmarks are required to measure progress. For instance, NLP has increasingly become **multitasks**, with the need to create new standards to assess their performance on a set of tasks. The **general language understanding evaluation (GLUE)** benchmark, established in 2018, evaluated models on nine sentence understanding tasks. Within only a year, models outpaced human benchmarks. Subsequently, **SuperGLUE**, a more challenging benchmark, was introduced in 2019. Once more, scores changed rapidly: by January 2021, new models from Microsoft and Google exceeded human baselines for that benchmark. To advance further research, another harder test, known as the **massive multitask language understanding (MMLU)** was presented between September 2020 and January 2021: the evaluation aims to measure the language model's multitask accuracy over 57 tasks, from mathematics and history to social sciences and law and problem-solving ability. The progress made against this new benchmark over the past 3 to 4 years has again been impressive as we can see for example with the evolution of the **generative pre-trained transformers (GPT)** family of models (Figure 1.4). This family of models has been released by **OpenAI**, an artificial intelligence research and deployment company that operates as a cappedprofit entity, with the mission to ensure that artificial general intelligence benefits all of humanity. Although GPT models are often cited as examples, it's important to note that there are numerous other large language models available in the market. More about GPT and the language models in Chapter 2 and Chapter 8.

Model	Number of Parameters	Release Month	MMLU Benchmark % (Average)
GPT-2	1.5 billion	Feb 2019	32.4
GPT-3	175 billion	May 2020	43.9
GPT-4	1.76 trillion (rumored)	Mar 2023	86.4

Figure 1.4: A simplified view of the progress made by the GPT family of models against the MMLU benchmark, for illustrative purposes only. More about GPT and the language models in Chapter 2 and Chapter 8.

These examples are provided solely to demonstrate the rapid pace of progress, but they no longer reflect the current state of the art. The interested reader will find an

---

9 Novel object captioning surpasses human performance on benchmarks – Microsoft Research.

updated list of more than 11,000 benchmarks for a very diverse set of more than 4,500 different AI related tasks supported by a broad range of AI techniques on the website "Browse the State-of-the-Art in Machine Learning," developed by **Papers with Code**, an open community project launched by a core team from Meta AI Research. The portal encompasses a wide range of AI techniques and benchmarks, spanning from computer vision, speech recognition, and wider natural language processing tasks to robotics. It also includes other AI fields such as diagnosis, theorem proving, knowledge bases, reasoning, etc.

Although the definition and threshold of **human parity** can be debated and benchmarks challenged, the rapid progress made by the AI techniques over the past few years is undeniable and the number of useful industrial applications has been exploding.

Not only have accuracy and models quality quickly improved over the last few years, but also the **speed to train** those has been drastically reduced, thanks to advancements in hardware, algorithms, parallel computing, and network architecture: as an example, the time to train convolutional neural networks (CNN), one of the deep learning techniques commonly used to classify images went down from 14 days in 2016 to 15 minutes at the end of 2017.

Note that the advancements in deep learning and NLP techniques, which have recently garnered significant attention, are emphasized here as illustrations, and shouldn't however make the reader forget about the importance and the complementarity of other AI disciplines, such as **planning, deductive reasoning, probabilistic reasoning, decision making**, etc. The diversity and richness of different AI techniques is an important aspect that will be highlighted and covered in this book.

1.2 Convergence of factors

Over the last years, a convergence of different factors mutually reinforced their effects, which have thereby accelerated the pace of innovation in artificial intelligence techniques. The widespread access to the internet with cheaper and simpler access to hyperscale computing resources in the cloud, an ever-increasing volume of accessible data, AI-specific innovations that ease the adoption by the industry, and the dynamics of the open-source software development model are some of the key ingredients that have together played a significant role in the rapid acceleration of the progress in AI, specifically in machine learning, deep learning and the natural language processing techniques, which heavily rely on deep learning.

1.2.1 The globalization and democratization of access to computing and AI resources

As of 2023, approximately 5 billion people, that is, 65 % of the world's population, probably more than 80 % of the developed world's population, have internet access. In today's digital age, it appears virtually unfeasible for businesses or organizations to operate without internet connectivity. The internet has enabled hundreds of millions of people, including over 25 million developers and a few million researchers, to tap into hyperscale computing resources. In a remarkably brief period, the ability to access and utilize cutting-edge technologies — once unimaginable just a few years ago — has become simultaneously more affordable, straightforward, and rapid. Let's take a couple of examples to illustrate those points.

Supported by **Moore's law** and hardware innovation, computing resources have become much cheaper to access. Storage capacity of a terabyte (10^12 bytes), which was worth $1,5b in the early 1980s, can now be rented for less than $10/month in the cloud. In the 1980s, only a few supercomputers had access to a GigaFLOPS,[10] estimated at about $45 billion; nowadays, one of AMD's[11] latest graphics processing units can deliver a GigaFLOPS for just $0.01...

Naturally, this has yielded and continues to yield advantages for the field of AI: the **cost to train** an image recognition system, for example, went down by around 150 times from 2017 to 2020. For data scientists, on-demand access to storage and computing resources in the cloud has not only become much cheaper but also much simpler and faster to get. Enabling extensive parallel processing and computing capabilities in the cloud, along with the latest data science tools and AI models, can now be accomplished within a matter of minutes. Cloud services also often include collections of machine learning templates contributed by the community. These templates can be then explored and reused by data scientists, streamlining the process of building machine learning solutions.

Furthermore, in an even simpler approach, software developers without any data science knowledge can benefit today from using advanced AI algorithms through simple application program interface **API**[12] **calls**: their software can initiate a request to another software module, which then executes an AI function and returns the result as a transaction. Thereby, access to powerful and advanced AI algorithms developed by the giants of the AI industry enable developers to develop rich AI based applications in

---

10 A GigaFLOPS is a billion numerical operations (e. g. addition, multiplication). It serves as a measure for the computing speed. See FLOPS – Wikipedia.

11 Advanced Micro Devices, Inc. (AMD) is an international American semiconductor company based in California.

12 Application programming interface, a type of software interface that allows applications to communicate with each other through a documented interface.

a matter of days. Big tech companies are routinely launching new APIs that developers can easily leverage in multiple categories of applications. For example, in:

- **speech services**, including transforming text-to-speech and speech-to-text, automated transcripts and translation services
- **image and vision services**, to detect and identify objects in pictures and videos, recognize handwriting, extract text from scanned documents or pictures
- **language services**, to automatically summarize content, extract intent predefined concepts, or answer questions from text
- **decision services**, to detect anomalies, suggest action to take, or learn from real-time user behavior and make recommendations
- and many more

AI based toolkits are even made accessible in the so-called *"low-code, no-code"* mode, an approach in software development that allows non-tech people to develop AI based applications, typically by connecting high-level, ready-made modules, without writing lower-level code.

1.2.2 Dematerialization, digitization, and exponential growth of data

Whether on the consumer side or on the business side, cheaper storage and processing power have paved the way to the explosion of digital content creation. The proliferation of social networks, easy ways to publish rich content online, connected objects sending data, physical tools and objects (such as books) replaced or mimicked by digital content or by their digital twins, etc. are all sources of exploitable data, which can then fuel the algorithms to improve, for example, the quality of predictions, classifications, or to grow some knowledge base when semantic meaning can be attached to the data. The **volume and diversity of publicly available data** on the internet (text, images, audio and video recordings) have boosted the take-up of deep learning and natural language processing techniques, as a free lunch for training their models. It also helps grow knowledge bases, such as for the semantic web, a vision of the World Wide Web, in which all the information may be linked to each other by the semantic meaning of the data (i. e., by their metadata[13]). This is exploited, for example, by search engines to provide answers based on the meaning of a text, rather than on the strings of texts.

According to the document "The Digitization of the World from Edge to Core," by IDC, a market intelligence and consultancy firm, the amount of data that was generated, stored, or replicated in 2018 was around 33 ZettaBytes (ZB). According to the same company, that amount almost doubled to reach 64 ZB in 2020, for about 6.7 ZB stored (IDC

---

13 data about data, structured data describing what the data is about.

report, Worldwide Global DataSphere Forecast, 2021–2025). 6.7 ZB is like giving a library of approximately 300 to 350 Netflix High-Definition quality movies on average to each internet user on the planet.

1.2.3 AI accelerators: AI techniques that speed up the adoption by the industry

Hardware and software innovation,[14] the availability of ever-increasing computing power and ever-increasing amount of data have supported the fundamental research in AI. Some specific AI techniques also contribute to the acceleration of the overall uptake by the industry. Without being exhaustive, here are a few examples.

Developing performant traditional machine learning systems require a pipeline of tasks with multiple choices and fine-tuning decisions to achieve the most optimal result (the predictors, the model itself with configuration and hyperparameters, etc. See Chapter 7). The search space to find the optimal parameters is sometimes complex with multiple dimensions, so that mathematicians would classify the data scientist's tasks as "high-dimensional combinatorial optimization" tasks. **automated machine learning (AutoML)** is an idea that emerged in the 1990s and whose objective is essentially to automate the generation and selection of the most performing algorithms and optimize their performance without the help of data scientists: a data scientist applying AutoML techniques would typically only require a couple of lines of codes to test multiple models with multiple hyperparameters in parallel, and let the algorithm select the best model under some defined quality metrics. These methods have progressively been making their way into standard commercial products (2018–2019) as a productivity tool that helps data scientists work faster and better. AutoML techniques clearly speed up the model development cycle, often with more performant models.

Self-supervised learning (see Chapter 7 and Chapter 8) is another type of accelerator that fastens AI application development. It is a form of learning that doesn't require a manually labeled dataset but generates the labels by itself as a pre-task, to train the algorithm with only a few labels as a downstream task. Self-supervised learning methods speed up the creation of models because they do not require human involvement for labor-intensive manual-labeling tasks (as in traditional **supervised learning**, for example).

A third accelerator that emerged over time is based on the concept of **transfer learning** (see Chapter 7 and Chapter 8): some of those models can be pretrained on extremely large datasets, so that developers can get accurate results even faster: instead of

---

14 AI algorithms are compute-intensive: progress in specific hardware chips specifically designed as "AI accelerators" (such as GPU, Google's TPU, ASIC, FPGA) and innovation in software development to distribute the tasks and the processing of big data with parallel executions, both with the ultimate goal to speed up computations.

starting the learning process from scratch, developers can start from patterns that have been learned from solving a related task (or even multiple tasks in recent developments) and customize the "last miles" of the learning to their specific tasks. **Pretrained NLP models**, for example, can be fine-tuned with just one additional learning layer to create models for a wide range of tasks, such as question answering. With the advancement of NLP techniques in recent years, companies can even now use state-of-the-art pretrained models that can handle different NLP tasks without requiring any fine-tuning to a specific task or any large datasets.

Finally, machine learning requires both a lot of data and data scientist expertise. So, another emerging AI field that can dramatically speed up model-building time is known as **machine teaching**. Machine teaching focuses on the efficacy of teachers, domain experts, to guide a learning algorithm. It seeks to leverage knowledge from domain experts in a more efficient way than using labeled training data alone. By designing the optimal training dataset to drive the learning algorithm to a target goal, subject matter expertise is used to help machine learning models find important hints about how to find a solution faster.

1.2.4 The dynamics of software, amplified by open-source: reuse, iterative improvement and innovation

The AI algorithms are developed in software written by people, sometimes by machines when AI techniques assist the developers to write code (especially generative AI techniques).[15] Software is easily copied and shared across networks and a digital medium, so fast reuse and iterative improvements are inherent characteristics of software. The reuse and iterative improvement of software have been further accelerated by the **open-source software** movement. Open-source authors make their source code available to users that, under open-source specific licensing terms, you can view it, copy it, learn from it, alter it, distribute it, thereby favoring sharing and an open collaboration: for example, the code used to create Linux, probably the most widely known open-source software, is free and available to anyone to view, edit, and contribute to it. Originating in the mid-1980s as a free software movement, open-source was later formally established by the Open-Source Initiative in 1998 and then also considered as a method. Open-source has become a component of corporate innovation strategy. Even if the software was created, or modified, and may or may not be made available for free,[16] such trend has accelerated improvement and innovation through open, shared, collaborative development

---

15 AI techniques can also assist software developers to write code faster, understand the comment and prepopulate software code.

16 "the term "free" in free open-source software refers to freedom, not monetary cost. Though most free open-source software is indeed free in price, the term "free" is referring to the freedom to use the software and source code as you please, as long as you attribute copyright to the person (or group) that

between communities of developers across the world. Software development isn't limited to company boundaries anymore, as it benefits from the power of collective work at scale.

The open-source movement as a shared, open, collaborative software development approach has favored virtuous circles of improvement and boosted algorithmic innovation. *"In an international software industry that is characterized by very short innovation cycles, open source projects have proven to be important incubators for new product lines and branch-defining infrastructures"* (Open-Source Projects as Incubators of Innovation: from Niche Phenomenon to Integral Part of the Software Industry, Jan-Felix Schrape, 2017).

The field of AI has made no exception to this. The open-source community has been a prolific source of innovation and progress in artificial intelligence. Whether originally developed by the open-source community (like the Apache Software foundation[17]), released by universities to the community or released by companies to the community, the source codes of many well-known AI software used in applications today have been made available for free in web-based repositories, such as GitHub,[18] commonly used to host open-source software projects supported by a community of more than 100 million developers. All types of AI techniques covered in this book have somewhere an open-source version. All at their disposal! Some industry examples in this book also leverage open-source codes and modules.

Beyond reusable libraries, databases of pretrained models also exist as open-source, allowing to easily discover, experiment, and contribute to newer, state-of-the-art models. These software libraries and pretrained models that are "ready to use" help speed up the progress of AI by reducing the time, the resources, and sometimes the expertise needed to create models from the ground up.

1.3 Opportunities and challenges for the industry

AI techniques are already widespread. They already power applications and use cases in the consumer space, in the enterprise market, and in the public sector around the world. AI technologies are already embedded in our daily life, in one or another form. When using a navigation system to find the fastest path to reach a destination, activating

---

created the software and the software stays free and open source when it is distributed to others." The value of free open-source software and collaborative communities | Opensource.com.

17 The Apache Software foundation is an all-volunteer community, established in 1999 to provide software for the public good.

18 GitHub is a web-based hosting and version control service for software development, based on the free open-source Git software, a distributed version control version for tracking changes in software development.

a spam filter, using a voice-enabled personal assistant, selecting music on Spotify or a movie on Netflix, searching for content on the web, getting personalized ads, playing a game against a machine, interacting online through chatbots, and using automated translation services, you are interacting with an application that uses some of the AI techniques.

While the technology companies, the telecom industry, and the financial services industry seem to lead AI adoption, AI applications already touch the business across all the industries and functions: whether to boost companies' productivity through the automation of their operations, to improve their decision making, to anticipate future outcomes with predictions, to better manage the risks and the cyber security threats, to get better customer's insights for better products or customer's experiences, or to enable new products and services. To name a very few examples only:

- **telecom companies** improve their customer support operation, reduce customers' churn, improve the reliability and security of their network with AI techniques
- **retailers** rely on AI-powered robots to run their warehouses, automatically reorder stock, optimize their supply chain and inventories, or improve customers' retention using AI
- **utility companie**s use AI to forecast energy consumption requirements, better balance the grid, or make more efficient use of renewable energy technologies
- **manufacturers** improve yield, throughput, or apply predictive maintenance to avoid machine and process downtimes or to extend the lifetime of their assets beyond the planned maintenance schedules
- fraud detection, risk analysis, and management are typical use cases in the **financial industry**
- lawyers and consultants in **professional services** companies are assisted by applications that automatically extract meaningful insights from documents, help summarize text or answer questions in natural language
- **news publishers** use AI to write simple stories, like financial summaries and sports recaps, without any human intervention
- AI powered computers assist **doctors** in their diagnosis, synthesize medical knowledge at scale, or scan and analyze medical images better than humans can
- AI algorithms aid **scientists** in the process of discovering new efficient drugs, and faster
- . . .

It even impacts more traditional sectors, such as **agriculture**, where crops, for example, are forecasted. AI applications also support generic, nonsector specific, corporate functions:

- **sales representatives** can receive assistance in their forecasting efforts, prioritizing their sales leads, and determining the next steps based on suggested actions

- **Human Resources** departments can benefit from AI in their recruitment process through automated resume screening, or in identifying potential churn in their company
- **service and call centers** can benefit from online virtual assistants, or chatbots, to handle the first customers interactions and their agents can take advantage of AI to automatically summarize and resolve some customers queries
- AI techniques can be utilized by **IT departments** to automate a variety of IT processes
- **knowledge workers** can benefit from Autonomous and Robotic Process Automation (RPA), a form of business process automation, that brings AI into the automation of routine tasks across different company processes
- last but not least, AI techniques and automation are also widely used to detect and respond to the increasing number of **cyber security** threats

This short enumeration is far from exhaustive. **Twenty-two case studies** in this book illustrate how AI techniques support applications in different industry sectors, from the banking and insurance sector to manufacturing and agriculture.

> — Before continuing with some statistics about the take-up of AI in the corporate world, we need to warn the reader: AI isn't covered only by one family of techniques, it is rather a broad set of different categories of techniques depending on the tasks to address. In recent years, the fields of machine learning, deep learning, and natural language processing have garnered most attention and investment. So, the term "AI" is often misused by overgeneralizing these specific AI fields to the much broader domain of AI. —

Those applications reflect how important AI techniques have become for industry. The results of McKinsey Global Survey on artificial intelligence released in November 2020 revealed that about 50 % of the 2,395 participants[19] confirmed their organizations had adopted AI[20] in at least one function ("The state of AI in 2020," McKinsey), and although a big part still struggle to scale it across their business, nearly 85 % of about 1,500 C-suite executives interviewed by Accenture, a multinational professional services company specialized in consulting and IT services, believe that they must leverage AI to achieve their growth objectives ("AI built to scale," Accenture, November 2019). For those companies that already adopted AI technologies, Deloitte, a multinational professional services network and accounting organization, concludes in a survey of more than 2,700 IT and business executives that the competitive differentiation brought by AI might even already soon diminish: *"Early-mover advantage may fade soon. As adoption becomes ubiquitous, AI-powered organizations may have to work harder to maintain an edge over their industry peers. An indicator of a leveling of the playing field: Most adopters expect that*

---

19 a random sampling across a broad range of regions, industries, company size.
20 with the same precaution about the meaning of the term "AI".

AI will soon be integrated into more and more widely available applications." (Thriving in the era of pervasive AI, Deloitte's State of AI in the Enterprise, 3rd Edition). This being said, business investments in AI, machine learning and NLP specifically, keep on increasing and fueling new innovation. According to Statista, a provider of market and consumer data, the total corporate investment in AI has grown over six times since 2016 to reach almost $92 billion in 2022.[21]

AI techniques also help to address **societal challenges**. It helps with medical diagnosis, aiding doctors with examining medical images or with recommending treatments, or in speeding up the drug creation process. AI techniques can also contribute to a more sustainable future by tackling climate change challenges: in climate predictions, in accelerating the discovery of new materials, in energy production and smart grids, in the industry and transportation system, and many more. The collaborative research document "Tackling Climate Change with Machine Learning," written by prominent scientists and engineers from 16 different organizations and universities, presents 13 solution domains where machine learning could bring a positive contribution to the climate challenges.

Therefore, artificial intelligence is increasingly essential for companies to survive, and the effects of those technologies are transformative for both the economy and the society.

At the same time, such rapid and widespread adoption of AI based applications also poses **challenges for the labor market** and brings **ethical risks** for the safety of consumers and the rights of citizens.

The impact of technology on jobs isn't new and has been constant, especially since the industrial revolution, going from hand production methods to machines. What is, however, unique is the accelerated speed of change, fueled by the convergence of multiple factors. Probably less visible than the automation of the manual work replaced by physical robots, the automation of knowledge work by AI systems in information intensive processes will also impact skilled and educated people. At the same time, technology and innovation are essential to **improving gross domestic product (GDP) and productivity growth**, which our economy is looking for. Research released by PricewaterhouseCoopers[22] in 2017 ("Sizing the prize: What's the real value of AI for your business and how can you capitalize") suggested that AI could boost global GDP by as much as 14 % in 2030 with an uplift initially coming from improved productivity factors, overtaken over time by consumption-side effects. Instead of just eliminating jobs, economic growth and wealth will also generate a need for new skills, new professions, supplementary and enhanced jobs. In its document *"The Future of Jobs 2018,"* the World Economic Forum even indicated a positive impact of AI technologies in favor of net jobs creation

---

21 Total global AI investment 2015–2022 | Statista.

22 also known as PwC. A global network of firms that provides professional services such as audit, tax, and consulting.

and reports *"Our analysis finds that increased demand for new roles will offset the decreasing demand for others. However, these net gains are not a foregone conclusion. They entail difficult transitions for millions of workers and the need for proactive investment in developing a new surge of agile learners and skilled talent globally."* While AI will cause **considerable upheaval in the labor market**, the more recent World Economic Forum report released in April 2023 still predicts a net positive overall. While forecasts remain uncertain, one can be confident to state that AI will increase productivity and economic growth while some people will need to switch jobs or upgrade skills, a **transition** that must be accompanied.

AI also presents **ethical risks**. Besides the intentional harmful uses that could impact people's privacy or safety, or that could influence public opinion by spreading false information on social media, AI systems also have their inherent limitations. As AI applications increasingly help to make decisions that affect people, ethical issues arise from those imperfections: for example, how someone is hired, whether someone receives a bank loan, or how people are treated. AI systems must be aligned with **fairness, transparency**, justice goals to avoid the negative outcomes of decisions taken or suggested by AI systems. Consensus emerges that AI systems should respect

- **privacy**,
- be **understandable** by humans,
- **fair and inclusive** to treat all people and groups the same way,
- **accountable** to provide people that have been harmed with a remedy,
- perform **reliably**,
- and be **safe** for humans.

The European Commission released in April 2021 a proposal for a regulatory framework, called the **"Artificial Intelligence Act,"** to ensure that AI systems are used in ways that respect fundamental rights and European values including human oversight, safety, privacy, transparency, nondiscrimination, and social and environmental well-being. At this time of writing, a draft text of the legislation serves as the negotiating position for talks between the member states and the European Parliament (see EU AI Act: first regulation on artificial intelligence).

In conclusion, those technologies present both fantastic opportunities and major socioeconomic and societal risks that require strong attention and action from leaders and governments to ensure both that the wealth created doesn't worsen a socioeconomic divide and that AI based systems remain human-centered and designed in a way that guarantees the ethical values that we want for our society.

In a world that's changing so quickly with many opportunities and risks, **education** is key to helping people and society to adapt: enlighten the leaders so that they can take informed decisions, educate law makers so that they devise the appropriate legal framework, assist individuals in adapting to changes, capitalizing on opportunities, and participating in democratic dialogues necessary for fair and ethical AI. It's widely recognized that education and democracy are closely linked, with studies indicating that nations with higher levels of education tend to uphold democracy more effectively (*Why does democracy need education? National Bureau of Economic Research*," https://www.nber.org/papers/w12128.pdf).

Unfortunately, skills and knowledge in AI are a shortage. O'Reilly, a global company that provides learning resources, found out in a survey of 3,570 business leaders that the first and most significant barrier to AI adoption is the **lack of skilled people** and the difficulty of hiring knowledgeable people in that field; a scarcity that has been reported for several years (O'Reilly, "AI Adoption in the Enterprise 2021").

1.4 Purpose and target audience of the book: enlighten the business practitioners

For about 70 years, a lot has been said and written about artificial intelligence: from academic books and research papers to business literature and computer science popularization. At the same time, the authors of this book also think that there is room for more efforts to connect the academic world and the corporate world, which still suffers from a lack of understanding AI: readings for business and political leaders focus on the economic benefits or social impacts but tend to oversimplify the concepts, sometimes to a point that leads to misconceptions and wrong conclusions; education materials for technical people in the field are often application or vendor-specific but don't necessarily cover the fundamentals. Training for software developers emphasizes practical hands-on methods but often neglects to explain the underlying principles of the AI tools they use.

The authors of this book care about clarifying some of the basic AI concepts and how or when they can be applied: the impressive results of certain AI applications shouldn't mislead leaders, technicians, or developers about the scope and limitations of the techniques. Business and political leaders could benefit from understanding the core issues at stake beyond the mere short-term economic impact: a better understanding of key notions should help them to not only better understand the opportunities but also to gain more strategic insights and mitigate the risks. Moreover, if the outcomes are incorrect, or if the advantages come with a loss of a bare comprehension of how they are achieved, it seems worthwhile revisiting some essential concepts. Developers that leverage AI more and more in their applications also need to understand some of those fundamentals and limitations behind the APIs that they use.

Acquiring a correct, even high-level understanding of those AI techniques can however represent a significant effort and time investment. Reducing that effort for the reader is the challenge that the authors of this book have decided to address. In an attempt to democratize the understanding of different AI fields and techniques, this book aims at giving a rather holistic, but nonexhaustive, view of AI while finding a digestible

middle ground between the academic theory and the oversimplified explanations that we may find outside of the academic world. Built as a **collaborative effort** involving a mix of **professors** teaching AI, researchers and **business practitioners**, this book demystifies what the core pillars of AI are made of by explaining their fundamental concepts and the core principles behind the mathematics and the algorithms that support those. Its originality lies in three key aspects:
– a balance between the complexity of the academic theory and an oversimplification,
– its approach to explain the concepts by answering common sense questions,
– the illustration of the theory by 22 real world industry examples.

With a didactic purpose in mind, the authors have strived to explain those technologies in simple terms, limiting when possible some mathematical and algorithmic developments,[23] yet preserving the scientific rigor to have a solid understanding of the fundamental concepts.

Readable **as a whole or by chapters**, this book is intended for **business practitioners**, leaders, or developers, who have a Bachelor or Master's degree outside of the field of computer science or AI but still want to understand the fundamental concepts of AI, their applications and limitations, in a relatively limited number of pages. Such reading can also be useful as a general introduction for **students** taking an MBA class, or similar.

The reader will find here a solid, yet digestible, overview of the different AI techniques supporting systems that **search and plan**, **reason with facts**, **with or without some uncertainties, learn and adapt, "understand" and interact**. All these terms are demystified in this book. It covers the two dominant and traditional paradigms in AI, which is also a coarse way to bring a holistic view of the domain:
1. the **statistical AI, or data-driven AI systems**, that learn and perform by ingesting millions of data points into learning algorithms, and
2. the "consciously modelled" AI systems, known as **symbolic AI** systems, that explicitly represent the world by means of symbols and are more deliberate in their actions.

Rather than opposing those two paradigms, the book also shows how those different fields can complement each other and can be combined for even richer applications.

Chapter 2 serves as an introduction to the rest of the book: it gives a first high-level overview of the different concepts, their applications, limitations, and complementarities. Chapters 3 to 8 form the core of the book as they address the key pillars that make up the backbone of most AI based applications. Those chapters are all structured in a pragmatic way that answers **common sense questions:**
1. *Why is this field of AI important within the broader AI domain?*
2. *What category of problems does this field solve?*

---

23 some sections are shaded for advanced readers that want to go deeper in their understanding.

3. *How are these problems solved?*
4. *What are the limitations of this field of AI?*

Each of the chapters also provides **concrete and real-world examples** coming from the industry. The theory supported by industry examples should give a clear understanding of the concepts, their applicability, limitations, and how they fit together. As it was briefly introduced, AI technologies also raise important ethical challenges. Chapter 9 explores some of the **ethical** dangers and techniques that can mitigate them. Chapter 10 shows examples of applications that combine different AI techniques to illustrate their complementarity.

Bibliography

AI built to scale, Accenture, November 2019, https://www.accenture.com/_acnmedia/Thought-Leadership-Assets/PDF-2/Accenture-Built-to-Scale-PDF-Report.pdf#zoom=50.

Apache Software foundation, http://www.apache.org/.

Browse the State-of-the-Art in Machine Learning, https://paperswithcode.com/sota.

Charts – Google Drive, https://drive.google.com/drive/folders/1sU7o-uPs2k5nBCNsU5sDZq0YmQyAOrG5.

EU AI Act: first regulation on artificial intelligence, https://www.europarl.europa.eu/news/en/headlines/society/20230601STO93804/eu-ai-act-first-regulation-on-artificial-intelligence.

FLOPS – Wikipedia, https://en.wikipedia.org/wiki/FLOPS#Hardware_costs.

IDC report, Worldwide Global DataSphere Forecast, 2021–2025, https://www.businesswire.com/news/home/20210324005175/en/Data-Creation-and-Replication-Will-Grow-at-a-Faster-Rate-Than-Installed-Storage-Capacity-According-to-the-IDC-Global-DataSphere-and-StorageSphere-Forecasts.

Microsoft researchers win ImageNet computer vision challenge—The AI Blog, https://blogs.microsoft.com/ai/microsoft-researchers-win-imagenet-computer-vision-challenge/.

Novel object captioning surpasses human performance on benchmarks – Microsoft Research, https://www.microsoft.com/en-us/research/blog/novel-object-captioning-surpasses-human-performance-on-benchmarks/.

O'Reilly, AI Adoption in the Enterprise 2021, https://www.oreilly.com/radar/ai-adoption-in-the-enterprise-2021/.

Schrape Jan-Felix, 2017, Open Source Projects as Incubators of Innovation: from Niche Phenomenon to Integral Part of the Software Industry, https://www.sowi.uni-stuttgart.de/dokumente/forschung/soi/soi_2017_3_Schrape.Open.Source.Projects.Incubators.Innovation.pdf.

Sizing the prize: What's the real value of AI for your business and how can you capitalize, https://www.pwc.com/gx/en/issues/analytics/assets/pwc-ai-analysis-sizing-the-prize-report.pdf.

Statista, https://www.statista.com/.

Tackling Climate Change with Machine Learning, https://arxiv.org/abs/1906.05433v1.

The Digitization of the World from Edge to Core, https://www.seagate.com/files/www-content/our-story/trends/files/idc-seagate-dataage-whitepaper.pdf.

The Future of Jobs, 2018, http://reports.weforum.org/future-of-jobs-2018/key-findings/.

The Microsoft 2017 Conversational Speech Recognition System, https://arxiv.org/abs/1708.06073.

The Stanford Question Answering Dataset, https://rajpurkar.github.io/SQuAD-explorer/.

The state of AI in 2020, McKinsey, https://www.mckinsey.com/business-functions/mckinsey-analytics/our-insights/global-survey-the-state-of-ai-in-2020.

The value of free open-source software and collaborative communities | Opensource.com, https:
//opensource.com/education/12/7/clearing-open-source-misconceptions.

Thriving in the era of pervasive AI, Deloitte's State of AI in the Enterprise, 3rd Edition, https://www2.
deloitte.com/us/en/insights/focus/cognitive-technologies/state-of-ai-and-intelligent-automation-
in-business-survey.html.

Total global AI investment 2015–2022 | Statista, https://www.statista.com/statistics/941137/ai-investment-
and-funding-worldwide/.

Why does democracy need education? National Bureau of Economic Research, https://www.nber.org/
papers/w12128.pdf.

OECD.AI, 2023, https://oecd.ai/en/data?selectedArea=ai-research&selectedVisualization=trends-in-ai-
application-areas-by-country.

Microsoft AI & Research, 2018, https://www.microsoft.com/en-us/research/uploads/prod/2018/03/final-
achieving-human.pdf.

Emmanuel Gillain

2 A holistic view of AI techniques, their limitations and complementarities

Academic review by Professor Hendrik Blockeel

The purpose of this chapter is to introduce the rest of the book by providing a simplified summary of the various AI concepts that are covered in this book, as well as their applications, limitations and synergies. It gives a general idea of some of the core concepts behind the AI methods that will be discussed more in the following chapters. As already noted in the introduction, this book is not meant to be exhaustive, nor to include the most recent and cutting-edge AI techniques, which evolve extremely fast since the beginning of this book writing process. The concepts covered in this book should however enable the reader to grasp some of the fundamentals of the newer techniques as well.

2.1 What is artificial intelligence (AI)?

AI is not one universal technology, it rather covers several categories of different techniques, that alone or combined together, add some form of intelligence to applications.

Many formal and informal definitions of **intelligence** have been proposed but there is no scientific consensus around any single definition. Rather than defining intelligence Alan Turing himself readdressed the question by a behavior that can fool human observers, the famous **Turing test** he proposed in 1950 (test of a machine's ability to exhibit a behavior that human benchmark can't distinguish). More recently, Legg and Hutter noted in a 2007 survey of intelligence definitions and evaluation methods: *"to the best of our knowledge, no general survey of tests and definitions has been published"* and the authors attempted to summarize no fewer than 70 definitions from the literature into a single statement: *"Intelligence measures an agent's ability to achieve goals in a wide range of environments.* (Shane Legg and Marcus Hutter, 2007).

The AI researcher François Chollet presents two different and contrasting views of intelligence in a research paper titled *"On the Measure of Intelligence"* (Chollet, 2019). The first one defines intelligence as a **collection of task-specific skills**, benchmarked by task specific performance and somehow similar to Minsky's[1] definition of AI (1968): *"AI is the science of making machines capable of performing tasks that would require*

---

1 Marvin Lee Minsky was an American cognitive and computer scientist, largely concerned with research of AI, cofounder of the Massachusetts Institute of Technology's AI laboratory, author of several AI books. He received multiple scientific awards, including the 1969 Turing Award, the greatest distinction in computer science.

https://doi.org/10.1515/9783111426143-002

intelligence if done by humans." The other vision considers intelligence as a **general learning ability**, an ability to acquire new skills through learning and handle situations (or tasks) that differ from previously encountered situations. Here again, generalization must be defined, something we'll cover later in this book. Chollet continues by proposing an interesting approach describing intelligence as **skill-acquisition efficiency**, with concepts that go beyond the scope of this book.

The major successes of AI have been so far in building special-purpose systems capable of handling well-described tasks, sometimes at above **human-level performance**: AI systems specialized in particular tasks in specific contexts, are classified as **narrow AI** (a term coined by Ray Kurzweil in 2005), as opposed to systems capable of broad generalization, adaptable to new tasks and environments without human intervention, broadly referred to as strong AI or **artificial general intelligence** (AGI), for which no single definition exists, despite a broad agreement on the general intuitive nature of AGI.

One approach to understanding general intelligence proposed is to look at the various **competencies and capabilities** that humans display: perception and actuation, memory, learning, reasoning, planning, attention, motivation, emotion, modeling, social interaction, communication, quantitative skills, building and creation skills (Ben Goertzel, "*Artificial General Intelligence: Concept, State of the Art, and Future Prospects,*" January 2014).

There seems to be a consensus that we're still far from reaching AGI. For example, in the survey titled *"When Will AI Exceed Human Performance? Evidence from AI Experts"* (Katja Grace et al., 2018), over 350 AI researchers across the world concur that at some point within the next 100 years, "*unaided machines will accomplish every task better and more cheaply than human workers,*" which again takes a task-oriented criteria. Some of them think that such an event may occur within 10 years, others a 100, with an aggregate forecast of 50 % chance that it happens around 2060.

At the same time, the current AI techniques of narrow AI systems will probably pave the way, in some shape or format, to reach such AGI. Therefore, a pragmatic way to look at AI is to consider the different categories of problems and tasks currently solved by the different techniques and the environment in which they operate (static or dynamic, deterministic, or stochastic). For our purpose, we'll therefore take a pragmatic approach to have a holistic, but not exhaustive, look at a series of AI techniques that can solve problems to assist human in decision-making, by

- *searching and planning* (*Chapter* 3),
- *reasoning with symbolic, logic-based, AI in deterministic settings* (*Chapter* 4 *and Chapter* 5)
- *reasoning with uncertainties in nondeterministic environments* (*Chapter* 6)
- *learning from data or from rewards in interacting with the environment* (*Chapter* 7)

Chapter 8 covers how natural language processing techniques help humans interact with the machines, that is, how human language can be either analyzed or generated by a machine. All those terms will be demystified by giving them a more precise meaning over the course of the book. Discussing some limitations of the different techniques will also give the reader a better sense of where the apparent magic stops and how the different techniques can potentially complement each other.

2.2 Decision by search and how an AI agent can construct a plan of actions

2.2.1 Overview and key concepts

The main purpose of **search** related techniques is to help take a decision to reach a specific goal; search agents built for that purpose are classified as **goal-based agents**. The decision must be understood here as: *"what are the series of actions to take in order to reach a particular goal state?"* The environment is modeled by so-called **states** and by legal **actions** to transition from one state to another one. In such a model where the set of states are discrete,[2] the environment can be represented as graph of a **state space**: nodes represent all possible states, edges representing a transition from a state to another, sometimes associated with specific weight or cost. For example, if we want to know how to reach a particular destination as quickly as possible, a position can be modeled as a state, the possible actions as "left, right, up, down," each of which action lands in another specific position. For this paragraph, we'll assume that the environment is **observable** (i. e., can be modeled), **known** (outcome of actions are known) and **deterministic** (transitions are certain). Some of those assumptions will be relaxed later.

The algorithms will then systematically explore the state space by progressively building a **search tree** on the state space graph. Once the goal state is found, the algorithm then simply follows the optimal path found. In cases where reaching an end-goal state with a minimum cost is all what matters, and not the path to reach it, we then talk about **local search** algorithms such as hill climbing or genetic algorithms, which can typically address **optimization problems**.

2.2.2 Industry applications and examples in the book

Applications of search-related techniques are typically found in **path-finding** related problems, navigation system like the GPS, to help robots or vehicles find their way while

---

2 Discrete as opposed to continuous. As a reminder, continuous variables have real number value and so aren't limited in the number of values they can take. Discrete variables have a finite domain.

avoiding obstacles or minimizing costs (time, distance, fuel, etc.), in road-building applications, or in other optimization and planning applications. Search-related techniques will be illustrated in Chapter 3 with **Microsoft Bing Maps**, which offers a set of routing tools to get directions and route information. Search techniques are also a **key pillar to support other AI techniques**: in propositional logic, for example, logic inference can leverage search techniques to conclude whether a logical statement is true in at least 1 model,[3] by systematically searching in a space of Boolean values that represent the knowledge of the environment, to search for finding a proof, or also as a way to solve more complex planning problems. See the section about SAT solvers in Chapter 3.

2.2.3 Moving to constraint search

Classical search techniques build a **planning** agent: a known environment is modeled, and the agent performs computations to anticipate possible moves until it finds the goal. So, the agent thinks ahead in a known environment that can be modeled, and whose transitions are **deterministic**. In its simplest form of modeling, search techniques solve problems by using atomic representations of the states: atomic in the sense that they are indivisible representations with no data structure associated to it. Such representations, however, are domain specific (including its heuristic to guide the algorithm to find the goal faster) and raises the challenges of a possible explosion of the number of possible states, which makes some problems too complex or even intractable.

Rather than using atomic representation for the states as in classical search, we can, and sometimes need to, have richer models, to represent the states with a set of variables, each with a set of **constraints** over those variables. Such richer representation, called a **factored representation**, will lead the reader to another section of Chapter 3 to solve **constraint satisfaction problems** (CSP). CSP techniques will associate a set of variables to each state, variables bound and linked by constraints: each state is defined by a possible assignment of values to variables (e. g., unary constraints to restrict the value of a single variable, binary constraints to restrict the combinations of 2 variables, etc.) and by explicit rules that define the legal combinations of the variable values. By identifying variable-value combinations that violate the constraints, large portions of the search space can then be eliminated, reducing the search space: such logic checks, based on constraints, are then combined with the classical search approach described in the previous section. CSP solving systems can be more efficient than pure state space searchers as they take advantage of the structure of states. In a stock management application, for example, a company could have the following constraints: a set of 10 possible products, available in 5 possible colors, with binary constraints that rule which color can be associated to which product, and ternary constraints that rule in which of their

---

3 A model in propositional logic fixes the truth value, true or false (Boolean value), to every symbol.

5 warehouses a specific product with a specific color can be found, like *"only blue or red (color) socks (product) are available in Brussels and London warehouses."* Constraint satisfaction propagation algorithms will then check the consistency of the variable values with the constraints, thereby reducing the number of possible values, and combine those constraint satisfaction checks with the classical search techniques (i. e., assignment of a value to the variables to search for the goal state) to find maybe the optimal way to deliver red socks from a warehouse to a retailer. Another example of constraints: if X can have the values $1, 2, 3$, and Y, the values $1, 2, 3, 4$, an additional binary constraint of $Y = X^2$ between X and Y would already limit the possible values of the search problem to the values $X = 1, 2$ and $Y = 1, 4$. As the reader can see, applying the constraints and propagating the consequences of those constraints checks reduces the overall search space to find a solution.

When the goal adds the objective to maximize or minimize an objective function, like a cost, it is referred as a **constraint optimization problem**. In cases where finding rapidly a good solution is more important than finding the best solution, then **local search** algorithms such as hill climbing or genetic algorithms, etc. can be applied. Chapter 3 will go through the details and illustrate some of those techniques.

2.2.4 Constraint satisfaction propagation: applications and examples in the book

Constraint satisfaction propagation algorithms can enable helpful applications, for example, in so-called **"configure, price, quote" (CPQ) systems** to help sellers quote complex and configurable products that follow production rules. **Maersk Container Industry A/S**, for example, has built a constraint-based product configurator to help its sales team generate valid customer and production-ready orders (Chapter 3). Besides the configuration of manufactured products, those techniques are also commonly used in **scheduling problems**, like job shop scheduling in manufacturing, where different jobs on the shop floor have to go through several manufacturing operations (drilling, painting, polishing, etc.) provided by a set of resources (machines, experts, tools, etc.).

2.2.5 Limitations of search-based techniques

As we have seen, search-based techniques rely on a simple representation of the world, but some planning problems might need a more complex way of representing the world to account for the consequences of the agent's actions. More elaborate environments might involve complex constraints with many states and variables that basic model-based representations can't handle. Factored representation with a set of values for attributes, such as CSP discussed above, improve the modeling but even more sophisticated relational representations of states and actions might be needed, using declarative languages[4] such as **first-order logic** briefly introduced later (Chapter 4), or even richer planning forms based on first-order logic.[5] They express **higher level concepts and relations** such as: "At(container, BrusselsAirport) ∧ At(Truck, Antwerpen)," could express that a container is at Brussels Airport, while a truck is in Antwerpen. The planning algorithms will then either search for a solution or prove the existence of a solution using **logic reasoning** techniques, typically using techniques described in Chapter 4. Robots, for example, also need task planning algorithms to sequence their actions toward accomplishing specific goals.

Besides dealing with the difficulties of a huge number of states and working at a higher level of abstraction, logic methods such as first-order logic, or variations, also have the benefit of having **general purpose solvers** that can use common solving methods to different particular domains. As we saw for the decision and planning problems above, richer expressiveness helps reduce the search space to reason or infer over more abstract concepts and reason in more general terms. That's a smooth transition to the next paragraph about **symbolic AI**.

2.3 Represent knowledge and reason with symbolic AI

Artificial intelligence has long been divided into two main paradigms: **symbolic AI**, or AI based on logic, and the **data driven**[6] AI, which is a coarse way to look at the AI field as a whole.

Despite some coexistence with research works and applications of neural networks (the first real-life application of neural networks dates from the early 1960s), **symbolic AI** has been the initial and most dominant paradigm of AI research for about 40 years. It defines **structured and symbolic representations** of the problem and its domain: domain experts model the world and the problem to solve with an explicit representation of a specific knowledge by means of facts and rules through symbols, symbols of different levels of expressiveness, according to the **knowledge representation** language chosen. Those symbols of facts and rules define a **syntax** to which domain analysts map some **semantics** to represent things, concepts, states, actions in the world.

---

4 Programming paradigm that expresses the "what," the logic of a computation (e. g., with Microsoft Excel® formulas), rather than the "how" like in imperative programming languages.

5 Planning Domain Definition Language is a popular language used in the planning community, and Answer Set Programming a general knowledge representation and reasoning language used in a variety of task planning problems.

6 Sometimes called statistical AI, the term data-driven AI is used here to ensure the reader doesn't confuse with probabilistic reasoning techniques, which not only applies statistics techniques and can learn from data but also associates some symbolic representation and models to solve problems.

Logic inference is then done by applying the rules and logical formulas to those symbols, at the syntax level. **Object-oriented programming** (OOP, such as C++, C#, Java), for example, can be seen as an example to create symbolic AI programs: OOP languages define classes, specify their properties, and organize them in hierarchies. Instances of these classes create objects, on which you can perform actions, rule-based instructions, queries that might read and change the properties of objects.

2.3.1 Overview and key concepts

One of the simplest forms of **knowledge representation** and of **logic reasoning** uses what is called **propositional logic**: knowledge composed of facts, axioms, and sentences, encoded by a combination of simple symbols and logical operators (such as and, or, not, equivalent, etc.), attached to Boolean values, true or false. The objective is then for logical systems to determine what logically works, reasons about the truth, or falsehood of logical expressions: either infer new logical sentences or determine whether a new logical sentence is valid or satisfiable[7] with the knowledge base. As explained in Chapter 3, the same product configuration problem of Maersk Container Industries A/S can also be solved by such techniques. More about propositional logic and this example in the section about Boolean satisfiability problems in Chapter 3.

Although useful to solve multiple problems, propositional logic can't represent the knowledge of complex worlds. It lacks the **expressive power** to represent an environment in a concise way: it only states the facts; it can't express relations among objects, and it is instance-specific. Therefore, more expressive knowledge representation languages, such as **first-order logic**, exist. First-order logic offers a powerful formalism to make more general statements about an environment: it assumes a world with objects, relations[8] among those objects, and functions on those objects, for example, the objects "employee," "skill," "task" with a binary relation "HasSkill(p,s)" to relate an employee and a skill, with a ternary relation "Assignment(e,s,t)" to assign an employee to a shift for a task and a function "Manager(e)" to map an employee to his/her manager. The facts or statements can be true, false, or unknown. First-order logic can also express concepts like "some" or general rules like "all" to express the properties of entire collections of objects, instead of enumerating them like in propositional logic. Such knowledge representation language facilitates a more concise and richer representation of the world. It not only allows queries to a knowledge base, but also allows to perform **automated logic reasoning** to infer new facts and prove theorems. See Chapter 4.

First-order logic can be **extended** to express some constraints with more flexibility, for example, to add numerical constraints such as "there should at least 2 people to be

---

7 A model in propositional logic fixes the truth value, true or false, to every symbol. Valid means true in all possible models; satisfiable means true in some models.

8 A set of t-uples that are related. Relationships might be unary, binary, ternary, n-ary.

married." A major limitation however of first-order logic though is that deductive inference is **undecidable**.[9] Some logical queries might not be answered in a finite number of steps. Other formal knowledge representation languages, such as **description logics** have been developed as decidable fragments or subsets of first-order logic. Those languages apply when the only objective is to enable the deduction of implicit knowledge by inference from a knowledge base. As explained in Chapter 5, many description logics are more expressive than propositional logic but less expressive than first-order logic. In contrast to the latter, the core reasoning problems for description logics are mostly decidable, and efficient decision procedures exist. Each family of description logic features a different balance between **expressive power and reasoning complexity**, defined by allowing or disallowing different logical operators in their language.

Often associated with first-order logic and description logics is the concept of **ontology**. An ontology is a formal specification, a convention to define an explicit specification of concepts, a declarative representation of what terms mean within a specific scope, with axioms and relations among terms, like binary predicates in first-order logic. Ontologies have strong ties with the semantic web initiatives, which illustrate quite well its practical applications. The **semantic web** is a vision of the world wide web, in which all information may be linked to each other. *"The term "semantic web" refers to W3C's vision of the web of linked data. Semantic web technologies enable people to create data stores on the web, build vocabularies, and write rules for handling data"* (Semantic web, W3.org). It is a vision of the internet being a single information model, instead of a vast meaningless collection of text embedded in web pages: text strings are related to meta-data, enabling the attachment of meaning to the strings; so rather than searching via a match on the text strings "nike," we can search on the associated meaning "shoes brand"="nike." This paves the way to the building of a semantic layer on the web. Semantic web technology is used by organizations in a wide variety of industries, essentially to harmonize and make the best possible use of their multiple and vast data stores. Ontologies not only bring a common framework to share and reuse data across different environments but also enable the application of logic reasoning techniques to answer complex questions about the domain. Indeed, when associated to description logics, for example, ontologies provide the semantics to enable systems to infer additional information based on the data explicitly provided. A simple example of how ontologies and logic can deduct new facts can be found on Wikipedia, at OWL: *"an ontology describing families might include axioms stating that a "hasMother" property is only present between two individuals when "hasParent" is also present, and individuals of class "HasTypeO_Blood" are never*

---

9 A logical system is decidable if there is an effective method for determining whether arbitrary formulas are theorems, i. e., logically valid formulas, of the logical system. Effective means here that it consists of a finite number of instructions, finishing after a finite number of steps, producing a correct answer. Propositional logic is decidable, because a truth-table method can be used to determine whether an arbitrary propositional formula is logically valid.

related via "hasParent" to members of the "HasTypeAB_Blood" class. If it is stated that the individual Harriet is related via "hasMother" to the individual Sue, and that Harriet is a member of the "HasTypeO_Blood" class, then it can be inferred that Sue is not a member of "HasTypeAB_Blood."

The use of ontologies and semantics also makes it possible to effectively separate the "what", i. e. the formal definition of the business problem and rules by business experts from the "how,", i. e. a technical implementation of the solution by the IT stack underneath the higher-level business logic. More about ontologies in Chapter 5.

2.3.2 Industry applications and examples in the book

Formally introduced in the mid-1960s, the so-called **expert systems** are a family of logical systems with an inference engine that deduces new facts from known facts of the knowledge base. They emerged in the 1980s as what some people saw as truly successful[10] application of symbolic AI and still have nowadays numerous applications. They are supportive and reliable decision-making systems used across industries: in medical diagnosis, in troubleshooting to infer possible malfunctions from facts, in finance to detect suspicious activities, frauds, or assist bankers before granting a loan, in multiple industries for configuring objects under constraints such as configuring devices in manufacturing, in planning tools or control systems,... Limited to knowledge bases that can be expressed as a set of specific Horn clauses[11] expert systems resolution methods use either forward chaining inferencing methods (generating a proof tree by following logical implications) or backward chaining inferencing methods (from the goal, chaining through rules to find known facts that support the proof).

Querying a database is actually another simple set of applications from logic reasoning: finding all instances that satisfy a certain logical formula represented by the user's query. Many successful database logics are actually based on first-order logic.

Another typical family of logic inferences focuses on **satisfiability checks**: is there an assignment of values for which a logical formula holds. Formal analysis, for example, can be performed to identify and match whether the production capabilities offered by some manufacturing assets can be compatible with the production tasks required by the production orders. Using formulas in first-order logic, **satisfiability solvers** not only

---

10 This may seem contradictory to the often-heard claim that the expert systems from the 1980s were a failure. This sentiment is mostly due to the fact that overly high expectations about expert systems "replacing experts" were not met: the term became associated with unrealistic expectations and was avoided as a result. The truth is that the research on expert systems has led to important scientific advances that are ubiquitously exploited today, though not under the denominator "expert systems" (Prof. Blockeel).

11 A Horn clause is a clause, that is, a disjunction of literals, with at most one positive, that is, unnegated, literal. A literal is an atomic formula. See Chapter 3 for details about literals and examples.

allow a richer modeling than propositional logic with their Boolean formula's (as with the example of Maersk Containers Industry A/S in Chapter 3) but also enrich the classical first-order logic with additional theories[12] of integers and numbers.

As mentioned above, logic-based inferences can also help with diagnostics in many complex industrial settings. Such industry example is covered in Chapter 4 with the example from **ASML**, a world's leading provider of lithography systems for the semi-conductor industry, where logic reasoning techniques are applied for the automated diagnostics of their lithography machines.

Automated theorem proving, another technique in the field of logic reasoning, provides higher level of inference and deduction. It automatically generates a proof, given a target theorem and a knowledge base of facts, all expressed in a formal language. It finds applications in mathematics to provide with mathematical proofs, to reason on the correctness of system properties, in integrated circuit design and verification or in software verification methods. Formal logic verification is especially a must for safety-critical systems to ensure they operate correctly and safely: systems whose correctness have a direct impact on the safety, such as autopilot systems in aircrafts, control systems in nuclear power plants, etc. Chapter 4 illustrates, for example, how theorem proving and **Imandra's** formal verification software can be leveraged to check the correctness of a simplified version of an autonomous controller found in drones and autopilot systems, such as the **Triton unmanned aircraft systems** of the US Navy. That example also high-lights another key advantage of automated theorem proving inference that can survey an infinite number of possible system behaviors through a finite computation.

As mentioned earlier, ontologies play a major role in the semantic web, where they are used to annotate web resources to perform semantic searches. Ontology-driven knowledge systems can also be found in various enterprise and corporate domains such as in life science, medicine, telecommunications, agriculture, astronomy, defense, resources, and energy management. In Chapter 5, **ONTOFORCE**, a company that helps organizations transform their data into insights in the fields of life science and health-care, provides with an example of how ontologies bring scattered data together to find relevant relations through linked semantical concepts and discover new knowledge in early-stage drug research.

2.3.3 Limitations of symbolic AI

From simple representations, like in propositional logic with knowledge encoded by simple symbols and rules defined by logic operators, to more complex representation and reasoning techniques, like first-order logic or other knowledge representation tech-niques, such as descriptive logics, symbolic AI approaches fundamentally enable hu-man engineers to bring **prior knowledge** into the applications: define facts or axioms,

---

12 a formal term.

rules, relationships, that explicitly represent a domain expert knowledge in an explicit declarative form. Those algorithms will then manipulate those symbols through different forms of inference, apply logic reasoning approaches to answer queries, make logical inferences and deductions. As mentioned earlier, richer expressiveness helps to **reason** over more abstract concepts and in more general terms. Even if sometimes tedious when there are a lot of complex rules, symbolic AI also generally offers the advantage of being explainable to humans. Symbolic AI models use symbols to represent and manipulate knowledge, which is similar to how humans think and reason.

Symbolic approaches work best on **well-defined structured problems**, wherein some structured knowledge is given to a system, which has to systematically follow search trees or apply logic rules to solve the problem: deductive inference (i. e., revealing new, implicit knowledge given a set of facts), consistency checking (i. e., detecting contradictions between facts), classification (i. e., generating taxonomies), reasoning-based decision support system, etc. Symbolic AI approaches however fall short when they are not directly programmed for a task, when the rules cannot be clearly defined, when obtaining knowledge is either too difficult or, when the size of the knowledge base increases to a level that the system faces computational difficulties. It therefore limits its applicability to microworlds and leads to the **commonsense knowledge** problem: "*If Peter is home, his head is also home*": human takes it for granted, computers don't. Coded knowledge must be explicit whereas humans infer meaning from sentences by using both explicit and implicit knowledge associated with concepts and relevant in a specific context. Also, areas that rely on procedural or implicit knowledge such as sensory or motor processes (can you define rules to drive a car?), are much more difficult to handle with the symbolic AI framework.

Some forms of logics such as first-order logic, classified as **monotonic**,[13] also bring the difficulty of **revising facts** once they are encoded, whereas not all the facts might be known at first: the more rules are added, the more knowledge is encoded in the system, but additional rules can't undo old knowledge: we can't add newer hypotheses, like exceptions, for example, that contradict the previous conclusions. Such logic doesn't leave room for default reasoning and unknown possible facts that would contradict the theory. Using a well-known example, if we state in monotonic logic that "*all birds fly*," we can't add penguin as a bird to our knowledge base later on: the addition of the premise would contradict the original conclusion, leaving the only alternative to list all possible exceptions, like penguin or ostrich are also birds. **Default reasoning**, a form of nonmonotonic logic, has been developed to formalize inference rules without explicitly mentioning all the exceptions. For example, some logic systems approach default reasoning by assuming that all positive information has been specified, and what isn't known to be true is false: they assume what is called a **closed world assumption** (CWA),

---

13 The monotonicity property requires that all derived conclusions remain valid after new facts are added to the knowledge base. A logic is nonmonotonic if some prior conclusions can be removed by adding more knowledge, if some conclusions can be invalidated by adding more knowledge.

like we typically see applied for databases. This considerably simplifies the representation since only positive information about the world need be explicitly represented in the database, negative information is inferred by default, so we don't have to list all the exceptions in the database. But still, this only works when databases are complete, when we have complete knowledge of the world. At the opposite, **open world assumption** (OWA), like used with ontologies and in the semantic web, assumes that what isn't explicitly specified is unknown, which allows to work with incompletely specified world and add knowledge as needed: statements about knowledge that are not included in or inferred from the knowledge explicitly recorded in the system may be considered unknown, rather than wrong or false. For example, if we have the fact: *"Piet is citizen of Belgium"* in our knowledge base and launch a query with the question *"is Piet citizen of Spain?"*, a CWA system will answer "no" but an OWA system will answer "unknown" since Piet can actually have dual citizenship.

Finally, although some **inductive learning** is possible with symbolic AI techniques (see **inductive logic programming** in Chapter 7 and the industry example in Chapter 10), they typically don't provide with mechanisms to learn and derive knowledge from nonmodeled data: symbolic logical systems don't make any association from raw data, they are inherently deficient to learn **correlations or associations** from data, and have no notion of proximity of concepts– all aspects that **statistical, data driven AI** and **machine learning** techniques address as we'll see next.

2.4 Learning with data-driven AI

2.4.1 Overview

Machine learning techniques are essentially inductive learning approaches that generalize from examples provided in the form of data. They infer statistical patterns that human might not see or know, to learn and extract information typically from very large data sets and raw signals. It's a broad field, typically known as the world of data science. The different families of techniques are typically classified based on the level of supervision[14] provided to the algorithm and the type of task to solve: **supervised learning, unsupervised learning, anomaly detection, reinforcement learning** to name the most important categories. Note that there is a wide range of settings where partial or indirect supervision is available, in semi-supervised learning settings. **Self-supervised learning** (where the labels are generated from the characteristics of the data itself removing the need for manual labeling), has also emerged as an approach in 2016–2017. **Automated feature engineering**,[15] also called **representation learning**, to replace the

---

14 supervision means that the ground truth value is given to the algorithm during the training phase.

15 Converting data from its raw form into numerical variables, called features, that are a meaningful representation for the problem to solve, a measurable property useful as an input for machine learning.

manual tasks of transforming raw data into feature vectors (i. e., some measurable property, numeric value that will then be used as representative elements of the data) is sometimes described as a specific category but essentially relies on supervised or unsupervised methods. More details are in Chapter 7.

2.4.2 Supervised learning key concepts, industry applications and examples in the book

Supervised learning techniques learn a function, a relationship, to map a given input to a given output based on provided examples as a guide for the algorithm. Training data, in the form of a set of input and correctly labeled output, is given to the algorithm so that it learns an optimal function that can then be used to predict the output associated with new inputs.

2.4.2.1 Classification

Predicting categories (i. e., categorical variables), like *"is it a cat or a dog?" "is that email a spam or not"* is called a **classification task.** Those classification algorithms will seek to optimize statistical metrics to reduce the error of the classification tasks, such as maximize the % images where actual cats are correctly classified as cats and minimize the % images wrongly classified as cats whereas they are dogs.[16] Classifications tasks serve a multitude of purposes, ranging from distinguishing between good and bad credit scores in credit rating systems, predicting market sentiment for product perception analysis or the benign or malignant nature of lesions in healthcare, to automatically classifying texts, documents, images, and objects in images. **Robovision**[17] and **Viu More,**[18] for example, developed a solution for **Veranneman Technical Textiles,** a European producer of woven and laid scrims (part of **Sioen Industries**), in order to automatically classify different classes of defects in the textile. See the industry example in Chapter 7.

2.4.2.2 Regression

Predicting a numerical variable is known as a **regression task.** Based on provided examples of input and output, regression algorithms will seek to minimize an error (defined from the difference between the predicted and the real value as a loss or a cost function) to fit the training data. Reducing that error will serve as a guide for the algo-

---

16 Known respectively as true positive rate and false positive rate.

17 Robovision is a company that started out as an AI consulting business and pivoted in 2021 to selling their AI platform, which helps create and manage vision AI models, as their core business.

18 Viu More specializes in building custom industrial solutions with image processing sensors and AI technology for various industries including recycling, technical textiles, food production, etc.

rithm to refine its parameters and fit the training data. For example, a linear regression model, one of the simplest forms of regression models, finds the line (or hyperplane for more complex linear combination) that most closely fits the data to minimize the sum of squared differences between the true given data and that line (or a so-called hyperplane when there are more variables). Of course, many other models have been created, ranging from the basic to the complex, to fit with the specific context of the problems. After training, the regression models will **generalize** and **predict** the output associated with new input, with some errors.

A widespread use case for supervised learning models is in creating predictive analytics systems to allow enterprises to anticipate certain results and forecast future opportunities and risks: optimize price points, predict power usage in an electrical distribution grid, predict the call volume in call centers for staffing, predict how many patients a hospital will need to serve in a time period, etc. For example, **KBC Group**, a Belgian integrated bank-insurance group[19] uses regression techniques to combine multiple variables (called **features**), such as the type of the building, the number of floors, or the surface in square meters to estimate property value and close insurance policies online (Chapter 7). Regression models can also typically be used to optimize manufacturing processes: an industrial company could create, for example, predictive models to assess the impact of the temperature in a nondeterministic process, to predict the remaining useful lifetime of industrial assets (**predictive maintenance**), to predict the quality of a particular output, etc. **ArcelorMittal**, the second largest steel producer in the world, for example, applies regression models to predict the iron quality at the output of a Direct Reduction[20] process. See the section industry examples in Chapter 7.

2.4.3 Unsupervised learning – key concepts, industry applications, and examples in the book

In **unsupervised learning** settings, data points have no labels and no ground truth value associated with them. They provide exploratory approaches to view data, group data, identify patterns in large volumes of data, describe its structure, or find relationships in data. They are typically classified in three main categories: **clustering, association**, and **dimensionality reduction**.

2.4.3.1 Clustering

Clustering techniques discover similarities and differences in data and are typically used for exploratory data analysis and customer segmentation: segment customers into

---

19 operating in Belgium, Bulgaria, Czech Republic, Hungary, and Slovakia.

20 A chemical process to transform iron ore (in the form of lumps, pellets, or fines) into iron.

groups by distinct characteristics, such as age and location, to better assign marketing campaigns, for example. A clustering technique is illustrated in Chapter 7 with **Microsoft Azure Form Recognizer**.[21] The latter uses, a. o., unsupervised learning to understand the layout and relationships between fields and entries in digital forms. The system clusters input forms by type, discovers what keys and tables are present in the digital forms, and associate values with keys and entries with tables. Another example is also provided in Chapter 7, where the **KBC Group** leverages unsupervised learning-based **anomaly detection** algorithms for detecting frauds in insurance claims.

2.4.3.2 Association rules

An **association rule** is a rule-based method for finding connections between variables in a given data set. These methods are often used for recommendation systems that do market basket analysis to understand how different products relate. Online retailers, for example, can use data from a previous purchase behavior to understand consumption habits of customers and develop efficient cross-selling strategies and make relevant add-on recommendations to shoppers. Or streaming services companies can make recommendations of movies based on users' history of interacting with movies; more about those techniques, and singular value decomposition, in particular, in Chapter 7.

2.4.3.3 Dimensionality reduction

Dimensionality reduction can be seen as an approach to summarize data: it's a technique used when the number of features or dimensions, in a given dataset is too high. Using data transformation techniques, it reduces the number of data inputs to a smaller number of dimensions by removing some of the statistical correlation between input variables, while preserving the integrity of the dataset as much as possible. It is commonly used as a preparatory step before applying other techniques. As part of its solution to classify defects (section 2.4.2.1), **Veranneman Technical Textiles**, for example, leverages a technique called **principal component analysis** (PCA) to extract the most relevant features in textile images data,[22] after which clustering techniques help them identify **outliers**,[23] considered as a defect and converted to a quality score for their fabric. See the industry example in Chapter 7.

---

21 The service now moved to a new service called **Azure AI Document Intelligence**, which added other learning solutions, so the customer can use whatever works best for their problem; see Chapter 7.

22 coming from the analysis performed by deep neural network techniques.

23 Data points or observations that significantly deviate from the majority of the data.

Autoencoders, a specific category of neural networks trained so that the output reproduces the input, also fall in this category: they try to learn a representation, or an encoding, as a sort of compression mechanism.

2.4.4 Anomaly detection – key concepts, industry applications, and examples in the book

Anomaly detection techniques are essentially methods to identify statistical **outliers**, rare events, or observations that significantly deviate from the majority of the data, with the remainder of that set of data. They indicate abnormal conditions in a specific task, which may cause a performance degradation or indicate a specific risk. The techniques for anomaly detection vary from problem to problem depending on the context and problem to solve. While **static rule-based** systems exist, identifying the rules can become a complex and subjective task. **Statistical** or **machine learning** based approaches to automatically learn the anomalies are then preferred to static rules. Some use supervised learning, trained as a classifier when access to normal and abnormal labels is balanced,[24] known and accessible. However, obtaining accurate and representative labels, especially for the anomaly class, is usually challenging. Techniques to learn from positive and unlabeled cases (a **semi-supervised** setting) assume that the training data has labeled instances but only for the normal class, hence more widely applicable than supervised learning techniques. When there is no prior knowledge of the data at all, **unsupervised learning** techniques can be used. They look for instances that seem to fit least to the majority of the dataset, by making the implicit assumption that normal instances are far more frequent than anomalies in the test data. If this assumption is not true, then such techniques suffer from high false alarm rate.

Detecting anomalies in **time-series data** (a series of data points listed in a chronological order, e. g., data samples of temperature over time) requires special care, as it often displays serial dependence. Serial dependence or autocorrelation,[25] occurs when the value of a datapoint at one point in time is statistically dependent on another datapoint in another time. However, this attribute violates one frequent assumption that data is statistically independent. On the other hand, autocorrelation is an ideal method for uncovering trends and patterns in time-series data that would have otherwise gone undiscovered. They require time-series and sequence modeling techniques such as the **hidden Markov model** and **recurrent neural networks**. See Chapter 6 and anomaly detection in Chapter 7.

---

24 When anomalous and normal classes are balanced.

25 Informally: intended to measure the relationship between a variable's present value and any past values.

Anomaly detection is applied across all industries. Financial institutions, for instance, utilize it as a mechanism for fraud and risk detection. In the manufacturing sector, it's employed to signal unexpected behaviors in production lines or machinery. The medical field leverages anomaly detection to identify medical irregularities. IT operations apply it to uncover unusual event patterns within their infrastructure. In cybersecurity, intrusion detection systems use it to pinpoint unexpected security behavior patterns. Even sales operations employ it to spot atypical sales trends... Those techniques can also be used to remove noise and labeling errors to improve the quality of data as a preprocessing step in machine learning.

An example in Chapter 6 shows how to move from rule-based approaches to detect anomalous activities in IT environments to more adaptive approaches: using an **unsupervised learning** approach and Markov chain model, **Microsoft Sentinel**, a security information and event management service in Microsoft Azure cloud, provide a modeling approach to help customers detect potential malicious activities in their IT environment. Another example is provided in Chapter 7, where the **KBC Group** leverages unsupervised learning-based anomaly detection algorithms for detecting fraud in insurance claims.

2.4.5 Reinforcement learning: key concepts, industry applications, and examples in the book

Originally inspired by psychological models of parts of the brain's reward system, **reinforcement learning** (RL) techniques fall into a different category of learning systems, both by their fundamental objective and by their learning methods. For their objective, those techniques belong to the family of **goal-based agents** that support complex sequential decision problems in stochastic environment: the transition from a state to another (from s_t to s_{t+1}) by taking an action a_t in the environment is subject to stochastic behavior, described by probabilities. The objective of reinforcement learning techniques is then essentially to learn to make a **series of decisions**, or **actions** a_t at time t, that will maximize an expected total reward [26] over those decisions (see Figure 2.1).

As for the learning part, RL is an AI paradigm whereby machines learn by **trial and error**, getting **rewards** (or penalties) at each step (r_t) from interacting with their environment; those techniques model unexplored territory and learns from their own experience by taking actions in the environment. To sum up, by trial and errors and getting rewards from the environment, reinforcement learning systems can simultaneously learn a model of the environment and use that model to decide on a course of

---

[26] Rewards are defined by the model designer. Total means that the agent factors in the expected rewards over all the time steps.

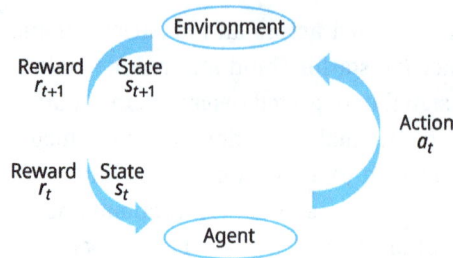

Figure 2.1: The agent takes a decision, or an action a_t, on its environment and lands in another state (s_{t+1}) where it receives a reward (r_{t+1}). From that new state, a new iteration starts by having the agent select again another action. The learning agent needs to learn the series of actions that will maximize the expected total rewards over all its decisions.

actions to maximize an **expected total reward**. More about reinforcement learning in Chapter 7.

Reinforcement learning finds its applications in a wide array of fields. These include robotics for task control, autonomous systems for safer and more reliable operations in real-world environments such as control systems, and gaming. It also plays a crucial role in decision-making processes such as determining whether to hold, buy, or sell stocks in trading systems. Furthermore, it can also aid in delivering a more personalized user experience on the web: for example in **recommendation systems**, where the RL system tracks the website reader's return behaviors and clicks form the basis of the rewards mechanism. The example in Chapter 7 illustrates how Microsoft delivers a personalized, relevant users experience through **Microsoft MSN news** website.[27] This personalization was implemented in 2016 with the aim of tailoring news articles to individual users. As a result, Microsoft reported a significant increase in the 'Click Through Rate' by 26 %.[28] An illustration of how reinforcement learning techniques are utilized in various sectors, including the manufacturing industry, can be seen in the case of **PepsiCo**. PepsiCo announced a deep reinforcement learning solution that monitors and adjusts the extruders that make Cheetos. The goal is to optimize the production line throughput, while maintaining the snacks quality for crunch, lightness, and shape. This approach reduces the time it takes to correct inconsistencies and allows operators to focus on parts of the line that require human expertise (See "More perfect Cheetos: How PepsiCo is using Microsoft's Project Bonsai to raise the (snack) bar").

---

[27] In 2022, Microsoft began phasing out MSN to Microsoft Start with news pages being moved to Start, and ads for the website appearing on the homepage.

[28] https://www.microsoft.com/en-us/research/blog/real-world-interactive-learning-cusp-enabling-new-class-applications/

2.4.6 What about deep learning?

Deep learning (DL), which relies on **deep neural network** (DNN), is a family of machine learning methods that leverage neural networks. **Neural networks** are typically comprised of nodes layers: an input layer, one or more hidden layers, and an output layer. Each node (or neuron) connects to another node in the next layer with an associated weight and threshold: the output of an individual node is simply a weighted sum of its inputs passed through an **activation function** sending the result to the next layer of the network (see Figure 2.2). Nonlinear activation functions allow such networks to compute nontrivial functions as they introduce a nonlinearity. They are called **deep** when many layers are stacked between the input and output layers. Learning the weights based on some loss or objective functions, deep neural networks can then represent **arbitrarily complex functions** to map raw input value to output value. More details in Chapter 7.

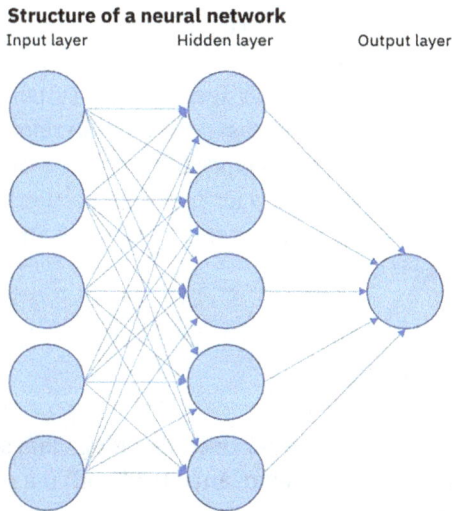

Structure of a neural network
Input layer Hidden layer Output layer

Figure 2.2: The node is a weighted sum of its input, passed through a nonlinear function at each node.

The concept of neural network isn't new: the **perceptron**, the first simplest neural network used as a binary linear classifier, was invented by Frank Rosenblatt in 1958. Evolution of multilayer perceptron in 1967, and the term **"deep learning"** was first introduced in 1986. Hardware innovations (like graphical processing unit), algorithmic innovations, and the increasing amount of available data, made neural networks progressively evolve from simple shallow architectures to more complex and bigger architecture with an increasing number of layers and parameters: LetNet5, one of the earliest convolutional neural networks,[29] used to recognize digit images had about 60,000 pa-

---

29 A type of deep neural network architecture.

rameters (1989). The deep learning-based **language prediction models**[30] developed by the OpenAI Research laboratory for **natural language processing (NLP)** went from 100 million parameters (GPT[31]-1, 2018) to 175 billion parameters in its third release (GPT-3, 2020). According to speculations, the GPT-4 model released in March 2023 would have more than 1.7 trillion parameters.

Deep learning techniques exist in various forms and architectures: standard **deep neural network**, **convolutional neural network** (known for image and video recognition, also applied to text classification and anomaly detections), **recurrent neural network** (typically used for temporal, time-series data, in language related tasks and NLP), **general adversarial network** and, more recently, **transformers**. Since their launch in 2017–2018, transformers have emerged as the leading technology in NLP, a status they maintain to this day (Chapter 8).

The success of deep learning is primarily attributed to its scalability with increasing data volumes. This scalability enhances performance as more data becomes available. Deep learning is also distinguished by its capacity for automatic **feature** or **representation learning**, which contrasts with traditional machine learning methods that rely on manual **feature engineering**. It allows deep learning systems to both learn the relevant features and use them to perform tasks; they are therefore much better to interpret unstructured data (i. e., data with no predefined data model, such as text, voice, image, video, etc.), creating opportunities for many applications that use speech recognition, image recognition, and natural language processing on text. Those two aspects are worth discussing in a bit more detail.

2.4.6.1 Bias-variance trade-off

In designing traditional models for **supervised learning**, data scientists have to compose with the so-called statistical **bias-variance trade-off**: the reducible errors in machine learning can be shown to be broken down into a **bias error** and a **variance error** of the parameters estimates (see Chapter 7). Unfortunately, it is typically impossible to reduce the total prediction error, for both the bias and variance error, beyond a certain point. Refer to the Figure 2.3 below, which illustrates the error in the function of the model complexity: it is impossible to both accurately capture the regularities in the training data (low bias, green dashed line), and also generalize well to new, unseen data (low variance, yellow line). This bias-variance trade-off causes the machine learning model to either **overfit** (variance error dominates) or **underfit** the given data (bias error dominates). So, even if techniques do exist, traditional machine learning (ML) techniques

---

30 A **language model** that uses deep learning to produce human-like text. See the concept of a language model in Section 2.7 and Chapter 8.

31 Generative pre-trained transformer (GPT).

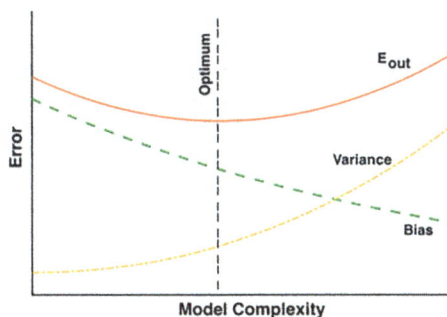

Figure 2.3: Underfit at the left side of the optimum, overfit with more complex models (right side of the optimum). Source: "A high-bias, low-variance introduction to machine learning for physicists" (Pankaj Mehta et al., 2019).

don't offer many approaches that just reduce the bias or the variance error without impacting the other one, and the model reaches some performance plateau.

The neural network size however takes better advantage of having more data to train and exploit more data than traditional ML algorithms: larger neural networks enhance the complexity of the model thereby **reducing bias**. Simultaneously, training on more data **decreases variance**, leading to improved generalization without affecting the bias.

2.4.6.2 Feature modeling

Traditional machine learning pipelines require some manual processing of the raw data, known as **feature selection**, to select good predictors, and **feature engineering**, to transform the raw data into feature vectors relevant to the specific task (e.g., to capture some characteristics of a sound, a text, or an image as a numeric vector that can be understood by the ML algorithms). Deep learning systems both learn the relevant features and use them to perform the specific task, eliminating the need for data scientists to handcraft the features: the raw input data is fed directly into the deep learning model and the model **automatically learns the features** from the input data, and the successive layers progressively extract higher level features from the raw input. So, deep learning systems will learn the patterns in the input data rather than having a data scientist to manually construct the features; they act therefore as **feature detectors**. For example, in the realm of **computer vision**, a neural network begins its analysis with raw input data, ie pixel values in an image; then, the network's lower layers identify basic visual elements like edges. Moving up to the intermediate layers, these edges are assembled into more complex shapes. Finally, the higher layers extrapolate these shapes to recognize concepts that are significant to humans, such as digits, letters, or faces. In **speech recognition**, the first layer captures the basic low-level waves, the intermediate ones capture basic sound units (phonemes) and words, while the last ones capture sentences in a complex assembly. In other **natural language processing** tasks, neural networks can link symbols to vectorized representations of the data, which are translations of raw data attached with some semantic meanings (see the notion of **embedding** for more details in Chapter 8).

Note that deep learning can be applied to various learning approaches, including supervised, unsupervised, and self-supervised settings.

2.4.6.3 Industry applications and examples in the book

Deep neural networks offer a range of applications. These techniques often surpass human benchmarks in tasks involving unstructured data, such as classifying or detecting objects in images within computer vision, and performing sentiment analysis, entity extraction, text classification, translation, and next-word prediction (known as language modeling) in natural language processing, among others. See Chapter 8 for more details.

The solution developed for **Veranneman Technical Textiles**, as mentioned in the classification-related paragraph, illustrates the use of **convolutional neural networks** in computer vision to automatically classify various defect classes in textiles (Chapter 7). Chapter 8 describes how other types of DNN architectures, such as **recurrent neural networks** and **transformers**, are successfully applied to natural language processing applications, like **Microsoft QnA Maker**. Microsoft QnA Maker helps to build knowledge bases by extracting Questions and Answers from manuals, FAQs, and documents.

Another major field of applications of deep neural networks belongs to the **deep reinforcement learning** family of algorithms. Deep reinforcement learning combines reinforcement learning and deep learning. Indeed, in many tasks, the **state** or **action-state space** that represents the environment in which the algorithm must make decisions (see paragraph about reinforcement learning) is either too large to be stored in tables and memory, or it is continuous: so, an approach that relies on look-up tables with discrete values isn't scaleable or doesn't even work. In theses cases, learning techniques that generalize from past examples (from seen states or state-action pairs) are required to estimate the values of new, unseen, states. Deep neural network can then be used as a **function approximator** to learn those complex functions. Autonomous driving is a typical application where deep reinforcement learning techniques are applied: following the recognition step to identify the surrounding environment (typically using DNN/CNN), and the prediction of environment dynamics (e. g., tracking an object), reinforcement learning facilitates the planning of decisions to avoid unwanted situations and ensure safe arrival at the destination, utilizing penalties or rewards.

2.4.7 Limitations of data-driven learning

Broadly speaking, machine learning techniques face two major challenges: the challenge of **generalization** and **knowledge engineering**.

Generalization
Machine learning essentially brings an inductive learning approach to generalize from data, based on examples. They infer statistical patterns from very large datasets and raw signals. Statistical learning is hindered by deviations from its assumptions and is limited by its specialization: such systems often work impressively well, sometimes better than human, when applied to the same environment on which they are trained but must be retrained if the environment differs, sometimes even in small ways. Once trained, a machine learning algorithms typically do what they learned to do. To make them perform a different task would generally require changing the parameters learned by the algorithms. It remains a localized generalization within a specific training domain, making it challenging to ensure its effectiveness in alternative circumstances, with new data that may not sufficiently resemble the previous training data: apply a trained model to a new domain may yield poor performance. Machine learning techniques are good at finding patterns in terms of data but can't broadly generalize. Another major source of bias is the **background knowledge** and the **preferences** of the data scientist, which both influence the choice and design of the learning models. Different types of models possess varying underlying assumptions and structures that mirror the data scientist's methodological preferences. As it will be discussed in Chapter 7, *learning without any inductive bias isn't possible.*

Also, some important domain concepts and laws cannot be learned from data alone: for example, **empirical laws**, such as Galileo's principle of inertia, are idealization that aren't observed in nature. It is practically impossible for any object to achieve a state of zero net force (i. e., the total force acting on an object is null) due to the presence of friction, air resistance, and various other forces constantly acting upon it.

Knowledge engineering
Machine learning techniques don't make any explicit and specific representations of knowledge, like with **symbolic AI** approaches do. The insight learned from data is a mapping based on a learned function, or an association, a pattern that fits the data but doesn't give meaning, nor causality by themselves. Machine learning techniques can assist in discovering insights within text, video or images. For example, they can extract and classify words in predefined categories, like recognize that "Paris" is a location or a person's name, or recognize a dog in an image. However, the underlying meaning of the category used to train the algorithm remains unknown to the machine. Unlike symbolic AI approaches, usual machine learning techniques lack a framework to represent the

knowledge of object and relations in a model and can't easily incorporate **prior knowledge** of the world. On the other hand, symbolic AI approaches—such as first-order logic and structured representation—provide data representations that are sufficiently expressive to encompass relational data. These methods can integrate some prior knowledge and are explainable to humans.

This brings us to techniques that combine learning with logical reasoning.

2.5 Learning in a logic framework

2.5.1 Overview and key concepts

Initially defined by Stephen Muggleton as the intersection of logic programming and machine learning (Stephen Muggleton, 1991), **inductive logic programming** (ILP) couples the area of logic programming with techniques from learning and essentially introduces the notion of generalization in a logic framework. As with other forms of learning, the goal of ILP is to induce a hypothesis that generalizes from training examples. Whereas machine learning represents hypotheses through specifically chosen learning models and learns functions, ILP represents hypotheses as a **set of logical rules** and learns new **logical rules and relations** from examples. In its simplest form, it learns new knowledge by induction: it forms hypotheses that both are consistent with the background knowledge and explain the provided examples (ground facts or rules). The new logical proposition must entail all the positive examples and none of the negative examples given to the algorithm. That hypothesis is then added to the knowledge base of the system. Generalizing that all birds can fly from seeing a duck and a sparrow that can fly is a very simple example. Given multiple examples of friendships between people, with some friends being female and others male, as well as examples of girlfriend relationships, an ILP algorithm could also infer the general rule by induction that having a girlfriend implies both being friends and being female.

In summary, ILP proceeds by forming hypotheses from provided evidence (somehow analog to label in supervised learning in the numerical data world) and that are consistent with the given **background knowledge**. See Chapter 7 for more details.

Having an approach to add explicit background knowledge to learn new knowledge makes ILP techniques quite unique. They combine generalization (learning) and specialization using both **inductive and deductive inference** rules within the same integrated framework: the examples, the hypothesis, and the added background knowledge are all expressed using the same symbolic representation. The induced, learned, hypothesis or rules are naturally incorporable in rule-based systems for deductive inference. ILP is differentiated from the other forms of machine learning both by its use of an expressive representation language and its ability to make explicit use of encoded background knowledge. This essential to consider notions like **cumulative learning**, where knowledge can be augmented and used to learn new things.

Finally, ILP techniques offers high interpretability of the results of the learning process: the learned hypotheses are represented in symbolic forms and therefore provide **transparency and explainability** to humans, while the learned knowledge can be remembered and explicitly stored in the **knowledge base**.

2.5.2 Industry applications and examples in the book

ILP is typically used to learn complex relational information, where the learned knowledge can then allow easy integration in expert knowledge system or deductive-based reasoning systems. ILP techniques have fueled applications in what is known as **Programming by examples** with the goal of automatically generating small programs synthesized from a few input and output examples. These include learning scripts, generating search query, or applying data transformation, extracting data from documents, all based on user-provided examples. **Microsoft Flash Fill**, a feature that automatically fills data when it senses a pattern, is a known example to automate repetitive string transformations in Microsoft Excel. As illustrated in the following Figure 2.4, once the user writes an instance of the desired transformation and proceeds to transforming another one, Flash Fill learns a program that automates the repetitive task.

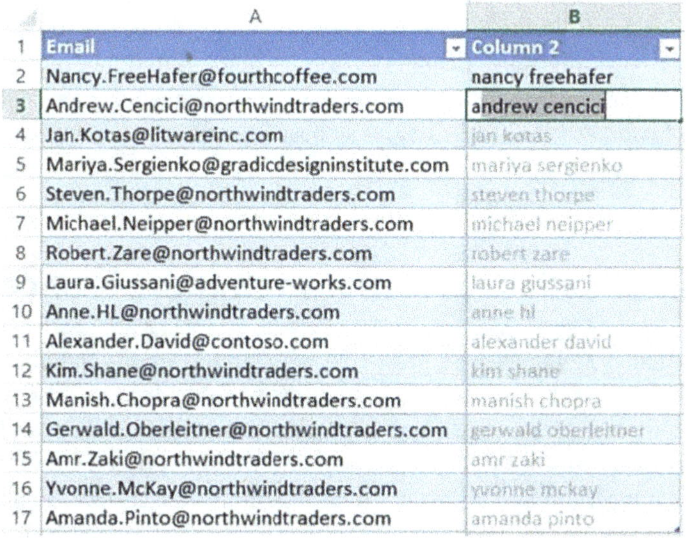

	A	B
1	Email	Column 2
2	Nancy.FreeHafer@fourthcoffee.com	nancy freehafer
3	Andrew.Cencici@northwindtraders.com	andrew cencici
4	Jan.Kotas@litwareinc.com	jan kotas
5	Mariya.Sergienko@gradicdesigninstitute.com	mariya sergienko
6	Steven.Thorpe@northwindtraders.com	steven thorpe
7	Michael.Neipper@northwindtraders.com	michael neipper
8	Robert.Zare@northwindtraders.com	robert zare
9	Laura.Giussani@adventure-works.com	laura giussani
10	Anne.HL@northwindtraders.com	anne hl
11	Alexander.David@contoso.com	alexander david
12	Kim.Shane@northwindtraders.com	kim shane
13	Manish.Chopra@northwindtraders.com	manish chopra
14	Gerwald.Oberleitner@northwindtraders.com	gerwald oberleitner
15	Amr.Zaki@northwindtraders.com	amr zaki
16	Yvonne.McKay@northwindtraders.com	yvonne mckay
17	Amanda.Pinto@northwindtraders.com	amanda pinto

Figure 2.4: Illustration of Microsoft Excel feature called Flash Fill. Suggestion made by the algorithm to fill-in the other cells in column 2, based on the first examples provided: extracts the first two words, converts them to lowercase, and concatenates them separated by a space character.

Other applications of ILP can be found in physics, for instance, where they solve problems from first principles and assist in discovering natural laws from collections of experimentally gathered data. Given physical models of the basic primitives, ILP systems can induce a target hypothesis that exhibits behavior derived from these primitives. Chemistry and molecular biology domains are also particularly appropriate for ILP due to the rich relational structure of the data, and the need to handle complex relational structures. Chapter 10 also illustrates how an international manufacturing company, known for its diverse construction products and innovative materials, utilizes ILP to parse tables in engineering diagrams and enables the creation of concise programs that deduce the labels of cells, based on the surrounding information in these tables (Chapter 10).

2.5.3 Limitations of inductive logic programming

Construction of the hypotheses can be a real challenge since the space of all hypotheses can be huge and complex, making it **sometimes intractable** if that space is not limited. It's therefore necessary to impose restrictions to constrain the search. This introduces a **bias to restrict the hypotheses space** to make the search tractable: restriction on hypotheses or fix how the space is being searched, for example. Next, to induce a hypothesis from examples, we need to provide an ILP system with suitable **background knowledge**. Finding the right balance of the appropriate background knowledge is a challenge: insufficient background knowledge may exclude a target hypothesis, while too much can degrade the performance. It can also be difficult and expensive to obtain handcrafted background knowledge from domain experts. Finally, training examples might be noisy with mislabeled examples, so it is difficult to find a hypothesis that is both complete and consistent.[32] Therefore, most approaches relax this definition and try to find a hypothesis that covers as many positive and as few negative examples as possible, with no perfect match.

2.6 Probabilistic reasoning when there is uncertainty

Our logical models so far, such as first-order logic and propositional logic, addressed facts and predicates with **certainty**, i. e., they were either true or false, possibly unknown. With those logical models and rules, we can express concepts such as "if A is true then B is true," but we didn't express **degree of belief** between true or false, either when we are not certain about the facts (nondeterministic facts like the probability of

---

32 Complete: works with all positive examples of the concept, consistent: works with all negative examples of the concept.

raining, the probability of having the flu if the body temperature is above 38 degrees), or when we face **stochastic behaviors** of the environment (uncertainty about the outcome of an action like a robot which takes a next step but might end up left or right due to the uncertain impact of the environment). **Probability theory** is the mathematical cornerstone used to express the degree of uncertainty or belief, as a value between 0 and 1. Dealing with uncertainty is not only important when we have **unpredictable facts and outcomes**, but also when we might **not have a complete knowledge** of the world to model (e. g., medical or fault diagnosis), or when it is practically impossible to compile an exhaustive list of all rules. This is the field of **probabilistic reasoning** techniques covered by Chapter 6.

2.6.1 Key concepts—uncertain facts and outcomes with probabilistic reasoning

The main objective of **probabilistic reasoning** is typically to answer queries such as the probability of an event, or a combination of multiple events given some evidence or observations. In what are called **causal models**,[33] for example, the objective is to estimate the probability of a cause given some observed effect or evidence. We observe some effects from the data and would like to assess the probability of a cause given these observed effects $P(\text{cause} \mid \text{effect})$, a question that is also called a **diagnostic question**. To achieve that objective, a key fundamental trick is to exploit **Bayes' theorem** properties:

$$P(\text{cause}|\text{effect}) = \frac{P(\text{effect}|\text{cause})P(\text{cause})}{P(\text{effect})}$$

and build a model in the opposite direction by rather determining $P(\text{cause})$, and $P(\text{effect} \mid \text{cause})$, easier for a subject matter expert to build since statistics are often available in that form and one can use available data for the evidence. See the first part of Figure 2.5a below. Once the model is generated and associated probabilities estimated, queries can then answer the diagnostic questions $P(\text{cause} \mid \text{effect})$ in the other direction. If subject matter experts don't know the $P(\text{effect} \mid \text{cause})$ and $P(\text{cause})$, a probability distribution is assumed, and the **likelihood of the observed data** is maximized so that we find the statistics that best fit the data. Such approach provides not only with the mechanisms to infer, or reason to answer diagnostic questions, but also the means to learn the probability distributions of the variables, and their dependencies, from the data. More generally than causal models, modern approaches to probabilistic reasoning in AI combine three fields:
1.   **probability theory** to address uncertainty,

---

33 Causal relationships among the involved random variables is assumed. The reader should remember that assumption. More about causality in Chapter 11.

2. **statistical, or machine learning** to learn the model parameters, or even the model, from data,
3. some form of graphical models, called **probabilistic graphical models (PGM)**, to represent the prior knowledge of experts and model the dependencies between the different variables. Probabilistic graphical models are a rich framework for encoding probability distributions over complex domains with multiple variables: joint multivariate[34] distributions over large numbers of random variables that interact with each other. They leverage concepts from probability theory, graph algorithms, machine learning, and more. The properties of those graphical models also ease the computation of the probabilistic queries.

Some key intuition about the general approach is developed here as an introduction to Chapter 6.

Any modeling exercise starts with a set of **random variables**[35] that are needed to model the environment: each random variable is described by a **probability distribution**[36] and the interactions between all the variables in the model are defined by what is called a **joint probability distribution**, which defines the probability of every possible combination of their values in the model, hence gives a complete specification of the model. For example, the definition of the $P(X = x_i, Y = y_i, Z = z_k)$ for the possible values x_i, y_i, z_k if we have 3 discrete random variables X, Y, Z in our model.

However, distributions over many variables, when used to model a real environment, can quickly become cumbersome to represent naïvely: defining the joint probabilities of n binary variables in a table already requires storing 2^n values. Techniques, therefore, seek to represent the joint probability distribution as product of local functions, each depending on a much smaller subset of variables. This approach exploits **prior knowledge** of the **conditional independence relationships** among the variables. While modeling the environment, domain experts incorporate their prior knowledge about the interactions of these variables by using a graph to build the model. The graphical model structures how objects with their variables are related: they capture relations among the variables as well as their uncertainties. Graphs conveniently represent probability distributions, detailing dependencies and independencies. They introduce structural assumptions about the joint probability distributions of variables, often reflecting independence assumptions among some variables. Representing independence between variables— by the absence of links in the graph—significantly

---

34 Multiple variables.

35 A variable whose value is uncertain.

36 A function which describes the dispersion of the values of the random variable, for every possible value of that random variable. Discrete probability distributions for discrete variables, probability density functions for continuous variables.

mitigates the combinatorial complexity of a full joint distribution table in the absence of prior knowledge about these independencies.

Important examples of such graphical models that support probabilistic reasoning techniques are:

- **Bayes networks** (also known as Bayesian networks) for modeling causal problems,
- **Markov random fields (MRF)** for modeling the global statistical distribution of the prior knowledge, in noncausal problems (when there are noncausal statistic dependencies),
- **Factor graphs**, which generalize the approach and describe the way in which a probability distribution p, such as $g(X_1, X_2, X_3)$, decomposes into a product of local functions, also known as a factorization.

Each model is illustrated by the three following pictures below (Figure 2.5).

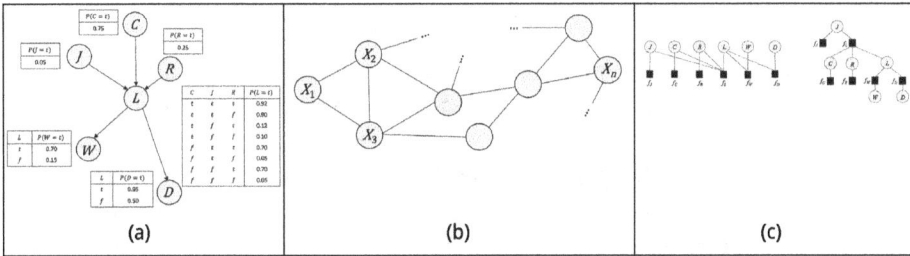

Figure 2.5: Three different probabilistic graphical models. (a) **Bayes network** for causal models as acyclic directed graphs. Nodes with capital letters represent the random variables of the model built by a subject matter expert. Tables are the associated probabilities and the conditional probabilities, given the parent(s) of the nodes. (b) **Markov random field (MRF)** for non-causal models. Edges represent the dependencies between the random variables, represented by the nodes, of a statistical distribution. $P(X_1, X_2, X_3, \ldots, X_n)$ is a product $\phi(X_1, X_2) \times \phi(X_2, X_3) \times \phi(X_1, X_3) \times$ etc. where ϕ is some function assumed in our prior model. (c) **Factor graph** decomposes the function $P(j, c, r, l, w, d)$ from Chapter 6 (function 6.27) into a product of factors: $P(j, c, r, l, w, d) = \underbrace{P(j)}_{f_J} \underbrace{P(c)}_{f_C} \underbrace{P(r)}_{f_R} \underbrace{P(l|j, c, r)}_{f_L} \underbrace{P(w|l)}_{f_W} \underbrace{P(d|l)}_{f_D}$.

Once the assumptions about variables interactions are established, that is, when the probabilistic model is constructed, we can compute the probabilities of specific events or joint events, potentially conditioned upon observed evidence. These computations are performed as operations on the graphs. Probabilistic questions are answered by performing **inference algorithms** within the graph. Graphical representation of a probabilistic model not only facilitates problem design and the expression of variable influences but also provides a structure conducive to efficient algorithmic computations that leverage the graph's topology.

If the probabilities or probability distributions aren't known, they can be **learned** from observed data. The first phase will be training, which aims to maximize the likelihood function given the observed data. This phase include:

–   for **discrete random variables**, for example, a simple statistical counting of occurrences of events
–   for **continuous random variables** with an assumed probability distribution parameterized by θ, inferring these parameters θ that best explain the observed data.

A more complex but possible task is to learn the structure of the network, i. e. the dependencies of the variables.

To summarize, the overall approach is to

1. **model**: model how variables are related to each other with some prior knowledge, exploiting some assumption of the independence between the variables,
2. **learn**: train the model to find the model parameters that maximize the likelihood of observed data if the probabilities of the model are unknown,
3. **reason or infer**: perform the query to answer some probabilistic question, usually conditioned on observed evidence by taking the value that maximizes its probability.

Such approach is also called **model-based machine learning** as the assumptions are made explicit in the form of a model, which includes the number and types of variables, which variables affect each other and how. A connection with classical **logic reasoning** is worth mentioning: the main idea of probabilistic reasoning is to find the relevant variables in the environment and build a probabilistic model of how they interact. Reasoning is then performed by applying evidence that sets specific variables to known states. Based on the observed values, we calculate the probabilities of interest. So, the rules of probabilities form a complete reasoning system, one that includes traditional **deductive logic as a special case**.

2.6.2 Industry application and examples from the book

Probabilistic reasoning techniques support decision-making in uncertain and complex environments. They have applications across various domains, including medical diagnosis, genomics, fault diagnosis, risk management, IT security, image analysis, social network models, environmental studies, and decision theory, among others.

Chapter 6 also illustrates how Bayesian inference techniques are used to learn the parameters of a model applied to **crop forecasting** in agriculture. Another example in the same chapter explains how Bayesian inference, factor graph and expectation propagation techniques are applied in the **Microsoft TrueSkill** ranking system, a skill-based ranking system for the **Xbox** network,[37] developed at Microsoft Research. The solution ranks the skills of gamers in order to match them into fair, competitive, matches.

---

37 online multiplayer gaming service, formerly Xbox Live, created and operated by Microsoft.

2.6.3 Probabilistic reasoning over time, or sequence to sequence

When we look at **dynamic processes**, we deal with temporal or **time-series data**, where states and observations depend on time: there is a **state** at a time t and **evidence** at time t. An on-going process can then, for example, be modeled by timestep sequences as

1. a **chain of states**, with a probabilistic **state transition model** to move from state at time$_{t-1}$ to another state at time$_t$ defined by $P(\text{State}_t|\text{State}_{t-1})$, and
2. a **chain of evidence** with an **observation model** to observe some evidence when landing in a state, defined by $P(\text{Evidence}_t|\text{State}_t)$.

Here as well, we typically build, then learn, the transition and observation models. The objective is typically to predict a state, the next state given all evidence so far $P(\text{state}_{t+1}|\text{evidence}_{1:t})$, or to find the most likely sequence of states that could have generated a sequence of observations $P(\text{sequence of states}_{1:t}|\text{evidence}_{1:t})$. In a similar way to the previous paragraph, different **graphical models** exist and **independence assumptions** are made in order to ensure tractable inferences. For example, with **hidden Markov model (HMM)**, a temporal probabilistic model with unobservable states of a single random variable State$_t$[38] and with observed variables Evidence$_t$ (e. g., evidence from sensors) is built as in Figure 2.6. The **Markov property** assumes that the current hidden state State$_t$ depends only on the previous value of the hidden variable state State$_{t-1}$: the values at time$_{t-2}$, and before, have no influence. Similarly, the value of the observed variable evidence Evidence$_t$ only depends on the value of the hidden variable state State$_t$, both at time t. Here again, if they aren't known, the transition and observation models can be **learned** from data.

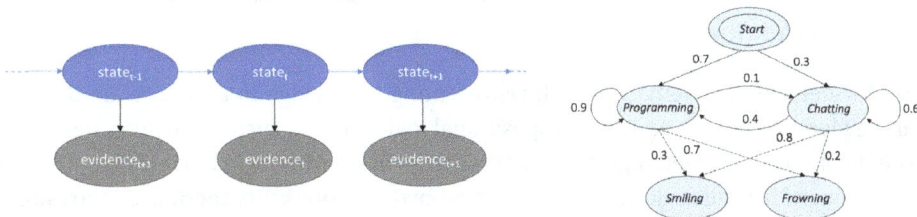

Figure 2.6: from Chapter 6. The first picture models a chain of steps over time with a hidden State$_t$ and the observable Evidence$_t$ at different time steps. Assuming that an observer tries to guess whether a developer is programming or chatting, only by looking at his or her facial expression, a simple model could define State$_t$ with possible value of programming or chatting, and Evidence$_t$ with possible value of smiling or frowning. The second picture, called a state transition diagram, illustrates the situation when the **transition probabilities** go from one state to another and the **observation probabilities** are known and stationary. Example: a probability of 0,1 to change from programming to chatting and a probability of 0,3 to smile when programming.

---

38 Or a composite of random variable as t-uples.

Typical family of techniques in this chapter are **hidden Markov model (HMM)** for discrete variables, **Kalman filters** for continuous variables and **dynamic Bayesian networks**, which are a generalization of HMM and Kalman filters. Just as in the previous static setting, learning can be done as a by-product of inferencing: inferencing the probabilities based on observed evidence provides a first estimate of probable states, and these estimates can then be used to update the learned model. More about this in Chapter 6.

2.6.4 Industry applications and examples from the book

Many of these models are well known in different branches of science, from physics to engineering. They are also applied in computer vision and natural language processing techniques (see Chapter 8), in bioinformatics to evaluate biological data sequences, in finance for financial predictions, in retail to anticipate consumer purchases, in marketing to enhance user conversion rates, etc. In robotics, factor graphs help autonomous systems make sense of the world and are essential to address the **simultaneous localization and mapping** (SLAM) problem in robotics, which is to construct and update a map of an unknown environment using the information coming from robot's sensors, while simultaneously keeping track of its location.

Chapter 6 features how Markov chain models are applied in **Microsoft Sentinel**[39] to help companies detect malicious activities and anomalous sessions in their IT environment.

2.6.5 Enrich the abstraction—unify probabilistic models and formal logic

Like for deterministic environment, representing knowledge in an uncertain domain can be modeled by using simple propositional logic models, such as in Bayes networks seen before, or by richer representations such as first-order logic models that allow to better generalize and scale, known as the **first-order probability model**. **Markov logic network** is such an example of probabilistic logic that generalizes first-order logic in uncertain environment: it applies the ideas of Markov networks and probabilistic graphical models (Chapter 6) to first-order logic, enabling uncertain inferences. This is also known as **symbolic-statistical modeling**.

---

39 A security information and event management (SIEM) and security orchestration, automation, and response (SOAR) solution.

2.6.6 Uncertain states or outcomes in the context of a decision

Probabilistic reasoning techniques not only apply to help predict a cause, or a state, given some evidence but also support the field of **decision theory**, where probabilities and utility theories are combined to make decisions under uncertainty. So, probabilistic reasoning also forms a core part of decision systems. In these problems, a **utility function**, which expresses user's preferences, is associated to the different possible outcomes, and a rational agent will make the decision that maximize the **expected utility** of the decision (the expected utility is the probability-weighted average of utility over all possible decision outcomes, given a set of random variables). Similar to Bayes networks, decision networks model the decision making problem at hand: **chance nodes** represent random variables with an associated conditional probability given their parents (e. g., patient's symptoms given a disease), **decision nodes** represent the decision variables that the decision maker sets (e. g., a treatment decision based on the symptoms, the test performed and the test result) and **utility nodes** that depends on parent variables (e. g., the utility may include costs of tests and treatments, the pain and inconvenience to the patient in the short term, and the long-term prognosis).

When a **sequence** of multiple, step-by-step, decisions or actions must be made in a stochastic environment –that is where transitions from a state to another occur with some degree of randomness– we are dealing with what is called **sequential decision problems**: at each step, a decision must be made after observing a certain state and receiving a reward. For a robot, for example,

- a **state** could refer to a position on a grid map,
- an **action** could refer to a valid move,
- the **transition** from a position to another based on the decision could be subject to some uncertainty (noise, failure, etc.), hence defined by some probabilities and a **transition model**,
- a **reward model** can be defined to reach a target destination as quickly as possible.

The objective is to assist the decision-maker in defining an optimal policy, ie a function that determines the best action to take in any given state. In this context, best is typically defined by summing the rewards over a state sequence, sum that the agent seeks to maximize as a utility function. If the model of the environment is known and the Markov property can be assumed, an agent with sufficient computational resources can perform computations using that model, plan ahead and design the optimal plan. That is, decide on an optimal course of actions by considering possible future situations offline before they are actually experienced live. The so-called **Markov decision process (MDP)** techniques help solve such problems: they define an optimal policy as a set of actions for optimal control in such a stochastic environment (i. e., given a transition model and a reward model). Applications that use MDP models exist in various domains: in gaming, robotics, manufacturing for scheduling machine maintenance or repairs, IT for server management, solving shortest path problems, optimizing planning to reduce queuing times, enhancing traffic control, among others.

Reinforcement learning (RL), which was briefly covered in the previous paragraph about learning, goes a step further. It can be considered as an extension of such MDP framework where the model of the environment (i. e., the transition and the rewards models) isn't known. RL uses the same conceptual framework of a MDP, but is typically a **model-free** approach, meaning that it learns the model by sampling: the RL-based agent learns on the basis of experience by interacting with the environment, from which it gets rewards based on successive trials. The goal is encoded by a **reward system** and the learning agent devises a strategy to optimize cumulative rewards based on the feedback received during its interactions with the environment: the algorithm employs a trial-and-error approach, testing various strategies, experiencing failures, and then gradually learning to optimize its decisions based on the rewards provided by the environment. It then exploits its findings for the best possible decisions. The algorithm balances between a so-called **exploration** and **exploitation** phase. So, to sum up, reinforcement learning techniques essentially study the problem of making **sequential decisions** in complex, unknown environments, with potentially long-term consequences. Examples of applications were mentioned already in the section related to learning, reinforcement learning.

2.6.7 Limitations of probabilistic reasoning

There are subjective and objective limitations. As the science of probabilities is a difficult topic for many, subjective limitations arise from errors or incorrect applications of the probability theory. Objective limitations arise due to the need for simplifications in the model, the trade-offs between accuracy and computing time in inference strategies, and the reliance on available data. Therefore, probabilistic reasoning used in AI shares many of the limitations of machine learning. More about these limitations in Chapter 6 and Chapter 7.

2.7 Interaction with machines in natural language

Natural language processing (NLP) is a field of AI focused on the interaction between the human and the machine. It studies how human language (speech, text, and by extension images) can be either analyzed (**natural language understanding**) or generated (**natural language generation**) by a machine. NLP encompasses various abilities such as:

- **recognizing speech**, converting **spoken language to text** that can be processed by a computer and transforming **written text into spoken audio**,

- **categorization** of documents,
- **extraction** of information, **relation** and **entities** from text (e. g., name of people, dates, locations, etc.) or images,
- **translation** of texts,
- **creation** of text, summaries, or images,
- the ability of **answering questions** from given text,
- building dialog systems, **conversational agents**, or chatbots,
- and many more.

Natural language is hard for computers to comprehend for multiple reasons: the **ambiguity** of multiple interpretations depending on the context, the **synonyms, the paraphrases**, etc. For example, does the term "bank," refer to a financial institution or the side of a river? Only the context can resolve such ambiguity, and furthermore, the context itself may vary depending on the circumstances. Capturing the **context** of human interaction is a challenge for AI systems since the foundation of NLP. Also, **commonsense knowledge** (such as "if Peter is home, his head is home as well," mentioned in the paragraph about the limitations of symbolic AI) is often assumed but not explicitly mentioned, hence not captured by the algorithms. Note that the use of the word "understanding" in understanding natural language is an abuse of language: it's worth stressing that there is no ground truth meaning but a translation into numerical representations that machines can handle and act upon accordingly. This might give the perception of understanding, to the extent that it can even fool a human.

2.7.1 Overview of some key concepts

While incorporating its own specificities, natural language processing techniques apply an assembly of techniques outlined in the preceding paragraphs. Although all the techniques still have their applications nowadays, the field has undergone 3 major stages of development: from **symbolic and rule-based** systems (utilizing dictionaries and lexical databases) to **statistical and machine learning** systems, and finally to advanced **deep learning** techniques. Let's delve deeper into the details of this evolution.

Early approaches to natural language processing applied existing knowledge of formal, language-specific, **linguistic theories.** These were used to model and design a pipeline of specific rules, which were supplemented by hand-crafted resources, such as dictionaries and ontologies. Despite the usefulness of this methodology for some applications, it is challenging to construct and maintain due to its complexity and the difficulty in determining the appropriate granularity of the definitions.

In **statistical and machine learning**-based systems, tasks typically follow a pipeline as well. After preprocessing, which includes morphological analysis to have canonical forms[40] and parsing to identify the structural relationship between words (the syntax and grammar), the process move to **semantic extraction**. In this step, a data scientist identifies key features that characterize the text and that are relevant for the NLP tasks to process. For example, determine the specific features applicable to perform sentiment analysis[41] and define a model that could make use of the adjective like good versus bad, or great versus poor for doing so. Although those methods remain in use, new NLP applications have, however, quickly adopted deep learning techniques with impressive results. To better understand why, a deeper look at two fundamental concepts in modern NLP systems is first needed: the concept of **embeddings** to determine a numerical representation of a word, a sentence, or a text, and the concept of a **language model** to determine the probability of the next word in a sentence.

2.7.1.1 Vectors and embeddings

Categorical features from text are typically represented by numerical value to be processed by machine learning or probabilistic models. Similarly, sentences or documents must also have their **numerical representation** to be analyzed. In modern approaches to NLP, such representation is done by means of **vectors**, by points in a n-dimensional space, called a **vector space**. A model that uses vectors to represent a vocabulary enables the transformation of words, phrases, sentences, or documents into vectors, which enable numerical operations for downstream computing and machine learning tasks. With 10 words in vocabulary, a very naïve and nonoptimal way, for example, could be to use 10 dimensions to represent each word with a vector of one 1 and 9 zero's (also known as **one-hot encoding** which represents the categorical variables as binary vectors).

A foundational idea in modern NLP is based on the intuition that similar words tend to appear in similar contexts and that the **meaning** of a word is determined by the words around it. *"It assumes that linguistic items that occur in the same contexts have similar meanings and that therefore some representation of the contexts in which a word occurs is a good meaning representation for that word"* (Chapter 8). Words like "apple" and "peer" will more probably appear together in a document about fruits than with words like "car" or "bus." That's where the concept of **embedding** kicks in.

Word embedding techniques learn to represent words as vectors of numbers in a way that captures semantic, or meaning-related, relations. It is a learned vectorial representation, where words that have the same meaning have a similar representation, that is, they appear close to each other in their numeric representations, as clusters.

---

40 Lemmatization and stemming are techniques to reduce words to their root form.
41 Determine the emotional flavor of a sentence or a text, whether positive, negative, or neutral, or identify even more advanced emotional states such as joy, anger, etc.

It gives a **semantic representation** to the words. "Cat" and "kitten" will, for example, appear close to each other in their representation because they are mentioned in the same context; these will be attached to vectors that are close to each other. The semantic similarity is then formally measured by the concepts of a distance, such as the distance between 2 vectors. For example, "dog" and "puppy" will be closer to each other in their representation than "cat" and "dog" will be closer to "cat" than "houses" (Figure 2.7), etc. While learning such semantic relations, the process of embeddings also ensures that the learned representations are **low-dimensional** and **dense**. Indeed, embedding reduces the size of the representation compared to the initial size of a naïve encoding of the dictionary, such as with the one-hot encoding example above. It represents the words in a much more compact representation than dealing with the original full-size dictionary: for example, an input dictionary can contain n words whereas the embeddings might be defined to limit the representation to a smaller number of dimensions. Figure 2.7 illustrates these different concepts: "man" is to "woman" what "king" is to "queen" (pri-

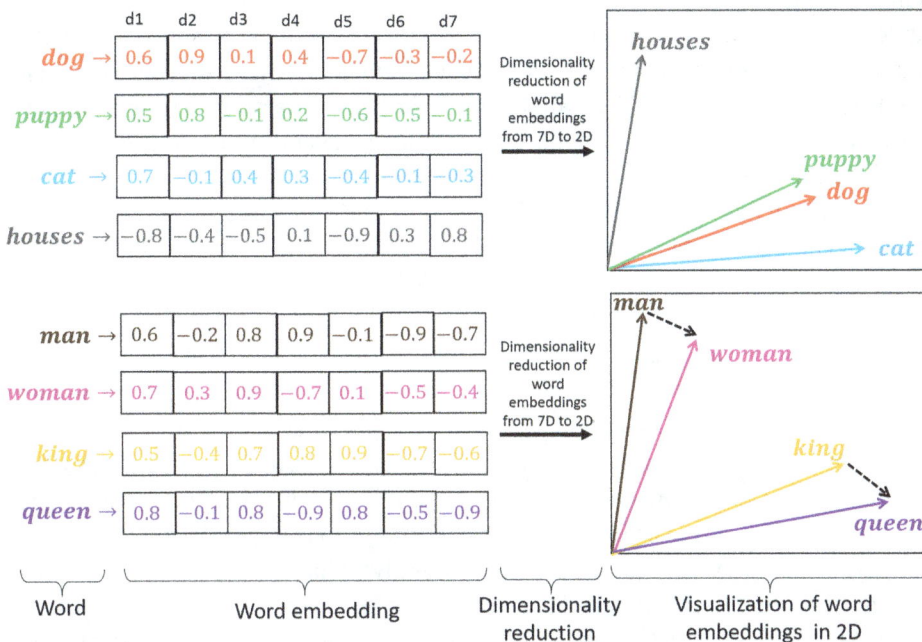

	d1	d2	d3	d4	d5	d6	d7
dog →	0.6	0.9	0.1	0.4	−0.7	−0.3	−0.2
puppy →	0.5	0.8	−0.1	0.2	−0.6	−0.5	−0.1
cat →	0.7	−0.1	0.4	0.3	−0.4	−0.1	−0.3
houses →	−0.8	−0.4	−0.5	0.1	−0.9	0.3	0.8

	d1	d2	d3	d4	d5	d6	d7
man →	0.6	−0.2	0.8	0.9	−0.1	−0.9	−0.7
woman →	0.7	0.3	0.9	−0.7	0.1	−0.5	−0.4
king →	0.5	−0.4	0.7	0.8	0.9	−0.7	−0.6
queen →	0.8	−0.1	0.8	−0.9	0.8	−0.5	−0.9

Word | Word embedding | Dimensionality reduction | Visualization of word embeddings in 2D

Figure 2.7: Dog, puppy, cat, man, woman, etc. are all part of an initial vocabulary, which might have n words. Seven dimensions (d1 to d7) are used in this example to attach a vector for the representation of these words. Those 7 dimensions capture the semantic relations between the words in the dictionary like "living being," "feline," or "gender." This dimension is then further reduced for a visualization in 2 dimensions. Source: Rozado David *"Wide range screening of algorithmic bias in word embedding models using large sentiment lexicons reveals underreported bias types"* 2020, https://doi.org/10.1371/journal.pone.0231189.

marily the gender), or "cat" and "dog" (living being) are closer to each other than they are to "houses". Note the reduction of the number of dimensions as well.

Embedding and similarity of meanings can also apply to sentences, documents, facts, or even pictures or audio, and then power multiple machine learning tasks for a diverse set of applications such as: text or image search based on semantics, text classification, question & answer systems, etc. In a question & answer application for example, the question *"When was John born?"* can be answered with *"John was born in 1970"* based on a higher similarity with the question than *"Peter was born in 1980"*. This method works by first mapping the database elements to their embeddings representation in the vector space, then mapping the user's query to its representation in the same space and finally choosing the element representation that is nearest to the query as an answer.

Basic word embedding techniques however are **static** in the sense that they don't capture the context: the word *"left,"* for example, will have exactly the same representation wherever it occurs in the text, even though it has 2 different meanings in the same sentence: *"I left my key on the left side of the table."* Given a word in the dictionary, a static embedding will always be the same, regardless of different context. Nowadays, capturing the **context** is addressed by more modern approaches with **contextual embedding**, typically generated by **transformers**, the latest generation of deep neural network architecture (see next section). These consider the entire sentence or the wider context before assigning an embedding. The embeddings generated for each word will then depend on the other words in a given sentence: *"left"* in our example will have two different embeddings. More about embeddings in Chapter 8.

2.7.1.2 Impact of deep neural network on natural language processing

Legacy machine learning-based approaches in NLP applications involve numerous manually intensive tasks. These include handcrafted engineering in the processing pipeline, such as preprocessing raw data, defining relevant text characteristics as features for the model, creating the learning models, and training them on specific tasks using manually annotated data sets. This is a manually intensive approach, which isn't realistic to address vast amounts of texts and documents. **Deep learning** has revolutionized the approaches by removing such chores: deep learning-based embedding algorithms process millions of documents on the internet to learn and produce condensed semantic representations of text or documents. These are **unsupervised learning** tasks performed at the scale of the internet. The data is then directly fed into these deep learning models, which learn the features and the language models (see next paragraph) from the data. So, it streamlines the end-end process by removing the need for **handcrafted feature engineering** and letting the deep neural network figure out the best features and models to use for the task at hand, which generally gives improved performance.

So, a deep neural network, trained on massive amounts of public data like Wikipedia in an unsupervised manner without labeling data, eliminates the need for complex, hand-crafted feature engineering and modeling. However, it comes at a cost as it is less transparent than traditional classical NLP approaches. It's also important to remember that these models inherit any **bias** present in the content used for training (Chapter 9).

2.7.1.3 Language models and transformers

After the embedding stage in the NLP processing steps, the objective of a **language model** is to predict the probability of a word (or a sentence,..., depending on the task) given the previous, or a set of other words around it. Language models are an essential part of the NLP system as they help create predictions for applications that have a language structure: predict the best answer to a question, write essays, summarize text, perform language translations, etc.

The techniques to make such predictions have evolved over time, from the classical statistical methods (such as simple counting, hidden Markov model, see Chapter 6) to the latest innovation in deep learning architectures: from **recurrent neural networks** to **transformer**-based language models, which rather process sentences as a whole using the mechanism of **attention** and additional type of **embeddings**,[42] concepts which will be covered in Chapter 8. Transformer-based architectures (Google, 2017) are at the heart of Google's BERT model and of the **generative pre-trained transformer (GPT)** models released by OpenAI research laboratory in 2018.

These models and their later versions have **pretrained language models** (an unsupervised pretraining step), which are then available for anyone to access and use. This approach has the main benefit that data scientists don't need to train a language model from scratch, which is expensive, computationally intensive, and needs huge amounts of texts. Transferring this learning (also known as **transfer learning**), the pretrained model is then possibly **fine-tuned** on downstream tasks (such as text classification, entity extraction, question answering) with a specific dataset and limited training needs.

And the progress has continued: the next generation goes even a step further by becoming **multi-tasks**, without requiring fine-tuning on downstream tasks. They can perform tasks for which they were not explicitly trained, with only a few, or even no training examples (this is respectively referred to as **few-shot learning** and **zero-shot learning**). Further details are available in Chapter 8.

---

42 Positional embeddings keep track of positions, segment embeddings keep track of structural components.

2.7.2 Industry applications and examples in the book

NLP applications can broadly be split by their tasks: they can **classify, extract, summarize, retrieve** information, or based on instructions, **generate information** into human-understandable text, credible images or even videos.

Textual NLP systems can determine the language of documents, translate, classify, and summarize documents, retrieve documents that are semantically relevant to a user query, and more. They can also analyze the content from documents: analyze the sentiment of texts, classify sections of text, extract information such as interesting **entities** (the instances of predefined categories such as people, date, location, such as "Brussels" as an instance of a city), and their relations[43] and derive key topics. NLP techniques also support **question answering** systems to pull answers from a collection of documents. All those capabilities are not only useful to automate company business processes (processing of orders, invoices, contracts, sales documents, etc) but also to automatically mine vast amounts of documents, perform advanced, semantic-based searches, develop applications to interact with documents in natural language and get much faster insight from these.

In generation related tasks, **natural language generation** can create sequences of tokens (e. g., words) based on a context, like suggesting the best next word to type in (e. g., when you type in a SMS) or automating the creation of human understandable content: generate responses in conversational AI agent, develop business reports, write news or articles, generate computer code, create credible images, and even videos.

More recently, NLP techniques apply in **multimodal** applications to understand and generate information across multiple modes of communication, such as text, speech, images, and videos. This allows, for instance, the automatic generation of text descriptions and summaries of what images or videos contain.

Legal Village, an **AXA**[44] business unit specialist in legal protection insurance, leverages a **knowledge mining** solution developed by **Pythagoria**, a company based in Luxembourg that offers services and solutions for knowledge management and text mining. Their solution to assist the lawyers in their daily tasks is illustrated in Chapter 8. The same chapter also describes how the procurement teams of **Daimler AG**,[45] a leading international automotive manufacturer based in Germany, are assisted by a solution developed by **Icertis**, an American software company that provides contract management solutions to enterprise businesses. The solution automatically parses contracts from different languages, sources and formats, identifies the contractual clauses, classifies them, then extracts relevant metadata and semantics. The information processed is

---

43 Information extraction (from unstructured information in texts into structured data) and relation extraction. So, NLP systems can help automate the creation of ontologies. See Chapter 5.

44 French multinational insurance company.

45 The official company name of Daimler AG is now Mercedes-Benz Group AG (February 2022).

then available for structured, rich semantics-based searches and reports. Finally, in the same chapter, we see how different NLP techniques, including transformer techniques, are used in **Microsoft QnA Maker** to create knowledge base by extracting questions and answers from manuals, FAQs, and documents. Microsoft QnA Maker is then used in chatbot applications to answer questions from users. Note that the techniques illustrated by the example of QnA Maker are no longer used in the new commercial applications: indeed, newer and more sophisticated methods have emerged since the book writing process began, but the basic ideas are still very applicable.

By adding a dialog management system and a response generator (using template or generative AI models), **conversational agents** or **chatbots** can create interactive systems that can converse in human languages: **KBC Group** offers an AI-enabled chatbot to automate several parts of the claim process in their car insurance offer. With half of the claims now being fully automated, their chatbot reduces the labor chore and offers a faster resolution time for their customers. See a description of the example in Chapter 10. **Amazon Alexa, Apple Siri, Microsoft Cortana, Google Assistant** are all well-known examples of voice activated personal assistants available on the market. As a result of breakthrough in NLP techniques, **ChatGPT**, developed and released by OpenAI in November 2022. **Microsoft Copilot**, was launched by Microsoft in the course of 2023. Both are examples of chatbots based on **large language models** (LLM, namely GPT-3.5 and GPT-4), fine-tuned[46] to handle conversations.

2.7.3 Limitations of natural language processing

Although they can generate high-quality text and display impressive useful results, NLP language models nowadays are fundamentally based on **probabilities**: they are not deterministic, still not fully reliable in the sense that they are still prone to so-called **hallucinations**, which refers to the generation of outputs that may sound plausible but are either factually incorrect or unrelated to the given context. These false outputs often emerge from the model's inherent biases, its lack of real-world understanding, the limitations of the training data, the model's propensity to guess based on statistical patterns rather than factual accuracy, etc. Although the rates of hallucination might be reduced and have improved with certain techniques, this could be detrimental in situations where deterministic and less error-prone answers are required. They also inherit the **errors** and **bias** introduced by their training material, which must therefore be carefully considered. An interesting paragraph about ChatGPT limitations, for example, can be found on the blog of OpenAI *"Introducing ChatGPT (openai.com)."* These aspects are important areas of current research. See Chapter 8 and Chapter 9.

How much of the **commonsense knowledge** is really picked up by those models from the available content on the internet is also debatable: will people explicitly ex-

---

46 With supervised learning and reinforcement learning techniques.

press on the internet what makes common sense to them? Is that only a question of time?

Besides limitations, new technologies also present the dangers of **misuse**. Performant NLP techniques such as GPT-3 and beyond, present potential harmful effects, raised by their inventors themselves. For example, such NLP systems can be used to automatically generate **fakes** that can hardly be distinguished from articles written by humans and, therefore, be used as weapons for misinformation, phishing, fraudulent writing, and influence the public masses. Artificial intelligence systems can generate text, audio, images, videos that are so realistic that humans may have a hard time distinguishing between outputs that are created by technology and those that are not. This again raises **ethical concerns** and the need for techniques that can address and mitigate those risks, such as fake detection techniques.

2.8 The importance of an ethical approach to AI

Trust is a prerequisite for people, companies, and societies to deploy and use AI systems. As usage of AI applications has grown, so has the awareness of the various risks raised by AI systems. Like any technology, they are susceptible not only to malicious usage but also to inherent imperfections. These imperfections can manifest as **errors** and **biases** in their predictions or classifications, leading to incorrect decisions being made. The effectiveness of trained systems is only as good as the data they were trained on and the models they employ, with errors. Learning systems can also sometimes operate in unexpected, undesirable, and opaque ways.

These imperfections thus raise **ethical concerns** as AI applications increasingly assist in taking decisions that impact people's lives whether, for example, someone gets a job, a loan, or something else ruled by an algorithm. Applications must be aligned with fairness, transparency, and justice goals to avoid the possible negative outcomes of decisions taken or supported by AI systems. While technology companies and public organizations have raised concerns and proposed different sets of guidelines, a couple of **ethical principles** seem to emerge: an analysis that aims at mapping the global landscape of existing guidelines for **ethical AI** systems reveal some commonalities: *"Our results reveal a global convergence emerging around 5 ethical principles: transparency, justice and fairness, nonmaleficence, responsibility and privacy, with substantive divergence in relation to how these principles are interpreted; why they are deemed important; what issue, domain, or actors they pertain to; and how they should be implemented."* (Anna Jobin, Marcello Ienca, and Effy Vayena, artificial intelligence: the global landscape of ethics guidelines, 2019). Consensus seems to emerge that **ethical AI** systems should adhere to the following principles:
- be **understandable** by humans,
- be **fair and inclusive** to treat all people and groups the same way,

- be **accountable** to provide people that have been harmed with a remedy,
- be **reliable and safe** for humans,
- respect **privacy**.

As a world's first, it's worth noting that the European Commission released in April 2021 a proposal for a regulatory framework, called the "**Artificial Intelligence Act,**" to ensure that AI systems are used in ways that respect fundamental rights and European values including human oversight, safety, privacy, transparency, nondiscrimination, and social and environmental well-being. At this time of reviewing, a draft text of the legislation serves as the negotiating position for talks between the member states and the European Parliament (see EU AI Act: first regulation on artificial intelligence).

Trustworthy AI systems require a holistic approach, conscious efforts from all the stakeholders to address the risks but it's outside the purpose of this book to consider all the aspects. Chapter 9 gives a glimpse at the risks and techniques to mitigate those, with a particular focus on two specific aspects: how to protect AI systems from bias (known as **fairness**) and have them more understandable (known as **interpretability**). Chapter 9 also illustrates how **EY** (also known as **Ernst&Young**, a multinational professional services firm that specializes in providing assurance, tax, consulting, and advisory services) applies some of those techniques to improve the fairness of loans decisions. It also shows how an airline company improved both the **transparency** and **fairness** of their fraud detection models, a common issue in customer loyalty programs.

Bibliography

Answer Set Programming, https://en.wikipedia.org/wiki/Answer_set_programming.

Chollet François. On the measure of Intelligence, 2019, https://arxiv.org/pdf/1911.01547.pdf.

EU AI Act: first regulation on artificial intelligence, https://www.europarl.europa.eu/news/en/headlines/society/20230601STO93804/eu-ai-act-first-regulation-on-artificial-intelligence.

Planning Domain Definition Language, https://en.wikipedia.org/wiki/Planning_Domain_Definition_Language.

Goertzel Ben. Artificial General Intelligence: Concept, State of the Art, and Future Prospects, January 2014.

Grace Katja, Salvatier John, Dafoe Allan, Zhang Baobao, Evans Owain. When Will AI Exceed Human Performance? Evidence from AI Experts, 2018, https://arxiv.org/pdf/1705.08807.pdf.

Introducing ChatGPT (openai.com), https://openai.com/blog/chatgpt.

Jobin Anna, Ienca Marcello, and Vayena Effy. Artificial Intelligence: the global landscape of ethics guidelines, 2019, https://arxiv.org/ftp/arxiv/papers/1906/1906.11668.pdf.

Legg Shane, Hutter Marcus. A collection of definitions of intelligence, 2007, https://arxiv.org/abs/0706.3639.

Mehta Pankaj et al. A high-bias, low-variance introduction to machine learning for physicists, 2019, https://www.sciencedirect.com/science/article/pii/S0370157319300766?via.

Microsoft Flash Fill, https://www.microsoft.com/en-us/research/wp-content/uploads/2016/12/popl11-synthesis.pdf.

More perfect Cheetos: How PepsiCo is using Microsoft's Project Bonsai to raise the (snack) bar, https://blogs.microsoft.com/ai-for-business/pepsico-perfect-cheetos/.

Muggleton Stephen, Inductive logic programming, 1991, https://www.doc.ic.ac.uk/~shm/Papers/ilp.pdf.
Robovision, https://robovision.ai/.
Rozado David. Wide range screening of algorithmic bias in word embedding models using large sentiment lexicons reveals underreported bias types, 2020, https://doi.org/10.1371/journal.pone.0231189.
Sioen Industries, https://sioen.com/en.
Viu More, https://www.viumore.com/.
OWL, https://en.wikipedia.org/wiki/Web_Ontology_Language.

Yves Deville

3 Solve problems by searching, including with constraints, a fundamental pillar

3.1 Why is solving problems by search important within the broader artificial intelligence (AI) domain?

Search techniques are important and efficient tools for decision-making and solving complex combinatorial problems. This chapter proposes different paradigms that can be integrated in search agents to handle various classes of problems. Classical search algorithms are used in goal-based agents that are looking for a series of actions leading to a specific goal. A classical application is route planning in a search algorithm, the environment is modeled as a state. At each step, the agent has to choose between different actions that change the current state of the environment. The set of possible actions lead to state space graph that has to be searched by the algorithm in order to find an adequate series of actions leading to the goal. Many AI applications are optimization problems looking for the best solution according to some cost function. Search algorithms can then be extended to ensure the finding of the best solution and integrate heuristics to speed-up the search.

The second searching paradigm is called constraint satisfaction problem (CSP). It generalizes classical search techniques as it proposes an expressive modeling through decision variables and constraints that must be satisfied by a solution. It also covers optimization problems by adding an objective function to be maximized or minimized. Two approaches are described for solving CSP. Constraint programming offers a propagation strategy that reduces the search space. Local search, although it cannot always find an optimal solution, is a pragmatic and well-used technique for solving complex problems. It ensures to rapidly find an approximation of the best solution. This chapter presents an application related to a configuration problem often present in product and software delivery. But CSPs can be used in many AI and operation research problems such as scheduling, timetabling, manufacturing problems, business, circuit design, configuration, etc.

Logic is an important tool in AI. First-order and other advanced logics can be used for reasoning and knowledge representation (see Chapters 4 and 5). This chapter shows how propositional logic, the simplest logic, can be used not only to model complex AI problems and CSPs but also to efficiently solve them by using SAT solvers.

https://doi.org/10.1515/9783111426143-003

3.2 Search algorithms

Search strategies are important methods underlying many approaches for problem solving. Search algorithms are also the basis for many optimization and planning methods.

3.2.1 What category of problems do search algorithms solve?

Search algorithms address problems for which the environment can be observed and modeled by a set of states, with known and deterministic rules of transition from one state to another. Using these rules, we look ahead for the best sequence of actions to take in order to reach from an initial state a specific goal, either a specific end state or a state that matches some conditions. The environmental knowledge that supports such decision is explicitly represented by states, and the decision taken by the agent involves the consideration of possible steps in the future, in order to reach the defined goal. So, the proposed techniques belong therefore to the category of "goal based" agent.

The use of search techniques requires an abstract formulation of the problem and of the available steps to construct a solution. By search, we mean the process of looking for a sequence of actions that leads to a goal starting from an initial state. For an optimization problem, one looks for the best sequence of actions, that is, the sequence with the lowest cost.

Example. The 8-puzzle, also called sliding blocks, is a simple but illustrative example. Starting from an initial state (see Figure 3.1), an action consists of moving a tile adjacent to the empty cell, also called blank cell, to the empty cell. The objective is then to find the (minimal number of) actions leading to the goal state from the initial state.

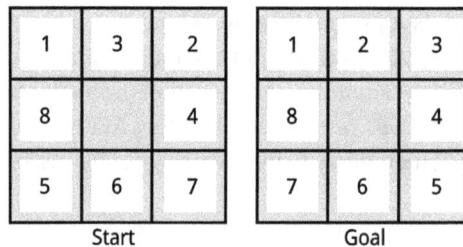

Figure 3.1: 8-Puzzle.

A state is an abstract representation of a possible stage of the problem. The initial state is a starting state for the search. The hypothesis is that the state is discrete, meaning that the number of states is finite, although usually very large.

An action transforms one state into another one by the application of an operator to the state. An operator can have preconditions on the given state to be applicable. The set of available operators depends on the given problem. The number of possible actions is an important factor on the complexity for solving the problem.

The goal can be an explicit state, but more generally, the goal is represented by a condition that must be satisfied by a state. In this case, the goal is a set of states.

In an optimization problem, a nonnegative cost is associated to each action. The cost of a sequence of actions, called a path cost, is then the sum of the cost of the actions in the sequence.

Example. A state for the 8-puzzle could be represented by a vector of 9 elements, listing the tiles, line by line, using the symbol B for the empty cell. The initial state is represented by $[1, 3, 2, 8, B, 4, 5, 6, 7]$. Of course, many other representations can be used, such as a $3{\times}3$ matrix.

An action for the 8-puzzle could be to move a (numbered) tile Up, Down, Left, or Right. Each action has a precondition that the move is possible (the target position is the empty cell). We thus have potentially $8 \times 4 = 32$ actions for a given state. A better design for the actions is to consider the move of the empty tile Up, Down, Left, or Right. We now only have 4 possible actions per state. As an example, $moveRight([1, 3, 2, 8, B, 4, 5, 6, 7]) = [1, 3, 2, 8, 4, B, 5, 6, 7]$.

The goal test of the 8-puzzle is the single state $[1, 2, 3, 8, B, 4, 7, 6, 5]$.

In the 8-puzzle, the cost of each action is 1.

3.2.2 Solving problems by search without heuristics

As illustrated by the above example, the overall approach to solve search problems requires essentially the following:
- modeling the problem as a search problem, defined by states, an initial state and a description of the possible transitions from a state to another one;
- a goal, defined by a condition which must be satisfied;
- search activities to identify the right sequence of actions, that is, choose the next action from several possibilities, in order to reach the goal.

Once a sequence of actions to reach the goal conditions is found, the agent can then execute that list of actions. The search activities essentially consist of growing a search tree directly on a "state space" graph until a solution is found. Different search strategies exist, and they can be uninformed or informed, with the notion of "heuristics."

The search for a solution, that is a sequence of actions, can be performed on a search tree derived from expanding the current state using the possible operators. This leads to tree-search algorithms that generate and traverse such a tree in order to find a state satisfying the goal.

A node in a search tree contains a state, a link to its parent node, the action applied on the state of the parent node yielding the state of the node, the depth of the node in the tree, and the path cost from the root to the node. In Figure 3.2, labels on the node correspond to different states. It clearly shows that in a search tree, the same state may appear in different nodes. Although the number of states is finite, the search tree can therefore be infinite.

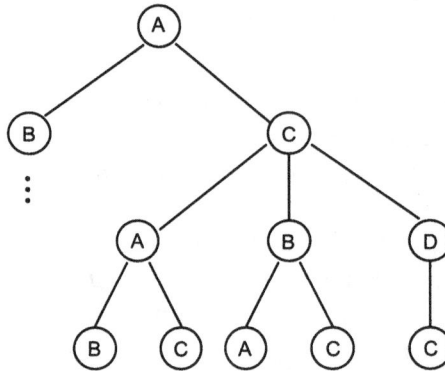

Figure 3.2: A search tree.

It is important to notice that the solution of the problem is not only the node with a state satisfying the goal, called a goal node, but the whole path from the root of the tree to a goal node defining a sequence of actions.

The parameters characterizing the complexity of a search problem are the branching factor, that is the average number of children of a node. In the 8-puzzle, the branching factor is about 1.732, assuming we do not consider the action to return to the state of the parent node. An example of a search tree for the 8-puzzle problem is shown in Figure 3.3.

3.2.2.1 A generic tree search algorithm

Figure 3.4 proposes a generic tree search algorithm. It has two parameters, the problem to be solved, and the frontier that should be empty when launching this algorithm. The frontier is an abstract data structure maintaining a set of nodes which ancestors in the search tree have all been goal-tested that have been visited.

The algorithm picks a node in the frontier and checks whether it has a state that satisfies the goal. If it is not the case, news nodes are created for each of the possible states reachable by an action from the state of the current node (the expand algorithm). These nodes are then inserted in the frontier and the search continues. At the beginning, the algorithm should start from the initial state. This is achieved by calling this tree

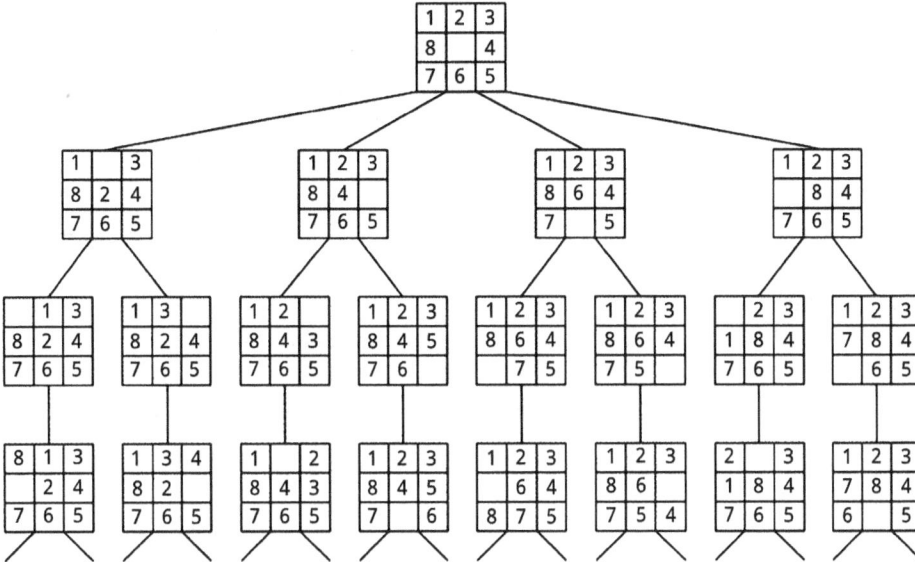

Figure 3.3: A search tree for the 8-puzzle.

```
tree_search( problem, frontier) : Node or Failure
        root := Node( problem.initial_state )
        frontier.insert( root )
        while not frontier.is_empty()
                node := frontier.pop()
                if problem.goal_test( node.state ) return node
                frontier.insert( expand( node, problem ) )
        return Failure
```

```
expand( node, problem ) : list of Nodes
        successors := empty_set
        for each operator in problem.operators such that operator.precondition( node.state )
                n := Node()
                n.state := operator.result(node.state)
                n.depth := node.depth + 1
                n. parent := node
                n.operator := operator
                n.path_cost := node.path_cost + operator.cost
                successors.add( n )
        return successors
```

Figure 3.4: Generic tree search algorithm.

search algorithm with the frontier containing only the root node with the initial state. The search will then start from the initial state.

The frontier is organized in such a way that the search algorithm chooses the next node to test and expand according to some preferred search strategy. Different organizations of the frontier lead to different versions of the algorithms. The frontier should be able to insert one or several nodes and should return the first element according to the internal organization of the frontier.

Figure 3.5 illustrates the first four iterations of the tree search algorithm on a small abstract example. The list of nodes in the frontier is successively (A), then (B, C), (C, D, E), and (D, E, F, G). In each step, the expanded node is highlighted by a marker.

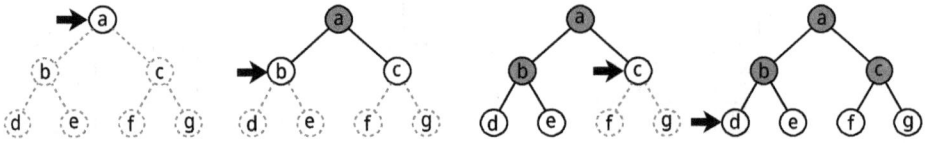

Figure 3.5: An example of execution.

The result of the tree-search algorithm is not only a node, but it is also implicitly the path from the root to the returned node, from which the sequence of actions to reach the goal from the initial state can be extracted.

The generic tree-search algorithm performed an uninformed search, also called blind search. The search is not goal oriented, nor does it exploit the path costs to guide the search to an optimal solution.

3.2.2.2 Breadth-first and depth-first search algorithms

Breadth-first search (BFS) is an instance of the generic tree search algorithm, depicted in Figure 3.6, where the frontier is organized as a FIFO queue (first in, first out). It means that after the visit of the root node, the algorithm will visit all the nodes at depth 1 before visiting the nodes at depth 2, etc. The order of the node visit is illustrated in Figure 3.7. It is assumed that the insertion of the nodes in the frontier realized by the frontier.insert() operation is done from left to right.

Depth-first search (DFS) is another instance of the tree search algorithm, also depicted in Figure 3.6. In DFS, the frontier is organized as a stack or LIFO queue (last in, first out). The order of the node visit is illustrated in Figure 3.8. It is assumed that the insertion of the nodes in the frontier realized by the frontier.insert() operation is done from right to left.

breadth_first_search_tree_search(problem, frontier) : Node or Failure
return tree_search(problem, *FIFO_queue()*)

depth_first_search_tree_search(problem, frontier) : Node or Failure
return tree_search(problem, *stack()*)

uniform_cost_tree_search(problem, frontier) : Node or Failure
return tree_search(problem, *priority_queue_G()*)

Figure 3.6: BFS, DFS and uniform cost algorithms.

The evaluation of search algorithms can be done according to different criteria. Does the algorithm find a solution if one exists, what is called completeness? Is it able to find the least cost solution, what is called optimality? What is its computation time and space complexities? Computation time and space complexities are measured in terms of the (maximum) branching factor (b) of the search tree, the depth of the least cost solution (d) and the maximum depth (m) of the search tree. The value b is assumed to be finite, but the value m can be infinite. The time complexity is measured in terms of the number of nodes visited.

BFS is complete; its time and space complexities are $O(b^{d+1})$, the number of generated nodes. It is also optimal when the cost of each possible action is 1. BFS is nice as it handles very well infinite search tree. Its space complexity is often a problem for large problem. No nodes can be deleted during the search as they may be part of the solution, that is a path from the root to a goal node.

BFS can also be seen as an instance of the generic algorithm where the frontier is organized as a priority queue ordered by the depth of the node. Uniform cost search is a standard variant of BFS, depicted in Figure 3.4, where the frontier is a priority queue ordered by the path cost of the node (denoted $g(n)$). The node with the lowest cost is thus explored first. Uniform cost is complete and optimal.

DFS is not complete as it can fall in an infinite branch of the tree when m is infinite. If m is finite, the time complexity is $O(b^m)$. Its space complexity is $O(mb)$, that is the maximum number of nodes in the frontier. One can observe that all the nodes in the frontier are all children of nodes in a single path in the tree. All the visited nodes without children in the frontier can thus be removed during the search. DFS is not optimal.

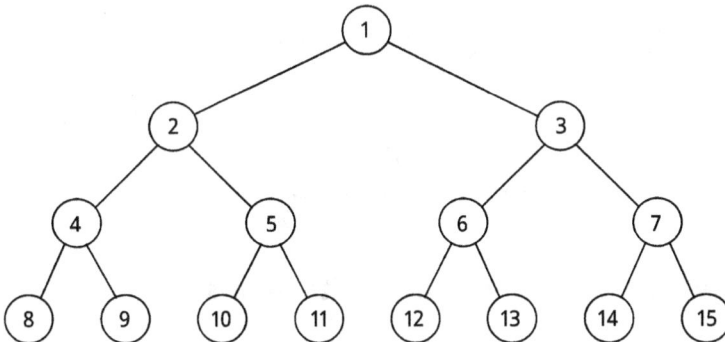

Figure 3.7: Order of node visit in BFS algorithms.

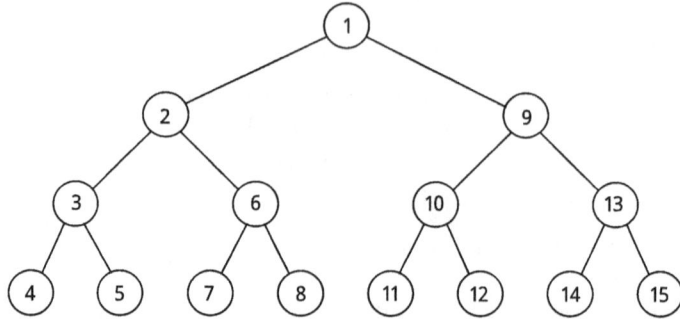

Figure 3.8: Order of node visit in DFS algorithm.

3.2.2.3 Iterative deepening

BFS has nice properties such as completeness and optimality but has a space issue. DFS has a nice space complexity but is not complete nor optimal. Iterative deepening combines the qualities of both algorithms. The idea is to apply DFS, but with a limited depth. The search tree is not expanded beyond the depth limit. Of course, DFS with a depth limit is not complete. Iterative deepening performs a succession of DFS with a increasing depth limit ($0, 1, 2, \ldots$) until a solution is found. It is complete as a solution will be found if there is one. The space complexity is $O(mb)$ as we restart a DFS at each step. Many nodes are visited multiple times, as illustrated in Figure 3.9. The root node is generated ($d + 1$) times, the d nodes at depth 1 are generated d times, ... the b^{d-1} nodes at depth $d - 1$ are generated twice, the b^d nodes at depth d are generated once. From a complexity perspective, the time complexity is thus $O(b^d)$. Iterative deepening is thus a suitable uninformed algorithm for tree search.

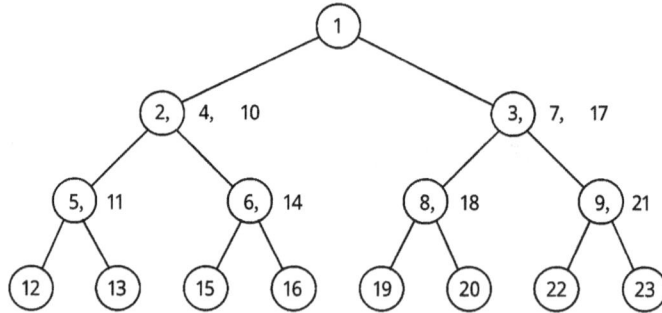

Figure 3.9: Order of node visit in iterative deepening algorithm.

3.2.3 Solving problems by search with heuristics

3.2.3.1 Evaluation and heuristics

BFS has nice properties. The idea of informed search is to exploit problem specific knowledge to find or deduce information about future states and future paths. This information is then used to make better decisions when traversing the search tree. Technically, informed search associates an evaluation function to each node n, denoted $f(n)$. This value models the quality of the node. Node with a low $f(n)$ value will be preferred when selecting a node in the frontier.

One usually decomposes the evaluation function $f(n)$ in two components:

$$f(n) = g(n) + h(n)$$

where $g(n)$ is the *exact* cost to reach node n from root node with the initial state, and $h(n)$ is an *estimated* cost, called the heuristic function, to reach a goal from node n. When no heuristic is used (i. e. $h(n) = 0$ for all n), the search reduces to an (uninformed) uniform cost search based on $g(n)$.

The objective of the heuristic function is to speed up the search from a node to a goal, by favoring nodes that are expected to be closer to a goal. A heuristic function $h(n)$ is called *admissible* if it never overestimates the real cost to reach a goal. The heuristic is thus optimistic. As a consequence, the value $f(n)$ never overestimates the cost of a solution through the node n. When n is a node with a goal state, $h(n) = 0$ if the heuristic is admissible.

3.2.3.2 The A* algorithm

The integration of a heuristic function leads to two different instances of the generic algorithm where the frontier is organized as a priority queue. In the greedy best first algorithm, the order is based on the heuristic function $h(n)$. In the A* algorithm, the order is based on $g(n) + f(n)$. In the greedy best first algorithm, the objective is to reach a goal node as soon as possible from the frontier. In the A* algorithm, the objective is to find the optimal solution as soon as possible. A* is thus particularly adapted to optimization problems.

Greedy best first algorithm is not complete as it can fall in an infinite branch of the tree when m is infinite. If m is finite, its time and space complexity is $O(b^m)$. It is also not optimal. But this algorithm is often better than DFS thanks to the heuristic function guiding the search.

A* is complete, with a time and space complexity if $O(b^d)$. Contrary to DFS, all the visited nodes must be kept in memory as they may be part of the solution. When the heuristics is admissible, then A* is optimal. It thus always returns an optimal solution. If the cost of the best solution is c^*, then A* will expand each node n with $f(n) < c^*$ but never expand a node n with $f(n) > c^*$. A nonadmissible heuristic can be useful to speed up the search, but A* does not longer grantees to find an optimal solution.

The idea of iterative deepening can also be applied on A*, leading to iterative deepening A* (IDA*). Instead of increasing the depth of the search from one execution of A* to the execution of A*, one increases the maximum value of the evaluation of nodes in the frontier that are expanded. Technically speaking, at each iteration, A* only expands nodes with an evaluation less than or equal to the smallest value of the evaluation function among all the nonexpanded nodes in the previous iteration. IDA* is complete, its time complexity is exponential (but hard to express) and its space complexity is linear. It is also optimal.

3.2.3.3 Examples

Example 8-puzzle
Different heuristics could be defined for the 8-puzzle problem.
- $h_1(n)$ is the number of misplaced tiles ($h(n) = 4$ in the start state in Figure 3.1).
- $h_2(n)$ is the sum of the Manhattan distance (horizontal + vertical distance) of each tile to its position in the goal ($h(n) = 6$ in the start state in Figure 3.1).

Both heuristics can be computed efficiently and are also admissible. They never overestimate the number of actions required to reach the goal. For any given node n, heuristics $h_2(n)$ will always be greater than or equal to $h_1(n)$. Heuristic $h_2(n)$ is said to dominate $h_1(n)$ as A* will always expand less nodes with $h_2(n)$ than with $h_1(n)$.

Example route planner
Finding the shortest path to travel from one position to another is a classical problem in combinatorial optimization. It can be modeled as finding the shortest path between a source node and a destination node in a weighted graph (**single-pair shortest path problem**), where the cost of an arc(A, B) is the distance (or time) to reach A from B and vice versa. An example is illustrated in Figure 3.10. The shortest path from A to D is A-B-F-C-D and has a cost of 18.

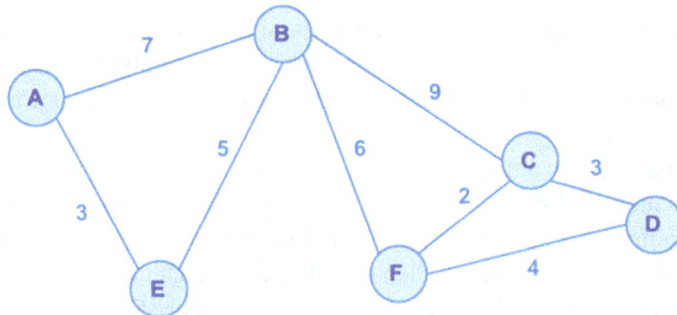

Figure 3.10: Shortest path problem.

The shortest path can be solved by the well-known Dijkstra algorithm. The complexity of this algorithm is $O((n + a) \log(n))$ using a priority queue, where n is the number of nodes and a the number of arcs in the graph. The Dijkstra algorithm actually computes the single-source shortest path problem, that is, finding the shortest path from a source node to all other nodes of the graph. For solving the single-pair shortest path problem, such a complexity can be unaffordable for very large graph as requires at worst to visit all the nodes and arcs of the graph.

The Dijsktra algorithm for solving the single-pair shortest path problem can be seen as the A* algorithm, but with a heuristic function $h(n) = 0$. By adding a heuristic, the performance of the search can be drastically improved. The standard heuristic used in travel routing problem is to use the Euclidean distance between a node of the graph and the goal node as an approximation of the cost of the path from this node of the graph to the goal node. This heuristic is admissible as it never overestimates the real cost: the Euclidian distance is the shortest distance between two points.

3.2.3.4 Designing and comparing heuristics

Designing a heuristic is not a simple task. A heuristic $h(n) = 0$ for all nodes is admissible but does not provide any information. On the other extreme, a heuristic function $h(n)$ yielding the exact distance to a goal is very informative but is computationally too costly as its evaluation requires itself a whole search. The basic idea is to relax the problem, for instance by reducing restrictions on possible actions allowing an efficient computation. In the 8 puzzle, heuristic $h_1(n)$ allows teleportation and $h_2(n)$ allows the superposition of tiles. In the shortest path problem, the heuristic removes the requirement to take roads.

How to compare different heuristics for a given problem? For a given instance of the problem, the shape of the search tree for the two heuristics could be very different and, therefore, not comparable. Comparing the number of expanded nodes in both trees is a possibility, but these numbers are specific to instances. In order to be independent from the instances, one introduces the *effective branching factor*. The idea is to reorganize a search tree into a complete tree in order to make them comparable. The branching factor b* of the complete tree is then the effective branching factor of the initial search tree. If we have a search tree with N nodes, the effective branching factor b* is the solution of the following equation where d is the depth of the goal node:

$$N = 1 + b^* + (b^*)^2 + \cdots + (b^*)^d$$

What is nice is that the effective branching factor is relatively independent from the depth of the solution and comparison can be made between different instances. An effective branching factor of 1 is ideal and means that at each step of the search, the right choice is made.

The effective branching factor of the 8-puzzle without heuristics is around 2.80 but reduces around 1.45 with heuristic $h_1(n)$ and around 1.30 with heuristic $h_2(n)$.

3.2.4 What are the limitations of search algorithms?

Solving by search proposes a series of algorithms that are more similar than different. They are all based on the same generic algorithm; they differ from the order of the nodes that are visited in the frontier. For solving a specific problem, one has to complete the chosen algorithm by designing a data structure representing a state and by developing methods encoding the goal test and the possible actions (operators) on a state. It thus requires some programming skill.

From an algorithm point of view, if the search is uninformed, IDA is a good choice among the described algorithms in this chapter. When it is possible to exploit a heuristic, A^* is a nice algorithm but has space complexity issue. IDA^* is a possible way, but there exist many other approaches, based on A^*, that offer a better space complexity than A^* (recursive best-first search, simple memory-bounded A^*).

Search algorithms are well suited when there is an initial state, a (set of) goal state(s), and the expected result is a sequence of actions leading to a goal. However, for many optimization problems, the objective is to find a goal and there is no need for a sequence of actions. The cost of a solution is then a function of the state rather than a sum of the cost of actions.

In all the proposed algorithms, the same state can be visited multiple times. This can be avoided by considering graph search instead of tree search. The idea is simple. In the generic tree search algorithm, a list of visited states, called closed, is maintained during the search. The extracted node in the frontier is expanded only if its state has never been visited. The generic graph search algorithm is depicted in Figure 3.11. All the variants of the tree search can also be applied for graph search. The time complexities are similar as they describe worst cases. However, the space complexity of graph search is always exponential as all the visited states have to be recorded.

```
graph_search( problem, frontier) : Node or Failure
        closed := emp yset
        root := Node( problem.initial_state )
        frontier.insert( root )
        while not frontier.is_empty()
                node := frontier.pop()
                if problem.goal_test( node.state )  return node
                if node.state  not in  closed
                        closed.add( node.state )
                        frontier.insert( expand( node, problem ) )
        return Failure
```

Figure 3.11: Generic graph search algorithm.

When solving a problem by a search algorithm, we need a procedural representation of the problem through a data structure and methods encoding the goal test and the possible actions. A procedural representation explicitly describes how to do or compute things. At the opposite, a declarative representation describes the characteristics and properties of the problem without explicitly proposing a computational way to update data structures not to solve the problem. Query languages to query databases like SQL, are typically declarative, for example. In order to exploit a declarative representation of a problem, one needs a generic problem solver such as constraint satisfaction or SAT solvers, developed in the next two sections. A generic problem solver is an algorithm that is able to solve any problem based on a given representation of the problem.

In an informed search, it is possible to exploit a heuristic, that is, properties of the problem to speed-up the search without changing the search algorithm. The heuristic is coded in a specific and external evaluation function. This approach has two limitations. First, it can only handle properties that can be encoded in a function evaluating the cost from a state to a goal. Second, it can only state problem dependent heuristics and no generic heuristics.

Search techniques only apply to deterministic and known environments that can be modeled by states and atomic actions. Limitations are that the modeling and the heuristics are problem specific, that the modeling might lead to an explosion of the number of states, and that although search techniques are able to perform some "look ahead" prediction through heuristics, but they don't infer nor deduce information, like "reasoning approaches" do.

The next sections will show how these limitations can be lifted and how a generic heuristic can be handled.

3.2.5 Industry example. The fastest route with search and heuristics, using Bing Maps Routes API

Yves Deville, Sanjib Dutta

3.2.5.1 Route planner

A route planner is a tool that finds the shortest path from one location to another location on an existing road network. While it sounds like a very trivial problem to solve, it has many challenges given the size of the road networks and the level of customization needed to support various user choices. For instance, preferences and constraints can be expressed to leave or arrive at a certain time, go through intermediate locations, change the cost function (time, distance, emission, financial), avoid some roads, combine different transport modes, or integrate schedules of public transportation. Route planning services are proposed by many tools and apps such as Google Maps, Route XL, Waze, MapQuest, Bing, and GraphHopper.

Unlike most of the other graph problems, a routing graph is dynamic. The cost of an arc can change dynamically based on the traffic condition at a particular time. A route, which is least cost at 12 AM in the night, may not be the best choice at 7 AM. Reversible road segments also reverse the direction of travel based on the time of the day; some turns also become illegal based on the time. Some roads segments may or may not be traveled based on the user's destination. For example, if the user's destination is a parking lot, the parking lot may be used for routing, but should not be used (even if using it leads to a better cost route) otherwise.

Performance is an important issue when planning a route from one point to some destination point. Because of the size of the road network, a simple 50-mile route may end up touching millions of edges before finding the optimal route. One of the techniques that is widely used to solve the performance issue is using shortcuts (precomputed small routes) that reduces the number of edges significantly, but shortcuts cannot be used with dynamic graphs or where user preference is considered.

A* with a good choice of heuristics can be used to considerably reduce the number of nodes needed to be examined during the search in a dynamic graph and user preference-based filtering.

3.2.5.2 Bing map routing using A*

Bing routing offers a set of tools to get directions and route information to and from anywhere. It provides an application programming interface (API) called Bing Maps Routes API to perform different tasks, including various routing problems. Bing router uses Euclidean distance (great circle distance) to the target location for the heuristic function to determine the cost of a node that is being explored next. To calculate the time required to travel this estimated distance, various options can be used. Examples are (i) incoming road speed, (ii) a fixed road speed based on the region (max speed in that region). Issues with (i) is that highways get prioritized too much over local roads and there is a risk that many legitimate local paths may not be explored. On the other hand, (ii) does not prioritize highways over the local roads. Bing map routing exploits a refined version of (ii) using road hierarchy. A higher hierarchy road will be given a little more preference unless it is too near to the source or the destination. Assuming there are 5 levels of road hierarchies, with hierarchy 1 higher than hierarchy 2, penalties can be assigned to lower hierarchy roads gradually, such as

$$\text{hierarchyModifiers} = [1.0, 1.05, 1.2, 1.3, 1.4].$$

Given a node n, the heuristic $h(n)$, that is, the estimated (time) cost to travel from n to the destination node is computed as follows. The estimated cost is first computed by dividing the Euclidian distance to the destination by a given fixed speed. If the node n is not too near to the source or the destination, this estimated cost is then multiplied by the hierarchy modifier corresponding to the road used to lead to node n in the search. In the above example, the estimated cost of nodes reached by a road with hierarchy 3 will be multiplied by 1.2.

This routing approach is illustrated in Figure 3.12 with a partial graph showing connected roads with two road hierarchies and length of the arcs in meter.
- Green: Hierarchy 3, Speed 36 km/h
- Purple: Hierarchy 1, Speed 54 km/h

The starting node is n3 and the destination node is n14. Table 3.1 shows the Euclidean distance between each node and the destination node n14. For each hierarchy level, the

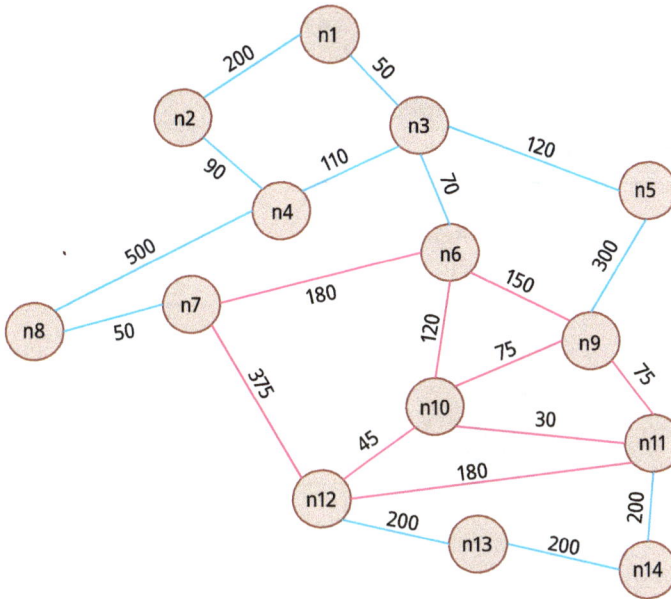

Figure 3.12: Finding a route from node n3 to n14.

Table 3.1: Table of distances and heuristics of nodes to destination n14.

	n1		n2		n3		n4		n5		n6		n7	
Dist	780		780		350		300		250		250		280	
Hrchy	1	3	1	3	1	3	1	3	1	3	1	3	1	3
$h(n,.)$	52	62	52	62	23	28	20	24	17	20	17	20	19	22

	n8		n9		n10		n11		n12		n13	
Dist	300		220		220		200		220		200	
Hrchy	1	3	1	3	1	3	1	3	1	3	1	3
$h(n,.)$	23	28	15	18	15	18	13	16	15	18	13	16

estimated cost is also provided. The used fixed speed here is 54 km/h. So, for node n1, we get the following values for $h(n1, i)$, expressed in seconds, where i is the hierarchy of the arc leading to node n1 in the search.

- $h(n1, 1) = 780 \, \text{m}/54 \, \text{km/h} * 1.0 = 52 \, \text{sec}$
- $h(n1, 3) = 780 \, \text{m}/54 \, \text{km/h} * 1.2 = 62 \, \text{sec}$

In this small example, hierarchy modification will be applied whatever the distance of the node to the source and the destination.

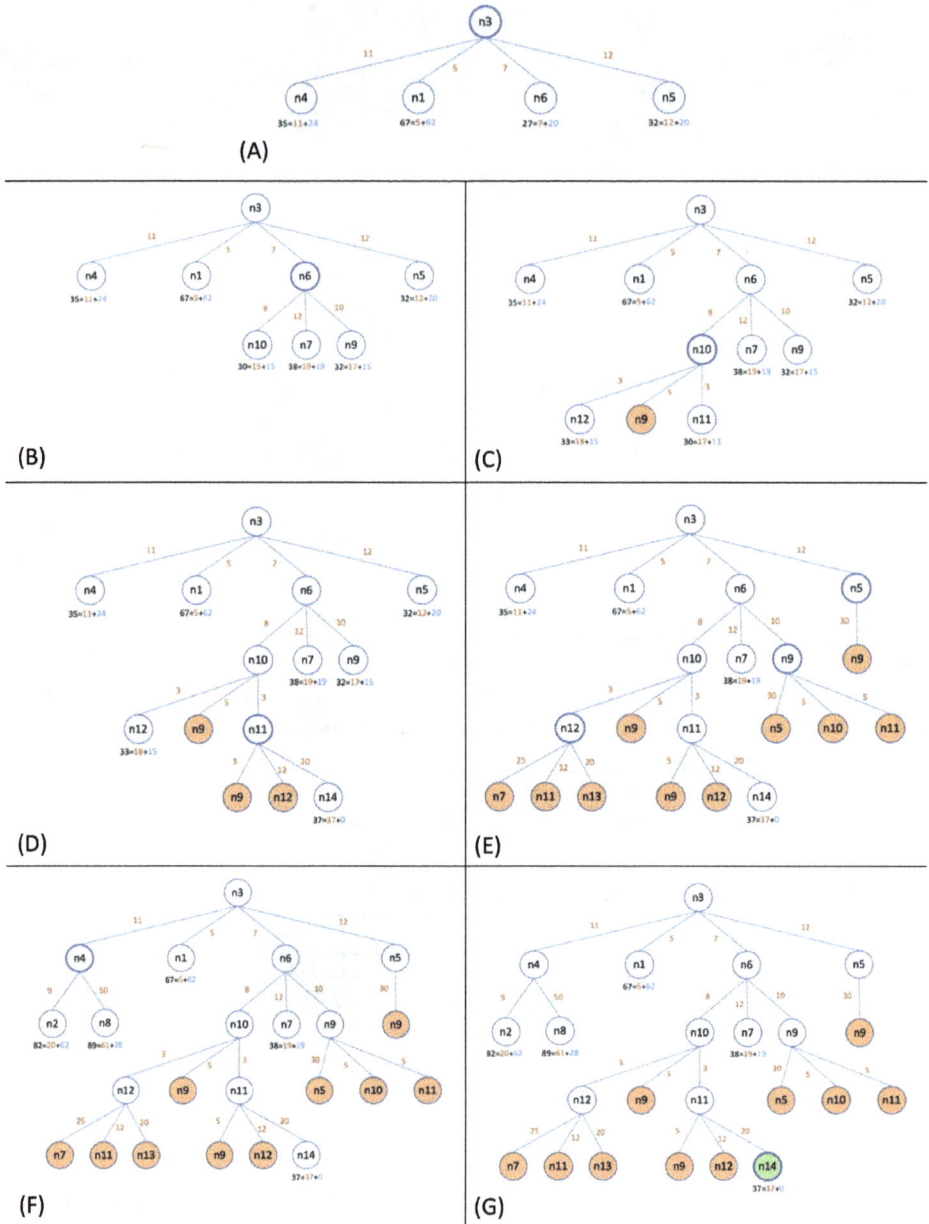

Figure 3.13: Steps in the search tree for a route from n3 to n14.

Figure 3.13 depicted the search tree in the different steps of the A* algorithm on the above graph to find an (optimal) route from n3 to n14. The frontier is initially the node n3.

- (A) Node n3 is extracted from the frontier. It is expanded and its successors nodes (n4, n1, n6, and n5) are added in the search tree and inserted in the frontier. Each node in the frontier is associated with the value of the evaluation function $f(n) = g(n) + h(n, i)$ where $g(n)$ is the *exact* cost to reach node n from root node n3, and $h(n, i)$ is an *estimated* cost described by the above heuristic function. For node n6, the values are computed as follows:
  - $g(n6) = 70\,m/36\,km/h = 7\,sec.$
  - $h(n6, 3) = 20$ as the hierarchy of the arc n3-n4 is 3
  - $f(n6) = 27 = 7 + 20$
- (B) Node n6 is extracted from the frontier as it has the smallest evaluation. It is expanded and its successor nodes (n10, n7, n9) are added in the search tree and inserted in the frontier. For node n6, the values are computed as follows:
  - $g(n10) = 120\,m/54\,km/h = 8\,sec.$
  - $h(n10, 1) = 15$ as the hierarchy of the arc n6-n10 is 1
  - $f(n10) = 27 = 7 + 8 + 15$
- (C) Node n10 is extracted from the frontier and expanded. The successor node n9 with an evaluation of $35 = 20 + 15$ is, however, not added in the frontier as this node is already in the frontier with a smaller evaluation (32). This is a basic optimization of the presented A^* algorithm.
- (D) Node n11 is extracted from the frontier and expanded. The successor nodes n9 and n12 are not added in the frontier as they are already in the frontier with a smaller evaluation. Node n14 is added with a cost $37 = 37 + 0$. This is a goal node, but the search continues as other nodes in the frontier with a smaller evaluation could perhaps lead to n14 with cost smaller than 37.
- (E) Nodes n5, n9, and n12 are successively extracted from the frontier. None of their successor nodes are, however, added in the frontier as they are already in the frontier with a smaller evaluation.
- (F) Node n4 is extracted from the frontier and replaced by its successors, nodes n2 and n8.
- (G) Node n14 is extracted from the frontier. It is recognized as a goal node and the search returns the path n3-n6-n10-n11-n14 with a cost of 37.

3.2.5.3 Key challenges

On large graphs, the A^* algorithm explores significantly a smaller number of edges than other algorithms such as Dijkstra. It, however, has a number of caveats.

A^* is not suited for routing problems involving constraints such as combining different transport or integration of public transport schedules.

The above routing algorithm uses a fixed speed to determine the estimated remaining cost. If this speed is an upper bound of the maximum speed, the heuristic (without hierarchy) is admissible and A^* always returns the best solution. When using a fixed speed lower than the maximum speed or introducing hierarchy modifiers, the heuristic is not always admissible as the estimated cost could be greater than the real cost. However, these cases are very rare and generally can be mitigated by continuing the search for some more time to check if better results exist.

3.3 Constraint satisfaction problems

When a state can be represented with a set of variables, each with a set of possible values, we move from an "atomic" representation to a richer "factored" representation that can take advantage of the structure of states and their constraints. This eliminates large portions of the search space and reduces the possible set of values by identifying variable/value combinations that violate the constraints. The SEARCH approach described in the previous section can then be combined with some form of INFERENCE based on the CONSTRAINTS. So, CSP solving systems can be more efficient than pure "state space" searchers as they take advantage of the structure of states.

3.3.1 What category of problems does constraint satisfaction solve?

A constraint satisfaction problem (CSP) is a problem that can be described by a set of (decision) variables, usually over a finite domain, and a set of constraints over the variables. More specifically, constraints specify the legal values and combination of values assigned to the variables: they can restrict the value of a single variable (un-ary), pairs of values (bin-ary), or more generally t-uples of values. Global constraints like "AllDifferent" can also be specified.

A solution of the problem is an assignment of values to the variables satisfying all the constraints.

A constraint optimization problem (COP) is a CSP with an additional objective function relating the variables. The goal is then to find a solution maximizing (or minimizing) the objective function.

The techniques for solving CSP and COP depend on the computation domain of the variables. Most approaches are handling finite domains, that is, each variable of the CSP has a finite domain. A particular case of finite domain is the domain $\{0, 1\}$, leading to Boolean CSP. Extensions of finite domains are finite sets and graphs, that is, where the value of a variable is a finite set (from a given initial finite set), or a graph (subgraph of an initial given graph). All these finite computation domains can be used to model combinatorial problems. Another class of problems are continuous CSP where the value of a variable is a real and has an infinite domain.

The description of a CSP or COP is a declarative representation of the problem. There are no data structures nor any algorithm in the description. CSP and COP cover a large class of problems such as vehicle routing, scheduling, warehouse location, planning, knapsack, configuration problems, and graph coloring.

3.3.1.1 Examples

Graph coloring
Given a graph and a set of colors, color the vertices of the graph so that two adjacent vertices are colored with different colors, as illustrated in Figure 3.14. This classical graph coloring CSP is known to be NP-complete, that is, a problem for which it is easy to check whether a given assignment of colors to the vertices is a solution, but finding a solution is very hard.

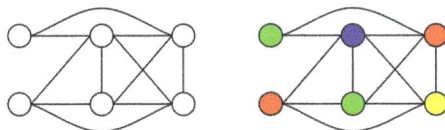

Figure 3.14: A solution of a graph coloring problem.

More formally, given a graph $G = (V, E)$, where V is the set of vertices and E is the set of edges, and a set of colors C, the (decision) variables are the vector colour[v], with $v \in V$. The domain of each variable color[v] is the set of colors C. For each edge $(v, w) \in E$, we have the constraint color[v] $\neq$ color[w].

The corresponding graph coloring COP is to find the minimum number of colors necessary to color the vertices of the graph so that two adjacent vertices are colored with different colors.

Graph coloring can be used to model various practical and industrial problems such as scheduling problems, pattern matching, or course schedule.

Cryptarithmetic puzzle
A classical cryptarithmetic puzzle is the SEND + MORE = MONEY problem, where each letter has to be replaced by a different digit so that this addition is correct, as illustrated in Figure 3.15. The unique solution is 9567 + 1085 = 10652. A possible model for this puzzle is to use the set of variables {S, E, N, D, M, O, R, Y} with domain {0, ..., 9}, and the additional four carry variables {C_1, C_2, C_3, C_4} with domain {0, 1}. A constraint

$$C_1 \ C_2 \ C_3 \ C_4$$

$$S \ E \ N \ D$$

$$+ \quad M \ O \ R \ E$$

$$M \ O \ N \ E \ Y$$

Figure 3.15: Cryptoarithmetic problem.

allDiff(S, E, N, D, M, O, R, Y) states that all the variables must have a different value. The constraints for the additions are then modeled as

$$S \neq 0; \quad M \neq 0$$
$$D + E = 10C_1 + Y$$
$$C_1 + N + R = 10C_2 + E$$
$$C_2 + E + O = 10C_3 + N$$
$$C_3 + S + M = 10C_4 + O$$
$$C_4 = M$$

Sudoku

Sudoku is a classical puzzle played on a 9×9 grid, divided in 9 squares made up of 3×3 cells. Each cell must be filled with a digit 1–9 such that each row, column and square, each having 9 cells, must have different digits. Some of the cells are given a specific value. An example is given in Figure 3.16, where the small digits are the possible values for the cells.

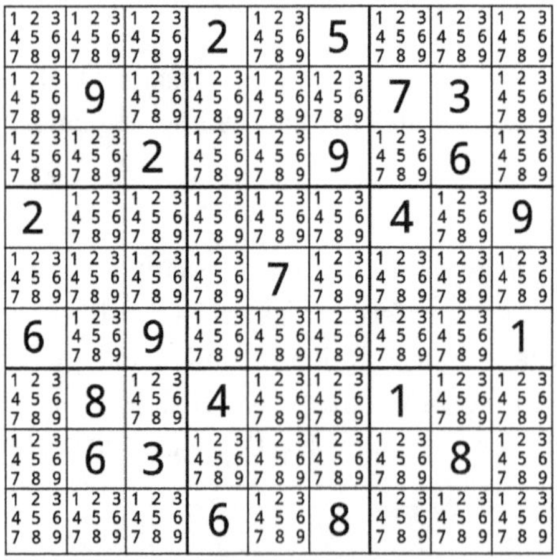

Figure 3.16: Sudoku.

The model of a sudoku has one variable per cell, that is, 81 variables, each with a domain $\{1, \ldots, 9\}$. There is an allDiff constraint for the variables of each row, of each column, and each square. The allDiff(X) constraints all the variables in X to be different. Then there are constraints imposing the specific value for the given cells of the instance of the sudoku.

3.3.1.2 Solving paradigms

There exist many approaches to solve CSP and COP. Classifying the different paradigms is helpful to their understanding and comparison. Two classification criteria will be used. The first one differentiates *perturbative* versus *constructive* approaches, while the second one distinguishes systematic versus incomplete search.

In a *perturbative* approach, a candidate solution is a complete solution (a value is assigned to each variable). New solutions are generated by modifications or perturbations of an existing one. The search is performed on a succession of candidate solutions. This is illustrated on graph coloring in Figure 3.17 where at each step, the color of one of the vertex is modified.

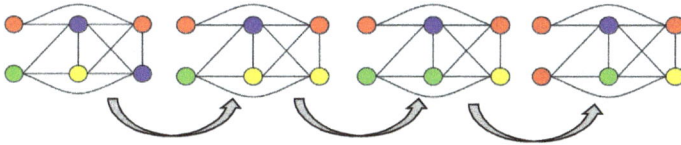

Figure 3.17: Graph coloring: perturbative approach.

In a *constructive* approach, a candidate solution is a partial solution (a value is assigned to some of the variables) and partial solutions are progressively extended. This is illustrated on graph coloring in Figure 3.18 where at each step, one new vertex is colored.

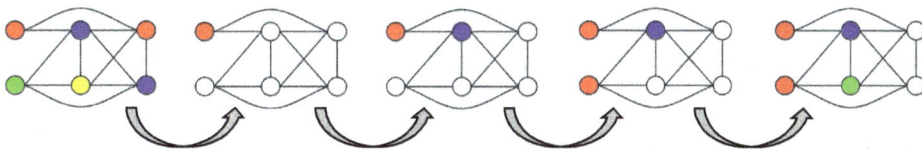

Figure 3.18: Graph coloring: Constructive approach.

In a *systematic* approach, the whole search space is systematically explored. As a consequence, the approach is complete; it finds an exact solution for a CSP and find a best solution for a COP. If no solution is found, this is a proof of nonexistence of solutions.

Of course, a systematic approach can be computationally expensive for large instances of problems.

In an *incomplete* approach, the exploration of the search space is based on a neighborhood function. The approach is incomplete in the sense that it may not find an exact solution of the CSP, but only an approximation of a solution where some of the constraints might be violated. For a COP, it will find a good solution that is not proved to be the optimal solution. An incomplete approach is computationally less expensive and usually allows us to find a reasonable (approximated) solution for large instances of problems.

Different paradigms for solving CSP and COP can be classified according to the above two criteria, as shown in Table 3.2. In this chapter, two paradigms will be developed: local search and constraint programming.

Table 3.2: Solving paradigms for CSP and COP.

	Systematic	Incomplete
Perturbative	Simplex	Local search
	Systematic local search	Evolutionary algorithms
Constructive	Dynamic programming	Greedy
	Branch and bound	Ant colony optimization
	Constraint programming	GRASP
	DPLL SAT solver	

3.3.2 Solving optimization problems by local search

Local search algorithms address problems for which only the solution state matters, not the sequence nor the path cost to reach it. In addition to finding a goal state, they are used to solve optimization problems, for which the objective is to find the values that minimize a cost function.

Local search covers a family of methods sharing common properties. The approach is incomplete and perturbative. It is based on iteratively improving the solution by finding a better one in the neighborhood of the current solution. When it stops, there is no proof the given solution is the best one.

3.3.2.1 From CSP to COP

In a CSP, one has to find a solution, that is, an assignment of the variables satisfying all the constraints. There is no given objective function to optimize. As local search is an optimization process, the CSP should first be transformed into a COP by introducing a cost function. The set of constraints is split in two sets: the hard constraints and the soft constraints. The set of hard constraints that can be emptycontains the constraints that must be satisfied by all the solutions considered by the local search. For each soft constraint, a cost or violation function measures how a given solution violates the constraint. The cost of a solution is then computed as the sum of the violation cost of all the soft constraints. The resulting objective function of this COP, usually called cost function in local search, is then used in the search for the best solution. When the cost of a given solution reaches 0, then it is a solution of the original CSP. If the search stops with a no zero cost solution, then the resulting solution is only an approximation as it still violates some of the soft constraints.

When the initial problem is a COP, the cost function used for the local search should combine the objective function of the COP and the violation function of the soft constraints, if any. The cost function usually has then the form,

$$\text{cost}(x) = \alpha.\text{obj}(x) + \beta.\text{cost\_soft}(x),$$

where x are the decision variables, $\text{obj}(x)$ is the objective function of the COP, and $\text{cost\_soft}(x)$ is the cost of the violation of the soft constraints and α, β are real values. Such a cost function combines optimality and feasibility. Usually, the β value is much higher than α to favor feasible solutions.

The iterative improvement of a solution needs to start from some initial solution. Such an initial solution is usually chosen randomly but respecting the hard constraints of the problem.

Graph partitioning

Given a graph with an even of vertices, divide the vertices in two equal-size sets of vertices such that the number of edges that go from one set to the other set is minimized as illustrated in Figure 3.19. More formally, given a graph $G = (V, E)$, where V is the set of vertices and E is the set of edges, find a partition (V_1, V_2) of V such that $\#V_1 = \#V_2$ (balance constraint). The objective is to minimize $\text{obj}(V_1, V_2) = \#\{(v, w) \in E | v \in V_1 \wedge w \in V_2\}$. This problem is known to be NP-hard.

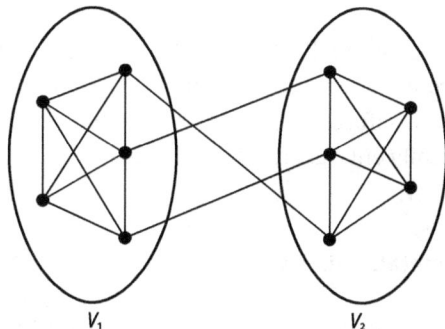

Figure 3.19: Graph partitioning: Example of a solution. The cost of this solution is 3.

3.3.2.2 Neighborhood

The neighborhood of a solution p is a set of solutions that are similar or close to p. The idea of proximity is usually measured in terms of computation. A solution in the neighborhood can thus easily be computed from the solution p. A neighborhood can be defined by one or more operators over a solution and yielding a set of solutions. Such operators must be computationally simple.

An important property of a neighborhood is called *connectivity*, stating that from each solution, there is a path to an optimal solution. This path is a succession of solutions p_1, p_2, etc. where each solution p_{i+1} is in the neighborhood of p_i. Of course, large neighborhoods lead to shorter paths to an optimal solution, but a large neighborhood needs more time to be explored to choose the next solution.

If the initial problem has constraints on the solution, should these constraints be respected by each solution in the neighborhood? If these constraints are seen as hard constraints, then the operators generating the neighborhood must be designed to fulfil the constraints. If these constraints are seen as soft constraints, then the neighborhood are larger as it may contain solutions violating these constraints, but the cost function would integrate the violation of these soft constraints.

Graph partitioning

In the graph-partitioning problem, if the balance constraint is hard, then the neighborhood of a (balanced) solution must only contain balanced solutions. A simple neighborhood could be defined by swapping two vertices in V_1 and V_2. If the balanced constraint is considered as a soft constraint, then a possible neighborhood would be to move a vertex from one set to the other set of vertices. The cost function should then integrate the cost of the soft constraint. In the following cost function for graph partitioning, the soft balance constraint measures the (square of the) distance of the solution to a balanced solution:

$$\text{cost}(V_1, V_2) = \alpha.\#\{(v, w) \in E | v \in V_1 \wedge w \in V_2\} + \beta.(\#V_1 - n/2)^2$$

These two neighborhoods are illustrated in Figure 3.20. The choice between a hard or soft constraint for a constraint of the problem is a design decision and is highly problem dependent. Choosing a soft constraint may reduce the length of the path to an optimal solution may drastically enlarge the search space.

3.3.2.3 Heuristics and metaheuristics

In local search, a heuristic defines how to choose the next solution in the neighborhood of the current solution. A heuristic uses local information, that is, the current solution

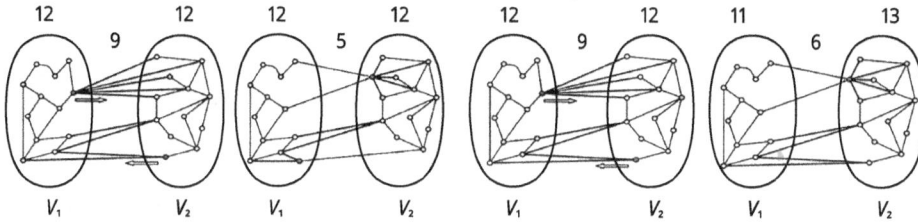

Figure 3.20: Graph partitioning: Solution in two definitions of neighborhoods.

and its neighborhood. A first property of a heuristic is whether it always improves the cost of the current solution or is it possible to have a degradation. A second property relates to the search of the neighborhood; is it an exploration or is it a random choice? The rationale behind improvement is to focus the search toward high-quality solution, what is called *intensification*. The rationale behind randomness and degradation is to direct the search toward other regions of the search space, what is called *diversification*. Different strategies to balance intensification with diversification have led to different types of heuristics. Some classical heuristics are presented hereafter.

Hill climbing explores the neighborhood and always improves the cost of the current solution. *Best-improvement* is hill climbing where the whole neighborhood is searched for the best solution while *first-improvement* stops as soon as a better solution is found. If no solution in the neighborhood improves the current solution, the local search is stopped to the current (local) optimum. Best-improvement is more costly because of the complete search of the neighborhood.

Random walk selects the next solution randomly in the neighborhood; the cost can therefore be degraded. *Random improvement* randomly selects a candidate solution; it is, however, accepted only if it improved the current solution. Otherwise, the current solution is unchanged. The *Metropolis* heuristic randomly selects a candidate solution. If it improves the current solution, the candidate solution is accepted as the new current solution. If it degrades the current solution, it is accepted with a small probability, usually depending on the severity of the degradation: a large degradation has a much smaller probability to be accepted.

While a heuristic focuses on choosing the next solution, a metaheuristic guides the search process toward a global optimum, hence trying to escape from local optimum. This usually collects information during the local search. Many metaheuristics have been defined for local search.

Simulated annealing is an analogy with metallurgy where heating causes molecules to move more freely to unusual location. During cooling, the movements are more restricted. In this metaheuristic, a temperature variable is reduced during the local search. This temperature is used in the Metropolis heuristic applied at every step; with high temperatures, the probability to accept the candidate solution is higher than with a low temperature. In the beginning of the local search degradation will be more often accepted, favoring diversification while at the end of the local search degradation will be less accepted, favoring intensification.

The idea of the *tabu search* metaheuristic is to select a neighbor that has not yet been visited. As it is difficult to keep track of all the visited solutions, tabu search maintains an abstraction of the last visited solutions. Such an abstraction can be based on the chosen operators or some properties of the solution. An evolving tabu list states criteria to reject candidate solutions. If the candidate solution is in the tabu list, then it is rejected. The abstraction of tabu list does not limit the rejection to already visited solutions. Tabu search often maintains the best solution found so far as the chosen heuristic may degrade the current solution. Tabu search is then often combined with an *aspiration* meta-heuristics overriding the tabu rejection if the candidate solution improves the best solution found so far.

In a *variable neighborhood metaheuristic*, a sequence of neighborhoods (of increasing size) is used during the local search. Guided *local search* or *dynamic local search* metaheuristics use a sequence of cost functions to escape from local optimum. *Adaptive local search* provides a more general framework where heuristics and metaheuristics are dynamically adapted during the search.

3.3.2.4 Termination

Local search is incomplete. It is impossible to know if the current solution is the best solution nor how far its cost is from the optimum. But whenever the search is stopped, a solution is provided. Termination is often defined through a time limit.

Some metaheuristics have their own termination criteria. In simulated annealing, for instance, termination occurs when the temperature reaches zero, according to a temperature schedule.

In order to favor diversification, it is interesting to do multiple local searches from different random initial solutions. That is what *Random restart* is doing, returning then the best solution. *Iterated local search* favor intensification by iterating the local search from a random initial solution close to the solution found in the previous iteration.

3.3.3 Solving problems by constraint programming

3.3.3.1 CSP as a search problem

A CSP can be solved using a search algorithm, such as depth-first search. The initial state is an empty assignment. The successor function assigns a value to some unassigned variable provided no constraint is violated. The goal test is to have all the variables assigned.

Constraint programming (CP) offers a general framework for solving CSP and COP. It can be seen as a general-purpose search algorithm, independent from the problem.

It can be applied to any CSP and COP. CP also provides generic heuristics. In practice, a CP language allows to state a model for a CSP or COP and proposes mechanisms to guide and enhance the search for a solution. Examples of CP languages are Choco, IBM cp optimizer, Gecode, OR-tools, Oscar, and JaCoP. SAS/OR and Cali Xpress also integrate CP modules.

The CP computation model is based on two steps. The *propagation* step reduces the domains of the (decision) variables while the *search* step decomposes the problem into simpler subproblems. These two steps are iterated until a solution is found or until the problem is proved to have no solution.

3.3.3.2 Propagation of the constraints

When the domain of a variable is modified, the objective of propagation is to use this information to (quickly) reduce the domain of the other variables. Propagation is made by applying a consistency algorithm that considers the constraints locally instead of considering all the constraints of the problem. The propagation aims at achieving some level of consistency. Different levels can be considered, each level offering more pruning than the previous level. Of course, a higher level of consistency requires more computation.

Any consistency algorithm must respect two properties. First, it never removes a value of a variable that is part of a solution of the CSP. This ensures that all the solutions will be found. Second, if the consistency algorithm is applied on a CSP where all the variables are assigned, then the consistency algorithm reduces to a satisfiability test. This ensures that when all the variables are assigned, the resulting assignment is a solution of the CSP.

Forward checking

When the search procedure assigns a value to a variable, forward checking only considered constraints involving this variable, but where all but one variable (say Y) is assigned to a value. Forward checking then removes all the values from the domain of Y that violate the constraint. It is clear that the removed value cannot be part of the constructed solution. Forward checking is illustrated on graph coloring in Figure 3.21.

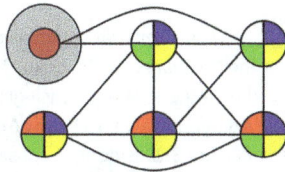

Figure 3.21: Forward checking in graph coloring. The value "red" is removed from the two neighbors to respect the constraints.

Domain consistency

Domain consistency, also called arc consistency, is easier to explain on a binary constraints $c(X, Y)$. Even if none of the variables are assigned, it is possible to remove values from the domain of X and Y. Domain consistency removes any value from the domain of X for which there is no value in the domain of Y that satisfies the constraint, and similarly for the domain of Y. It is clear that all the removed values cannot be part of the constructed solution. When all these impossible values have been removed, the constraint is said to be domain consistent. More precisely, a binary constraint $c(X, Y)$ is domain consistent when for each value a in the domain of X, there exists a value b in the domain of Y such that $c(a, b)$ holds. Domain consistency is illustrated in Figure 3.22, on the binary constraint $X \geq Y + 2$.

$$X \geq Y + 2$$

$$X \in \{\cancel{1},\cancel{2},\cancel{3},4,5\} \qquad Y \in \{2,3,\cancel{4},\cancel{5},\cancel{6}\}$$

Figure 3.22: Domaon consistency for $X \geq Y + 2$.

Bound consistency

Bound consistency is a weaker form of domain consistency where the domain of a variable is approximated by an interval [min Domain(X), max Domain(X)]. Only the bounds of the domains are verified to be part of a solution. Values are thus never removed in the middle of the domain; only the bounds are moved. The pruning is weaker but for some constraints, such as arithmetic constraints, this pruning can be achieved very efficiently.

Fixpoint computation

Propagation requires iterating the process until no more value can be removed. This process is known as a fixpoint computation. If we consider the propagation achieved by domain consistency on the constraint $X \geq Y + 2$ in Figure 3.22, the domain of variable X (and also Y) has been modified. All the other constraints involving X should then be (re)considered because the removal of value $\{1, 2, 3\}$ in the domain of X could imply the removal of values in the domain of another variable. When no more value can be removed, the fixpoint is reached and the CSP is domain consistent, that is, all its constraints are domain consistent, assuming domain consistency is the chosen level of consistency. A consistency algorithm is an algorithm achieving some consistency for CSP.

Global constraint

A global constraint such as allDiff($X_1, \ldots, X_n$), is a constraint with a variable number of arguments. This is illustrated in Figure 3.23, on one of the 9 squares of a sudoku. The left part is a given state where 2 variables (X_4 and X_9) are assigned, and the others have a nonsingleton domain. The constraint on this square is that these variables must have a different value. This can be modeled by a set of thirty-six $X_i \neq X_j$ constraints. In that case, the left state of Figure 3.23 is domain consistent as any of these inequalities is domain consistent. However, when these constraints are modeled by an allDiff($X_1, \ldots, X_9$) constraint, the left state of Figure 3.23 is not domain consistent. For instance, it is not possible to find an assignment of

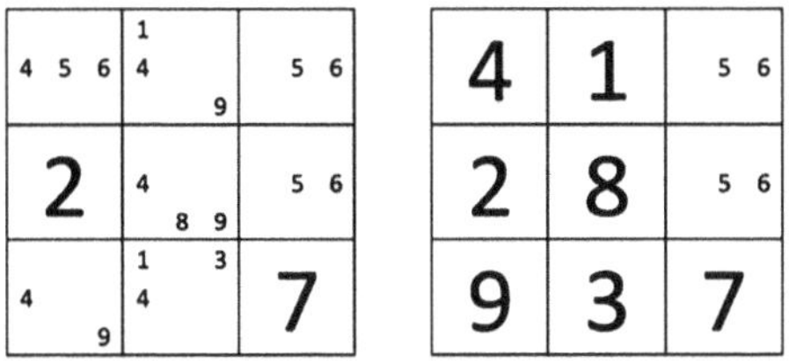

Figure 3.23: The allDiff global constraint on a sudoku square.

these nine variables with $X_1 = 5$ satisfying this allDiff constraint because the values 5 and 6 must be taken by the variables X_3 and X_6. Similarly, the values 4 and 9 must be pruned from the domain of X_5 because the value 8 occurs only in the domain of this variable. The right part of the figure is the corresponding domain consistent state. The achieved pruning is very effective.

In the case of allDiff, the constraint can be expressed as a set of simpler constraints, but as a global constraint, it achieves a much better pruning. Of course, the computational complexity is higher. Other global constraints cannot be expressed with simpler constraints, and thus increases the expressivity of the constraint language. Each global constraint requires a dedicated algorithm to achieve the expected consistency level. For the allDiff constraint, the algorithm achieving domain consistency is based on matching theory in bipartite graphs.

3.3.3.3 Search

The search implicitly develops a search tree. Each node of the search tree contains a CSP that only differs from the initial CSP by its domain of values. The root contains the initial CSP. Search and consistency checks are interleaved. At a given node, the search algorithm first applies a consistency algorithm. If the domain of a variable is empty, then the CSP of this node has no solution and is a failed node. If all the variables are assigned, this assignment is a solution of the CSP and the node is a solution node. In any other case, the search algorithm chooses a variable with a nonsingleton domain and creates children nodes for the different values of the chosen variable. The CSP in a child node is a copy of the CSP of the parent node, but with one new variable assigned. The search tree is finite and is usually traversed using a depth-first strategy. An example of (part of) a search tree for the graph coloring CSP is depicted in Figure 3.24. The root is the initial CSP where each of the six variables has four colors in its domain. In a child node, the chosen variable and its chosen value (its color) are highligthed. In the child node of the root, the top left variable has been chosen and set to the red color. The domain consistency algorithm is then applied and removed the red color from its two neigbhors variables. At the left bottom of the tree, two nodes are failed nodes as they contain a CSP with variables with an empty domain. The bottom right node is a solution node as all variables are assigned.

The basic search algorithm illustrates that a CP search is made of two components: the definition of the search tree and the use of an exploration strategy for traversing the

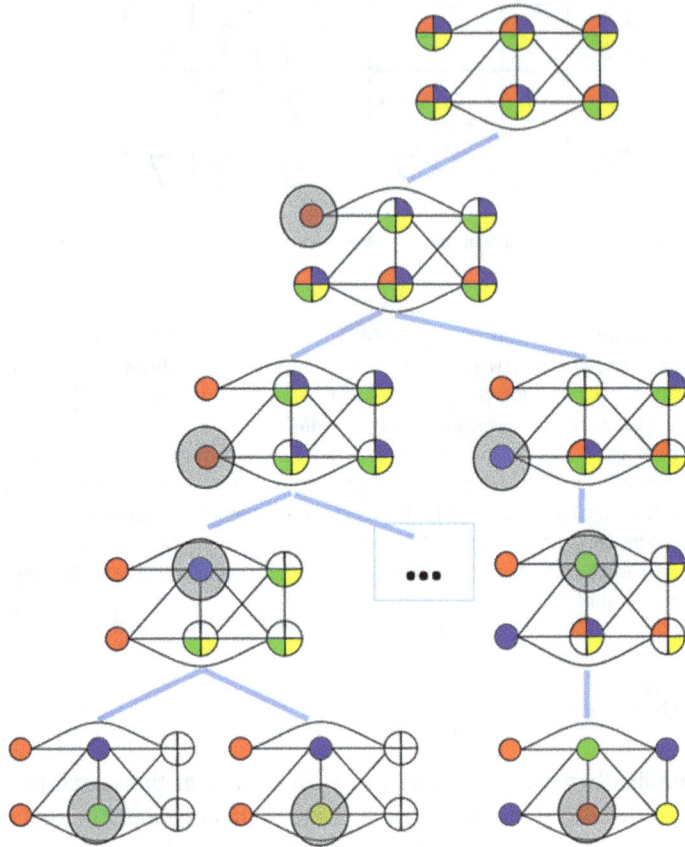

Figure 3.24: Search tree for the graph coloring CSP.

search tree. The different exploration strategies defined for searching problems could be used in this context. DFS is the standard approach, although best-first search can be applied using some measure of quality of the nodes in the fringe.

The definition of the search tree is induced by the branching strategy, that is, how are generated the children nodes of a given node. In the standard branching strategy, called labeling, a variable is chosen and there is a child for each of the value in the domain of that variable. There exists various heuristics guiding the choice of the variable to label. The *domain* heuristic chooses the variable with the smallest domain as such a choice minimizes the size of the search tree. The *degree* heuristic chooses the variable involved in the largest number of constraints. Such a choice will quickly drive the search to a fail node if the CSP has no solution.

Optimization problems

For COP, that is, CSP with an objective function $f(x)$ defined over the variables of the CSP, a branch-and-bound approach can be used to find an optimal solution. As soon as a solution (with value f^* for the objective function) is found, this solution is stored as the current best solution and an additional constraint $f(x) < f^*$ is added to the CSP. With this new constraint, the only solution improving the value of the objective function will be found, thus pruning the search space from nonoptimal solution. When a better solution (with value f^{**}) is found, then it replaces the best current solution and the additional constraint is replaced by the constraint $f(x) < f^{**}$, and so, for each better solution found. When the search is completed, that is, it is not possible to find yet another better solution, then the current best solution is known to be the best solution of the COP.

Large neighborhood search

CP is a systematic approach. It is thus complete but requires exploring the whole search space what can be very costly, especially for COP where proof of optimality is not always necessary. Large neighborhood search (LNS) is a local search metaheuristic that can be applied to many frameworks, including CP. This approach is incomplete but orients the search in promising areas of the search space. LNS iteratively applies two steps: relaxation and reconstruction. Relaxation relaxes parts of the best solution found so far, that is, some random variables are chosen to become uninstatiated. The relaxed CSP is simpler than the original one and contains parts of the best solution, and thus looks promising. The reconstruction attempts to solve the relaxed CSP obtained. If the reconstruction finds a better solution, a new iteration is done. If the reconstructions fail to find a better solution in a fixed amount of time, then a new iteration begins from former solution with a different relaxation.

3.3.4 What are the limitations of constraint satisfaction problems?

CSP and COP offer a declaration description of problems. They can then be solved by different techniques. A systematic approach ensures to find a solution if one exists or to prove there is no solution. It is, however, computationally more expensive. An incomplete approach is computationally less expensive but does not ensure the finding of a solution.

Thanks to its perturbative and incomplete approach, local search always provides an (approximated) solution, whatever the chosen computation time. For many problems, an approximate solution is often sufficient, and certainly better than having no solution at all. Local search usually does not provide an estimation of the distance from the cost solution to the optimal solution. Local search requires designing and programing the neighborhood and the cost function for the soft constraints. Solving a CSP by local search usually requires some significant tuning of the parameters of the chosen heuristic and metaheurisfic. Local search covers a large variety of methods, each of them having its advantages on different classes of problems. All local search methods are based on randomness, a key feature in incomplete approaches.

Constraint programming (CP) is constructive and perturbative, hence always provides a solution if one exists. CP only requires the description of the CSP or COP and does not require any programming. CP also proposes generic heuristics to speed up the search. Solving practical CSP with CP could however require some tuning in the modeling of constraints and in the search procedure.

Local search and constraint programming are only two of the various methods for solving CSP and COP but illustrate various interesting features and concepts.

3.3.5 Industry example: containers configuration using CSP at Maersk Container Industry A/S

Yves Deville, Karen Veldeman, Dennis Conrad

3.3.5.1 Product configuration

The need to configure products to meet special and custom requirements is becoming increasingly important for companies in different industries. A product configuration tool is used to guide a user to configure a product or process, while considering different constraints. Although product configuration tools are used in different industries, most manufacturers and automotive producers rely on them to support them in creating the right products or processes. They are often employed as design tools that allow manufacturers to constrain product lines or even end customers to codesign their own products. Once the product specifications and constraints are modeled by a subject matter expert into a product configurator, this configurator guides an employee or a customer to express his demand by proposing favorable combinations of attributes and ruling out impossible ones. It thereby removes the need for the expert to be in the specific product configuration process, while ensuring compliance with the specifications. For many manufacturers, relying on product configurators helps them to lower their distribution costs, accelerate their production, react faster to customer inquiries, and reduce overproduction of general products.

Product configurators exist in many shapes and forms. Depending on the reasoning techniques used, they are usually classified as rule-based or constraint-based systems. Rule-based configurators are used in the earliest systems and derive solutions in a forward-chaining manner, by evaluating all rules and eliminating rules that are not applicable to the configuration of the product. This approach anticipates all possible configurations by specifying the configuration using restrictive "if... then..." statements. By using constraint-based systems, companies can build product models composed of modular entities.

Use case

Maersk Container Industry A/S (MCI) manufactures containers. At MCI, a product configurator is used to translate customer specifications into a container configuration. For example, a customer can choose the type of condensers to control the temperature within

Figure 3.25: Example of a MCI container with built-in cooling system.

the container, the type of battery pack or the type of panels outside the container. An example of a MCI container is illustrated in Figure 3.25. Based on the specification of these attributes, the configured container can serve as input to generate the quote that will be shared with the customer, as well as a bill-of-materials (BOM) of raw materials and components that will be shared with the Maersk production department. The MCI product configurator is developed in Microsoft Dynamics 365 Supply Chain Management, an enterprise resource planning (ERP) system, featuring a constraint-based product configuration tool based on Z3, a solver and theorem prover, open sourced by Microsoft Software Research in 2015.

3.3.5.2 Modeling product configuration as a CSP

The need to configure products to meet special and custom requirements is becoming increasingly important for companies in different industries. A product configuration tool is used to guide a user to configure a product or process, while considering different constraints. The definition of the underlying product configuration model has been designed by product experts. They first defined product attributes (the variables of the CSP) and their domain of possible values. For example, the Maersk container attributes integrate the type of condensers to control the temperature within the container, the

Table 3.3: Some variables and their domain in the MCI product configurator.

Variables (attributes)	Domain
Atmosphere control (AC)	{Manual ventilation (MV), Automatic ventilation (AV), Automatic ventilation (AV+), Automatic ventilation (AV+O2)}
Fresh air module (FAM)	{Manual ventilation (MV), Automatic ventilation (AV)}
Sensor module (SM)	{No sensors (No), CO2 sensors (CO2) CO2 and O2 sensors (O2)}

type of battery pack, the type of fresh air exchange module or sensor module, as illustrated in Table 3.3.

After defining the product configuration variables and their domains, the experts need to define how these attributes can be combined into one product of service by defining constraints. These constraints can make sure that a user can only configure containers that can be sold and produced. For example, not all types of fresh air exchange modules and sensor modules can be combined into one single container. These constraints are expressed in the form of a table expressing the valid combinations between one or more attributes. Each column of the table is assigned to a variable of the CSP. Each row defines a valid combination of the values for the variables. Other constraints can also be specified in the CSP. For example, the MCI configurator uses a table constraint illustrated in Table 3.4 to express that some combinations of fresh air exchange modules and sensor modules cannot be combined into one single container.

Table 3.4: A table constraint in the MCI product configurator.

Atmosphere control (AC)	Fresh air module (FAM)	Sensor module (SM)
Manual ventilation (MV)	Manual ventilation (MV)	No sensors (No)
Automatic ventilation (AV)	Automatic ventilation (AV)	No sensors (No)
Automatic ventilation (AV+)	Automatic ventilation (AV+)	CO2 sensors (CO2)
Automatic ventilation (AV+O2)	Automatic ventilation (AV+O2)	CO2 and O2 sensors (O2)

The constraint in this example ensures that a user of the product configurator cannot assemble a container with automatic ventilation (AV+) but without CO2 sensor. Once the user has chosen to configure a container with automatic ventilation (AV+), the user will only be able to select the automatic ventilation and CO2 sensor option. The configurator leverages the constraint solver to determine the feasibility of each value to help the user in selecting feasible combinations of values. If the feasible domain of an attribute is reduced to a single value, this value is automatically filled in to make sure all configured containers can be sold and produced.

3.3.5.3 Solving product configuration

After modeling the product configuration as a CSP with its variables, domains and constraints, a solving paradigm has to be chosen to solve the different product configurations for the end users. After processing the input provided by the end use, the solver needs to share which product configurations match all specifications of the product model and its constraints.

The MCI product configurator integrates a constraint programming solver. The illustrated table constraint allows propagating the information on values on variables to reduce the domain of other variables. Propagation here achieves domain consistency. This is illustrated in Figure 3.26, where on the left, the three variables are symbolically related by the table constraint. On the right, we assume that the value MV is chosen for the variable AC and the propagation of the table constraint reduces the domains of the two other variables. Since there is only one value left in the domain of the variables, a solution is found $\{AC = MN, FAM = MV, SE = No\}$.

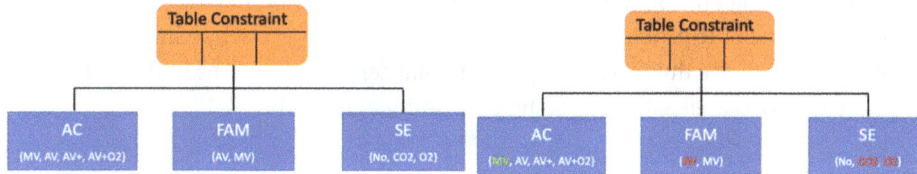

Figure 3.26: Example of propagation using the table constraint in the MCI product configurator.

3.4 SAT solvers

3.4.1 What category of problems does SAT solvers solve?

The Boolean satisfiability problem (SAT) is the problem of deciding whether a logical formula of Boolean variables is satisfiable, that is, whether a specific assignment of true and false values to those variables leads to that formula evaluating to true. It relies on propositional logic.

SAT solvers are typically helpful in problems where there are many options and dependencies that can be encoded as Boolean formulas: correctness of hardware design, automatically test, and detect connection defects, which may cause circuit failure in integrated circuits, software verification, planning and resource allocation problems in product configurations problem to ensure that product instances and configuration options satisfy the component dependencies and customer's restrictions, etc.

When a constraint satisfaction problem is defined over Boolean variables and the constraints are expressed as propositional formulas, the CSP can then be solved using a SAT solver, that is, a solver aiming at finding a Boolean assignment of the variables, satisfying the constraints.

Graph coloring

The graph coloring problem has been described as a CSP. The objective is to color the vertices of the graph so that two adjacent vertices are colored with different colors. This problem is NP-complete. This problem can also be modeled as a Boolean CSP.

Given a graph $G = (V, E)$, where V is the set of vertices and E is the set of edges, and a set of colors C, the Boolean variables are the array of color$[v, c]$, with $v \in V$ and $c \in C$. An element color$[v, c]$ is true if the color of vertex v is c; it is false otherwise. For each edge $(v, w) \in E$, we have the constraint $\neg(\text{color}[v, c] \wedge \text{color}[w, c])$ preventing adjacent vertices to have the same color. For each $v \in V$, the disjunction of color$[v, c]$ over all the colors $c \in C$ ensures to have at least one color for each vertex. For each $v \in V$, the conjunction of $\neg(\text{color}[v, c1] \wedge \text{color}[v, c2])$ for each different pair of colors (c1,c2) ensures that each vertex has at most one color.

Finding a Boolean assignment to the Boolean variables that satisfies these formulas yields a solution of the problem. Such an assignment can be found by using a SAT solver.

In this section, we thus solve any problem that can be modeled as finding a Boolean assignment to variables satisfying some propositional formula.

3.4.2 Solving problems with SAT solvers

3.4.2.1 Propositional logic

Propositional logic is a very simple logic centered around propositions, that is statements that can be either true or false, such as *"Task T1 must be performed before task T2."* Logic is described by two components: the syntax defines what are the well-formed sentences or formulas in the logic, and the semantics determines the meaning of sentences.

The syntax of propositional logic

A propositional formula is formed by using the following symbols: the logical constants True and False, propositional variables (represented by string of letters), logical connectives ($\neg$ for negation, $\wedge$ for conjunction, $\vee$ for disjunction, $\Rightarrow$ for implication and $\Leftrightarrow$ for equivalence), and parentheses.

A logical constant and a propositional variable are propositional formulas. If p and q are propositional formulas, then $(\neg p)$, $(p \wedge q)$, $(p \vee q)$, $(p \Rightarrow q)$ and $(p \Leftrightarrow q)$ are propositional formulas.

The semantics of propositional logic

The semantics define the meaning of a propositional formula. More precisely, the semantics of a formula assigns a truth value to each formula in the context of some world where some facts are *true* and others are *false*. A possible world, called *interpretation*, for a given formula is an assignment that assigns a truth value (*true* or *false*) to each of the propositional variables of the formula. For a given interpretation I, it is then possible to decide the truth value of a formula following the following rules:

- The truth value of the logical constants True and False are *true* and *false*, respectively.
- The truth value of a propositional variable A is the truth value of A in the interpretation.
- The truth value of a formula of the form $(\neg p)$, $(p \wedge q)$, $(p \vee q)$, $(p \Rightarrow q)$, or $(p \Leftrightarrow q)$ depends on the truth value of the formulas p and q, following the classical table.

p	q	$\neg p$	$p \wedge q$	$p \vee q$	$p \Rightarrow q$	$p \Leftrightarrow q$
true	true	false	true	true	true	true
true	false	false	false	true	false	false
false	true	true	false	true	true	false
false	false	true	false	false	true	true

Example. Given an interpretation with the assignment ($A = true, B = false, C = false$), the truth value of the formula $(A \vee B) \wedge (B \Rightarrow \neg C)$ in this interpretation is *true* as ($true \vee false$) $\wedge$ ($false \Rightarrow (\neg false)$) reduces to ($true \wedge (false \Rightarrow true)$) and to ($true \wedge true$), which yields *true*.

Given a formula, there are interpretations where the formula is true and interpretations where the formula is false. An interpretation where the formula is true is called a *model* of the formula. The meaning of the formula, that is, its semantics is then defined by the models of the formula. A propositional formula is *satisfiable* if it has at least a model; otherwise, it is *unsatisfiable* (no model). A formula is *valid* (also called a tautology) when it is true in all interpretations. We obviously have that a formula p is valid if and only if $\neg p$ is unsatisfiable. Two formulas with the same propositional variables are equivalent if they have the same models. A (sub)formula can always be substituted by an equivalent (sub)formula without changing its semantics. Classical examples of equivalent formulas are presented in Figure 3.27.

$\neg\,(\neg\,p)$	is equivalent to	p	(double negation)
$p \wedge q$	is equivalent to	$q \wedge p$	(commutativity of $\wedge$)
$p \vee q$	is equivalent to	$q \vee p$	(commutativity of $\vee$)
$p \Leftrightarrow q$	is equivalent to	$(p \Rightarrow q) \wedge (p \Rightarrow q)$	($\Leftrightarrow$ elimination)
$p \Rightarrow q$	is equivalent to	$\neg p \vee q$	($\Rightarrow$ elimination)
$p \Rightarrow q$	is equivalent to	$\neg q \Rightarrow \neg p$	(contraposition)
$\neg\,(p \vee q)$	is equivalent to	$\neg p \wedge \neg q$	(de Morgan)
$\neg\,(p \wedge q)$	is equivalent to	$\neg p \vee \neg q$	(de Morgan)
$p \vee (q \wedge r)$	is equivalent to	$(p \vee q) \wedge (p \vee r)$	(distributivity)
$p \wedge (q \vee r)$	is equivalent to	$(p \wedge q) \vee (p \wedge r)$	(distributivity)

Figure 3.27: Examples of equivalent propositional formulas.

The satisfaction problem

The satisfaction problem aims at deciding whether a given formula is satisfiable. The set of satisfiable propositional formulas has been formalized as SAT. Deciding whether a formula belongs or not to SAT is an NP-complete problem, hence requiring an exponential complexity (assuming P $\neq$ NP). If the formula has n propositional variables, the number of interpretations is 2^n. Deciding if a formula is satisfiable can be done by enumerating all the possible interpretations of the formula and evaluating the truth value of the formula in this interpretation. The formula is then satisfied as soon as the truth value of the formula is *true* in one of the interpretations. This naïve approach is only practical when the number of variables is small.

Example. The formula $(A \vee B) \wedge (A \vee \neg B) \wedge (\neg A \vee B) \wedge (\neg A \vee \neg B \vee \neg C) \wedge (\neg A \vee C)$ is unsatisfiable. For each of its 8 possible interpretations, this formula evaluates to false.

3.4.2.2 SAT solvers

A SAT solver takes a propositional formula as input and either returns a model of the formula (i. e., a assignment of its variables), or output that the formula is unsatisfiable. Most SAT solvers restrict their input to specific propositional formulas called conjunctive normal forms (CNF).

Conjunctive normal form (CNF)

A *literal* is either a propositional variable (A) or its negation ($\neg A$). A *clause* is a disjunction of literals. A formula is in CNF when it is a conjunction of clauses. All propositional formulas can be transformed into an equivalent that is CNF.

Example. The formula $(A \vee B) \Leftrightarrow (\neg C \wedge D)$ can be rewritten as the following CNF formula $(\neg A \vee \neg C) \wedge (\neg A \vee D) \wedge (\neg B \vee \neg C) \wedge (\neg B \vee D) \wedge (A \vee B \vee C \vee \neg D)$. The size of the resulting CNF is, however, exponentially large. For instance, a disjunction of n subformulas, each being a conjunction of m variables leads to a CNF formulas of size $O(n.m^n)$. Tseytin transformation avoids this exponential explosion by introducing new variables for subformulas. For instance, the formula $(A \wedge B) \vee (C \wedge D)$ is first transformed as the conjunction of the following three formulas X1 $\Leftrightarrow$ $(A \wedge B)$, X2 $\Leftrightarrow$ $(C \wedge D)$, and X1 $\vee$ X2. These formulas are then transformed into CNF.

Basic backtracking algorithm

The basic algorithm for a SAT solver explores a search tree where each node contains a partial assignment of the variables, and the leaves are all the possible interpretations. The traversal is performed by a backtracking algorithm. It chooses a variable, assigns it a truth value and recursively checks if the resulting formula is satisfiable. If it is unsatisfiable, the other truth value is chosen. If one of the truth values leads to satisfiability, then the algorithm returns the model, otherwise it returns unsatisfiability. The basic case of the recursive algorithm is when one of the clauses evaluated to *false* under the current assignment, or when all the clauses evaluated to *true* under the current assignment. Specific data structures are used to simplify these tests. This approach is complete as it explores the whole search space. It is also constructive; it starts from an empty partial assignment and extends it until an assignment satisfying all the clauses is found.

DPLL: adding unit propagation and pure literal elimination

When a clause contains a single unassigned variable, the value of this variable can be fixed to satisfy this clause. The other truth value should not be explored. This is called *unit propagation*. If the occurrences of a variable in the different clauses are always positive (*A*), one could assign the value *true* to this variable; the value *false* should not be considered in the search. A similar reasoning can be made when the occurrences are always negative (¬*A*). This is called *pure literal elimination*.

Both unit propagation and pure literal elimination can be applied iteratively at each step of the recursive algorithm, reducing the exponential growth of the search tree. The integration of unit propagation and pure literal elimination into a backtracking algorithm leads to the DPLL algorithm (Davis–Putman–Logemann–Loveland), which is the basis of many efficient SAT solvers.

CDCL: introducing clause learning

The traversal of the search tree with DPLL is depth first. Backtracking is chronological as the search goes back to the parent node in the search tree. However, when a partial assignment is found to lead to unsatisfiability, no information is deduced from the origin of this unsatisfiability. Conflict-driven clause learning (CDCL) improves DPLL in a significant way by introducing a nonchronological backtracking when the search reaches a dead end. A precise description is out of the scope of this chapter. The effect of CDCL is to build a new clause that is implied by the original clauses and blocks the current branch together with as many branches that share the choices leading to unsatisfiability. This is achieved by building an implication graph on the partial variable assignment and introducing a new clause with the negation of the assignments that lead to the conflict. Backtracking can then be made directly to the variable involved in the new clause that was first assigned in the search tree.

SAT solving through local search

Both DPLL and CDCL are complete and constructive approaches to the satisfiability problem. Although these methods are quite efficient, SAT is NP-complete and there will always be a complexity barrier for large instances. There are methods of SAT solving based on local search, that is, using an incomplete and perturbative approach. Starting from a (random) initial assignment to the propositional variables, at each step the value of one variable is flipped. The cost function is the number of unsatisfied clauses, and a solution is found when the cost reaches zero. This optimization problem is called max-SAT as it aims at optimizing the number of satisfied clauses, what is also interesting for unsatisfiable formulas. There exist many algorithms that solve MaxSAT through a local search. For instance, GSAT and WalkSAT methods are based on a MinConflict CSP local search, that is, choosing a variable, which flip minimizes the cost function. WalkSAT also introduces some perturbation to enhance diversification and avoid local minima.

3.4.2.3 Modeling problems

Problems that can be modeled and described as a SAT problem using propositional formulas. They can be solved with SAT solvers. The resulting solution, a specific assignment of Boolean values to the variables, can then be expressed as a solution to the initial problem. SAT solvers can therefore be helpful in a variety of problems where options and dependencies can be encoded as propositional formulas, as with the following example illustrating a planning challenge aiming at having a drink. A state of the world is here modeled by five Boolean status, represented by five propositional variables: "Is there a clean glass" (CG), "Is there a bottle of beer" (BB), "Is there a beer in a glass" (BG), "Am I thirsty" (T), and "Am I happy" (H). Initially, there is a clean glass (CG), a bottle of beer (BB), and the person is thirsty (T). There is no beer in a glass ($\neg BG$) and the person is not happy ($\neg H$). There are two actions. The action MakeDrink() requires CG and BB as precondition, and BG, $\neg CG$, $\neg BB$ as the effect. The action Drink() requires T and BG as a precondition, and H, $\neg T$, $\neg BG$ as the effect. The objective is to reach a state where the person is happy. What is a suitable combination of actions to reach the goal state?

This planning problem is now modeled as a SAT problem. To simplify the presentation, it is assumed that the goal can be reached by a succession of two actions. There will then be three considered states: the initial state, the state after the first action, and the state after the second action, which will be the final state. Each propositional variable describing a state will be subscripted by 0, 1, or 2 to model these three states. For each level of action, the two propositional variables $MakeDrink_i$ and $Drink_i$ state depending on whether the corresponding action is performed at level i.

The formula *PLAN* for this planning problem has then the following form:

$$InitialState \wedge DescriptionAction_1 \wedge DescriptionAction_2 \wedge Goal$$

where *InitialState* is the formula $CG_0 \wedge BB_0 \wedge T_0 \wedge \neg BG_0 \wedge \neg H_0$, *Goal* is the formula H_2, and *DescriptionAction_i* are formulas modeling which action is performed as the first and second actions in the plan.

The *DescriptionAction_i* should model the following elements:
- The precondition of the action must be satisfied by the preceding state:

$$(MakeDrink_i \Rightarrow CG_i \wedge BB_i) \wedge (Drink_i \Rightarrow T_i \wedge BG_i) \quad (i = 1, 2)$$

- The effect of the action must be satisfied by the next state:

$$(MakeDrink_i \Rightarrow BG_{i+1} \wedge \neg CG_{i+1} \wedge \neg BB_{i+1}) \wedge (Drink_i \Rightarrow H_{i+1} \wedge \neg T_{i+1} \wedge \neg BG_{i+1}) \quad (i = 1, 2)$$

- After an action, the state propositional variable not involved in the effect of the action are unchanged after this action:

$$(\text{MakeDrink}_i \Rightarrow (H_i \Leftrightarrow H_{i-1}) \wedge (T_i \Leftrightarrow T_{i-1})) \wedge (\text{Drink}_i \Rightarrow (CG_i \Leftrightarrow CG_{i-1}) \wedge (BB_i \Leftrightarrow BB_{i-1})) \quad (i = 1, 2)$$

- At least one action must be performed at each level

$$\text{MakeDrink}_i \vee \text{Drink}_i \quad (i = 1, 2)$$

- At most one action can be performed at each level

$$\neg(\text{MakeDrink}_i \wedge \text{Drink}_i) \quad (i = 1, 2)$$

Finding a model for the formula *PLAN* ensures to have a plan that starts from the initial state and reaches the goal states after two actions. To find a model, the formula *PLAN* can be transformed into a CNF form and a SAT solver algorithm can then be used. The following variable assignment is a model of the formula *PLAN*:

$$CG_0, BB_0, \neg BG_0, T_0, \neg H_0, \quad \text{MakeDrink}_1, \neg\text{Drink}_1, \quad \neg CG_1, \neg BB_1, BG_1, T_1, \neg H_1,$$
$$\neg\text{MakeDrink}_2, \text{Drink}_2, \quad \neg CG_2, \neg BB_2, \neg BG_2, \neg T_2, H_2$$

The conclusion is that the first action is MakeDrink and the second action is Drink.

For a real problem, the propositional formula can be automatically generated from a formal description of the planning problem, such as in planning domain definition language (PDLL), a family of languages that allow to define a planning problem using logic. The propositional formula can also be enriched by additional clauses expressing possible mutual exclusions between actions and/or literals that can be algorithmically derived.

3.4.3 What are the limitations of SAT solvers?

SAT solvers are used to solve Boolean CSPs or any problem that can be modeled as a satisfaction problem in propositional logic. One of the advantages of propositional logic is its simplicity. It is also compositional, meaning that different formulas can be composed without affecting the meaning of each of them. Although SAT is NP-complete, modern SAT solvers are quite efficient, even for large problems. The main limitation of propositional logic is its inability to represent objects and relations between objects. It also lacks quantifiers allowing stating conjunctions or disjunctions over different objects. In the planning problem, it is impossible to have a formula about all actions in a given level. As a consequence, a propositional model of a real problem may involve tens of thousands of variables and hundreds of thousands of clauses. Despite these limitations, SAT solvers are competitive for solving complex problems.

SAT solvers have been extended to satisfiability modulo theories (SMT) solvers, such as z3. Such solvers do not only solve SAT, but also formulas over richer logics, such as arithmetic or uninterpreted functions. SAT solvers have also been combined with constraint programming solvers, leading to CP-SAT that can be used in operational research applications. The CDCL technique is a basis for efficient SMT solvers as well as for CP-SAT.

3.4.4 Industry example: containers configuration using SAT, at Maersk Container Industry A/S

Yves Deville, Karen Veldeman, Nikolaj Bjorner

3.4.4.1 Modeling product configuration as a SAT problem

The industrial example on product configurators in Section 3.3.5 is further developed. The product configuration model has been described as a CSP using variables with a finite domain, and a set of constraints defining the possible configurations of containers. The MCI product configurator also integrates a SAT solver. The SAT solver relies on encoding all constraints using Boolean variables only in the form of clauses. On the surface, this may seem very low level and potentially losing structure, but this representation allows for efficient propagation using clauses in an efficient SAT solver.

An encoding of the sensor module attribute (SM) and its domain into SAT can be achieved using three Booleans SM.No, SM.CO2, and SM.O2, where the variable SM.x is true when the value of SM is set to x. To ensure that in an assignment of the Boolean variables, the SM attribute has a value, and the following clause is added to the CSP: SM.No $\vee$ SM.CO2 $\vee$ SM.O2. Other formulas ensure that this attribute is not assigned more than one value: $\neg$(SM.No $\wedge$ SM.CO2) $\wedge$ $\neg$(SM.No $\wedge$ SM.O2) $\wedge\neg$(SM.CO2 $\wedge$ SM.O2). The resulting clauses are then also added: $\neg$SM.No $\vee$ $\neg$SM.CO2, $\neg$SM.No $\vee$ $\neg$SM.O, $\neg$SM.CO2 $\vee$ $\neg$SM.O2. Similar clauses should also be added for the other attributes.

The table constraint illustrated in Table 3.4 expresses that some combinations of fresh air exchange modules and sensor modules cannot be combined into one single container. This table constraint should be transformed into clauses. A first idea would be to encode each line of the table as a conjunction (line 1 as AC.MV $\wedge$ FAM.MV $\wedge$ SM.No), and relate these formulas with disjunction. The size of the resulting formula when transformed in CNF is, however, exponentially large. This exponential explosion can be avoided by using Tseytin transformation that introduces a new Boolean variable to each row of the table (T1, T2, T3, and T4). A set of formulas relates these new variables with the attribute variables in the table constraint. For the first row, the formula T1 $\Rightarrow$ (AC.MV $\wedge$ FAM.MV $\wedge$ SM.No) relates T1 and the values of the attributes in this line. The resulting clauses are then AC.MV $\vee$ $\neg$T1, FAM.MV $\vee$ $\neg$T1, SM.No $\vee$ $\neg$T1. Similar clauses are added for the other rows. Tseytin transformation traditionally should also add formulas such as T1 $\Leftarrow$ (AC.MV $\wedge$ FAM.MV $\wedge$ SM.No) for the first row. It is, however, better to reexpress these formulas by other equivalent formulas. For each value of each attribute in the table, a formula relates the value to the corresponding row in the table. For the value number of the attribute SM, the formula is SM.No $\Rightarrow$ (T1 $\vee$ T2), leading to the clause ¬SM.No $\vee$ T1 $\vee$ T2.

The proposed encoding is very efficient. When applying DPLL on the above SAT encoding of the problem, thanks to unit propagation, it actually achieves a pruning equivalent to domain consistency. For instance, as soon as the variable AC.MV is set to true, the variable T1 has to be set to true by the clause ¬AC.MV $\vee$ T1. Then the variables FAM.MV and SM.No have to be set to true by the clauses FAM.MV $\vee$ ¬T1 and SM.No $\vee$ ¬T1.

Bart Bogaerts

4 Reasoning with first-order logic

4.1 Why is reasoning with first-order logic important within the broader artificial intelligence (AI) domain?

In this chapter, the general goal we focus on is the development of a so-called *knowledge-based* agent, which can reason based on a rich body of knowledge, stored in what we will call a *knowledge base*. While later chapters focus on *learning*, in this chapter, we assume that the knowledge is *provided* to the agent, most likely by a domain expert.

What this knowledge represents can depend on the context. For instance, it can be a symbolic representation of regulations relevant for its users, or it can be a representation of the causal mechanisms that govern the world in which the agent operates, or of strategic plans of a company, etc.

When representing knowledge, there are a couple of concerns to keep in mind. A first is the principle of *elaboration tolerance*; that is, when the world changes (e. g., new laws are introduced, company policies shift, the environment is changed), the agent's knowledge base should be easy to update, bring up to speed. To achieve this, it is important that this knowledge is presented in a *clear, understandable*, and preferably *modular* way. A second concern is that the underlying knowledge should also only be *represented once*. If the agent relies on its domain knowledge for multiple purposes, it should be able to use a single representation of that knowledge to guarantee consistency. In other words, the *representation of the knowledge is independent of the task to be solved*.

Once the agent is equipped with said knowledge base, we want it to perform various types of tasks, to solve problems that arise in the problem domain. As such, this chapter is concerned with two main tasks: *(1)* how to represent knowledge in a way that is understandable both for humans and computers and *(2)* how to exploit such represented knowledge for different forms of (deterministic) reasoning.

For the former, we study first-order logic (FOL), a much richer logic than propositional logic from the previous chapter; for the latter, we will show that in fact a theory in first-order logic allows for a wide variety of reasoning mechanisms.

First-order logic is chosen here as the base language because of its historic importance, as well as the fact that its connectives are simple, and have a crisp and clear informal semantics (by this, we mean that the formal meaning of statements in first-order logic is often what one would expect; as an example, first-order logic contains a quantor $\forall$ informally read as "for all"; as one can expect formulas of the form $\forall x : \phi(x)$ will hold in case ϕ is true for all possible objects). However, many other knowledge representation languages exist, taking different criteria into account (efficiency of reasoning methods, expressivity of the language, etc.); we discuss alternatives later in the chapter, as well as in Chapter 5, where we discuss knowledge representation languages known as "description logics," which are designed specifically with the purpose of making deductive reasoning decidable.

In the previous chapter, we studied propositional logic and saw how modern SAT solvers can solve the satisfiability problem of propositional logic with remarkable ease. However, from the perspective of *representation*, propositional logic falls short: the knowledge it represents is always instance-specific; it lacks the expressive power to represent knowledge of a problem domain in a concise way, independently of the instance.

Throughout this chapter, we will use *hospital scheduling* as a running example. In this context, a very natural piece of knowledge is that everyone can only be assigned a single task during each shift (no one can do two tasks at the same time). While it is very easy to specify this in natural language, it is not easy (in fact, even impossible) to do this in propositional logic independently of a concrete instance.

Of course, it *is* possible to encode this natural language constraint, given a concrete set of employees, in propositional logic, but this encoding will have the following undesirable properties:

- it is instance-specific: if the set of employees changes, so does the encoding; and
- it is hard to read, and hence difficult to understand or debug.

On top of that, it is inherent to propositional logic that when the instances get larger (more personnel or more tasks), so does the representation of the knowledge that everyone can do at most one job at a single time. For instance, when considering the set of employees $A(lice)$, $B(ob)$, and $C(harlie)$ with two tasks T_1 and T_2 and two shifts S_1 and S_2, we could use propositional variables p_{ETS} with E an employee, T a task and S a shift, where p_{ETS} has the intended interpretation (the intended interpretation of a symbol is the concept in the real world it represents) that p_{ETS} holds if and only if E is assigned task T during shift S, one can craft the following encoding:

$$\neg(p_{A11} \wedge p_{A21}) \wedge \neg(p_{A12} \wedge p_{A22}) \wedge$$
$$\neg(p_{B11} \wedge p_{B21}) \wedge \neg(p_{B12} \wedge p_{B22}) \wedge$$
$$\neg(p_{C11} \wedge p_{C21}) \wedge \neg(p_{C12} \wedge p_{C22})$$

While this formula correctly characterizes the given natural language constraint for the given instance, it raises challenges of understandability (it can hardly be called transparent and convincing someone of its correctness is not an easy job, especially if the instances get larger), as well as maintainability (if the knowledge of what constitutes a valid schedule changes, or even if the set of tasks changes, the entire formula needs to be updated). Furthermore, even if one could convince someone of the correctness of this encoding, nothing could be concluded about other instances. These limitations of representation in propositional logic are in stark contrast with the simplicity of a natural language expression

"everyone can do at most one job at a single time."

To address the understandability problem, it would make sense to not use the actual propositional theory, but rather use a *schema* of how to generate the propositional theory. Concretely, here this could be something of the form

$$\neg(p_{EXS} \wedge p_{EYS}) \quad \text{for each employee } E, \text{ each shift } S,$$
$$\text{and each two distinct tasks } X \text{ and } Y.$$

While this representation solves the issue of understandability, it has other weaknesses, most notably that this notation is informal and not computer-readable. This is exactly where representation languages with a higher level of expressivity (such as first-order logic) come into play: first-order logic provides us with the means to express the knowledge underlying the scheduling problem compactly, in a formal way that is independent of the instance. Concretely, in first-order logic, we would make use of a relation *Assignment* with intended interpretation that *Assignment*(e, t, s) holds if employee e is assigned task t at shift s. The constraint that every employee can be assigned at most one task during each shift then becomes

$$\forall e, \forall s, \forall t_1, t_2 : (t_1 \neq t_2) \Rightarrow \neg(Assignment(e, t_1, s) \wedge Assignment(e, t_2, s)),$$

or, equivalently,

$$\forall e, \forall s, \forall t_1, t_2 : Assignment(e, t_1, s) \wedge Assignment(e, t_2, s) \Rightarrow t_1 = t_2,$$

which intuitively states that if a single employee e is given the tasks t_1 and t_2 during a given shift s, then it must be that t_1 and t_2 are actually the same. Moreover, providing a problem-independent, formal description of this knowledge, provides us with the means to use it for various types of reasoning. Indeed, the knowledge about "what constitutes a good scheduling" is independent of any problem to be solved; while it is often associated to it, this knowledge is not inherently linked to the problem of *finding schedules*. It could also for instance be used for *proving properties that all valid schedules satisfy*, that is, to reason *independently of a specific instance*.

4.2 What category of problems does reasoning with first-order logic solve?

The focus of this chapter lies on problems that arise in *knowledge-intensive problem domains*. These are domains about which background knowledge is available, either explicitly (for instance, in the form of regulations, policies, etc.), or implicitly (e. g., in the mind of a domain expert). In Chapter 7, we will study how to make intelligent agents in case the knowledge is *not* available explicitly, but instead we want to *learn* it from data. In such domains, we are concerned with problems whose solutions can *unambiguously*, and *deterministically* be defined in terms of this background knowledge. In Chapter 6, we will see other forms of reasoning that work *probabilistically*, for example, taking uncertainty about the world into account.

Some examples of such knowledge-intensive problem domains include *decision management* (where the logic underlying everyday business decisions can be made explicit), *scheduling* (where the knowledge about what constitutes a valid scheduling can be made explicit), *product configuration*(where the information about what constitutes a good configuration can be made explicit), legal reasoning (where laws or regulations can be made explicit), etc.

In such a knowledge intensive domain, many different problems can arise; and part of the research in logic-based reasoning is about classifying such problems as more generic problem patterns. These generic patterns are defined independently of the problem domain at hand, directly in terms of the logical representation. Some examples of such generic inference methods include:

– *Model checking*: Given a complete specification of a state-of-affairs, check whether this state of affairs indeed satisfies the knowledge that was made explicit. For instance, in case the knowledge describes "what are valid schedules" and the state-of-affairs is one concrete proposed schedule (that might have been crafted manually), this task boils down to checking whether the manually crafted schedule indeed satisfies the scheduling constraints.

– *Model expansion*: Given a partial description of the state-of-affairs, complete this to a *complete* description *in such a way that* the given knowledge is respected. In the scheduling example, the partial description of the state-of-affairs could be information about the available rooms and personnel and shifts, but not contain information about the actual schedule. The complete information would then also contain the actual schedule. That is, in that case the problem-specific task would be *searching* for a schedule. This inference is sometimes extended to *optimal model expansion*, where one is not just interested in searching *any* extension of the given input but one that is *optimal* with respect to a certain criterion. For instance in the scheduling domain, one might want to optimize conformance to explicated preferences of employees.

– *Querying*: Given a representation of the world, finding all instances that satisfy a certain logical formula. This type of querying corresponds exactly to querying a database (and in fact, many successful database languages are based on first-order logic).

– *(Finite domain) Propagation*: Given partial information about the world, derive more information based on the explicit knowledge. For instance in case a partial schedule is constructed, an automated reasoner could already infer that a given

employee could not do a certain task at a certain moment (because they already have another task at that moment, or because regulation forbids to fill more than two consecutive shifts, or ...)
- *Deduction*: Checking whether a certain logical sentence follows from the formalized knowledge independently of the instance. For instance, in the scheduling use case, one could use this to verify whether the constraints used for scheduling guarantee that the local fire safety regulations are respected (given a representation of the fire safety regulations in first-order logic). By proving this independently of the instance, we obtain the guarantee that all previous and future schedules created based on this knowledge indeed satisfy those requirements.

All the methods presented in this chapter will be based on an explicit representation of domain knowledge. This, however, does not mean that *all* knowledge needs to be represented formally. Indeed, again consider the hospital scheduling application. In such an application, a lot of knowledge will be easy to formalize (e. g., constraints imposed by the hospital, as well as constraints imposed by the government), but other parts of the knowledge might be *tacit*, that is, expert schedulers might know about certain sensitivities: who can work well with whom, and which kind of schedules are deemed fair by employees. This kind of knowledge is often hard to formalize. In such cases, we can still use the formalization of the "hard" knowledge, not to *solve* the complete scheduling problem, but to *assist* the expert scheduler, that is, the knowledge on scheduling policies can be used to develop a *decision support system*. Such a system can for instance use *propagation* to derive consequences of choices made the expert scheduler and even *explain* him/her why certain (undesired) consequences follow from other assignments. In situations like this, multiple inference can be used on the same knowledge; the idea of doing this is known as the *knowledge base system paradigm* (Denecker and Vennekens, 2008).

4.3 How are those problems solved?

In order to solve problems in knowledge-intensive domains, as described above, we need:
- A language in which domain knowledge can be expressed. This language should be understandable both by a computer (for reasoning) and by humans (for transparency, as well as to enable maintenance/updates).
- Reasoning engines that can exploit this knowledge.

As for the language, this chapter focuses on first-order logic, because of its historic importance in different domains, its expressive connectives, and its intuitive informal semantics. However, many other knowledge representation languages exist, each of them designed with specific goals in mind. A nonexhaustive list of some examples follows.

For instance, Chapter 5 is concerned with the study of description logics (Baader et al., 2003), which are essentially *limitations* of first-order logic that enable efficient deductive reasoning. On the other hand, in other languages the expressivity of first order is *extended* instead of reduced, for example, by adding second-order features (such as one for instance in ProB (Leuschel and Butler, 2003) and Alloy (Jackson, 2002)).

Statements in first-order logic are in essence *objective*, they are either true in the world or not. However, in some cases one might want to express knowledge not about the actual world, but about the *state of mind of an agent*. For instance, rather than reasoning about what is true, in some cases it might be useful to reason about *what another agent knows*. For this purpose, a variety of so-called *epistemic logics* has been developed (Hintikka, 1962; Fagin et al., 1995; Halpern and Moses, 1990).

First-order logic is a *monotonic* logic. By this, we mean that if one can make a conclusion from a limited set of statements, adding more statements can never undo the conclusion, can only result in us making *more* conclusions. Certain natural language statements do not have this property. For instance, if I tell you that "Birds can usually fly," and "Tweety is a bird," it is natural to assume that Tweety can fly. However, when given more information, such as, for example, that Tweety is in fact a penguin, the conclusion might become invalid. This kind of logics is studied in the field of *nonmonotonic reasoning* (Reiter, 1980; Moore, 1985; McCarthy, 1986; Gelfond and Lifschitz, 1988).

Other types of languages, often with a basis in first-order logic, include
- languages to express dynamic domains, in which the world is modeled in a temporal setting and actions can change the state of the world (Reiter, 2001; Mueller, 2007; Reiter, 1991);
- languages to express *causal information* (Pearl, 2000; Pearl and Mackenzie);
- languages to express *spatial relations* (Cohn and Renz);
- domain-specific languages, to represent knowledge about a particular application domain. For instance when taking the example of modeling business logic and decision management, several domain-specific languages have been proposed (Group, 2008; Governatori, 2005; Abdelsalam and Shoaeb, 2016).

4.3.1 Representing knowledge in first-order logic

To represent knowledge in first-order logic, we will discuss three important concepts. The first is a *vocabulary*; it specifies the set of symbols we will use. By choosing the vocabulary, and agreeing on the informal semantics (the meaning in the real world) of each of the symbols, we determine which concepts in the world we can express knowledge about. Second, a *structure* consists of a domain (all the objects in the world), as well as an interpretation of the symbols in the vocabulary; it is an abstract representation of the world of interest; it has a *domain* (the set of all objects in the world) and assigns a value (of the right type, see below) to each of the symbols. Third, a *formula* then represents the actual knowledge to be formalized; it specifies how different symbols from the vocabulary are related. Formulas should follow the *syntax* of first-order logic. The *semantics* of first-order logic is used to determine whether or not a formula is satisfied in a given world (structure).

Vocabularies

While propositional logic starts from a set of *propositions,* statements that can be true or false in the world, first-order logic takes the view that the world consists of (different sets of) objects (often called *domain elements*), with *relations* and *functions* between them. Uncoincidentally, this assumption is very similar to the one made in the context of relational databases (with a strong focus on relations there). Of course, it is debatable whether the real world is indeed made up of such objects and relations, but knowledge representation also starts from the idea that a certain abstraction of the actual world needs to be made. Thus, we will abstract away certain details of the world to be modeled and end up with a set of objects with relations and functions between them. This immediately brings us to the first challenge to be tackled when using knowledge representation techniques: *finding the right level of abstraction for representing a problem domain.* Often, an informal specification of the problem domain can give us an idea of what this level of abstraction should be. Take, for instance, the problem of hospital scheduling. A specification of what constitutes a valid scheduling will most probably mention *nurses* (a set of "objects" we cannot take abstraction of) and shifts (suggesting that using entire days as building blocks is probably too coarse, but scheduling per minute on the other hand is probably a too detailed representation). Such a specification will also mention that there is a relation between nurses, time points (abstracted in shifts) and assignments, etc.

The exercise described above does not just establish a level of abstraction, it also gives us an idea what the objects and relations between them will be. In other words, this exercise determines the *ontology* to be used in the representation of our domain knowledge. In first-order logic, the concept of a *vocabulary* is used to represent this information: a vocabulary is a collection of relation symbols (often called *predicate symbols*[1]), and *function symbols*, each with an associated arity—the number of arguments they take. Nillary function symbols (function symbols that take no arguments) are often called *constants.* For instance, in our running example on hospital scheduling, it is very likely that one needs:

– Constant symbols, such as *CEO*, to refer to specific persons (or other objects of interest in the domain).

---

1 First-order logic is also often called *predicate logic.*

- A set of nurses; this will become a *unary* (one input argument) symbol *Nurse*, meaning that every object in the abstracted world, either is a nurse or is not. For instance *Nurse(CEO)* will (as we see below) express that the chief executive officer (CEO) is a nurse, which can either be true or false.
- A set of employees containing all employees of the hospital; again, in the vocabulary this will correspond to a unary predicate symbol, for example, called *Employee*. It is very likely that all the nurses in the domain of interest are employees; we will see below how to express this.
- A unary predicate *Surgeon* representing a set of surgeons.
- A set of shifts (e. g., "Monday 6am–4pm" could be a shift); again, this will be represented by a unary predicate *Shift*: every object in the domain of discourse either is a shift or it is not.
- A set of possible qualifications, again represented by a unary predicate symbol, for example, *Qualification*.
- A relation between employees and qualifications stating who has which qualification; since this is a relation between two objects (for every employee e and every qualification q, either e has the qualification q or e does not have it), it is represented by a binary predicate symbol (e. g., *HasQualification*) in the vocabulary. The intended interpretation of this symbol is that *HasQualification*(e, q) holds when e has the qualification q (e. g., a $C1$ qualification for the English language).
- A set of tasks to be performed, represented by a unary predicate *Task*.
- A relation stating who does what at which moment. Since this is a relation between three objects (a person, a task, and a shift), this is modeled by a ternary relation in first-order logic, for example, a relation *Assignment* with intended interpretation that *Assignment*(e, t, s) holds if employee e is assigned task t at shift s.
- A relation stating who is on call during each shift: *OnCall*(e, s) means that e is on call during shift s.
- A function *Manager* with intended interpretation that it maps every employee to its manager. This would be a unary function, taking one input argument; every function in first-order logic produces one output.

In the description of this vocabulary, already a couple of important points show up. First, these symbols are independent of which task one wants to solve in the context of hospital schedules: it might be used for checking whether a schedule satisfies the hospital's rules, for generating a work schedule, or for proving that the hospital's rules conform to the state regulation. Second, the choice of symbols, and thereby level of abstraction, determines which kind of information we will be able to express, and how easy it is to express that information. Third, with each symbol here we specified its *intended interpretation*; this is an often overlooked, but extremely important thing to do. Similar to documentation of procedural code, the intended interpretation of symbols is a means to communicate, which objects and relations in the world one will make claims about and hence is crucial for making sure these claims are indeed well understood. Also similar to programming, it is important that the relations have a meaningful name, again to ensure readability of the formalized knowledge.

We already associated to the symbols an *intended* interpretation. However, symbols can also have an *actual interpretation*: symbols can have a value; for constant symbols, this is an element of the domain; for predicate symbols this value will be a relation, for function symbols, this will be a function. For instance, in the real world, everyone either has a certain qualification or not; this determines the actual value of the symbol *HasQualification*. Also, in a solution to the scheduling problem, each person is either assigned a task during a certain shift or not, thereby determining a value for the symbol *Assignment*. Such an interpretation of the symbols is called a *structure* and is discussed next.

Structures

The basic semantic object of first-order logic is a *structure*, often denoted S. It is a possible state of the (abstracted) world. A structure consists of two parts. On the one hand, it specifies a *domain D*: the set of all objects in the world. On the other hand, it also specifies for each symbol in the vocabulary an *interpretation*. The interpretation of a function symbol of arity n is a function from D^n to D; the interpretation of an n-ary predicate symbol is a set of n-tuples with elements in D, that is, an n-ary relation over D. Thus, in short, a structure determines

- a set of objects (its domain) D,
- for each constant symbol c, a domain element c^S (i. e., each constant symbol denotes an object).
- for each n-ary predicate symbol p a relation $p^S \subseteq D^n$;
- for each n-ary function symbol, a function $f^S : D^n \to D$.

For instance, in our scheduling example,
- The domain (the set of all objects) could be $\{Ann, Bob, Charlie, Q_1, Q_2, \dots\}$
- The interpretation of *Nurse* should be a set of 1-tuples, that is, elements of the domain. It could for instance be $Nurse^S = \{Ann, Bob\}$.
- The interpretation of *HasQualification* should be a set of 2-tuples indicating who has which qualifications. It could be $HasQualification^S = \{(Ann, Q_1), (Ann, Q_2), (Charlie, Q_1)\}$.
- The interpretation of *Manager* should be a function, mapping each employee to their direct manager. It could map *Ann* to *Bob* (i. e., $Manager^S(Ann) = Bob$), *Bob* to *Charlie* and *Charlie* to himself (assuming *Charlie* is the CEO).

It is important to remark that in our informal discussion, we only specified how the function *Manager* should behave *on employees*. In standard first-order logic, however, functions are assumed to be *total*. That is, they are supposed to map every tuple of values to some value. In practice for us, this will mean that the function that *Manager* denotes will also assign values to qualifications, assignments, etc. Most likely, we will not be interested in such values. To deal with this in a more natural way, first-order logic is often extended either with *partial functions* (functions that do not need to map every object to a value) or with *types*. In a typed logic, we partition the domain of interest in different sets (the types), and in this setting, each symbol is not just given an arity but also a typing, for example, one could there state that *Manager* is a function *from employees to employees*, and thus get rid of the redundant information.

Terms and formulas

First-order logic now allows us, using the symbols from the vocabulary, to write complex expressions, as detailed below. In first-order logic, *terms* are expressions that denote an object. There are three types of terms that can denote an object of the world, constant symbols, function symbol applications, and variable symbols. For instance, if *HeadNurse* is a constant symbol, then it denotes an element of the domain in every structure; the interpretation of this symbol specifies which element it denotes; let us assume it denotes *Bob* in an extension of the structure described above. Another type of terms is the application of a function symbol (of arity n) to n terms. For instance (with $n = 1$), *Manager(HeadNurse)* is a term as well. In our structure, it denotes *Charlie* since the expression *HeadNurse* denotes the object *Bob* and the interpretation of *Manager* maps *Charlie* to *Bob*. A last type of terms are *variables*, for example, x, y, z. These symbols are local to some quantification and their value is not part of a structure.

As can be seen, the semantics of terms is *compositional* in the sense that it works by giving meaning to an expression in terms of the meaning of its subexpressions, and *unsurprising* in the sense that it seems like the only reasonable semantics to be given to it. This is the case for the entirety of first-order logic. We now discuss all the language constructs that are used to form formulas in first-order logic. A formula in first-order logic denotes a truth value (true (**t**) or false (**f**)) in the context of a structure and assignment of values to its (free) variables. The basic building block are atomary formulas; they either equality atoms (equality between two terms) or a predicate symbol applied to the right number of terms. For instance,

$$Manager(HeadNurse) = HeadNurse$$

is true if and only if the head nurse is his/her own manager and

$$Employee(Manager(HeadNurse))$$

is true if and only if the manager of the head nurse is an employee.

First-order logic then makes use of the same Boolean connectives we know from propositional logic (Chapter 3): if φ_1 and φ_2 are formulas, then so are $\neg\varphi_1, \varphi_1 \wedge \varphi_2, \varphi_1 \vee \varphi_2$ (and $\varphi_1 \Rightarrow \varphi_2$ and $\varphi_1 \Leftrightarrow \varphi_2$). Their semantics are as expected, for instance,

$$\neg Manager(HeadNurse) = HeadNurse$$

is true if and only if

$$Manager(HeadNurse) = HeadNurse$$

is false, that is, if the head nurse is not his/her own manager. Similarly,

$$Manager(HeadNurse) = HeadNurse \wedge Employee(Manager(HeadNurse))$$

is true if two claims hold: *(1)* the head nurse is their own manager, and *(2)* the head nurse's manager is an employee. The semantics of $\Rightarrow$ is a bit more subtle. This operator is to be understood as follows: $\varphi_1 \Rightarrow \varphi_2$ states that *if* φ_1 is true, then φ_2 must be true as well. Otherwise, no claim about the value of φ_2 is made and the proposition is always satisfied. For instance, consider the formula

$$Nurse(x) \Rightarrow Employee(x).$$

Taking abstraction for a moment of where the value of the x comes from, this formula states that *if* x is a nurse, then x is an employee as well. This proposition should surely hold for all xs that are nurses, but it actually holds for all nonnurses as well. To see this, assume x denotes an office chair. In that case, this statement holds as well: *if* my office chair is a nurse, then it is an employee as well. While this might seem a bit counter-intuitive at first, this kind of construct is actually used constantly in natural language. Consider someone saying to their friend "If we meet each other at the conference, we go for a drink together, I promise." Now assume the two people do not see each other at the conference (and hence of course, do not go for a drink), would that make the first person a liar? Most people would agree not. Hence, the logical implication (conditional) formalizes this sort of "if ...then ..." from natural language; however, it deserves to be mentioned that in natural language we often use "if ...then ..." to mean other things as well (causation, definition, etc.).

We now turn our attention to a language that really sets first-order logic apart from propositional logic. That is *quantification*. That is, we can say that a certain property holds for *all* or for *some* objects. Concretely, if, as discussed above, we wish to express that every nurse is an employee, we can express this as

$$\forall x : Nurse(x) \Rightarrow Employee(x).$$

The semantics of such a formula is defined as follows: $\forall x : \varphi$ is true if φ is true for all assignments of domain elements to x. Similarly, $\exists x : \varphi$ is true if φ is true for some (one or more) assignments of domain elements to x, allowing us for instance to express properties such as

$$\exists x : Employee(x) \wedge Manager(x) = x$$

stating that there exists an employee who is their own manager (e. g., the CEO).

While each of these language constructs are quite simple, combining and nesting them allows us to express complex constraints such as

$$\forall x : (\mathit{Nurse}(x) \wedge x \neq \mathit{HeadNurse}) \Rightarrow \mathit{Manager}(x) \neq x,$$

which states that all nurses different from the head nurse must have a manager that is different from themselves (i. e., no nurse except for the head nurse can be their own manager).

Remark 4.1. There is one subtlety that deserves some attention. When communicating (in natural language), humans rarely quantify over "everything" (i. e., all objects in the world). Instead, we usually quantify over restricted subsets. For instance, "all men are human" or "all lectures should be scheduled between 8am and 6pm." That is, universal quantification (and the same holds for existential quantification) in natural language is usually of the form

"All P's are Q's"

This kind of construct is sometimes called *binary quantification* but is not present in first-order logic. The way this will typically be written in FO is $\forall x : P(x) \Rightarrow Q(x)$. This statement states that *for all objects in the world*, if they are a P then they should also be a Q (and otherwise no restriction is imposed on them). Existential quantification in natural language takes the form

"There is a P that satisfies Q" or "Some P is a Q"

for instance "(at all times) there is a surgeon on call," where P is "surgeon" and Q is "is on call." In first-order logic, this would be expressed as $\exists x : P(x) \wedge Q(x)$, which immediately highlights an asymmetry between universal and existential quantification. Indeed, for conditional/binary quantification, a universal quantifier is paired with an implication while an existential typically occurs with a conjunction. Both connectives are visible when formalizing the previously mentioned statement "at all times, there is a surgeon on call" as

$$\forall sh : \mathit{Shift}(sh) \Rightarrow \exists s : \mathit{Surgeon}(s) \wedge \mathit{OnCall}(s, sh).$$

This rule-of-thumb can be of great value when debugging knowledge expressed in first-order logic: expression of the form $\exists x : \varphi \Rightarrow \psi$ or $\forall x : \varphi \wedge \psi$ are rarely correct.

The formal semantics of the different type of formulas is summarized in Table 4.1; the informal reading of different formulas is summarized in Table 4.2. Often, one will not be interested in a single arbitrary formula, but rather in a set of so-called *sentences*: formulas in which all variables are quantified (for instance, $\mathit{Nurse}(x) \Rightarrow \mathit{Employee}(x)$ is not a sentence; it is not clear what this expresses (the existence of such an x or that this

Table 4.1: Semantics of formulas in first-order logic.

formula	interpretation in structure S
$t_1 = t_2$	**t** if t_1 and t_2 have the same interpretation in S; **f** otherwise.
$P(t)$	**t** if $t^S \in P^S$; **f** otherwise.
$\phi \wedge \psi$	**t** if both ϕ and ψ are true in S; **f** otherwise.
$\phi \vee \psi$	**t** if at least one of ϕ and ψ is true in S; **f** otherwise.
$\phi \Rightarrow \psi$	**t** if ϕ implies ψ in S (i. e., if ϕ is true, so is ψ); **f** otherwise.
$\forall x : \phi(x)$	**t** if $\phi(d)$ is true for all the elements d in the domain of S; **f** otherwise.
$\exists x : \phi(x)$	**t** if $\phi(d)$ is true for at least one element d in the domain of S; **f** otherwise.

Table 4.2: Informal interpretation of first-order logic.

formula	informal interpretation
$t_1 = t_2$	t_1 and t_2 have the same value
$P(t)$	t is in the interpretation of P
$\phi \wedge \psi$	ϕ and ψ are both true
$\phi \vee \psi$	ϕ is true, or ψ is true, or both
$\phi \Rightarrow \psi$	if ϕ is true, so is ψ
$\forall x : \phi(x)$	ϕ holds for all possible values of x
$\exists x : \phi(x)$	ϕ holds for some possible value of x

holds for all such x?), while $\forall x : Nurse(x) \Rightarrow Employee(x)$ is a sentence). Such a set is often called a *theory*, or *knowledge base*.

This concludes our introduction to the type of knowledge representable in first-order logic. An overview of the most important syntactic objects in first-order logic can be found in Table 4.3. To increase natural expressivity, the operators discussed here are often extended, for example, to include aggregates (to naturally express "at least 5"), types (to make quantification more natural and make functions only range over the relevant domain elements), and many more constructs.

Table 4.3: Overview of the most important elements of the syntax of FO.

type of expression	(type of) value in a structure	examples
constant symbol	domain element	*CEO, HeadNurse*
function symbol	function	*Manager*
predicates symbol	set of tuples	*Employee, OnCall*
variable symbol	–	*x, y*
term	domain element	*CEO* *Manager(HeadNurse)*
formula	true or false	$\forall x : Nurse(x) \Rightarrow Employee(x)$. *Manager(HeadNurse) = CEO*

We now turn our attention to how this knowledge can be used for various forms of reasoning and how this relates to many other fields.

4.3.2 Reasoning with first-order logic

Now that we have presented the syntax and semantics of first-order logic, we are ready to present a variety of different forms of reasoning that use the knowledge expressed in first-order logic. In fact, the reasoning methods we present (often called *inference methods*) are not just applicable to first-order logic, but are applicable as well to other logics with a *model semantics*, that is, where the semantics is defined by formally stating which states of affairs satisfy expressions in the logic and which do not, that is, which structures are *models* of a theory. In FO, a structure S is called a *model* of a theory T if it satisfies all sentences in T (if all the knowledge in T is indeed true in S). Formally, S is a model of T if for all sentences ϕ in T, $\phi^S = \mathbf{t}$.

Some of the inference methods discussed below make use of a *partial structure*. A partial structure, like a normal structure has a domain and interprets symbols. The difference with a regular structure is that in a partial structure part of the information may be "unknown." For instance, it can be unknown who the CEO is, or it can be unknown whether or not "Alice is a nurse." Partial structures can be compared in *precision*: S_1 is *more precise* than S_2 if there are fewer things unknown in S_1 than in S_2, but everything that has a value in S_2 has the same value in S_1. In case S_1 is a normal structure more precise than S_2 we call S_1 an *expansion of* S_2. There are different ways to define such partial structures formally, the simplest one to think of, for now, is at the granularity level of symbols: *a partial symbol interprets only some symbols in the vocabulary, and leaves the value of the other symbols open/unknown.*

Inference method: **Model checking**

Given: a (finite) structure (an abstraction of the world) and a theory (the formalized knowledge)
Decide: whether or not all sentences of the theory are satisfied in the given structure

---

The first inference method discussed is quite a simple one: for *model checking*, one is given a structure and a theory (the formalized knowledge) and the problem at hand is deciding whether or not all formulas in the theory are satisfied in the structure, the abstraction of the world. That is, in the context of the running example, this inference method would boil down to checking whether a (for instance, manually created) schedule satisfies all the scheduling constraints. This inference method can be implemented very efficiently (in so-called *polynomial time* in terms of the size of the world). One way to implement this would be to directly apply the definition of the semantics, for instance, for the sentence $\forall x : Nurse(x) \Rightarrow Employee(x)$, given a structure S, we can check for each d in the domain whether $Nurse(d) \implies Employee(d)$ holds. As soon as we find one instance for which it does not, we know the original sentence is not satisfied. If we finish this loop over the domain without finding one such instance, we know that it holds.

Inference method: **(Optimal) Model expansion**

Given: a (finite) partial structure (a structure that interprets some symbols, but not all of them) and a theory (the formalized knowledge) and an optimization criterion that specifies which worlds are preferred

Find: an expansion of the partial structure that satisfies the theory (and that is optimal with respect to the given criterion)

The second task, model expansion, is in fact a generalization of model checking. Here, the world is not completely given, but parts of it are to be searched. For instance in nurse scheduling, it would be a realistic assumption the parts of the structure given include which rooms/shifts/etc. there are, while the nongiven part includes the actual schedule. As such in this case, the problem of model expansion is the problem of searching a schedule that satisfies all the constraints (and that is optimal with respect to a certain criterion, e. g., how well it respects preferences of the employees or fairness). This problem is typically solved by the techniques from the previous chapter, such as SAT solving. To do this, the combination of the structure and theory is first reduced to a SAT problem by a technique called *grounding*, which eliminates all quantifications, intuitively by replacing a quantifier

$$\forall x : \varphi(x)$$

by

$$\varphi(d_1) \wedge \varphi(d_2) \wedge \cdots \wedge \varphi(d_n)$$

where the d_i are all the elements of the domain. Similarly, an existential quantifier

$$\exists x : \varphi(x)$$

would be replaced by

$$\varphi(d_1) \vee \varphi(d_2) \vee \cdots \vee \varphi(d_n).$$

When this procedure is applied recursively, for example, translating $\exists x : \forall y : \varphi(x, y)$, to

$$(\varphi(d_1, d_1) \wedge \varphi(d_1, d_2) \wedge \ldots) \vee (\varphi(d_2, d_1) \wedge \varphi(d_2, d_2) \wedge \ldots) \vee \ldots,$$

the resulting formula will have no more quantifiers or variables. In case there are no function symbols, the result will be a *theory in propositional logic,* and hence the SAT solving techniques from the previous chapter are directly applicable to find satisfying assignments, which can then subsequently be translated back into a first-order structure. In case the original theory had function symbols, there are two options. The first option is to replace the function symbols by predicate symbols by a technique that is known as graphing (a function f is replaced by a predicate G_f such that $G_f(x, y)$ holds if and only if $f(x) = y$), to again, arrive in propositional logic. The second option is to keep the function symbols, but in this case, after replacing all variables by all their instantiations, not only propositional symbols will remain, but also *finite domain variables*. In that case, a *constraint solver* can be used to solve the remaining problem, using all the techniques from the previous chapter.

Inference method: **Querying**

<u>Given</u>: a (finite) structure (an abstraction of the world) and a formula
<u>Find</u>: assignments to the free variables of the formula for which it is satisfied

---

The query inference is used mainly in the context of databases, where the structure is represented by a database, and the query typically by an structured query language (SQL) statement. However, it deserves to be mentioned that first-order logic (there often referred to as the *relational calculus*) lies at the basis of SQL. The goal of the query inference is to evaluate a certain formula (with free variables) in a given structure. For instance, in case a complete structure of the scheduling vocabulary (i. e., a complete schedule) is found, one might want to ask questions about it such as "which surgeons are on call during the July 11 Saturday AM shift?", for instance, to inspect the schedule or to develop applications that visualize it. In first-order logic, this would be expressed as

$$Surgeon(s) \land OnCall(s, \text{"Jul11} - SatAM\text{"})$$

or as

$$\{s \mid Surgeon(s) \land OnCall(s, \text{"Jul11} - SatAM\text{"})\}$$

in case the variables whose instantiations is to be found is made explicit. For reference, the same query in SQL would be written as

SELECT person FROM *OnCall* NATURAL JOIN *Surgeon*
WHERE shift = "Jul11-SatAM" ;

Another example would be a query that searches for the set of all people who worked together with a certain individual who tested positive for COVID-19, which would be expressed as

$$\{p \mid \exists t, s : Assignment(p, t, s) \land Assignment(\text{"PersonX"}, t, s) \land s < Today\}$$

where *Today* is a constant interpreted as the current day (to only select shifts that have already passed, not shifts that are planned in the future). Computing the solutions to such a query efficiently is extensively studied in the field of databases. One very important technique is *join reordering*, which when translated into first-order logic, essentially boils down to using the fact that for all formulas α, β, and γ, for instance,

$$(\alpha \wedge \beta) \wedge \gamma$$

is equivalent to

$$(\alpha \wedge \gamma) \wedge \beta,$$

and hence that in order to compute all instances that satisfy each of these three formulas, we can first compute all instance that satisfy any of the two and afterwards intersect with those instances satisfying the last formula. Join reordering techniques often make use of the size of the interpretation of certain symbols in order to reorder the query so that the internal processors will compute the result much faster.

Like model expansion, querying is a generalization of model checking; in case all occurrences of variables are quantified, the query becomes a so-called *Boolean query* and the task reduces to checking whether a formula is satisfied in the structure or not.

Inference method: **Finite domain propagation**

<u>Given</u>: a partial structure (an abstraction of parts of the world) and a theory (the formalized knowledge)

<u>Find</u>: a more precise structure (one that interprets more symbols) that is a consequence of the input

---

The task of propagation takes as input some partial information about the world and produces more refined (possibly still partial) information based on the logical theory. That is, it will only derive consequences of the theory given the current information. To define this formally, if the partial structure in the input is S and the theory is T, the output S' should be such that for every model M of T more precise than S, M is also more precise than S' (i. e., "no models are lost"). For instance, in the hospital scheduling application, this type of inference could be used in a support system for expert schedulers who make the schedule by hand, but in doing so interact with the assistant. For instance, if the scheduler has assigned a surgeon to be on call during a certain shift and the theory contains a constraint stating that exactly one surgeon should be on call during every shift, the system can automatically derive that none of the other surgeons should be on call. Or in case a nurse is scheduled to work a certain number of shifts in a period of time, the system could (depending on the actual regulations holding for that particular hospital) decide that by law the nurse cannot work any more shifts in that period of time. Such propagations can help the expert planner keep an overview of the consequences of their previous actions, or detect inconsistencies early on. In this kind of setting, a formalization of the knowledge underlying the scheduling problem is useful even when the scheduling problem is not solved fully automatically.

There are different ways in which propagation can be implemented. One way would make use of *grounding*, as described in the section on model expansion, to reduce the first-order problem to a propositional problem or a finite-domain constraint problem. Afterwards, the propagation can be done by the techniques seen in the previous chapter, for example, *unit propagation* or *constraint propagation*. Another solution would be to use model expansion to compute *all models* of the theory that are more precise than the given partial interpretation. Everything that is true in all those models is then a consequence of the given input. This second method is computationally much more demanding than the first, but also produces more precise information. Which of the two is preferred depends on the size of the problem (for small problems, computational cost is not an issue), and the required precision.

Inference method: **Deduction**
Given: two theories
Decide: if the first theory entails the second

Inference method: **Satisfaction checking**
Given: a theory (the formalized knowledge)
Decide: if the theory has a model

The last two types of inference methods are discussed together since they are two sides of the same coin. What is important to notice about these two inference methods is that they do not take a structure as input. As such, these inference methods operate purely on the formalized knowledge in an instance-independent way. Such types of reasoning are what really sets first-order logic apart from satisfiability solving, where conclusions are always for a specific instance. Let us discuss these methods in a bit more detail. The deduction inference methods takes two theories as input and checks whether the first entails the second, that is, whether all models of the first are also models of the second. In the nurse scheduling application, this could be used as follows: Suppose theory $\mathcal{T}_1$ contains the scheduling constraints given to an automatic scheduler, or to a system that supports an expert in creating a schedule. Now assume that $\mathcal{T}_2$ contains a formalization of a novel labor law hospitals are supposed to adhere to. In this case, we could wonder whether the "old" scheduling constraints already enforced conformance to the laws or not. If that is the case, we are sure that

- every schedule ever generated by the software remains valid, taking the new law into account, and
- no updates to the knowledge base used by the software are needed.

This can be checked using the deduction inference. Alternatively, we could also solve this problem using satisfaction: we can create a novel theory $\mathcal{T}_3$, which states that the all the "old" scheduling constraints are satisfied while the new law is violated, and check whether this theory has a model. If it does, it means that the $\mathcal{T}_1$ does not entail $\mathcal{T}_2$. On the other hand, if it does not have a model, $\mathcal{T}_1$ does entail $\mathcal{T}_2$. The satisfaction can in a similar way be reduced to the deduction problem.

This stronger form of reasoning is the main focus of the research domain on *automated theorem proving*. As the name suggests, when deciding satisfiability of a logical theory, automated theorem provers will not just provide a yes/no answer, but in case the theory is unsatisfiable, they will also produce a *proof of unsatisfiability*. An often-used technique to create such proofs is the method of analytic tableaux. Here, a specified set of inference rules is used to construct a tree-shaped proof (called the *tableau*). This tree has the property that if each of its branches contains an inconsistency, the original formula is unsatisfiable. An example of a semantic tableau for the formula

$$(\forall x : P(x)) \wedge (\exists y : \neg P(y) \vee \neg P(f(y)))$$

is depicted. Several types of inference rules are applied here: the $\wedge$-rule allows reducing a formula $\alpha \wedge \beta$ to two formulas α and β. The $\exists$-rule transforms an existential statement $\exists x : \phi(x)$ into $\phi(C)$ where C is a new constant name, often called a *Skolem*. Intuitively, what it does is given the knowledge that some existential statement is true, it gives a name to an instance satisfying it. The $\forall$-rule simply instantiates universally quantified formulas: if a certain formula holds for all domain elements, then we can fill in any term and it should also holds for that term. The $\vee$-rule splits a branch in two branches: if it is known that $\alpha \vee \beta$ holds, and we can assume that both options lead to a contradiction, then the whole is unsatisfiable:

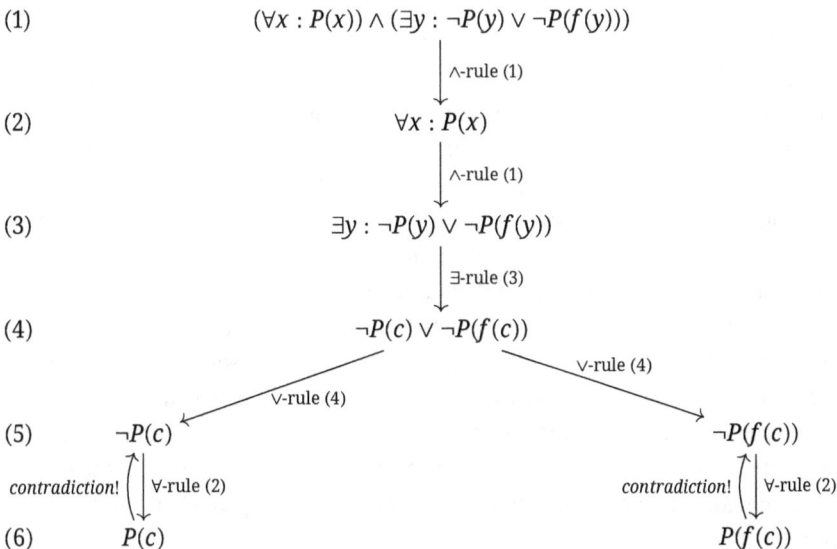

Another domain concerned with the satisfication problem is *SAT modulo theories*. Solvers in this field combine the efficient search methods of SAT solvers with rich modules supporting theories in first-order logic; specifically, for various theories, such as one axiomatizing certain forms of arithmetic, special-purpose propagators are developed that can check satisfication efficiently and translate their findings (lazily) into clauses for the SAT solver to avoid duplicate work.

4.4 What are the limitations of reasoning with first-order logic

The approach described in this section starts from the assumption that a large body of **domain knowledge is available and can be represented explicitly.** In case the assumption is satisfied, this is often considered a strength, since it gives you a lot of control, and typically results in high confidence in correctness of the conclusions of the system. However, there are many cases in which this assumption is not satisfied. To explicate knowledge about a problem domain, humans need not just be able to make correct decisions, but also need to be able to provide rational arguments about why certain decisions are made, and need to be able to align exactly when the same decision is to be made. Instead, humans often rely on *tacit knowledge*—knowledge that is difficult to transfer to another person by means of writing it down or verbalizing it—to make decisions, making it hard to replace the human completely by a knowledge-based system. For instance, in the context of scheduling, an expert scheduler might take personal relationships and preferences that are hard to formalize into account, and might, from experience have already learned that certain combinations work better than others. Furthermore, even when the knowledge is not tacit, the process of extracting it from experts possessing the knowledge is often not a one-shot procedure: it is not easy to provide a complete formalization of the knowledge used in decision-making. Instead, when doing so, the experts will often only realize that they use certain laws or rules in case they are presented with a situation that requires applying it. In Chapter 7, we will study different approaches that instead of starting from an explicit representation of knowledge start from data, for instance, historic decisions and *learn* from that data what the desired behavior is.

Another weakness is that the types of knowledge and reasoning studied in this chapter are all **deterministic**. That is, a black-and-white view on the world is taken: a structure is either possible (if it is a model, i. e., if it satisfies all the sentences in the theory) or impossible (if it is not a model) according to a given logical theory. In certain cases, however, this type of black-and-white representations does not suffice, for instance in case probabilistic knowledge (e. g., when rolling a die, there is a one in six chance of rolling a one) is relevant for the application. Such probabilistic approaches will be studied in Chapter 6.

Finally, first-order logic was presented here as a knowledge representation language, because of its natural informal semantics and historic importance. However, it is far from perfect (in fact, we believe that there is no such thing as the perfect knowledge representation language for all applications). There are two important points to be discussed here. The first is its **inability to express certain concepts (naturally)**. Certain concepts cannot be expressed in first-order logic. One example is the claim that one graph is the *transitive closure* of another graph. Formally, this means that in the one graph, there is an edge from node a to node b if and only if in the other graph, there is a *path* from node a to node b. Another example is in general *nonobjective information*, such as, for example, claims about the knowledge of another agent, for example, the other agent knows that I have either the King or the Queen of Spades. Certain concepts can be expressed, but cannot be expressed naturally. In this case, extensions are needed to improve the language. We already encountered this issue when discussing the lack of types: for instance, the fact that variables always range over "everything in the world," is an unnatural assumptions that does not match with how quantifications occur in the wild (in natural language). This is often solved by using a multisorted extension of first-order logic. Another limitation would be for instance constraints of the form "at most 7 people satisfy a certain restriction" (e. g., at most 7 people can be in a room together at the same time). While this is expressible in FO, it would take several lines to do so and would be very error-prone. For this reason, FO is sometimes extended with *aggregates*. A second important point to mention that has historically been a source of great criticism, **undecidability of deductive inference**. It is well known that deduction for first-order logic is not decidable in general. Hence, **if** one wants to use their knowledge only for deductive reasoning, it makes sense to only consider fragments of the language. This observation gave rise to the field of description logics, as discussed in the next chapter.

4.5 Industry examples

4.5.1 Automated design-driven diagnostics for lithography machines at ASML

Pieter Van Hertum, Thomas Nagele

4.5.1.1 Introduction

ASML is the world's leading provider of lithography systems for the semiconductor industry, manufacturing complex machines that are critical to the production of integrated circuits or chips. A typical semiconductor manufacturing process consists of a number of steps, as can be seen in Figure 4.1. These are the 5 major steps:

1. **Deposition**: Semiconductors are made with silicon wafers (extremely smooth discs of 99.99 % pure silicon). In the deposition step, thin layers of conducting, isolating, or

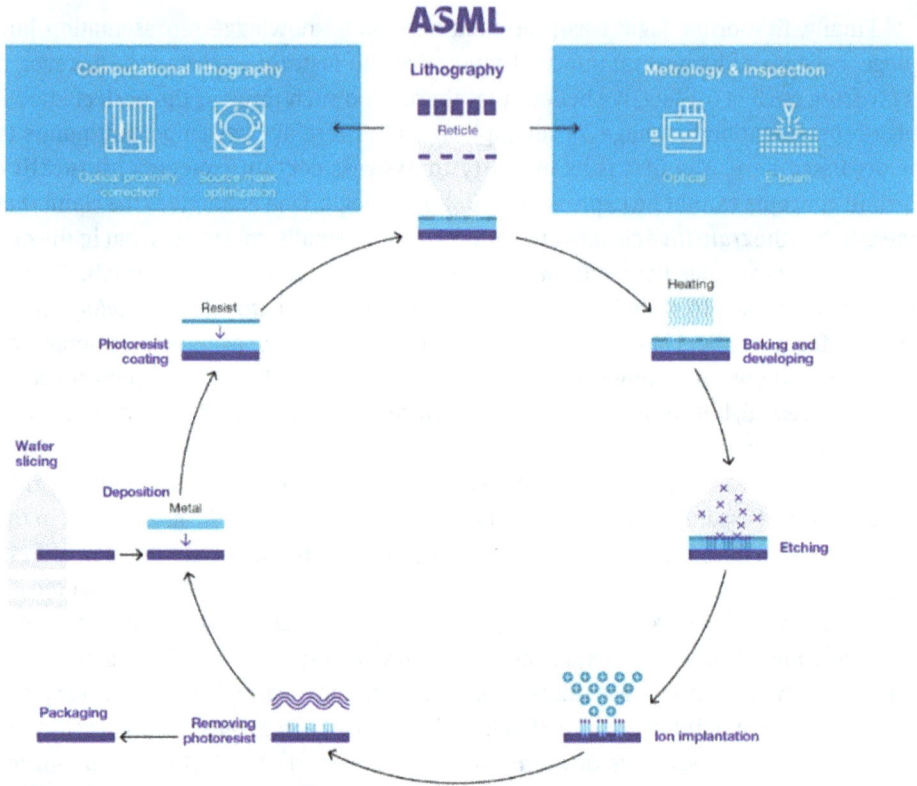

Figure 4.1: The semiconductor manufacturing process.

semiconducting material are deposited on the wafer to enable layers to be printed on.

2. **Photoresist Coating:** On top of this material layer, a layer of light sensitive coating, the photoresist, is deposited. This layer enables the subsequent step of printing patterns onto the wafer. A photoresist layer changes chemical structure when exposed to light, for example, becoming more soluble.

3. **Lithography:** The lithography step is the step in the manufacturing process where the actual patterns are transferred from a blueprint (a *reticle*) toward the wafer, using light that shines through the reticle. With the pattern encoded in the light, the pattern is shrunk by the system's optics and focused on the photosensitive wafer, chip by chip.

4. **Developing and etching:** When the pattern has been transferred, the next step is to remove the degraded photoresist. By etching, the redundant material is removed to reveal the intended 3D pattern. Baking and developing is done to fix the structure permanently.

5. **Ionization:** By bombarding the wafer with ions (positive or negative), the electrical conducting elements of the silicon are modified, in order to create the transistors. After the ionization step, the remaining layers of photoresist are removed to prepare the wafer for the next layer.

This process is repeated hundreds of times to create a wafer full of microchips, every step crucial and sensitive. Think of the resolution, focused and perfect positioning of the lithography step, the perfect depth that has to be etched or the sensitivity to the type of material in the deposition and coating steps. After this process, the wafer is cut into its individual chips (ranging anywhere from 10's to 1000's per wafer) and it is packaged and placed onto its baseboard.

To support this lithography step, ASML also develops tools to optimize and finetune the lithography machine and different types of measurements on the produced layers on the wafers.

4.5.1.2 Diagnostics of a lithography machine

A lithography machine is a complex piece of equipment, consisting of 10,000's different parts interacting to print patterns at nanometer scale. Sometimes, interactions between these parts or the aging of certain components can cause a machine to work suboptimally. ASML's service organization focuses on optimizing performance, and, in case of performance issues, on getting a machine back up to specification as quickly as possible.

The high physical complexity and nanometer resolutions lead to many interactions between different parts and modules, making system diagnostics a big challenge. Next to training customer support engineers to diagnose, maintain, and repair the machines in the field at our customer sites, ASML builds diagnostic software that supports them in this process. The tools automatically analyze data and suggest potential causes or tests the engineer can execute.

As a consequence of this high complexity, these tools need to combine domain expertise with the data to make sense of the machine structure. When diagnosing a root cause of a failure or performance problem of a lithography machine, it is this combination of data and knowledge that allows for causal reasoning in this complex domain.

4.5.1.3 Strategies for automated or supported diagnostics

To diagnose a complex machine, many data sources on that machine provide relevant information, coming in the form of software loggings and traces of physical sensors. Since software logs are introduced into the system by the domain experts in design, they are often the first source to use for diagnostics. For more complex issues, often caused by complex interactions between many modules that were not foreseen during design, these logs are not sufficient and data traces have to be used. The data traces coming from the system sensors contain all types of measurements and need additional interpretation before becoming useable for diagnostics. Models are used to help the engineers to understand data that comes from the system. Such models can be based on the machine design or machine learning techniques being applied on historical and other data sources.

While data from events and interpreted sensor traces give a good overview on the current state of the machine, it does not—on its own—offer an understanding of what causes a certain issue, and what action can resolve it. To do this, the expert working on diagnosing and repairing the machine needs domain knowledge, which can be their own built-up expertise or through documentation. Here, we are looking to automate or support this process by incorporating this knowledge into the diagnostic tools themselves. The knowledge can originate from data, from engineer experience (feedback) or directly from the machine design.

When working with knowledge from expert experience, support engineer feedback or data, the goal is to leverage known and historical cases and experiences to learn relations between patterns of symptoms and root causes. This works well for issues that were identified in advance or appear multiple times. By specifying or learning the relation of a certain failure of a machine to a number of symptoms, a machine shows (sensors exceeding thresholds, specific loggings appearing, performance metrics dropping, etc.), it becomes possible to automate part of the diagnostic process.

Due to the large number of components, and consequently, the large number of sensors, symptoms, and potential causes, specifying every possible problem with their symptoms becomes impossible to scale. Learning these relations directly from the data and comparing machines to others is therefore applied more and more in industry context, and ASML is also incorporating this in its diagnostic landscape. Data-driven diagnostics is a booming field, and there is much to learn from the other machines in the field. This approach is strongest when you have many other samples to learn from (i. e., when diagnosing lightbulbs, coffee machines, or cars). Typically, a certain type of lithography machine has at most a few 100 machines (which even function in different configurations, making it more challenging). This problem of low sample size, together with the high complexity inspired to investigate a third solution: design-based diagnostics.

Instead of directly linking symptoms to causes, another option is to specify the intended behavior of a system and use this knowledge to reason backwards to the most likely root causes of a certain issue, by deducing and excluding. In this approach, the intended behavior of the system is first captured in logical formulae. When a failure occurs in one of the systems, its observations—such as sensor data or error logging—can be inserted in the logic structure, after which a diagnostic engine will compute possible explanations for what was observed.

4.5.1.4 An example

To control the system, it is filled with electrical and hydraulic circuits. Electricals are used to control the different subsystems, gather the sensor information and hydraulics for the supply of liquids, and for the distribution of cooling water over several other subsystems. To illustrate both our modeling approach and diagnostic method, we will apply the approach on an example of such a hydraulic system, which is a simple example based on a hydraulics subsystem of an ASML lithography machine. The hydraulic system consists mostly of interconnected pipes that transfer the water. Figure 4.2 shows a schematic representation of the hydraulic system. The components from the hydraulic domain are shown in blue and the electronic domain is shown in green. A pump pushes the water via a pipe through a manifold, which splits the water over two branches. Each of the branches can be manually closed by a valve at the end of the branch. A manifold combines the water coming from the branches again toward the final pipe. The water pressure is measured with pressure sensors at the end of each branch. The pump and the pressure sensors are powered by two separate power supplies.

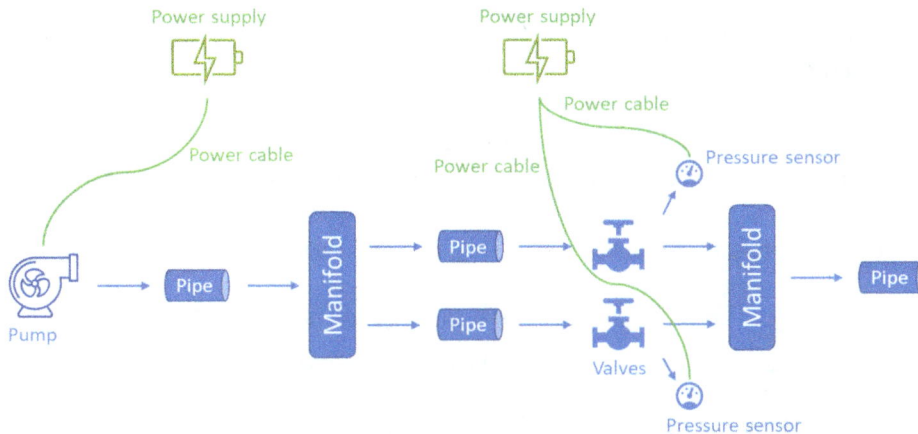

Figure 4.2: A schematic representation of the hydraulic system. The hydraulic domain is shown in blue, and the electronic domain is green.

Modeling behavior

The schematic shows the components of the system and how these are connected to each other, but it does not provide the behavior yet. The intended behavior is specified per component, after which the system behavior is the composition of those individual behaviors by connecting the components to each other. The behavior of each component is specified as the relation between inputs and outputs. These inputs and outputs represent the (physical) variables at either side of the component in a discrete way to abstract away from the highly detailed physics domain. For example, relevant properties of the water flowing through a system can be pressure and flow. The intended behavior is that the water entering the pipe will also exit. If there is water going in, but nothing exiting the pipe must leak and if pressure is applied to the water, and it is not flowing, it must be blocked. This expression of both normal and failure behavior can be translated to a logic expression.

A logic specification

To specify the behavior of pipes, we would need following symbols:[2]

-   A type **Pipe**, used to specify the collection of pipes we have in a system.
-   A type **HealthStatePipe** stating the possible health states of a pipe, containing *HealthyPipe, BlockedPipe, LeakingPipe*.
-   A predicate **Connected(Pipe, Pipe)** to specify the connection between two pipes.
-   A predicate **FlowIn(Pipe)** specifying the flow coming into a certain pipe.
-   A predicate **FlowOut(Pipe)** specifying the flow coming out of a certain pipe.
-   A predicate **PressureIn(Pipe)** specifying the pressure at the beginning of a certain pipe.
-   A predicate **PressureOut(Pipe)** specifying the pressure at the end of a certain pipe.
-   A function **StatePipe(Pipe)** : **Pipe → HealthStatePipe** specifying the health state of a certain pipe.

These symbols are stored in a Vocabulary V_{pipe}.

A theory T_{pipe} is used to specify the relations between these symbols, detailing out pipe behavior.

-   *For a healthy pipe, the input state of the fluid always equals the output state.*

$$\forall p[Pipe] : StatePipe(p) = HealthyPipe \Leftrightarrow PressureIn(p)$$
$$= PressureOut(p) \land FlowIn(p) = FlowOut(p)$$

-   *If no fluid can pass through (pressure but no flow), the pipe is blocked.*

$$\forall p[Pipe] : StatePipe(p) = BlockedPipe$$
$$\Leftrightarrow PressureIn(p) \land \neg FlowIn(p) \land \neg PressureOut(p)$$

-   *If there is fluid going in, but not coming out, the pipe is leaking.*

$$\forall p[Pipe] : StatePipe(p) = LeakingPipe \Rightarrow FlowIn(p) \land \neg FlowOut(p)$$

---

2 Note that we are using a typed extension of first-order logic, where the domain is split into a number of subsets, called types. Predicates, functions are defined over those types instead of over the entire domain. When writing a theory, we allow quantification over these types.

This specification now depicts all possible behaviors of the pipe, together with a health state. This health state can identify the state of the pipe and later be used to diagnose specific issues. Together with the following theory containing a few *commonsense* constraints, inferences can be executed on this specification to build diagnostic solutions.

Theory T_{common}:

- *Flow cannot be created:*

$$\forall p[Pipe] : \neg FlowIn(p) \Rightarrow \neg FlowOut(p)$$

- *Pressure cannot be created:*

$$\forall p[Pipe] : \neg PressureIn(p) \Rightarrow \neg PressureOut(p)$$

- *No flow without pressure:*

$$\forall p[Pipe] : FlowIn(p) \Rightarrow PressureIn(c)$$
$$\forall p[Pipe] : FlowOut(p) \Rightarrow PressureOut(c)$$

- *The meaning of "Connected":*

$$\forall p1[Pipe]p2[Pipe] : Connected(p1, p2) \Leftrightarrow FlowIn(p2)$$
$$= FlowOut(p1) \wedge PressureIn(p2) = PressureOut(p1)$$

In similar ways, specifications for the functioning of valves, manifolds, pumps, power supply units (PSUs) can be created by domain experts. By combining these specifications and supplying information of the known state of the world (the specific pipes, valves, pumps, etc., their connections and sensor measurements) diagnostic reasoning can be done.

Inferences
Using the model for diagnosis

Once the behavior of the complete system has been captured in a logical specification, it can be used to help the service engineer in finding the defective part. The diagnosis is based on what is observed from the system, a combination of the data coming from sensors and software logging. An input structure for the model can be compiled from the observed data, after which model expansion inference can be used to find explanations for the observations.

For example, given the pipe specification above, the following structure could be compiled from design information and observations:

$$S_1 = \{$$
$$Pipe = \{p1, p2\}$$
$$Connected = \{(p1, p2)\}$$
$$FlowOut = \{p2\}$$
$$\}$$

By performing model expansion on this theory, with this partial structure, the following complete structure is calculated:

$$S = \{$$
$$Pipe = \{p1, p2\}$$
$$Connected = \{(p1, p2)\}$$
$$FlowIn = \{p1, p2\}$$
$$PressureIn = \{p1, p2\}$$
$$PressureOut = \{p1, p2\}$$
$$FlowOut = \{p1, p2\}$$
$$StatePipe = \{p1 \rightarrow HealthyPipe, p2 \rightarrow HealthyPipe\}$$
$$\}$$

This structure (the only full structure extending the partial structure and satisfying the theory) states that both pipes are performing healthy and as such, the fluid is flowing through the system, and the pressure is propagated as expected.

However, when starting with the following structure:

$$S_2 = \{$$
$$Pipe = \{p1, p2\}$$
$$Connected = \{(p1, p2)\}$$
$$FlowIn = \{p1 \rightarrow Flow\}$$
$$FlowOut = \{p2 \rightarrow NoFlow\}$$
$$\}$$

the model expansion inference creates many different possible structures that extend this partial structure and satisfy the theory, for example, the first pipe can leak, or the second pipe can leak. This approach shows its value when reasoning over larger specifications, such as shown in Figure 4.2. While only expecting a domain expert to specify simple behavior of components and how they are interconnected (which can be further simplified through a graphical interface), quite complex possible scenarios can be calculated with incomplete input information.

When handling more complex scenarios with limited input information, there are often many possible structures that satisfy a theory extending the input. For example, in the Figure 4.2 specification, where there is flow coming into the first pipe and not coming out of the last pipe, it could be that the first pipe is leaking, or it could also be a failure of both parallel valves. When reasoning, it is often good practice to assume that failures happen rarely, so a model with less failures is more likely than a model with more failures. To this end, an optimization inference can help. By adding a term $t = \#\{p \mid StatePipe(p) \neq HealthyPipe\}$ and minimizing that term, we encode the assumption that in most situations components are behaving as we expect.

Assessing diagnosability during design
The model can also be used during the system design to assess its diagnosability. Instead of using observations coming from the real system as input to the model, one can also run the inference for an assumed failure. For this, the observability configuration should be known, which comprises a list of variables in the system for which you know you can observe its value, either via a sensor or software logging.

Through inference, one can assess the values on the observable variables when all components are healthy to understand what the observables of the running system will tell when everything is working as expected. This set of readings on the observables is referred to as a *signature*. This assessment can also be done for every single failure mode in the system. For this, all other components are assumed to behave normally, while only one component has a problem. Each of these failure simulations provides one or more signatures, which are the computed failure signatures.

Repeating this assessment for all failures or possibly even failure combinations results in a set of failure signatures. Depending on the numbers and locations of observables in the system, multiple failures may have identical signatures. Based on this insight, the designer may add more observability to the design to reduce the number of failures with identical failure signatures, or procedures could be formulated to help the service engineer in the field to find out what failure really caused the issue.

4.5.1.5 Conclusion

When doing diagnostics, machine learning can help us to interpret data sources and sensor data. However, to separate cause from effect, and optimally make use of the available knowledge, reasoning systems can help. By specifying system design and interpreting software loggings and sensors, inferences can be used to support diagnostic tooling, or to support system design by analyzing observability.

In order to enable this for entire systems, the scalability of building these specifications is crucial. Good (graphical) interfaces and tooling can help immensely for domain experts to build these specifications. On top of this, it is important to build tools and study techniques that can (semi)automatically build these specifications directly from design documents.

4.5.2 Modeling and verifying simple vehicle controller, such as the Triton unmanned aircraft systems of the US Navy: using Imandra system and first-order logic

Djordje Markovic, Bart Bogaerts, Grant Passmore

4.5.2.1 Introduction

Designing complex systems is quite an extensive and expensive process, and sometimes mistakes are just not affordable. In these cases, before developing the desired product, it is essential to model it and prove its specific properties. Formal verification methods often use mathematical logic and symbolic AI (artificial intelligence) for designing and analyzing engineering artifacts such as software and hardware. The difficulties are that complex systems strive to have infinitely many different possible behaviors. This seemingly miraculous feat— surveying an infinite number of possible system behaviors through a finite computation—is made possible using symbolic mathematics and mechanized techniques for logical inference.

Formal verification has a tremendous use-value for safety-critical systems, that is, computer systems whose correctness have a direct bearing on the safety of others. For example, autopilot systems in aircraft, control systems in nuclear power plants, and collision avoidance controllers in drones are directly related to public safety.

In practice, designing and tuning functions like the controller is a considerable challenge, and formal verification is necessary to ensure they operate correctly and safely. Consider that you are creating an algorithm for controlling some aspect of an aircraft; would you be able to trust it and be a passenger on a test flight? So, before giving it a test ride, how is the safeness of the controller verified? The first step is often simulation. That is, we may first gain some primary assurance of its safety and correctness through simulating its behavior in many different situations. However, no matter how many unique runs are done through a simulator, only a finite number of scenarios can be observed. But this controller can, in general, be in an infinite number of possible situations. How can this gap be bridged? How can we ensure that the algorithm is correct? Formal verification provides an answer. The key is to use logic and reason symbolically about its possible behaviors to prove that the controller follows desired safety and correctness properties.

In this section, we shall model a *simple autonomous vehicle controller* dynamic system and verify its correctness. This controller may be seen as a simplified version of a controller found in modern day drones and autopilot systems, such as the Triton UAS of the US Navy.

The system is modeled in first-order logic and proofs are explained accordingly. Nevertheless, we give a few examples using an Imandra[3] syntax side-by-side with first-order logic statements, illustrating the connection between reasoning about programs and first-order logic. These parts are more suitable for advanced readers, and they are not relevant for the understanding of this section; therefore, less experienced readers can safely skip them.

The upcoming text is split into two parts: first, the discussion of modeling of a simple autonomous vehicle controller and, second, detailed analysis and proof of the two safety properties of the system.

4.5.2.2 The domain knowledge

This section introduces the problem of a *simple autonomous vehicle controller* and essential elements needed for modeling such a system. The analyzed controller is originally described in the article (Boyer et al., 1990) by Boyer, Green, and Moore.

The goal of a vehicle controller is to take care of wind changes and keep the vehicle on the course. The brief specification of the system follows: The system is restricted to only one space dimension (y-component). The time is abstractly represented as a sequence of discreet time points. The vehicle can move with a certain velocity in the positive or negative direction of the y-axis, and wind can blow with a particular speed (in the positive or negative direction of the y-axis). Wind cannot change more than one unit between two time points. The drone controller can increase vehicle velocity at any time point for an arbitrary value. The controller has an insight into all values at a certain time point.

---

3 Imandra is a formal verification environment that facilitates the design and verification of safety-critical algorithms. More on the official website: https://www.imandra.ai/

Representing the world (state of affairs)

Formal modeling of any system usually starts with the choice of an adequate ontology. Ontology serves to represent the *states of affairs* in the world we are formalizing. Naturally, this choice is of immense importance because it has a strong impact on formalization.

First, it is important to notice that first-order logic does not have the concept of time, and yet we would like to model a dynamic system. The common approach is to interpret time as natural numbers,[4] zero being the *start* point in time and *next* being a function mapping time point n to time point $n + 1$.

Once we have the abstract representation of the time, we can start representing the simple vehicle controller system. The system is described by wind speed, vehicle speed, vehicle position, and wind change at any time point. As we abstract away units, all these values are represented by integers. Knowing that values are unique per time point, it is natural to represent them as temporal functions mapping time points to their values.

Finally, vehicle *controller* should update vehicle speed based on two consecutive time points. However, we shall see later the abstract version of the controller that takes as input two integers and returns relative speed change. Table 4.4 supplies first-order logic vocabulary suitable for a simple vehicle controller system.

Table 4.4: Vehicle controller— first-order logic ontology.[5]

First-order logic:	
Temporal functions describing state:	
– $w : Time \rightarrow \mathbb{Z}$	– Wind velocity
– $y : Time \rightarrow \mathbb{Z}$	– Position of the vehicle
– $v : Time \rightarrow \mathbb{Z}$	– Velocity of the vehicle
Temporal function describing wind change:	
– $dw : Time \rightarrow \mathbb{Z}$	– Wind change
Controller function:	– Binary controller function
– $controller: \mathbb{Z} \times \mathbb{Z} \rightarrow \mathbb{Z}$	
Time vocabulary:	
– $start : () \rightarrow Time$	– Starting point
– $next : Time \rightarrow Time$	– Next function

---

4 For the simplicity of the example, we assume that every structure in first-order logic contains two types, integers $\mathbb{Z}$ and Time. Time is interpreted by natural numbers, where "start" is a constant designating 0 in every structure and "next" is a function mapping each time point to the following (i. e., mapping n to n+1).

5 We use (w, y, v) to represent a state in the later text, dw is not part of the state, and it can be treated as a parameter of the system.

State transition

The system that we want to model is dynamic; the vehicle can move, and the wind can blow in one direction of the y-axis with a certain velocity. The vehicle controller is supposed to control vehicle speed and balance the wind impact. An essential property of this dynamic system is that it is deterministic, that is, for any state of a system (including the wind change), there is exactly one next state. Here, we are going to model this transition between two states.

For a given current state and wind change, we can compute the next state of the system. Wind velocity changes with respect to the wind change. Vehicle position changes concerning the previous position, vehicle velocity, and new wind velocity. The exciting part is how the controller updates the vehicle velocity, but let's keep it abstract for a moment.

The next state function is defined as: For a given state (w, y, v) and for a wind change dw, the new state (nw, ny, nv) is

- Wind change: $nw = w + dw$
- Vehicle position is changed: $ny = y + v + w + dw$
- New vehicle velocity depends on the controller: $nv = v + controller(w, y, v, dw)$

The controller from (Boyer, 1990) is more abstract and considers a sign of the new and old position of the vehicle. So, we can model it as a $controller((sgn(ny)), (sgn(y)))$.

Table 4.5 shows the first-order logic statements expressing this state transition. It is important to notice that each statement starts with universal quantification "For each time point t..." So, we are saying something about time. Let us translate the first statement to the natural language statement given standard informal semantics for first-order logic and expected interpretation for w – wind and dw – wind change. The first statement expresses: *"For each time point t, wind at the next time point is equal to the wind at time point t augmented with wind change at time point t."* This translation clearly states the informal interpretation of the first statement, and it precisely describes our thoughts of this dynamic system.

Table 4.5: Vehicle controller—first-order logic specification.

First-order logic theory – T_{vc} (Theory – vehicle controller):

$\forall t : w(next(t)) = w(t) + dw(t)$.
$\forall t : y(next(t)) = y(t) + v(t) + w(t) + dw(t)$.
$\forall t : v(next(t)) = v(t) + controller(sgn(y(t) + v(t) + w(t) + dw(t)), sgn(y(t)))$.

Table 4.6 represents the same function in the Imandra syntax. The syntax is clear, and the representation is very readable and compact. This function returns a new state vector that represents the state after applying wind change dw to the state vector s. One

Table 4.6: Vehicle controller—Imandra specification.

Imandra:

next_state dw s = {
  w = s.w + dw;
  y = s.y + s.v + s.w + dw;
  v = s.v + controller (sgn (s.y + s.v + s.w + dw)) (sgn s.y)
}

can see that statement talks about two consecutive time points, but time is not mentioned explicitly.

Controller

Let's assume that the controller function is defined as

$$\forall x, y : controller(x, y) = (-3 * x) + (2 * y).$$

Given such a controller, we may first gain some elementary assurance of its correctness through simulation and testing. Let us look at one particular use case when the wind increases in the positive direction of the y-axis for three consecutive time points and then stays constant for four time points.

The scenario from the Table 4.7 is also graphically represented in Figure 4.3. One can observe that after four time points of constant wind, the vehicle is back at the course. We can try this many times with different setups, and the vehicle always gets back at the center. Furthermore, we are not able to find counterexamples. In an analogous way, one can notice that the drone never strays further than 3 units from the center. All these suggest that these properties always hold. But how can we prove them?

Table 4.7: Vehicle controller simulation example.

Time	0	1	2	3	4
w	0	0 + 1 = 1	1 + 1 = 2	2 + 1 = 3	3 + 0 = 3
y	0	0 + 0 + 0 + 1 = 1	1 − 3 + 1 + 1 = 0	0 − 1 + 2 + 1 = 2	2 − 4 + 3 + 0 = 1
v	0	0 − 3 = −3	−3 + 2 = −1	−1 − 3 = −4	−4 − 1 = −5
dw	1	1	1	1	0

Time	5	6	7	8
w	3 + 0 = 3	3 + 0 = 3	3 + 0 = 3	3 + 0 = 3
y	1 − 5 + 3 + 0 = −1	−1 + 0 + 3 + 0 = 2	2 − 5 + 3 + 0 = 0	0 − 3 + 3 + 0 = 0
v	−5 + 5 = 0	0 − 5 = −5	−5 + 2 = −3	−3 + 0 = −3
dw	0	0	0	0

Vehicle position in time

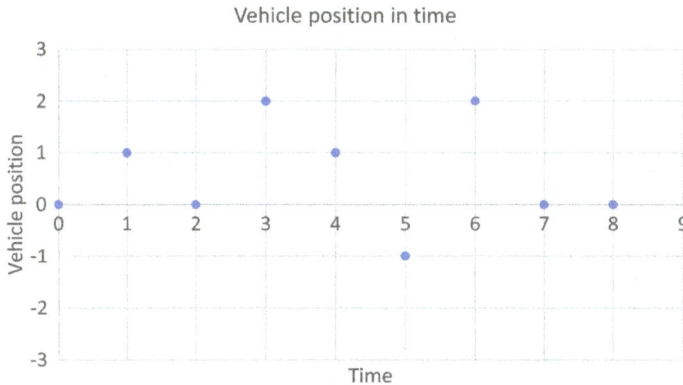

Figure 4.3: Controller behavior when wind remains steady for four time points.

4.5.2.3 Reasoning and theorem proving

This section opens with a concise introduction to theorem proving. Suppose that there is a theory T and a formula φ both expressible in first-order logic. Proving that T entails φ would be as simple as formalizing both and using an adequate automated theorem prover. However, these provers could struggle to prove the complex properties of dynamic systems. The main reason for these problems is the *time*, which is interpreted by natural numbers. One typical way to solve these problems is induction, which naturally can be conducted on time: *First, we prove that the property holds at the first time point, and then we prove that the property is preserved by the transition function.*

Since induction is a crucial concept in the following text, it deserves a brief explanation. Induction proofs in the context of dynamic systems are usually applicable for proving single state properties. A single state property ($\forall t : \varphi[t]$) is a formula quantifying over *time* and the only time point mentioned in the body of the formula is the quantified one (t). Consider a case of proving that theory T entails some property $\forall t : \varphi[t]$, the induction proof would consist of

– *Base case* – Proof that property φ holds initially: $T \models \varphi[Start]$.
– *Induction case* – When φ holds at some time point, it will also hold at the next time point: $T \cup \{\varphi[t]\} \models \varphi[Next(t)]$.

Proving these two entailments ensures that property φ always holds. It is important that φ is a single state formula since otherwise, we cannot split it easily into base and induction cases. A more detailed explanation of this approach can be found in the paper (Bogaerts et al., 2014).

Let's look at the theorems (properties) we would like to prove.

Theorem 4.1. *If the vehicle starts at the initial state $w = 0; y = 0; v = 0$, then the controller guarantees the vehicle never strays farther than three units from the y-axis.*

Theorem 4.2. *If the wind ever becomes constant for at least four sampling intervals, then the vehicle returns to the 0 of y-axis and stays there as long as the wind is still.*

Proving these theorems formally ensures that the vehicle appropriately stays on the course under each of the infinite number of possible wind change sequences. Let us prove our theorems.

4.5.2.4 Discussion and proof of Theorem 4.1

No matter how the wind behaves, we want to prove that if at the beginning the system is in the state $(y = 0, w = 0, v = 0)$, then the controller guarantees the vehicle never strays farther than three units from the y-axis, or more formally,

$$w(Start) = y(Start) = v(Start) = 0 \Rightarrow \forall t : -3 \le y(t) \le 3.$$

This theorem states that vehicle position is always in some range if some preconditions are met. The part that states conditions on each state looks like a suitable candidate for induction, but the problem is that the entire formula is not a single state.

For clarity, let's name *if part* of the theorem φ_{cond} (*condition*) and *then* part φ_{Inv_1} (*invariant*[6]). Now the theorem can be abbreviated as $\varphi_{cond} \Rightarrow \varphi_{Inv_1}$. Our goal is to show that $T_{vc} \models \{\varphi_{cond} \Rightarrow \varphi_{Inv_1}\}$. Using sequence calculus, we can transform this question to equivalent one $(T_{vc} \cup \{\varphi_{cond}\}) \models \{\varphi_{Inv_1}\}$. This transformation allows us to *merge* vehicle controller theory and *condition* of the theorem leaving single state formula on the right-hand side. Now the problem is entailment of a single state proposition, which allow us to use induction to prove it.

Induction proof starts with the *base case*, the goal is to show that initially vehicle position is in an adequate range, or formally $-3 \le y(Start) \le 3$. This is trivial since the start condition (φ_{cond}) is added to the main theory, and hence the vehicle position at the *start* time point is always 0.

The more difficult part is the induction case. Here, the goal is to show that at any time point t if the vehicle position is between 3 and -3, it will still be at time point $t + 1$ (this is known as induction hypothesis). As the whole idea is to eliminate time from the theory so we can use theorem provers, we must transform our theory to consider only two consecutive time points. This can be done by introducing constants instead of functions for vehicle position, wind speed, and vehicle speed (technical details are available in the provided formalizations). As these two time points stand for an arbitrary segment of time, the start constraints (φ_{cond}) are not applicable for the first time point (t). Since the first time point can be any possible state, the induction hypotheses start to sound a bit too optimistic. Consider a state $(y = 0, w = 100, v = 0)$, no matter how the

---

6 In the context of dynamic systems, "Invariant" denotes a single state property that always holds.

wind and vehicle speed change in the next time point, vehicle position will certainly be somewhere around 100. The problem is that connection with the starting point is lost, it is probably not possible to reach such a state from the state $(0, 0, 0)$.

To handle this issue, we can strengthen the induction hypotheses by adding the information to it. In the paper (Boyer, 1990), the notion of a good state is introduced for this purpose. A good state is a class of states represented as a pair $(y, w + v)$ reachable from the state $(0, 0, 0)$. The good states are represented in the Table 4.8.

Table 4.8: Good state— class of states reachable from the state $(0, 0, 0)$.

y	−3	−2	−2	−1	−1	0	0	0	1	1	2	2	3
$w + v$	1	2	1	3	2	−1	0	1	−2	−3	−1	−2	−1

The new invariant would look like $\varphi_{Inv_2} = \forall t : GS(y(t), w(t) + v(t))$. Note that this invariant entails the old one, since in any good state vehicle position is always between 3 and −3. The new invariant is stronger because the first time point in the induction case must be a good state, and hence reachable from the initial state. The final shape of the entailment to be proven is represented in Table 4.9.

Table 4.9: Theorem 4.1— induction schema in first-order logic.[7]

First-order logic:

$T_{vc}[t] \cup T_{as}[t] \cup \{\varphi_{Inv2}[t]\} \models \{\varphi_{Inv2}[Next(t)]\}$
Where:
- T_{vc} – Is a simple vehicle controller theory.
- $T_{as} = \{\forall t : -1 \leq dw(t) \leq 1.\}$ – Assumption that wind change is between −1 and 1.
- $T[t]$ – is a temporal theory applied to the time point t

We use the IDP system (De Cat et al., 2018) to provide specification of simple vehicle controller example using first-order logic. IDP is a knowledge base system for the FO(·) language[8] and it supports a multiple forms of inference methods, among others also proving invariants in dynamic system specifications.

---

7 Note that we didn't express the base case of induction since it is trivially true. Also, note that wind change constraint appears as an assumption in the theorems.

8 FO(·) is an extension of first-order logic with types, aggregates, inductive definitions, bounded arithmetic, etc.

Detailed proof procedure for Imandra is available here,[9] while the solution using the IDP system based on first-order logic can be found at this link.[10]

Good state discussion

There is something puzzling about the notion of a good state, namely how to compute it, and if we can compute it, does it not prove Theorem 4.1? The difference is in one subtle puzzle piece. To compute the set of good states, we can bound the system and use finite domains. Because of this restriction, the computed set is not enough to be considered as proof of Theorem 4.1. Therefore, induction comes in to prove that this set of good states is the maximal one.

One can try as an exercise to remove extremes from the good state relation and retry the induction proof.

4.5.2.5 Discussion and proof of Theorem 4.2

The second property to prove an autonomous vehicle controller is "Whenever the wind is constant for four consecutive time points, vehicle will be back to the course and will remain there as long the wind is constant." This statement talks about at least five different states, which makes it harder to prove. Here, we explain some ideas on how to simplify this statement and how to prove it. We are not going into details about this theorem, rather we sketch how the proof can be constructed.

It is important to notice that the theorem is composed of *two* parts, first expressing that the vehicle will come back to the course after four consequent time points of steady wind, and second expressing that the vehicle remains there if wind remains steady. This suggests that the theorem can be split. The next two formulas stand for these new theorems expressed in first-order logic. Note that we abuse the notation and write $GS(t)$ where it should be $GS(y(t), w(t) + v(t))$. Also, $GSw(s)$ stands for the state after four consecutive time points of steady wind:

$$\forall s : GS(s) \land (\forall t : 0 \leq t \leq 4 \Rightarrow dw(s + t) = 0) \Rightarrow y(s + 4) = 0$$
$$\forall s : GSw(s) \land dw(s) = 0 \Rightarrow y(s) = 0$$

Intuitively, after the vehicle comes back to the center, to stay there (if the wind does not change) the wind and vehicle speed should cancel each other. Speaking in terms of *good states*, this is the state $(0,0)$, and we will refer to this state with $GS_0(t)$ for a time

9 https://docs.imandra.ai/imandra-docs/notebooks/simple-vehicle-controller/

10 The full solution is explained at https://djordje.rs/posts/svc.html. The raw IDP files can be retrieved from https://gist.github.com/dmkoder/6c39aa5768a9fb3305745f7f999285f4 and https://gist.github.com/dmkoder/2a1c564c7a3eba07b5b54f2ed3799e9a.

point t. So, the $GSw(s)$ in the second part of the theorem is actually $GS_0(t)$. To restore the connection between the two theorems, the consequent of the first part should be strengthened to $GS_0(t)$. The new theorems would look like the following:

$$\forall s : GS(s) \land \forall t : (0 \le t \le 4 \Rightarrow dw(s + t) = 0) \Rightarrow GS_0(s + 4)$$

$$\forall s : GS_0(s) \land dw(s) = 0 \Rightarrow y(s) = 0$$

The second part satisfies all preconditions for induction to be applied, and hence we will not discuss it further here as it is the same as in Theorem 4.1. However, the first part still talks about four different time points relative to s. Keeping in mind that a *good state* stands for a class of states, quantification over states is a bit redundant here, and hence can be eliminated:

$$GS(Start) \land \forall t : (Start \le t \le 4 \Rightarrow dw(t) = 0) \Rightarrow GS_0(4)$$

Finally, we have a statement that talks about exactly four consecutive time points. Now we can drop the time (natural numbers) from the theory by simply introducing fresh new constants for each time point and defining the transition between each of them. This method is sometimes called forward-chaining.

This idea can be automated, and that is what the Imandra system is doing. We don't show details of this idea here, since they are too technical, but they are available in supplied full specifications of both Imandra and IDP solutions.

4.5.2.6 Conclusion

The focus of this section was on understanding the pragmatic importance of formal systems as first-order logic in industrial use cases. We have shown these on the example of the simple vehicle controller, but the same ideas could be applied to other dynamic systems, perhaps much more complex.

The first-order logic allowed us to go deep into the essence of problems of proving invariants of dynamic systems and to analyze them. It is important that in higher-level tools that we could provide more automated procedures for the problems that we have discussed; the underlying principles are the same as in this section. Hence, to be a profound user of such systems, one should keep these ideas in mind.

Bibliography

Abdelsalam Hisham M., Shoaeb Amal R. S., and Elassal Magy M. Enhancing decision model notation (DMN) for better use in business analytics (BA). In *Proceedings of the 10th International Conference on Informatics and Systems*, INFOS '16, page 321–322, New York, NY, USA, 2016. Association for Computing Machinery. ISBN 9781450340625. https://doi.org/10.1145/2908446.2908514.

Baader Franz, Calvanese Diego, McGuinness Deborah L., Nardi Daniele, and Patel-Schneider Peter F., editors. The Description Logic Handbook: Theory, Implementation, and Applications. Cambridge University Press, 2003. ISBN 0-521-78176-0. URL http://www.cambridge.org/asia/catalogue/catalogue. asp?isbn=9780521876254. Second edition, 2007.

Bogaerts Bart, Jansen Joachim, Bruynooghe Maurice, De Cat Broes, Vennekens Joost and Denecker Marc. Simulating dynamic systems using linear time calculus theories. Theory and Practice of Logic Programming, 14(4-5):477–492, 2014.

Boyer Robert S., Green Milton W. and Moore J Strother. The use of a formal simulator to verify a simple real time control program. In Beauty Is Our Business. Springer, 1990.

Cohn Anthony G. and Renz Jochen. Qualitative spatial representation and reasoning. In Handbook of Knowledge Representation van Harmelen et al. (2007), pages 551–596, 2007. ISBN 0444522115, 9780444522115. https://doi.org/10.1016/S1574-6526(07)03013-1.

De Cat Broes, Bogaerts Bart, Bruynooghe Maurice, Janssens Gerda and Denecker Marc. Predicate logic as a modeling language: the IDP system. In Declarative Logic Programming: Theory, Systems, and Applications (pp. 279–323), 2018.

Denecker Marc and Vennekens Joost. Building a knowledge base system for an integration of logic programming and classical logic. In María García de la Banda and Enrico Pontelli, editors, ICLP, volume 5366 of LNCS, pages 71–76. Springer, 2008. ISBN 978-3-540-89981-5. URL http://dx.doi.org/10.1007/978-3-540-89982-2_12.

Fagin Ronald, Halpern Joseph Y., Moses Yoram, and Vardi Moshe. Reasoning About Knowledge. MIT Press, 1995. URL http://library.books24x7.com.libproxy.mit.edu/toc.asp?site=bbbga&bookid=7008.

Gelfond Michael and Lifschitz Vladimir. The stable model semantics for logic programming. In Robert A. Kowalski and Kenneth A. Bowen, editors, ICLP/SLP, pages 1070–1080. MIT Press, 1988. ISBN 0-262-61056-6. URL http://citeseer.ist.psu.edu/viewdoc/summary?doi=10.1.1.24.6050.

Governatori Guido. Representing business contracts in *RuleML*. International Journal of Cooperative Information Systems., 14(2-3):181–216, 2005. https://doi.org/10.1142/S0218843005001092.

Halpern Joseph Y. and Moses Yoram. Knowledge and common knowledge in a distributed environment. Journal of the ACM, 37(3):549–587, July 1990. ISSN 0004-5411. https://doi.org/10.1145/79147.79161.

Hintikka Jaakko. Knowledge and Belief. Ithaca: Cornell University Press, 1962.

Jackson Daniel. Alloy: A lightweight object modelling notation. ACM Transactions on Software Engineering and Methodology (TOSEM'02), 11(2):256–290, 2002.

Leuschel Michael and Butler Michael. ProB: A model checker for B. In Keijiro Araki, Stefania Gnesi, and Dino Mandrioli, editors, FME 2003: Formal Methods, LNCS 2805, pages 855–874. Springer-Verlag, 2003. ISBN 3-540-40828-2.

McCarthy John. Applications of circumscription to formalizing common-sense knowledge. Artificial Intelligence, 28(1):89–116, 1986.

Moore Robert C. Semantical considerations on nonmonotonic logic. Artificial Intelligence, 25(1):75–94, 1985. https://doi.org/10.1016/0004-3702(85)90042-6.

Mueller Erik T. Event calculus. In Handbook of Knowledge Representation van Harmelen et al. (2007), pages 671–708, 2007. ISBN 0444522115, 9780444522115. https://doi.org/10.1016/S1574-6526(07)03017-9.

Object Management Group. Semantics of business vocabulary and business rules (sbvr). OMG document number formal/08-01-02, January 2008. Version 1.0.

Pearl Judea. Causality: Models, Reasoning, and Inference. Cambridge University Press, 2000.

Pearl Judea and Mackenzie Dana. The Book of Why: The New Science of Cause and Effect. Basic Books, Inc., USA, 1st edition, 2018. ISBN 046509760X.

Reiter Raymond. Knowledge in Action: Logical Foundations for Specifying and Implementing Dynamical Systems. MIT Press, 2001. ISBN 9780262264310. URL http://books.google.be/books?id=exa4f6BOZdYC.

Reiter Raymond. A logic for default reasoning. Artificial Intelligence, 13(1-2):81–132, 1980. https://doi.org/10.1016/0004-3702(80)90014-4.

Reiter Raymond. The frame problem in situation the calculus: A simple solution (sometimes) and a completeness result for goal regression. In Artificial Intelligence and Mathematical Theory of Computation, pages 359–380. Academic Press Professional, Inc., San Diego, CA, USA, 1991. ISBN 0-12-450010-2.

van Harmelen Frank, Lifschitz Vladimir, and Porter Bruce. Handbook of Knowledge Representation. Elsevier Science, San Diego, USA, 2007. ISBN 0444522115, 9780444522115. URL https://www.elsevier.com/books/handbook-of-knowledge-representation/van-harmelen/978-0-444-52211-5.

Isabelle Linden

5 Knowledge representation and engineering with ontologies

5.1 Why are knowledge representation and engineering with ontologies important within the broader AI domain?

Knowledge representation with ontologies is the branch of artificial intelligence (AI) that studies languages that model domain knowledge and support reasoning and queries answering on complex and often partial models. First generations of knowledge representation languages, as semantic networks and frames, were not free of ambiguities. Ontologies, for their part, integrate a formal semantics based on first-order logic (FOL). Thanks to this formalization, several checks and queries on the models can be automatically performed.

This ambition of providing knowledge representation associated with automated reasoning engines is already addressed in the previous chapters. Propositional logic (Chapter 3, SAT) states facts, using atomic representations to describe the world, and simple logic rules, such as Boolean algebra, to do the inference. FOL (Chapter 4) enriches the modeling with structured representations (e. g., facts, objects, relationships) and quantifiers by representing it in general rules. As such, it gains in expressive power by applying logic reasoning techniques in the modeled environment. Many extensions of FOL have also been proposed to further extend its expressiveness. However, as mentioned in the conclusion of the previous chapter, a limitation of FOL lies in its undecidability, that is, the fact that some queries cannot be answered in a finite amount of time. This is a well-known problem in the field of knowledge representation (KR): the balance between expressiveness and complexity. Studies on "fragments" of FOL (fragments are a subset of a logical language, resulting from syntactical restrictions within that language) aim to explore this trade-off.

When representing knowledge with ontologies, a knowledge base is not assumed to provide a complete and very detailed representation of the studied domain, but rather to offer a synthetic and structured presentation of the relevant elements and features as far as they are known to the domain experts. As such, ontologies support the formal description of specific domains while integrating their semantics as well. The definition of a knowledge model is done by both the objects of the domain themselves and by their intrinsic relations. Further logic reasoning is supported at two levels: specific entities (i. e., instances) of the domain and the categories that further structure it.

Ontologies with their associated reasoning engines are particularly used in the development of semantic web applications that support knowledge management, semantic data integration, and information retrieval. The ontology web language (OWL[1]) is certainly one of the most well-known and biggest successes of description logics (DLs). This standard defined by the World Wide Web Consortium (W3C) is one of the key elements of the semantic web. Much more widely, ontologies are also already integrated in ontology-driven systems in various domains as medicine, telecommunications, agriculture, astronomy, biology, defense, and natural resources and energy management.

Indeed, whatever application domains or languages and AI techniques used to develop an intelligent system, it is valuable to have a model of the domain shared by all engineering stakeholders. In particular, such a model, expressed in a language independent of technological choices, offers an important support for
-   ensuring alignment between the developers' understanding and the meaning given by the domain experts and/or users,
-   guaranteeing interoperability and achieving integration of diverse components,
-   integrating data from heterogeneous sources,
-   maintenance and evolution of the developed system.

There are two different approaches that support ontology-driven knowledge base integration: conceptual graphs and description logics. Both involve a wide variety of specific languages. *Conceptual graphs*[2] represent knowledge as labeled bigraphs (i. e., graphs with two types of nodes). Their nodes are either individuals and concepts, or relations between them. Reasoning mainly relies on graph homomorphism (i. e., retrieving a subgraph in the graph whose structure corresponds to that of the query). Conceptual graphs offer efficient algorithms for reasoning on assertions, which are statements on specific entities. *Description logics*, on the other hand, are a family of logical languages limited to binary relations, which offer large sets of constructors allowing to express rich properties of the different categories of objects. As such, description logics emphasize classification reasoning (i. e., belonging of one entity to specific categories) and reasoning on the domain structure (i. e., inclusion hierarchies between the categories of the domain).

The current chapter focuses on description logics (DLs) and how it addresses the challenges of knowledge representation from a dual perspective:
-   on the one hand, preserving (a certain degree of) human readability. Therefore, the important notions of the domain to model are expressed using concept descriptions (i. e., definitions of sets of elements) and role definitions (i. e., relations between these elements),

---

1 www.w3.org/OWL/

2 *Knowledge graphs* is sometimes used as a synonym for *conceptual graphs*, but it also commonly refers to multiple different families of graph-based knowledge representations supported or not by formal semantics. Since 2012, the *Google Knowledge Graph* introduction makes the use of the term even more ambiguous. Therefore, we limit our considerations to *conceptual graphs*, which are well formalized and equipped with formal semantics.

–   on the other hand, offering a language equipped with formal semantics and support-
    ing decidable reasoning, (i. e., expressible queries that can be answered by a finite
    process). First-order logic offers these requested semantics formalization. Unfortu-
    nately, as noted in Chapter 4, it is undecidable. In order to preserve decidability,
    it requires then limiting the expressiveness of the language with respect to first-
    order logic. The variety of languages in DLs' family explores this trade-off between
    expressiveness and complexity of the performed reasoning.

The rest of this chapter is organized as follows:
–   the next section specifies the notion of ontology and formalizes the problems solved
    using these ontologies,
–   then Section 5.3 explains how these problems are solved by using description logics
    to implement ontologies and further reason upon them. Therefore, the attributive
    language with complement ($\mathcal{ALC}$) is used as a representative example of the de-
    scription logics family to introduce the formalism and reasoning processes,
–   after that, Section 5.4 discusses the limitations of ontologies and description logics,
–   and finally, Section 5.5 draws some conclusions and further outlooks.

5.2 What category of problems do knowledge representation and engineering with ontologies solve?

Unlike the techniques presented in most other chapters, ontologies do not have as their
main challenge to solve business problems. Their key objective is to offer a strong domain. They thus offer an essential support to knowledge engineering and to the development of knowledge-based systems in any technology. The knowledge that they pro-
pose to model mainly concerns the organization of the domain into categories of objects,
called *concepts*, and the *relations* that exist between these. In addition, ontologies also
enable the characterization of specific domain objects as members of these concepts.

Such domain models provide representations at a conceptual level, that is, at a level
that is independent of the technological choices: a model could, for example, describe
a domain involving entities classified as `administration`, `company`, `employee`,...by ab-
stracting the specific implementation choices. Similarly, models can express relations
and constraints between the classes and properties of these entities without any indication of how these can or should be implemented. As such, this level of abstraction is very
useful to support validation of the model by any business or process domain expert. On
the IT-side, formalizing the ontology of the domain is a powerful preliminary step for the
use of many of the techniques presented in this book, as well as for the conception and
development of systems, intelligent or not, that integrate heterogeneous technologies.

Models with formal semantics like ontologies furthermore enable the execution of analyses and queries to be carried out with a guarantee of validity supported by any underlying logical theory.

A first set of analyses consists of ensuring the quality of the model. On the one hand, a *concept satisfiability check* ensures that all the defined classes, that is, the *concepts*, really make sense, and on the other hand, a *consistency check* verifies that there is no contradiction between the model's assertions. Most of the time inconsistency diagnosis reveals modeling errors. Indeed, as a knowledge base models a real state of the world, it should not be contradictory. This consistency diagnosis is critical to ensure the usability of the model and the validity of the deduced statements. Indeed, in the worst case, any statement could be derived from an inconsistent knowledge base.

More typically, ontologies allow for the analysis of the hierarchy of concepts that structures the domain, that is, the multiple relations of inclusion, called *subsumption*, between the concepts of the domain. Figure 5.1 illustrates such a hierarchy for a world that contains two classes of instances represented by, for example, the concepts of Organization and Person. Among the instances that belong to the Organization concept, two particular subgroups are identified, for example, the Company and the Administration concepts. Among the instances that belong to the Person concept, a particular subgroup stands out, covered by, for example, the Employee concept.

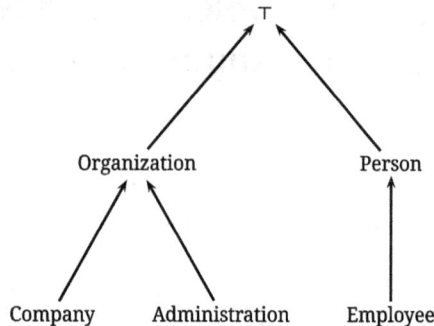

Figure 5.1: Graphical representation of a subsumption hierarchy.

Finally, ontologies support queries at instances level, too. These can either check the property of a given entity (e. g., "Does John work for an administration?") or retrieve all entities that satisfy a given property (e. g., "Who is working for MySBCompany?").

Computer sciences literature commonly refers to Gruber's definition of an ontology as "an explicit specification of a conceptualization," which specifies "the objects, concepts, and other entities that are presumed to exist in some area of interest and the relationships that hold among them" (Gruber, 1993). The knowledge model expressed by an ontology can be seen as a logical theory organized into two main pieces: a set of *logical axioms* ($\mathcal{O}$), which describe general knowledge on the domain, and a set of *assertions* ($\mathcal{A}$) that concern specific entities of that domain.

The logical axioms represent an abstract representation of the domain knowledge and its structure. They specify the concepts and relations of the domain under consideration with their corresponding semantics. In other words, they define the vocabulary of the domain and formalize its meaning. This is achieved by providing a vocabulary and a set of logical formulas that constraint the acceptable models for that domain. For example, completing the vocabulary introduced above, a logical axiom could express that "being a `civil-servant` means being an `employee` who works for an `administration`."

Using the vocabulary defined by the logical axioms, the assertions make statements on individual entities of the domain as "Mary works for MySBCompany" or "MySBCompany is a company." Some authors call $\mathcal{O}$ an abstract ontology, and the pair $(\mathcal{O}, \mathcal{A})$ a concrete ontology. Others consider $\mathcal{O}$ as the actual ontology, and call $\mathcal{A}$ domain assertions. In the current chapter, we call the pair $(\mathcal{O}, \mathcal{A})$ an ontology or knowledge base, as is commonly done in description logics and semantic web literature.

Expressed in specific languages, logical axioms and assertions can respectively be seen as *rules* and *facts* in standard logics. If we compare ontologies with databases, we can see the logical axioms as the definition of the database schema, and the assertions as instances within the database. But the reasoning context differs on an important point. Databases rely on a *closed world assumption*, that is, the information stored in the databases is supposed to be complete. Any statement, which is not in the database, is supposed to be false. Whereas knowledge bases expressed through ontologies deal with an incomplete description of the world. As such, the associated reasoning process involves a so-called *open world assumption*.

This difference, that is, the support of reasoning on partial knowledge, makes the manipulation of negations much more delicate, and consequently, commonly more limited. For example, if the model doesn't provide any information about Nathan, it can neither be inferred that he works for MySBCompany, nor that he does not. However, if the model states that no MySBCompany employees work for BBCorp, knowing that Mary works for MySBCompany, we can deduce that she does not work for BBCorp.

5.3 How are those problems solved?

In order to develop an ontology that models a specific domain, it is necessary to have a language that allows to describe
- the *concepts*, that is, the categories that organize the domain, for example, *Person, Employee, Client, Company*, etc.
- the *relationships* that exist between (some) elements of the concepts, for example, *works_for, buys*, etc.
- the *axioms*, that is, subsumption relations between concepts, for example, employees are persons, clients are persons that buy some products, etc.
- the *assertions*, that is, statements on specific object of the domain, for example, John is an employee, etc.

To further support reasoning, this language must also have, beside a well-defined syntax, a formal semantics.

The multiple formal languages developed to model ontologies are distinct from one another by the statements they allow to express. Technically, this difference consists in the different objects and relations considered as well as different sets of constructors[3] offered by these languages. At the semantical level, the differences impact the respective expressiveness of these ontology languages.

The family of *description logics* (DLs) involves multiple formal languages, supporting the description of *concepts* and *roles* from *concepts names* (unary predicates) and *role names* (binary predicates). Most of these logics are decidable fragments of the first-order logic, but they differ by their respective subset of constructors.

This section first introduces the formal definition of description logics by considering one of the most famous basic ones called attributive language with complement ($\mathcal{ALC}$), and then shows how ontologies reasoning problems are addressed in this language.

5.3.1 $\mathcal{ALC}$ description logic formal definition

The basic elements of $\mathcal{ALC}$ are concepts and roles, which provide the vocabulary of the domain. They are introduced first. Then the domain knowledge is formalized by a knowledge base made of a *terminological part*, called the TBox, and an *assertional part*, called the ABox. They correspond respectively to the abstract ontology $\mathcal{O}$ and the concrete ontology, or domain assertions, $\mathcal{A}$ introduced above. Finally, considerations are drawn on the semantics of $\mathcal{ALC}$.

Concepts and roles

The definition of a knowledge base within a DL uses a vocabulary defined from two sets: the *concept names* and the *role names*. The set of concept names is commonly named **C** and its elements, denoted by A, B, etc. represent sets of elements of the domain. In our example, they are Company, Administration, Person, etc. In $\mathcal{ALC}$, the basic concepts are these provided in **C** plus the universal concept, $\top$, involving all elements of the domain and the bottom concept, $\bot$, that does not contain any element.

The set of role names, denoted by **R**, involves binary relations r, s, etc. Typical relation names in our business example are works_for, sales, is_client_of, etc. They can be seen as binary predicates.

---

3 *Constructors* are symbols or syntactic structures used to build (*construct*) acceptable complex objects and formulas from basic ones (technically called *well-formed formulas*).

Concepts represent sets, called their *extension*. They can be further combined to define new sets, called compound (or complex) concepts thanks to syntactic structures called constructors. The first ones, Boolean constructors follow quite naturally set theory. In our example,

- if we want to talk about female employees, we use the concept defined by the conjunction of concepts Female and Employee and denoted Female ⊓ Employee,
- all workers, whether employed or self-employed, are represented by the new concept Worker defined as a union (also called disjunction) Worker ≡ Employee ⊔ Self-employed,
- unemployed people are people that are not workers: ¬Workers ⊓ Person.

An important precision lies in a syntactic detail that it is important to observe here. The concept ¬Worker would denote all the elements of the domain, which are not workers, involving elements which are not persons as companies, administration, etc. The restriction to persons is obtained by putting the conjunction with the Person concept.

A powerful way to define other concepts in DL is offered by bringing the roles into play with either existential or universal restrictions. So, for example, using existential restriction,

- ∃works_for.⊤ denotes the set of individuals that work for at least one individual (instance of ⊤),
- civil servants, as employees who work for administrations, can be described by the concept Employee ⊓ ∃works_for.Administration.

Universal restrictions are a bit more delicate to manipulate. For example, to represent the set of individuals that work exclusively for administration(s), a concept could be defined, using the universal restriction, as

$$\forall\text{works\_for.Administration}$$

The members of this concept respect one condition: all the individuals with which the member is in a works_for relationship (i. e., all his employers) are administrations. This may appear as counter intuitive, but this condition is satisfied by the individuals (instance of ⊤: organizations and persons) that do not work for anyone. Indeed, if Gabriel is unemployed, the set of his employers is empty and, technically, each of his employers satisfies the constraint of being an administration.[4] So, the precise concept to denote employees that actually work for at least one administration and only for administrations is

$$\text{Employee} \sqcap (\exists\text{works\_for.administration}) \sqcap (\forall\text{works\_for.administration})$$

---

4 For the reader familiar with FOL, remind that whatever is the predicate p, $\forall x : x \in \emptyset \Rightarrow p(x)$.

Formally, *compound concepts* are inductively build from the vocabulary using the constructors, which are specific to each DL. In $\mathcal{ALC}$, the syntax of compound concepts is described by the following grammar:

$$
\begin{aligned}
C, D \quad \rightarrow \quad & A \mid \perp \mid \top \mid \\
& \neg C \mid \\
& C \sqcup D \mid \\
& C \sqcap D \mid \\
& \exists r.C \mid \\
& \forall r.C
\end{aligned}
$$

where A is a concept name, C and D are (either atomic or complex) concepts, and r is a role.

By common abuse of language, we call "concept of $\mathcal{ALC}$," or even "concept" if the reference to $\mathcal{ALC}$ is obvious, the concept descriptions of $\mathcal{ALC}$. As defined by this grammar, a concept is just a well-written formula, still meaningless, or at least without any formal semantics for now. As for other logics, the formal semantic of a language is defined by referring to the notions of both interpretation and model. Given the FOL presentation in Chapter 4, we limit here the formalization of the semantics to the presentation of the first-order logic formulae corresponding to the $\mathcal{ALC}$ concepts.

A concept C represents a set of elements called its *extension*, which can be seen as an unary predicate $C(x)$, which is true for every value of x corresponding to elements in the extension of C. Similarly, a relation r corresponds to a binary predicate $r(x,y)$. Given this convention, the semantics of complex concepts is given by the corresponding formulae as defined by the following mapping functions τ_C and τ_R:

$$
\begin{aligned}
\tau_C(C, x) &= C(x) \\
\tau_C(\top, x) &= \textit{true} \\
\tau_C(\perp, x) &= \textit{false} \\
\tau_C(C \sqcap D, x) &= \tau_C(C, x) \wedge \tau_C(D, x) \\
\tau_C(C \sqcup D, x) &= \tau_C(C, x) \vee \tau_C(D, x) \\
\tau_C(\neg C, x) &= \neg \tau_C(C, x) \\
\tau_R(r, x, y) &= r(x, y) \\
\tau_C(\exists r.C, x) &= \exists y [\tau_R(r, x, y) \wedge \tau_C(C, y)] \\
\tau_C(\forall r.C, x) &= \forall y [\tau_R(r, x, y) \Rightarrow \tau_C(C, y)]
\end{aligned}
$$

Given a so-defined specific description language ($\mathcal{ALC}$ in this case), the knowledge on the domain is organized in two sets of assertions, that is, the TBox that involves axioms on the concepts and the ABox that involves assertion on individual names.

Terminological knowledge: the TBox

A TBox $\mathcal{T}$ is a set of *ontologic axioms* that express inclusion ($\sqsubseteq$) or equivalence ($\equiv$) between concepts (possibly compound). The terminological knowledge represented in Figure 5.1 can be modeled by the following TBox $\mathcal{T}$:

$$\mathcal{T} = \{\, \text{Company} \sqsubseteq \text{Organization}$$
$$\text{Administration} \sqsubseteq \text{Organization}$$
$$\text{Employee} \sqsubseteq \text{Person}\}$$

This TBox could also involve for example, the definition of a civil servant and unemployed persons as introduced above:

$$\text{Civil-servant} \equiv \text{Employee} \sqcap (\exists \text{works\_for}.\text{Administration})$$
$$\text{Worker} \equiv \text{Person} \sqcap (\exists \text{works\_for}.\top)$$
$$\text{Unemployed} \equiv \text{Person} \sqcap \neg\text{Worker}$$

The mapping of axioms to first-order logic formulae is obtained by completing the mapping defined above with the τ mapping defined as follows:

$$\tau(C \sqsubseteq D) = \forall X(\tau_C(C, x) \Rightarrow \tau_C(D, x))$$
$$\tau(C \equiv D) = \tau(C \sqsubseteq D) \wedge \tau(D \sqsubseteq C)$$

Domain axioms: the ABox

An ABox $\mathcal{A}$ involves a set of assertions that express the knowledge about specific individuals of the domain. These individuals are represented by the elements (john, mary, MySBCompany, etc.) of a set **I**. Given a, b belonging to **I**, C a concept and r a role name, the elements of an *ABox* are either
- *concept assertion*: $a : C$,
    for example, mary:Employee and mySBCompany:Company state, respectively, that Mary is an employee and that MySBCompany is a company,
- *role assertion*: $(a, b) : r$,
    for example, (mary,mySBCompany):works-for states that Mary works for MySBCompany.

The mapping of axioms to first-order logic formulae is obtained by completing the mapping defined above with the τ mapping defined as follows:

$$\tau(a : C) = \tau_C(C, x)[x|a]$$
$$\tau((a, b) : r) = \tau_R(r, x, y)[x|a, y|b]$$

About the semantics

Now let us outline the main aspects taken into account by tools that compute reasoning on ontologies.

First, it is to note that the links between individual symbols in **I** and the objects of the real world is not specified by the ontology. mary and ann are just syntactic identifiers without any intrinsic meaning. So, mary and ann could be names for persons as well as for organizations. They could also be two different designations for the same person. Indeed, DL does not make a unique name assumption.

Second, ontologies do not pretend to fully describe the state of the world, which is commonly referred to as the open world assumption. As the knowledge is partial, there are potentially many different states of the world compatible with the ontology. The more axioms there are in the ontology, the fewer compatible states of the world there are. An ontology is said to be consistent as long as there is at least one state of the world, which matches all the assertions in the knowledge base.

Given this, powerful reasoning on ontologies are still possible. In particular, an assertion α is a consequence of an ontology $\mathcal{O}$ if α holds in all the states of the world compatible with $\mathcal{O}$. One then says that $\mathcal{O}$ *entails* α ($\mathcal{O} \vDash \alpha$). The following subsection presents the main reasoning problems supported by DLs.

5.3.2 Queries and reasoning problems

Besides offering a machine readable format for human knowledge representation, description logics also offer large reasoning capabilities. Built on top of basic checks, more advanced queries offer real support and services to further knowledge engineering.

Among the questions addressed by DLs, one distinguishes these only concerning the structure of the domain, the TBox, and also these addressing the global knowledge base, the TBox, and the ABox.

Concept satisfiability check

Given a, possibly empty, Tbox $\mathcal{T}$, a concept C is said to be *satisfiable* with respect to $\mathcal{T}$, if one can imagine a world that satisfies all the constraints expressed in $\mathcal{T}$ in which the extension of C contains at least one element.[5] A *satisfiability check* of a concept ensures that the concept C, as it is defined, makes sense in the world described by $\mathcal{T}$. A negative answer to the satisfiability check most frequently reveals modeling trouble either in the definition of C or in the knowledge base $\mathcal{T}$. In the knowledge engineering process, a global check of the satisfiability of all the concept names A defined in $\mathcal{T}$ offer

---

5 Formally, if there exist a model of $\mathcal{T}$ in the domain of which at least one element belongs to the interpretation of C.

powerful debugging support. Indeed, common sense rarely leads to the provision of a sophisticated description of a concept, which doesn't have any instances in any configuration of the world.

In our example, we could consider extending the model by introducing the definition of a job seeker as an unemployed worker by adding the concept:

$$\text{Job-seeker} \equiv \text{Worker} \sqcap \text{Unemployed}$$

Since we have defined the concept of unemployed as disjoint from that of worker (Unemployed ≡ Person ⊓ ¬Worker), the concept satisfiability check will then reveal that the new concept Job-seeker is empty and will thus manifest a necessary adjustment of the model.

Subsumption hierarchy

Combining the statements set in the axiomatic knowledge on the structure of the domain as expressed in $\mathcal{T}$ induces implicit relations between the concepts, too. Given a, possibly empty, Tbox $\mathcal{T}$, and two (atomic or complex) concepts C and D, DLs offer reasoning support to check the *subsumption* of concept C by D with respect to $\mathcal{T}$ ($\mathcal{T} \models C \sqsubseteq D$) and *equivalence* of concepts C and D with respect to $\mathcal{T}$ ($\mathcal{T} \models C \equiv D$).

The subsumption relation between concepts as defined by the axioms of a TBox induces a preorder between them, that is, the subsumption is a reflexive and transitive relation on the set of concept names. Building the *subsumption hierarchy*, and offering a visualization of the subclass-superclass global relationship between concepts, supports the capability to easily overview the knowledge domain model. The *classification* of a TBox, that is, the calculation of the subsumption hierarchy, is one of the reasoning tasks most used in knowledge engineering.

The computation of the subsumption hierarchy on the TBox described above provides the result as illustrated in Figure 5.2. It reveals that the intuitive inclusion of employees among workers is not present in the current model. This suggests an adaptation of the model, for example, by defining Employee as Person⊓∃works_for.organization, which will then be recognized as being subsumed by Worker.

Consistency checking

Let's consider now a complete knowledge base $\mathcal{K}$ made of a TBox $\mathcal{T}$ and a ABox $\mathcal{A}$. The first question that arises is the insurance that all the assertions involved in the knowledge base are compatible. This task is called *consistency check* or *knowledge base satisfiability check*. It finds any contradictory statements in a knowledge base, and thus preciously supports debugging and validation of knowledge bases.

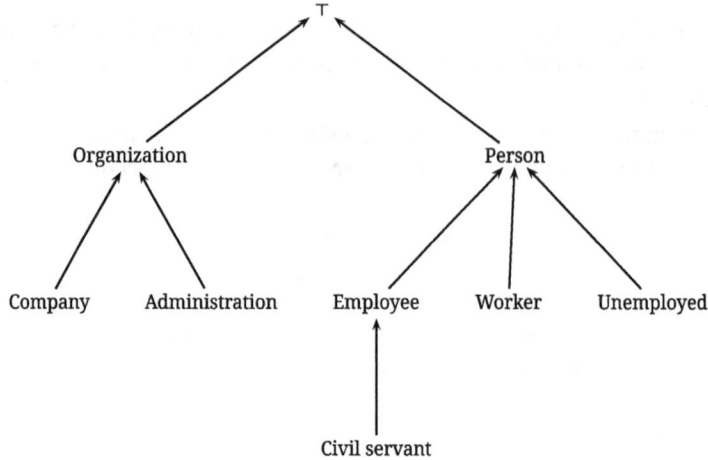

Figure 5.2: Graphical representation of the computed subsumption hierarchy.

Instances queries

Given a knowledge base $(\mathcal{T}, \mathcal{A})$, the focus can be put on a specific element a. Checking if a has some specific property can be rephrased as checking it is belonging to the set of elements having this property, or in other words, to a concept, let's say C, that has this set as extension. DLs offer this reasoning of *checking whether a is an instance of C* with respect to $(\mathcal{T}, \mathcal{A})$ $((\mathcal{T}, \mathcal{A}) \models a : C)$.

Here, it is important to understand that the instance checking answers the question *does the model guarantee that $a : C$?* In our example, the checks `mary:Worker` and `mary:¬Worker` return respectively yes and no, while as the model does not involve any information about `nathan`, both queries `nathan:Worker` and `nathan:¬Worker` will return no, indicating that neither statement can be derived from the model. This behavior manifests the open world hypothesis, that is, the model being partial, there are statements that can neither be confirmed nor denied.

The analysis of a given element can be generalized to the complete set of concepts defined in $\mathcal{T}$. This task is called the *realization* of an individual with respect to a knowledge base $\mathcal{K} = (\mathcal{T}, \mathcal{A})$. For each concept name A occurring in $\mathcal{T}$, it tests whether a is an instance of A with respect to $\mathcal{K}$ and returns the set of those concept names for which the test is positive. In our adapted model, if `michaël` is declared to be a `Civil-servant`, its realization will indicate that he is also member of the `Employee`, `Worker`, and `Person` concepts.

Finally, probably the most expected query is *instance retrieval*. It aims to retrieve elements that satisfy a certain description, that is, that belongs to a specific concept. Given a concept C and a knowledge base $\mathcal{K}$, this task checks for each individual name a occurring in $\mathcal{K}$ whether it is an instance of C with respect to $\mathcal{K}$ and returns the set of those individual names for which the test is positive.

Reasoning problems reduction

Before turning to the implementation of the reasoning problems, let's observe the relationships between them. Without going into the details of theorems and proofs, an intuitive presentation of the main results helps to understand how problems can be *reduced* to other ones, that is, rephrased as a problem of another type. The study of reduction leads to the identification of a minimal set of problems that can be used to solve all other reasoning problems, and consequently, of the minimal set of reasoning for which an implementation has to be realized.

Let's consider first equivalence and subsumption between concepts C and D. By its definition, the equivalence check $\mathcal{T} \models C \equiv D$ can easily be reduced to the two subsumption problems $\mathcal{T} \models C \sqsubseteq D$ and $\mathcal{T} \models D \sqsubseteq C$. The reader familiar with set theory,[6] won't be surprised that the subsumption problem $\mathcal{T} \models C \sqsubseteq D$ is equivalent to the satifiability check of $C \sqcap \neg D$ with respect to $\mathcal{T}$.

Now, as the satisfiablity check of a concept C with respect of $\mathcal{T}$, intuitively corresponds to the identification of one element belonging to the extension of the concept, one can easily guess that this problem can be reduced to the consistency check of the knowledge base formed by adding to $\mathcal{T}$ an ABox including the assertion $b : C$, that is, the consistency check of $(\mathcal{T}, \{b : C\})$.

Combining these results suggest that all the reasoning problems introduced in the previous subsection can be reduced to mere consistency checking. Complementary results enhance that, provided the TBox does not involve cyclic concept definitions, checking a knowledge base definition can be reduced to checking the consistence of knowledge bases with an empty TBox. This follows, on the one hand, from the fact that checking whether a is an instance of C with respect to $(\mathcal{T}, \mathcal{A})$ $((\mathcal{T}, \mathcal{A}) \models a : C)$ is equivalent to checking the consistency of $(\mathcal{T}, \mathcal{A} \cup \{a : \neg C\})$ and, on the other hand, of the definition of an unfolding operation, which turns an acyclic TBox into an ABox.

Reasoning implementation

$\mathcal{ALC}$, like most DL, is a decidable fragment of first-order logic. The consistency check of an ontology (and consequently all the above mentioned problems) can be realized through an algorithm, which is
- sound: any time the algorithm declares a model consistent, the model is actually consistent (no false positive),
- complete: if the model is consistent, the algorithm declares it consistent (all positives are identified),
- terminating: the algorithm stops after a finite amount of steps on any input.

---

6 Reminding that for all sets A and B, $A \subseteq B$ if and only if $A \cap \neg B = \emptyset$.

In $\mathcal{ALC}$, the concept satisfiability and the ABox consistency check are PSpace-complete for an acyclic TBox, and ExpTime-complete for the general TBox.

A complete discussion of the numerous results and implementations is far beyond the scope of this current chapter. However, *tableau algorithms* are considered an important milestone in the DLs reasoning systems implementation, and are still being used in numerous tools. This subsection introduces the main elements of these algorithms focusing on the consistency check, whose importance has been outlined in the previous subsection.

Basically, the algorithm proceeds in a knowledge expansion from the original knowledge base until it meets a contradiction, called a *clash*, or no more expansion is possible. To get the essence of the algorithm, we illustrate a naive implementation on the following ontology:

$$\mathcal{T} = \{\text{ClassBcomp} \sqsubseteq \text{Comp} \sqcap \neg\exists\text{sell}.(\text{Far} \sqcap \text{Phytosan})\}$$
$$\mathcal{A} = \{\text{bbCorp} : \text{ClassBcomp} \sqcap \text{Comp} \sqcap \exists\text{sell}.\text{Far},$$
$$\text{apple} : \text{Product} \sqcap \text{Phytosan},$$
$$(\text{bbCorp}, \text{apple}) : \text{sell}\}$$

The TBox, $\mathcal{T}$ defines Class B companies as companies that do not sell any products that are both distantly sourced and produced using phytosanitary products. Assertions in the ABox (A) state that (i) bbCorp is a Class B company that sells products distantly sourced, (ii) apples are products produced using phytosanitary products, and (iii) bbCorp sells apples.

The algorithm is processed in two steps:

Step 1

All concepts are reduced to their *negation normal form*, that is, equivalent concepts where negation is applied only to concept names. This is achieved by applying an *nnf* transformation summarized in Figure 5.3. In our example, the concept on the right of the rule in the *TBox* is rewritten as follows:

$$nnf(\text{Comp} \sqcap \neg\exists\text{sell}.(\text{Far} \sqcap \text{Phytosan})) = \text{Comp} \sqcap nnf(\neg\exists\text{sell}.(\text{Far} \sqcap \text{Phytosan}))$$
$$= \text{Comp} \sqcap \forall\text{sell}.nnf(\text{Far} \sqcap \text{Phytosan})$$
$$= \text{Comp} \sqcap \forall\text{sell}.(\neg\text{Far} \sqcup \neg\text{Phytosan}).$$

Other concepts do not involve negation, so they do not have to be transformed.

$$nnf(C) = C, \text{if C is atomic}$$
$$nnf(\neg C) = \neg C, \text{if C is atomic}$$
$$nnf(\neg\neg C) = C$$
$$nnf(C \sqcup D) = nnf(C) \sqcap nnf(D)$$
$$nnf(C \sqcap D) = nnf(C) \sqcup nnf(C)$$
$$nnf(\neg(C \sqcup D)) = nnf(\neg C) \sqcap nnf(\neg D)$$
$$nnf(\neg(C \sqcap D)) = nnf(\neg C) \sqcup nnf(\neg D)$$
$$nnf(\forall r.C) = \forall r.nnf(C)$$
$$nnf(\exists r.C) = \exists r.nnf(C)$$
$$nnf(\neg\forall r.C) = \exists r.nnf(\neg C)$$
$$nnf(\neg\exists r.C) = \forall r.nnf(\neg C)$$

Figure 5.3: Negation normal form transformations.

Step2

The knowledge expansion is realized by building a tree similar to the tableau algorithm presented in Chapter 4. The $\mathcal{ALC}$-specific expansion rules are presented in Figure 5.4. The expansion of our example knowledge base $(\mathcal{T}, \mathcal{A})$ is further given by the tree illustrated on Figure 5.5. If all the branches of the tree conclude with a clash, the knowledge base is inconsistent. In our case, (at least) one branch of the tree (a tableau) is clash-free and no more rule can be applied, as such the knowledge base is consistent.

$\sqcap$-rule   if $a : (C \sqcap D) \in \mathcal{A}$ and $\{a : C, a : D\} \not\subseteq \mathcal{A}$,
         then $\mathcal{A}' = \mathcal{A} \cup \{a : C, a : D\}$

$\sqcup$-rule   if $a : (C \sqcup D) \in \mathcal{A}$ and $\{a : C, a : D\} \cap \mathcal{A} = \emptyset$,
         then $\mathcal{A}' = \mathcal{A} \cup \{a : C\}$ and $\mathcal{A}'' = \mathcal{A} \cup \{a : D\}$

$\exists$-rule   if $a : \exists r.C \in \mathcal{A}$ and there is no b such that $\{(a, b) : r, b : C\} \subseteq \mathcal{A}$,
         then $\mathcal{A}' = \mathcal{A} \cup \{x : C, (a, x) : r\}$, for a new individual $x \notin \mathcal{A}$

$\sqcap$-rule   if $a : \forall r.C \in \mathcal{A}$ and $(a, b) : r \in \mathcal{A}$, but $b : C \notin \mathcal{A}$,
         then $\mathcal{A}' = \mathcal{A} \cup \{b : C\}$

$\sqsubseteq$-rule   if $a : A \in \mathcal{A}, A \sqsubseteq C \in \mathcal{T}$ and $a : C \notin \mathcal{A}$,
         then $\mathcal{A}' = \mathcal{A} \cup \{a : C\}$

$\equiv_1$-rule   if $a : A \in \mathcal{A}, A \equiv C \in \mathcal{T}$ and $a : C \notin \mathcal{A}$
         then $\mathcal{A}' = \mathcal{A} \cup \{a : C\}$

$\equiv_2$-rule   if $a : \neg A \in \mathcal{A}, A \equiv C \in \mathcal{T}$ and $a : nnf(\neg C) \notin \mathcal{A}$
         then $\mathcal{A}' = \mathcal{A} \cup \{a : nnf(\neg C)\}$

Figure 5.4: Tableau expansion rules for $\mathcal{ALC}$.

$$bbCorp:ClassBcomp \sqcap Comp \sqcap \exists sell.Far$$
$$apple:Product \sqcap Phytosan$$
$$(bbCorp,apple):sell$$

↓ ⊓–*rule*

$$bbCorp:ClassBcomp$$
$$bbCorp:Comp$$
$$bbCorp:\exists sell.Far$$
$$apple:Product$$
$$apple:Phytosan$$

↓ ∃–*rule*

$$(bbCorp,x):sell$$
$$x:Far$$

↓ ⊑–*rule*

$$bbCorp: Comp \sqcap \forall sell.(\neg Far \sqcup \neg Phytosan)$$

↓ ⊓–*rule*

$$bbCorp:\forall sell.(\neg Far \sqcup \neg Phytosan)$$

↓ ∀–*rule*

$$apple :(\neg Far \sqcup \neg Phytosan)$$

⊔–*rule* (left branch) ⊔–*rule* (right branch)

$$apple :\neg Far$$

↓ ∀–*rule*

$$x:\neg Far \sqcup \neg Phytosan$$

⊔–*rule*

$$x:\neg Far$$
XClash

$$x:\neg Phytosan$$
consistent KB

$$apple :\neg Phytosan$$
XClash

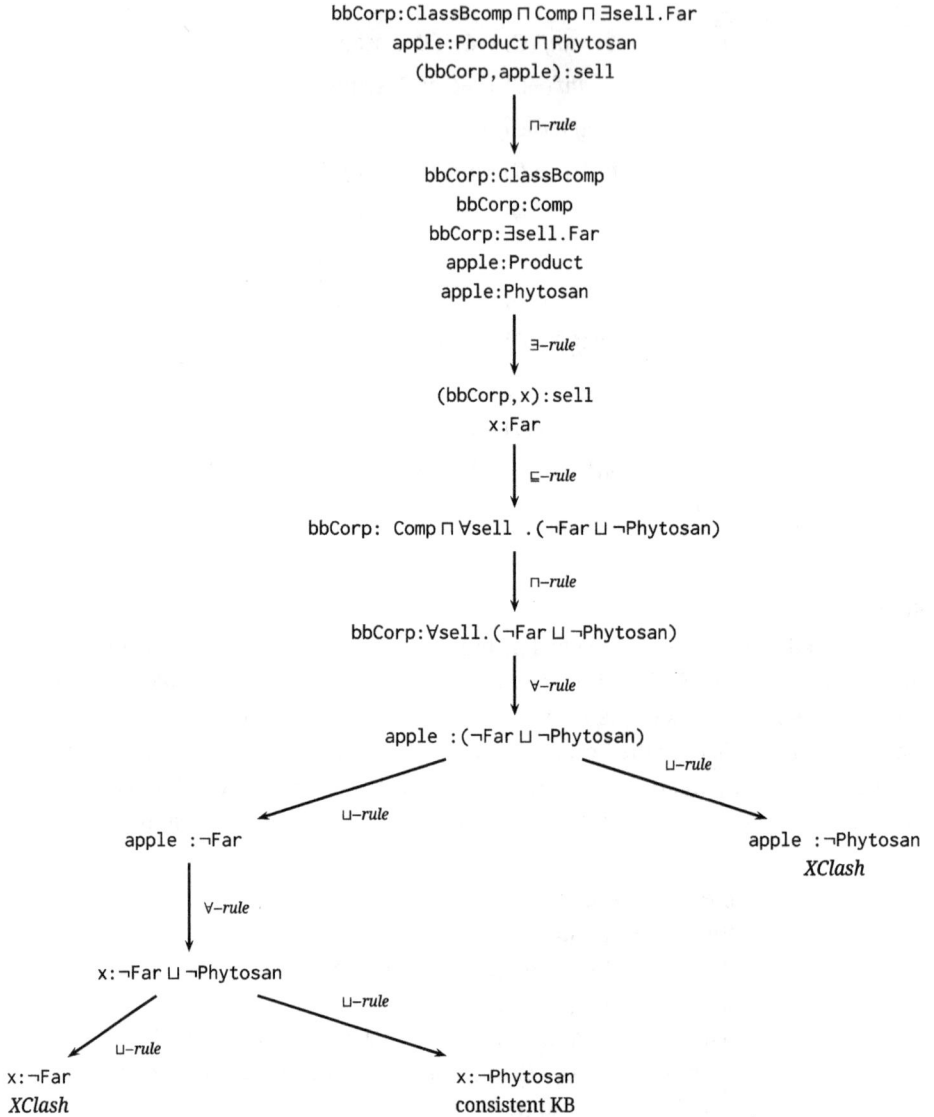

Figure 5.5: Knowledge base consistency proof.

Note that the naive version of the algorithm presented here is nondeterministic, as the order of the rules (the two branches of ⊔-rule) can be applied in any order. Mind you that this does not impact the conclusion of the algorithm, but an optimization of the implementation can further take advantage of defining an exploration strategy.

5.4 What are the limitation of knowledge representation and engineering with ontologies?

The previous section presented the DL $\mathcal{ALC}$ as a representative of this family of languages. However, there are multiple DLs. This section first introduces their diversity in Subsection 5.4.1. Then Subsection 5.4.2 addresses the common limitations of DLs and 5.4.3 the challenges faced when DLs are being used for knowledge engineering. Finally, extending somewhat the view on ontologies covered in this chapter, the semantic web ontology language, OWL, is shortly described in Subsection 5.4.4 and Subsection 5.4.5 introduces knowledge graphs.

5.4.1 Expressiveness and complexity of DLs

The previous section stated that if, on the one hand $\mathcal{ALC}$ loses expressiveness compared to FOL; on the other hand, it gains decidability on the consistency check. Moreover, the complexity of the concept satisfiability and the ABox consistency check is PSpace-complete for an acyclic TBox, and ExpTime-complete for a general TBox.[7]

The history of the DL family is a kind of dance back and forth between language expressiveness and algorithm complexity. In parallel, particularly with the semantic web and semantic data management applications, another movement back and forth is from queries on a TBox to queries on an ABox. The following subsection therefore introduces the variety of DLs issued from these explorations.

Descriptive logics with low expressiveness

Targeting the reduction of the complexity of reasoning, one may focus on two minimal description logics: $\mathcal{EL}$ and $\mathcal{FL}_0$. The set of constructors allowed in $\mathcal{EL}$ is limited to the conjunction ($C \sqcap D$), existential restriction ($\exists r.C$), and the top concept ($\top$). In this very limited language, the subsumption problem, even with general TBoxes, can be solved by a consequence-based reasoning, which remains polynomial. In contrast, in $\mathcal{FL}_0$, despite a set of constructors only limited to the conjunction ($C \sqcap D$), value restriction ($\forall r.C$), and the top concept ($\top$), the subsumption check remains ExpTime-complete. It is to note that the satisfiability check loses its interest in both languages as their constructors cannot cause unsatisfiability.

---

7 For the reader nonfamiliar with complexity theory, mind that the complexity of a problem estimates how the resources quantity (time and space) required to reach the answer (by the best possible algorithm) in the worst case grows with the size of the problem. Problems are classified according to the following hierarchy: $PTIME \subseteq NP \subseteq PSPACE \subseteq EXPTIME \subseteq NEXPTIME$.

DL-Lite is another remarkable DL with limited expressiveness. It is one of the DLs proposed to support *query answering in the presence of ontologies*. This type of queries concerns search in huge data sets issued from heterogeneous data sets (as web sources), where ontologies are used to compensate the incompleteness of information by providing the domain knowledge. Given the constructors' restrictions, for any TBox, queries on the knowledge base (typically conjunctive queries) can be rewritten into first-order queries[8] on an ABox that is equivalent. The translation and the execution (being both polynomial) of the resulting query ensures polynomial execution of the query in a polynomial time.

Beside concept names, DL-Lite further has as *basic concepts* the top concept ($\top$), unquantified existential restriction ($\exists r$, to be understood has $\exists r.\top$), and unquantified existential restriction on the inverse role ($\exists r^-$). The TBox axioms are either positive basic concept inclusions ($B_1 \sqsubseteq B_2$), negative basic concept inclusions ($B_1 \sqsubseteq \neg B_2$), or role inclusions ($r \sqsubseteq s$, where r and s are concepts names).

Concept constructors and the ALC family

A variety of DLs results from the introduction of new concept constructors, that is, their names, notation, and symbols are provided in Table 5.1 together with their corresponding logic formulae.

Table 5.1: Advanced concepts constructors corresponding formulae.

Sym	name	notation	Corresponding logic formula
$\mathcal{F}$	functionality	($\leq 1r$)	$\#\{y\|r(x,y)\} \leq 1$
$\mathcal{N}$	unqualified number restriction	($\geq nr$)	$\#\{y\|r(x,y)\} \geq n$
		($\leq nr$)	$\#\{y\|r(x,y)\} \leq n$
$\mathcal{Q}$	qualified number restriction	($\geq nr.C$)	$\#\{y\|r(x,y) \wedge C(y)\} \geq n$
		($\leq nr$)	$\#\{y\|r(x,y) \wedge C(y)\} \leq n$
$\mathcal{O}$	nominals	a	$x \in \{a\}$

These $\mathcal{ALC}$ extensions are called $\mathcal{ALCX}$ where X denotes the selection of constructors among $\mathcal{FNQO}$ where $\mathcal{F}$ is dropped when $\mathcal{N}$ is included, and $\mathcal{FN}$ are dropped when $\mathcal{Q}$ is included.

Introducing (any subset of) these concept constructors in the concept definitions preserve the PSpace-completeness of the concept satisfiability and the ABox consistency check on acyclic TBoxes. With general TBoxes, the problems remain ExpTime-complete for any subset of concept constructors.

---

8 First-order formulae that use only unary and binary predicates and no function symbols or constants.

Role constructors

The basic DLs considered so far have the concepts as their main focus. Among the broad diversity of DLs, many of them provide constructors supporting the advanced definition of roles and include constraints on roles in the knowledge base. The complete study of all DLs go far beyond the scope of this introductory chapter though. However, we introduce here the most well known of these tools.

Among the role constructors, we find *inverse role* ($\mathcal{I}$), role intersection, union, (atomic or full) negation, role composition, and reflexive-transitive closure corresponding to the operations on relations. Among the possible constraints on roles the following ones are the most referenced ones:

- *Role hierarchy* ($\mathcal{H}$), denoted $r \in s$, expresses inclusion,
- *Role transitivity* ($\mathcal{S}$), denoted $Trans(r)$, expresses transitivity,
- and *complex role inclusion* ($\mathcal{R}$) allows the expression of role inclusion of a relation obtained by combining other ones.

As previously, languages are named by adding the letters of the features to $\mathcal{ALC}$. For advanced languages (including $\mathcal{S}$ or $\mathcal{R}$), the $\mathcal{ALC}$ prefix is commonly dropped. Except for inverse, denoted by $\mathcal{I}$, the selection of role constructors is specified in parentheses after the name of the language.

Although there is no fundamental reason to distinguish the *relational axioms*, that is, the axioms formulated using these constraints on relations from the TBox, they are commonly grouped in a third set of axioms called the RBox.

Without going into the details of all languages' inclusion and the respective complexity of problems in each of the DLs, which gave rise to multiple publications, it is important to note that some gains in expressiveness have important costs toward the complexity of the problems. For example, in $\mathcal{ALCOI}$, satisfiability and subsumption are ExpTime-complete, even on empty TBoxes. As it happens, these same problems become NexpTime-complete in $\mathcal{ALCOIQ}$. For some extensions, the reasoning even becomes undecidable.

Despite these theoretical limitations, one has to remind that these complexity studies consider the worst possible cases. In real case scenarios, the combination of constructors used rarely reach these levels. Most implemented DLs reasoners have acceptable performance on ontologies of a reasonable size.

5.4.2 DLs expressiveness limitations and challenges

Most DLs are equivalent to fragments of FOL. This formal limitation makes it impossible to express certain properties that are easily expressed in FOL. This is obviously the case for properties, which require constructors not involved in the specific studied DL. For all DLs, this is the case with properties, which require an *n*-ary predicate with $n \geq 3$. For example, it is not possible to express a generic property modeling a generic *provides*(x, y, z) relation between company x, product y, and country of origin z. Similarly, reasoning with natural numbers, addition, or multiplication is known to be undecidable, and consequently, can not be modeled with DLs.

Finally, DLs presented above also face limitations similar to these of FOL, that is, they are not well suited to model knowledge involving time, modalities, or fuzzyness. To date, researchers attempt to include these concerns into DLs similar to the various logics extending FOL. The key challenge for future DL research is to extend the expressiveness while preserving decidability.

5.4.3 Knowledge engineering

Consistency of a model can be ensured by automated reasoning and this is precious for knowledge engineering. However, that does not guarantee that the model is adequate with respect to the reality, or does actually fit the needs of the developers, whatever they are. Many challenges remain in the hands of the knowledge engineers as, for example,

– arbitrate between an elegant, very abstract model, which is more easily maintainable and a more tricky one, which reflects all the details of the real world,
– reconcile in one ontology the possibly multiple users' conceptions of the domain,
– deal with the ambiguity and the polysemy of natural language,
– make explicit as much as possible of implicit knowledge,
– and deal with the fuzzyness in knowledge.

5.4.4 OWL and the semantic web

Ontologies expressed in DLs are used to support development and maintenance of systems in numerous heterogeneous languages and application domains. One of the most famous application domain is certainly the semantic web. The ontology web language (OWL[9]) allows the definition of ontologies by means of *classes* (ontology concepts) and *properties* (ontology roles). OWL adopts a web-friendly syntax relying on preexisting extensible markup language (XML[10]) and resource description framework (RDF[11]) standards.

Without going into the details of the RDF language,[12] let us remind that it relies on a triples' structure $< subject, predicate, object >$. So, for example, the assertion that "John

---

9 www.w3.org/OWL/

10 is a markup language that defines a set of rules for encoding documents in a format that is both human-readable and machine-readable.

11 RDF is a standard that uses a graph data model to describe web resources and facilitate data interchange on the internet.

12 www.w3.org/RDF/

works for mySBC" can be expressed by a triple $< John, worksFor, mySBC >$." Similarly, this triple structure allows to naturally express ABox assertions, too, that is,

- a role assertion $(a, b) : r$ can be expressed by a triple $< a, r, b >$, called in OWL *property assertion*,
- a concept assertion $a : C$ can be expressed by a triple $< a, rdf : type, C >$, called in OWL *class assertion*, where $rdf : type$ is a predified predicate denoting the "instance of" relationship.

The RDF schema extension of the basic RDF vocabulary (RDFS[13]) provides a predicate $rdfs : subClassOf$ denoting the class inclusion relationship. Thanks to this predicate, a *TBox* axiom $C \sqsubseteq D$ can be written as $< C, rdfs : subClassOf, D >$.

This adoption by OWL of the RDF vocabulary and syntax facilitates interactions with information available in the RDF format, in particular, with the semantic web and the linked data (data formatted according to the web semantic best practices[14]).

The description of the formal semantics of OWL 2[15] further enhances its compatibility with $\mathcal{SROIQ}$. The standardization of the language has led to the development of multiple tools that allow the development and manipulation of ontologies via user-friendly interfaces,[16] that is, Protégé[17] was one of the first and best known open source tools to manipulate RDF data sources.

5.4.5 Knowledge graphs

This chapter has highlighted how DLs support the implementations of ontologies in a way that is both human-readable and allows queries to be executed with respect to a formal semantic.

An alternative family of models can achieve the same goals: graph-based knowledge models. Like DLs, they are the subject of a large literature that studies their operations and expressiveness (see Chein and Mugnier (2009) for a study of the foundations of these works).

With the notoriety of Google Knowledge Graph, the world of the semantic web has turned to specific models of this family: the knowledge graphs (KGs). KGs provide human-readable visualizations in the form of node-relationship diagrams whose nodes represent entities (objects, events, or concepts) and the labeled links denote the relationships between them.

---

13 www.w3.org/TR/rdf-schema/

14 https://www.w3.org/standards/semanticweb/data

15 www.w3.org/TR/owl2-direct-semantics/

16 www.w3.org/wiki/Ontology_editors

17 protege.stanford.edu/

The business case below details how large data sets can be stored in such a knowledge graph. Then it illustrates how ontology-driven queries can be computed on this graph and offer support to decision in a drug development process.

5.4.6 Conclusion

Ontologies introduced in this chapter aim to offer formal specification of domain knowledge that support reasoning on partial model. Their technology independent formulation offers support to various knowledge engineering related tasks from the specification to code maintenance going through the data and code integration.

Among the diversity of languages that implement ontologies, this chapter introduces description logics. In these languages, the variable free syntax preserves a reasonable human readability. Thanks to their formal semantics, they offer both support for model checking and query answering. DLs are equivalent to decidable fragments of first-order logic and, therefore, the associated decision engines proceed requests in a finite time. Even if the complexity is high in worst cases scenarios, most pragmatic queries are answered within a limited amount of time.

The emergence of the semantic web has largely contributed to the notoriety of ontologies. However, they are used in a wide variety of application domains. Decision support tools in health domains are among the systems that most extensively integrate ontology-based knowledge representation. The drug development business case presented below illustrates how a tool supporting research in the conceptually complex domain of pharmaceutical industry can integrate rich knowledge bases by using ontologies. It enhances how efficient semantical queries can be performed.

The reader who wishes to delve deeper into the subject will be interested in the following references.

- Probably the most quoted definition of ontologies in computer sciences literature comes from the paper by Gruber (1993).
- For more detail on description logics and the associated theoretical results and proofs, the reader could refer to Krötzsch et al., Baader et al. (2017), Baader et al. (2007).
- A browser presenting the multiple DLs and their respective expressiveness and complexity is available on http://www.cs.man.ac.uk/ezolin/dl/.
- For a study of knowledge representation and reasoning with graph-based models, refer to Chein and Mugnier (2009).
- A simple comparison of DLs and conceptual graphs can be found in Leclère et al. (2014).
- The W3C OWL standard is on www.w3.org/OWL.

5.5 Industry example: drug development using ontologies and semantic search, the ONTOFORCE example

Erik Mannens, Filip Pattyn

5.5.1 Context

Life sciences is one of the fastest evolving domains and is fueled by the technological evolutions that marked the twentieth century such as electronics and computer sciences. There is a tremendous amount of information available ranging from observations and insights in fundamental biology up to the clinical data of millions of individuals. The newest technological developments in wearables and sensors have created openings for data generation in another order of magnitude. As a result, one is capable of diagnosing, treating, curing, and eradicating an extensive number of illnesses and improving the quality of life for billions of people. Nevertheless, our profound understanding of biology has thought us that there are still lots of pieces in the puzzle missing. The quest to continue to strive and deeply understand biology, to discover, and to innovate has turned the domain into a massively data-driven business. The pharmaceutical industry, and more specifically novel drug development, is therefore characterized by a fierce competition to find the next big hit that boosts the business and brings value to patients.

5.5.2 Data is ubiquitous but siloed

Developing a new therapy is a labor intensive and long process with a high risk of failure. The drug development process can take up to 13 years until a new therapy gets market approval and is characterized by a succession of specific phases (see Figure 5.6). All phases are very data-driven and require bringing together public, licensed, and internal data to find new insights or to stay ahead of the competition.

Although the digital transformation is running behind in biopharma compared to other industries, the related scientific domains of biology and biomedicine were one of the drivers and early adopters of the semantic web and linked data technologies and principles. Since the last two decades, a lot of public initiatives to capture and structure subsets of biological data have mushroomed. In contrast, the industry remained very protective of their growing amounts of internal data. Data is very fragmented per operational or business unit or drug development phase. One of the counterreactions to stimulate data exchange and interoperability is the creation of many standards and ontologies.

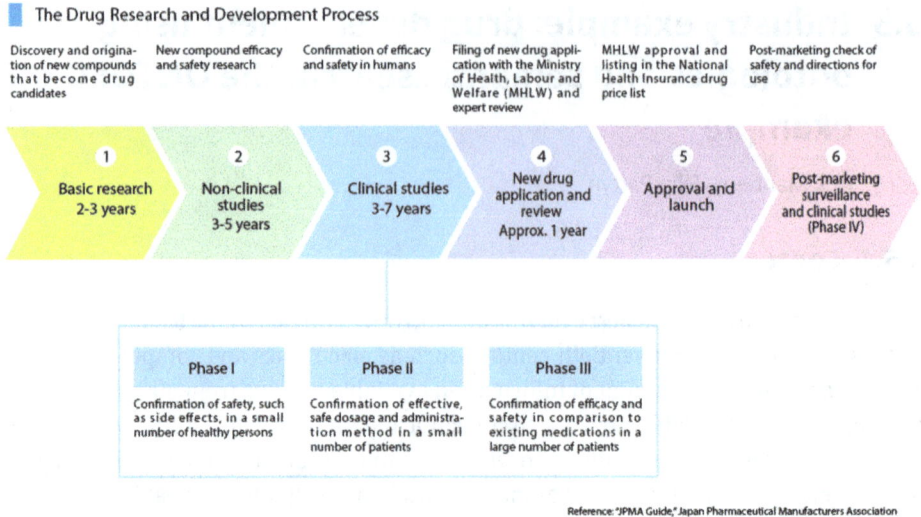

Figure 5.6: The drug development process.

5.5.3 ONTOFORCE

It is in that atmosphere that ONTOFORCE was founded; a Belgian SME with offices in Ghent, Belgium, and Cambridge, MA. Their primary ambition is to develop a technology that eases the pain of bringing siloed data together and creating links between concepts in order to allow finding information faster, revealing new insights and stimulating collaboration. The technology is coined DISQOVER and is built on the principles of *semantic web technologies* and *linked data* and is embracing the *FAIR Data Principles* (Wilkinson et al., 2016).

It contains a data ingestion engine with a graphical user interface that allows to build data processing pipelines that are able to transform and integrate a wide variety of data into a *knowledge graph* stored in a proprietary storage engine. On top of that, the platform is equipped with an intuitive user interface that is able to search, navigate, traverse a knowledge graph, and allows to visualize subsets of the graph as customizable dashboards used for visual analytics. The DISQOVER platform is built specifically for life sciences and delivers value in use cases acccross the drug development cycle.

A lot of the public domain knowledge in life sciences is stored in databases or is available as a thesaurus, taxonomy, hierarchy, or ontology. Transforming a number of these data sources and storing this data as a semantic knowledge graph is the starting point for solving multiple use cases based on data interoperability and reuse. Therefore, a DISQOVER version consisting of public data is created and managed by ONTOFORCE. It currently holds more than 100 data sources and is available via a publicly accessible user interface (http://www.disqover.com) and serves as a hub for enterprise data federation and synchronization services (see Figure 5.7).

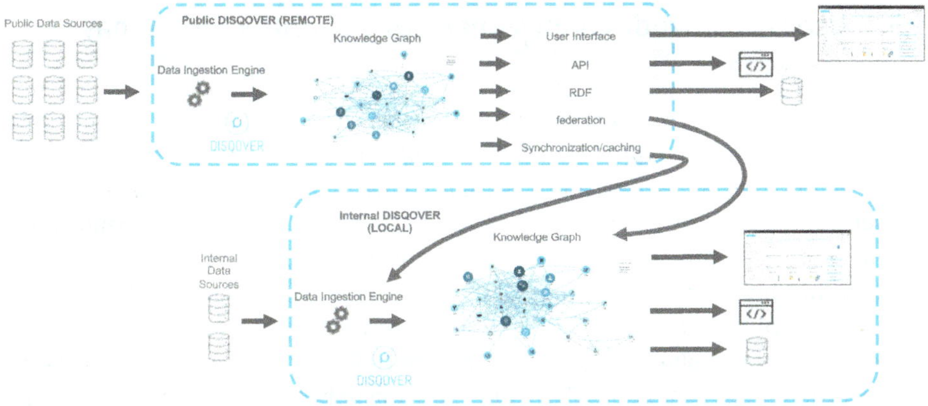

Figure 5.7: The DISQOVER technology.

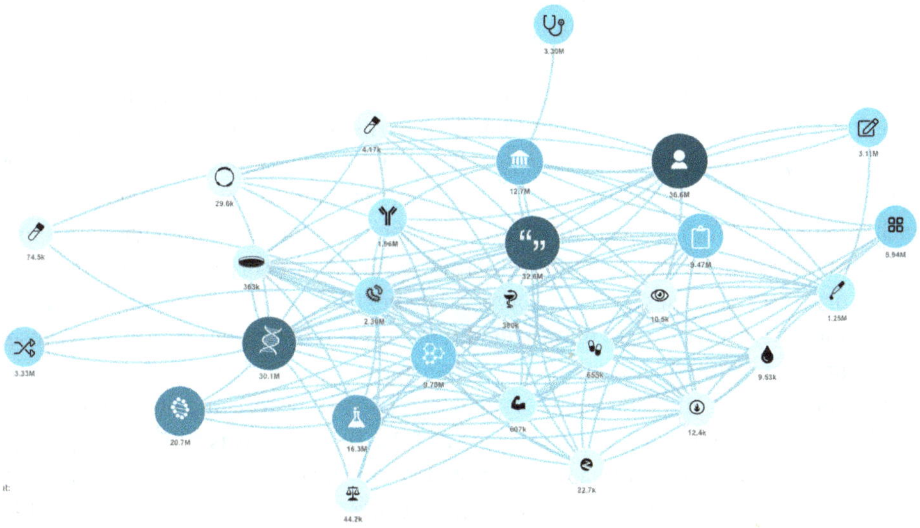

Figure 5.8: The DISQOVER data type ontology graph showing the semantic links between the major data types. The largest data types are Person (36.6M), Publication (32.6M), and Gene (30.1M).

The core of a DISQOVER *knowledge graph* is the overarching data type ontology graph (see Figure 5.8). Specific data types can be configured (via *constructors*), which allows to group, define, or distinguish object classes available in existing data sources. The configuration of a data type entails the definition of data properties, filter properties, and typed relations. At any moment, the provenance of the data is maintained in *OWL rules* and can be highlighted in the user interface. The public data knowledge graph spans different subdomains of the life sciences and can be applied for a multitude of use cases. The most elementary ones are in basic research.

5.5.4 Use case introduction: semantics and ontologies in basic research

One of the strategies in drug development is to start with finding a potential biological target for intervention. Typical targets are biological entities that are involved in biochemical processes such as specific types of proteins. Biological entities come in different kinds and create extremely complex networks of relations. Two fundamental interactions or relationships are: a gene, specific segment of Deoxyribonucleic acid (DNA[18]), is transcribed by a biochemical process into messenger ribonucleic acid (RNA) (mRNA), which is in turn translated into a protein. The transcription and translation processes are regulated via complex networks of proteins and RNAs such as signal transducers and members of binding complexes. There are approximately $3.0 * 10^{13}$ individual cells in a human body (Sender et al., 2016) and each of these cells contains the genetic code consisting of $3 * 10^9$ letters and encoding for 30,000 genes. Cells have specialized functions and execute them via activating specific combinations of genes. They are also interacting with each other. Moreover, humans are not exact copies and have a genetic makeup, which is highly similar between two individuals—this makes us human and members of one species—but shows an extremely wide variability. Currently, more than 900 million different human gene variations are identified in the human population.[19] This is just the beginning of a description of the complexity of biological systems. Most of this information is stored in public databases of which some of them contain cross-references to each other. In addition, organizations are continuously generating internal data about biological interactions and variations in order to find direct correlations between biological processes, diseases, and phenotypes. The next step is finding the agents that interact efficiently with the right disease modifying target (efficacious drug), which causes a minimal of unwanted adverse effects (safe drug). In an early phase of the inception of the semantic web and linked data, researchers started to store and distribute biological data in a semantic representation such as RDF. This process is still ongoing and at different speeds is accompanied with a sprawl of new *taxonomies, hierarchies, and ontologies* making it often even more complex to bring it to practical applications. Most importantly, this evolution hasn't fully proved its value to business users.

5.5.5 Use case scenario: semantics and linked data at work for knowledge discovery in early-stage drug research

One of the main objectives of ONTOFORCE is to create value for business users by bringing this scattered data together and to enable finding relevant relations or associations

---

18 It's a molecule that carries the genetic instructions used in the growth, development, functioning, and reproduction of all known living organisms and many viruses.

19 NCBI Allele Frequency Aggregator (ALFA): https://www.ncbi.nlm.nih.gov/snp/docs/gsr/alfa/ALFA_20201027095038/

in a user-friendly way. In DISQOVER, *semantic relations* between instances such as a disease, a publication and a gene are included when available. In addition, the labels, abbreviations, alternative labels, or symbols of all instances in the graph can be used for semantic search expansion (see Figure 5.9). This allows to perform a basic string search, which is easily extendable by including other labels or synonyms retrieved from different data sources.

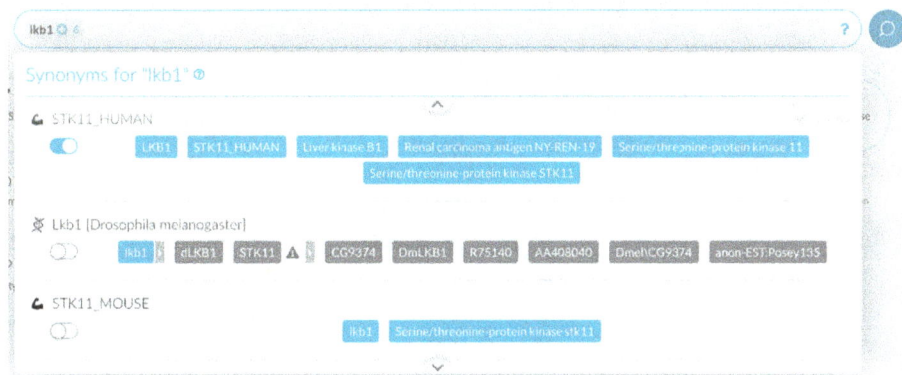

Figure 5.9: Semantic search expansion in DISQOVER. The human protein STK11 is selected together with all known synonyms and labels.

When selecting the "protein" data type—not the "exact hits" box—a new dashboard displays an overview of all matching hits (both semantic as search string hits) (see Figure 5.10). This dashboard allows to use filtering facets (based on a combination of materialized *OWL rules* on top of *elastic search indexes*) that can be used for visualization in combination with more specific visualization or analysis facets.

The dashboard panes and facet contain different subsets of data—here and, for example, in the SPARQL pseudocode to retrieve a few subsets (see Figure 5.10).

When clicking on a specific hit, you'll retrieve the detailed properties of a single concept combined with the semantic links to other concepts grouped per data type (see Figure 5.11). The data provenance can be visualized to ascertain the origin of every property. In addition, the different public or internal URIs of that concept can be visualized. This is useful to explore the data when preparing for advanced and automated searching via scripting or preparing new data sources to integrate. By clicking on the circle "Disease," one gets guided via the semantic links (by using the RDF descriptions together with RDFS and/or OWL rulings) to the 11 diseases associated to this protein. By adding a critical mass of relevant public data sources and combining this with internal data, DISQOVER can become an environment for knowledge discovery leading to improved, evidence-driven hypothesis generation and decision making.

```
1   Variables:
2   $searchStringPreviousStep1 = "lkb1" OR "STK11_HUMAN" OR ... (See Fig below)
3   $selectedDatatypePreviousStep2 = :Protein (Step not shown)
4   $selectedConceptInFacetOrganism = :HomoSapiens (Step executed on the active dashboard)
5
6   Result list (See Fig 5, right pane):
7   GET Properties from Concepts
8   ?URIConcept, ?prefLabelConcept, ?descriptionConcept,
9   [?URIDatatype ?prefLabelDatatype], [?URIConceptClassConcept, ?prefLabelConceptClassConcept]
10  GET Provenance from Properties
11  ?URIDatasource , ?prefLabelDatasource
12  FILTER
13        ?prefLabelConcept OR ?alternativeLabelConcept MATCHES $searchStringPreviousStep1
14        AND  ?URIDatatype EQUALS  $selectedDatatypePreviousStep2
15        AND ?URIConceptClassConcept EQUALS $selectedConceptInFacetOrganism
16
17  Facet Counter (See Fig 5, middle pane):
18  COUNT ?URIConcept  from Concepts
19  FILTER
20        ?prefLabelConcept OR ?alternativeLabelConcept MATCHES $searchStringPreviousStep1
21        AND  ?URIDatatype EQUALS  $selectedDatatypePreviousStep2
22        AND ?URIConceptClassConcept EQUALS $selectedConceptInFacetOrganism
23
24  Facet Organism (See Fig 5, middel pane)
25  GET Properties from ConceptClassConcepts
26        ?URIConceptClassConcept , ?prefLabelConceptClassConcept
27  FILTER
28        ?prefLabelConcept OR ?alternativeLabelConcept MATCHES $searchStringPreviousStep1
29        AND  ?URIDatatype EQUALS  $selectedDatatypePreviousStep2
30        AND ?URIConceptClass EQUALS :Organism
```

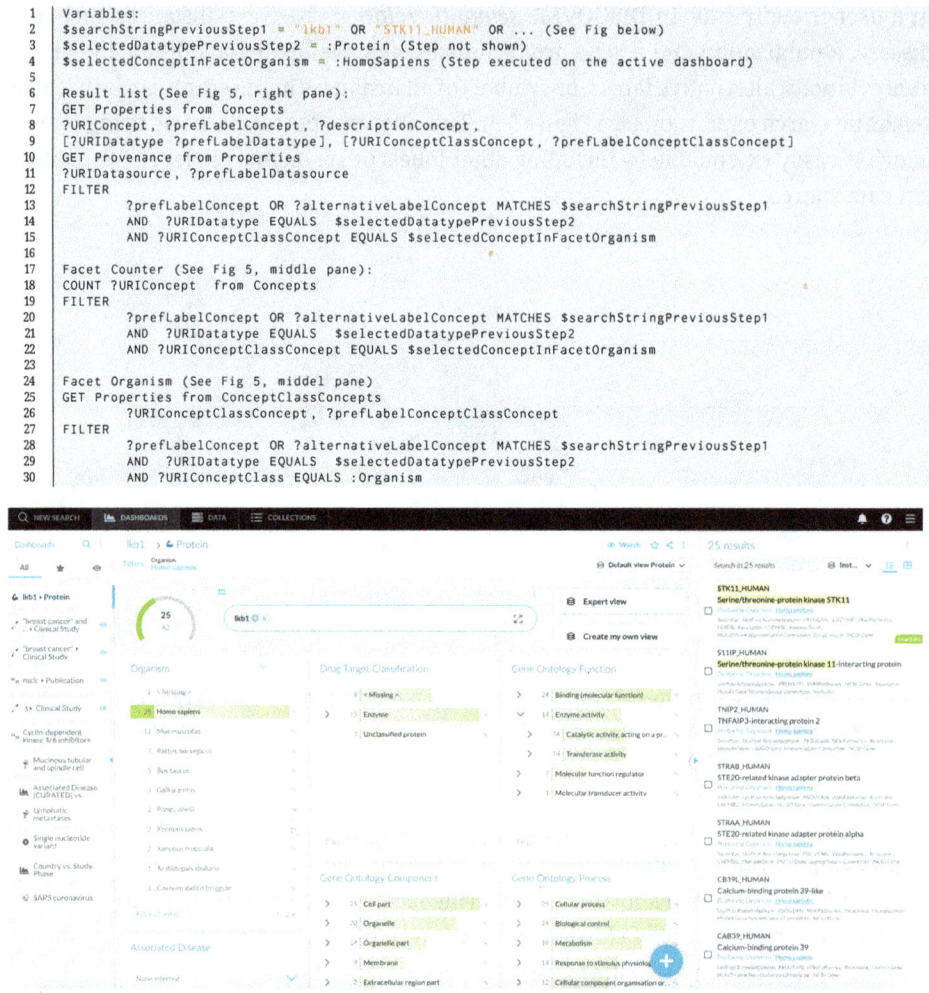

Figure 5.10: Dashboard displaying a protein-centric search. The left pane shows a history of previous searches, the middle pane is reserved for filtering and visualization facets, and the right pane shows a list of semantic hits on top.

5.5.6 Application of early-drug research knowledge for translational research use case

Under the hood, a lot is done via *DL reasoning* to calculate state changes in the *knowledge graph*. The above example of the *STK11 protein* is elaborated on as a possible treatment for kidney cancer by introducing the *semantic description of state changes*. The introduction of background knowledge and target description also uses examples from known kidney cancer treatments. Figure 5.12 shows the semantic description of a *STK11 protein therapy* in the domain of kidney cancer.

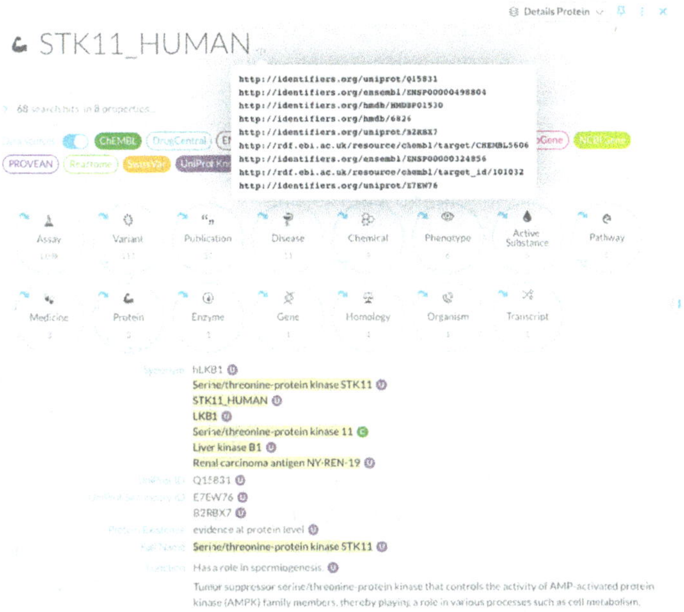

Figure 5.11: Detailed view of a single protein concept.

```
1   PREFIX math: <http://www.w3.org/2000/10/swap/math#>
2   PREFIX xsd: <http://www.w3.org/2001/XMLSchema#>
3   PREFIX DQV: <http://quirkdirk.org/eye/reasoning/DQV/DQV-schema#>
4   PREFIX action: <http://quirkdirk.org/eye/reasoning/DQV/action#>
5   PREFIX sct: <http://snomed.info/id/>
6   PREFIX therapy: <http://quirkdirk.org/eye/reasoning/DQV/therapy#>
7   PREFIX care: <http://quirkdirk.org/eye/reasoning/DQV/care#>
8
9   {care:Kidney_cancer                                    #Map
10    DQV:description (
11    {?patient care:tumor_size ?size.
12     ?patient care:metastasis_risk ?risk.}              #From
13    {?patient DQV:therapy therapy:Serine/Threonine-protein-Kinase11.}
14    {?patient care:tumor_size ?new_size.
15     ?patient care:metastasis_risk ?new_risk.}          #To
16     action:Serine/Threonine-protein-Kinase-11          #Action
17     "20D/ML"^^xsd:desiredHourlyRate                     #Dose 60ML/H
18     14147                                              #Cost
19     0.9                                                #Belief
20     0.4                                                #Comfort
21    )} <=
22   {?patient care:diagnosis sct:254915003.              #Kidney cancer
23    ?patient   care:tnm_t ?t_value .                    #Tumor-Metastasis
24    ?t_value math:greaterThan 2 .
25    ?patient care:tumor_size ?size.
26    (?size 0.7) math:product ?new_size.
27    ?patient   care:metastasis_risk ?risk .
28    (?risk 0.5) math:product ?new_risk. }.              #Condition graph
```

Figure 5.12: Sample state change representation of "hoax" kidney cancer therapy.

In general, this sample description indicates that by taking the action of administering a certain STK11 protein dose, the size of the tumor is expected to shrink to 70 % of its original size. Line 9 indicates the specialized domain of the action, in this case, it is *care:Kidney_cancer*. We use "Map" to indicate domain information of each specific medical domain, mimicking a map that provides different paths. We use the concept "Map" to separate different domain knowledge, so that *domain experts* can focus on *creating rules in their own expertise.*

The graph stated in the *from* section (lines 11–12) indicates the state before the action is applied. It will be retracted once the action is started. In this case, the current *tumor size* and *metastasis risk* of the patient will then be retracted. Line 13 contains the transition state. It indicates that during the state transition, the patient is receiving some dose of the STK11 protein. The graph stated in the transition section will be asserted when the action is started. It will be retracted when the action of the administration of some dose of STK11 protein is finished. Lines 14–15 indicate the target state. When the state transition is finished, new values of tumor size (*?newsize*), and metastasis risk (*?newrisk*) will be asserted. The new size and new risk are reflecting the expectation of the treatment. In reality, the new size and new risk might differ as the emphconfidence of reaching the target is indicated by the parameter *belief.* Line 16 indicates the action to be taken in the state transition is receiving *some dose of the STK11 protein.* Lines 17–20 indicate the weights of the state transition. Dose indicates the action will be administering 60 ml/hour at a rate of 20 drops/min. *Cost* indicates the cost of the action is 14147 Euros. It is believed 90 % chance the target can be reached, and the comfort level of this action is 40 %.

Both *Belief* and *Comfort* are initially subjective values based on known knowledge base inputs of physicians. Nevertheless, they can also be based on existing studies harvested from other linked open data sets, as well as being updated following the outcome of evaluating the actual outcome of the state transition. Lines 22–28 form the section of *Condition.* Lines 22–24 indicate the premise of carrying this action, that is, a patient is diagnosed with kidney cancer (line 22), and the tumor is reaching more than two layers of the kidney (lines 23–24). The new size of tumor is calculated in lines 25–26; it will be 70 % of the original size. The new risk of metastasis is calculated in lines 27–28; it will be 50 % of the original risk. The current calculations of the target values are simplified for demonstration purposes.

In clinical practice, the new metastasis risk can be calculated based on the detailed status of a patient, even including factors such as genetic variants or extrapolating the results via external machine learning services. In this way, the newly acquired knowledge is updated in the DISQOVER knowledge graph, taking into account all previous provenance information, that is, "one learns from the past." As such, earlier less effective treatments can be taken into account, and thus "old" patients could proactively get better treatment.

Bibliography

Baader F., Calvanese D., McGuinness D., Nardi D., and Patel-Schneider P. The Description Logic Handbook, 2nd ed., 2007.

Baader F., Horrocks I., Lutz C., and Sattler U. An Introduction to Description Logic, Cambridge University Press, 2017.

Chein M. and Mugnier M.-L. Graph-based knowledge representation, Computational Foundations of Conceptual Graphs, Springer, 2009.

Gruber T. R. A Translation Approach to Portable Ontology Specifications. Knowledge Acquisition, 5:199–200, 1993.

Krötzsch M., Simančík F., and Horrocks I. A Description Logic Primer. CoRR, abs/1201.4089, 2012.

Leclère M., Mugnier M.-L., and Rousset M.-C. Raisonner avec des ontologies: logiques de descritpion et graphes conceptuels in Panorama de l'intelligence Artificielle Vol. 1, Chapter 5, 2014.

Sender R., Fuchs S., Milo R. Revised Estimates for the Number of Human and Bacteria Cells in the Body. PLoS Biol. 14(8):e1002533, Aug 19 2016. doi: https://doi.org/10.1371/journal.pbio.1002533. PMID: 27541692; PMCID: PMC4991899.

Wilkinson MD, Dumontier M. et al.The FAIR Guiding Principles for scientific data management and stewardship. Sci Data 3:160018, Mar 15 2016, doi: https://doi.org/10.1038/sdata.2016.18. Erratum in: Sci Data 6(1):6, Mar 19 2019. PMID: 26978244; PMCID: PMC4792175.

Aleksandra Pižurica

6 Probabilistic reasoning: When the environment is uncertain

6.1 Why is probabilistic reasoning important within the broader AI domain?

In the real world, agents as well as people must often deal with *uncertainty* due to partial observability, lack of complete domain knowledge, nondeterministic aspects of the environment, or various adversaries. Logical reasoning provides us with sound methods and tools to organize knowledge, perform inference, and construct plans but we need other mechanisms, too, to enable efficient decision making and acting under uncertainty. For example, an agent navigating through a partially observable environment and relying on its noisy sensor readings may never know for sure at which exact location it is and even less so in which state it will lend after a sequence of actions.

Problem solving and purely logical reasoning approaches can deal with some levels of uncertainty by maintaining the set of all possible states that the agent might be in—so-called *belief state*—and generating a contingency plan for every such possibility. This approach can work in scenarios with relatively few possible random outcomes of the actions and/or relatively few possible explanations for sensor readings at each stage. In reality, however, such a belief state tends to grow rapidly while incorporating many unlikely instances. Similarly, the corresponding contingent plan becomes prohibitively complex to support reasonably the deliberation process.

We thus need a reasoning mechanism that supports decision making in uncertain and complex scenarios. This also means being able to rank in some consistent manner the worthiness of actions or plans even when none of them is guaranteed to succeed. Suppose the goal is to be on time for an important meeting in a different part of the city while leaving the office as late as possible. Plan A is to drive along a ring around the city, Plan B take a shorter road through the small streets in the city center, and Plan C to go by bicycle. Plan A is likely to fail if there is a road accident or a lane is closed for road works. For Plan B, a blocked street and detour through one-way streets can be detrimental while increased intensity of rain or pants caught in bike chain can be devastating for Plan C. A sudden disturbance in the ionosphere caused by a geomagnetic storm can invalidate your GPS navigation system and get you off the optimal route. None of these obstacles can be excluded—no matter how unlikely they are—and many others may arise that we could not think of or did not want to bother listing explicitly. Still, we need a consistent way to rank the merits of the various plans in uncertain situations. This leads us to the concept of *probabilistic reasoning*, a. k. a. *reasoning under uncertainty*.

https://doi.org/10.1515/9783111426143-006

The importance of probabilistic reasoning is often illustrated with diagnostic tasks. Making medical diagnosis, determining the cause of a mechanical failure, or a given effect in any other context involves almost always uncertainty. Think, for example, of diagnosing the cause of chronic fatigue. The list of possible causes is almost infinite, ranging from lack of good sleep, not enough activity or unbalanced diet to anemia or even heart problems, cancer, or covid infection, just to name a few. Furthermore, in some cases one has to make a diagnosis while lacking complete knowledge of the domain (theoretical ignorance) and/or being unable to perform all the relevant tests due to various practical reasons and limitations (practical ignorance).

It is therefore often said, after Russel and Norvig (2021), that probabilistic reasoning succeeds where purely logical reasoning fails due to our "laziness" (avoiding exhaustive lists of all possible causes or consequences) and ignorance (both theoretical and practical). Figure 6.1 highlights this concept.

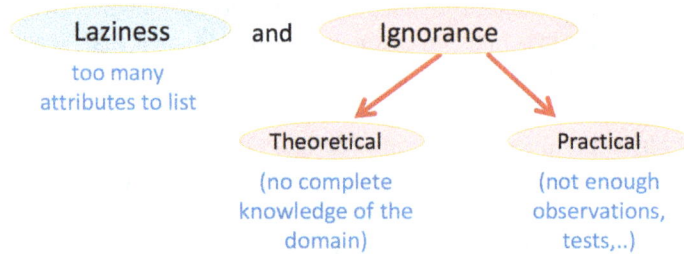

Figure 6.1: Sources of uncertainty that trouble purely logical and motivate probabilistic reasoning.

In essence, probabilistic reasoning provides us with means to evaluate *degrees of belief* for the success of different outcomes (e. g., from a sequence of actions). This way, it also constitutes a crucial component in designing *rational agents*. Modern artificial intelligence (AI) often qualifies intelligence as *acting rationally*, that is, choosing the action that *maximizes the expected utility* given the available evidence. A decision-theoretic agent does so by combining the probability theory (to evaluate the likelihood of each outcome) and the utility theory (to express preferences regarding different outcomes), as it is highlighted in Figure 6.2. Hence, probabilistic reasoning forms also a core part of decision systems, like Markov decision processes (MDP) and their natural extension to reinforcement learning (RL).

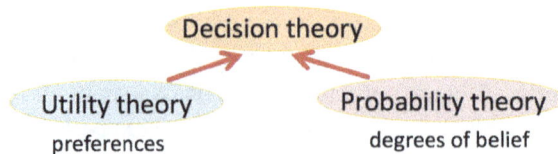

Figure 6.2: Decision-theoretic agents combine probability theory and utility theory.

6.2 What category of problems does probabilistic reasoning solve?

The fundamental problems addressed by probabilistic reasoning often boil down to answering questions of the type *what is the probability of an event* (or a combination of events, or a sequence of states) *given some evidence.* Technically, we say that probabilistic reasoning finds the probability of a *query proposition* (like "I'm having flu") given some evidence (e. g., symptoms like cough and fever). Similarly, we might be interested in the probability that "it will rain tomorrow and the picnic won't be canceled" given some meteorological data over the past days and some evidence about my friends' habits. Another type of problem that we encounter in probabilistic reasoning is that of finding an explanation or a sequence of most likely states of a given system, given a sequence of some observed features. In this context, the terms evidence, observations, observable variables, measurements, data, and (observed) features have the same role in the reasoning process and are thus often used interchangeably and typically denoted generically as *evidence* in technical descriptions. The problems that are being addressed by probabilistic reasoning can be categorized in different ways:

- *Static* vs. *dynamic* environment. A task environment is said to be static when it doesn't change while the agent is deliberating. It means that the states of all the entities of interest for a given problem remain fixed (each random variable has a fixed value) in the time span in which the agent performs inference and decides on the action. For example, the fact that the alarm went off (or not) does not change while inferring the probability of burglary, and the state "there was burglary" or "not" also remains fixed all this time. In a dynamic environment, the states and the observations keep constantly changing while the agent is deciding about what to do next. Think of self-driving cars: the car itself and other cars keep moving, and overall, the traffic situation is changing while the driving algorithm decides on what to do next. To deal with dynamic problems, we need reasoning over time, where we account for probabilities of various transitions between the states of the environment in subsequent time intervals. This type of problems will be addressed later in this chapter.
- *Causal* vs. *noncausal* problems. Some models assume "causal" dependencies between variables, the influence of a variable over another one, represented in a graphical model by directed links. With directed graphical models, we are describing the dependencies of the type "parent-child" where one influences the other in a particular direction, for example, having flu causes with some probability headache, so *Flu* has a causal influence on *Headache* and this is expressed by a directed link between the two nodes in the model. Bayesian networks are often used to model this type of problem. In noncausal models, the interdependencies among the involved random variables are not directional, but are characterized symmetrically as neighboring relations. For example, in a digital photo, neighboring

pixel intensities are statistically dependent. If pixel values are modeled as random variables, we can express their statistical dependencies with an undirected probabilistic graphical model, like Markov random field (MRF) or conditional random field (CRF), which are covered in the subsequent sections.

- *Discrete* vs. *continuous* (or *hybrid*). The involved random variables can either be discrete or continuous. A discrete random variable can take only a finite number of distinct values like being *true* or *false* or the number of free seats in a bus, while a continuous random variable follows some continuous distribution (e. g., current speed, body temperature, average price of a house, etc.) In some cases, we have a mix of continuous and discrete random variables, which are then modeled by a hybrid probabilistic graphical model (e. g., a hybrid Bayesian network). Furthermore, if we are dealing with dynamic environments, the discrete/continuous characterization can also refer to how the time is handled: is it divided in some discrete steps or considered as a continuous variable?

In dynamic scenarios (*probabilistic reasoning over time*), the problems we address can be grouped in the following categories:

(i) Estimating the probability of a given state in a particular time instant, given the evidence available *before* (prediction) or *up to* (filtering) or *beyond* that time instant (smoothing). Say we don't know if the concentration of a given air pollutant, like PM2.5, in our city is above the standard but we can try to infer it based on some observations (like dense fog).
    - Prediction makes the inference a step ahead. (e. g., "Will the concentration of PM2.5 exceed the threshold *tomorrow* based on the observations until today?")
    - Filtering makes the estimate for the present with all the evidence so far ("Is PM2.5 *today* above the threshold based on the evidence so far?").
    - Smoothing improves the estimate in the past based on the new evidence that arrives in the meantime ("Was PM2.5 above the threshold *yesterday* based on the evidence we received until now?").

(ii) Finding the most likely *sequence* of states given a sequence of observations—also called the *most likely explanation* (e. g., inferring for 10 days in a row whether PM2.5 was above the threshold or not based on the fog observations in those 10 days).

These concepts relating to reasoning in dynamic scenarios will be explained in more detail in the section "Probabilistic reasoning over time."

After introducing some basic background concepts that we need for solving any of these categories of the problems, we will first address probabilistic reasoning in a static setting. There, we will start from causal problems, which are well described by Bayesian networks and then we will turn to noncausal problems modeled by Markov random fields. In both of these, we will exemplify discrete and continuous or hybrid models. We will subsequently unify causal and noncausal models within a factor graph representation and we will introduce inference methods by belief propagation, Markov chain

Monte Carlo (MCMC) samplers and briefly comment on some other inference mechanisms. Then we will turn to probabilistic reasoning over time, where we will start from basic inference problems in a dynamic setting, and explain how they are solved by the belief propagation-type of methods. Then we will introduce the concepts of hidden Markov models, dynamic Bayesian networks, and approximate inference by particle filtering.

6.3 How probabilistic reasoning problems are solved?

A minimum knowledge of the probability theory is needed to understand the key concepts of this chapter. Therefore, the reader will find some useful reminders in the appendix of this chapter. These reminders cover:
1. random variables, conditional probabilities, independence, and the Bayes' theorem,
2. estimators used in probabilistic reasoning. Estimators are based on statistics, so a function of observed data used to estimate the value of unknown parameters. They are key in machine learning since those express one of the approaches to "learn" from data.

See 6.6.1 Appendix: Basic Concepts and 6.6.2 Appendix: maximum likelihood estimation (MLE) and Bayesian estimation.

In many cases, the information of interest is not changing while the agent is deliberating on the action to take or while it infers the cause of a given manifestation, true configuration of some phenomenon, or the likelihood of a particular event given the available evidence. In this case, we are in a static environment: we'll first show how we can model causal problems using Bayesian networks (Section 6.3.1), then we turn to MRF (Section 6.3.2) and CRF (Section 6.3.3) for noncausal problems. A unifying representation model using factor graphs will then be introduced in Section 6.3.4. After the modeling, Section 6.3.5 will cover a couple of techniques that are used to answer the inferencing questions. Finally, Section 6.3.6 briefly addresses the probabilistic reasoning over time, when we are in a dynamic environment with random variables that change over time.

6.3.1 Modeling causal problems with Bayesian networks

Often, we need to infer probability that some events arise given *causal* relationships among the involved random variables. Consider a somewhat simplified scenario from the Introduction to this chapter that we describe in the example below.

Example. *Late to meeting.*
Sharon has to leave for a meeting where she gives a demo. She can go either by car or by bicycle. By car, she will likely be on time unless there is huge traffic jam. It is even safer

to arrive timely by bicycle, unless there is sudden heavy rain in which case she will have to look for a shelter on the way or at least do something about her hair before entering the meeting. The chairman typically decides herself to wait a while until everyone is present, but not always. Sharon's boss might decide to give the demo instead of her even if she is on time, and almost for sure if Sharon is late.

A possible way to model this problem is to use the Bayesian network given in Figure 6.3 together with some reasonable prior and conditional probability tables (CPTs). All the involved random variables here are discrete and, moreover, Boolean. $P(C = true)$ models the probability to take the car, $P(J = true)$ models the probability of a traffic jam, $P(R = true)$ the probability of raining, while $P(L = true)$ models the probability that Sharon arrives late. Observe that random variables C, J, and R influence the value of L, while the opposite is not true. Hence, C, J, and R are *parents* of L, that is, they provide *causal support* for L. Similarly, L is the parent of W and D.

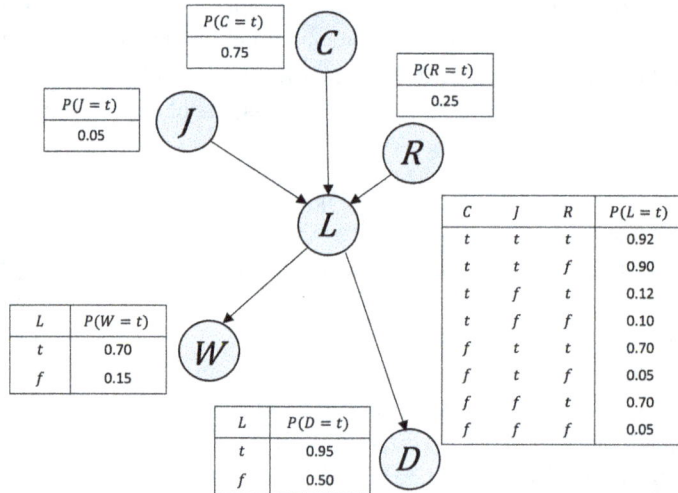

Figure 6.3: Bayesian network for the *late to meeting* problem. Random variable C denotes choice to go by car (if the value is true (t) or by bicycle if the value is false (f)). J denotes traffic jam, R heavy rain, L being late for the meeting, W that chairman suggests to wait, and D that the boss decides to give the demo.

In constructing a Bayesian network for a particular problem, we need to make some reasonable assumptions about the independence and conditional independence relationships between the different variables in the model, in order to simplify the probabilistic representations of the environment. For example, in our model we assume that traffic jam and rain are statistically independent, but we could devise a more complex model stating that rain might impact the probability of traffic jam.

In general, a Bayesian network is a *directed* and *acyclic* probabilistic graphical model, expressing causal, that is, *parent–child* relationships among the involved random variables. The joint probability of a Bayesian network $P(X_1 = x_1, \ldots, X_n = x_n)$ denoted for brevity by $P(x_1 \ldots x_n)$ is

$$P(x_1, \ldots, x_n) = \prod_{i=1}^{n} P(x_i \mid parents(X_i)) \qquad (6.1)$$

where $parents(X_i)$ denotes the values of the parents of the node X_i that appear in $x_1, \ldots, x_n$. Observe that a node is *conditionally independent of its nondescendants given its parents*. For example, in our network from Figure 6.3, D is conditionally independent of W given L, and D is also conditionally independent of J, C, and R given L.

If we wouldn't know anything about the conditional independences among the nodes, we could only express the full joint probability, which we can always rewrite using the *chain rule*: $P(x_1, \ldots, x_n) = P(x_n \mid x_{n-1}, \ldots, x_1) P(x_{n-1} \mid x_{n-2}, \ldots, x_1) \ldots P(x_2 \mid x_1) P(x_1) = \prod_{i=1}^{n} P(x_i \mid x_{i-1}, \ldots, x_1)$. We thus make use of the conditional independence assertions to reduce the full joint probability to the expression (6.1), where we in fact reduced the set $\{x_{i-1}, \ldots, x_1\}$ to $parents(X_i)$ only.

For the Bayesian network from Figure 6.3, we have that

$$P(j, c, r, l, w, d) = P(j)P(c)P(r)P(l \mid c, j, r)P(w \mid l)P(d \mid l) \qquad (6.2)$$

Observe that when dealing with Boolean random variables, by convention we denote $P(J = true) = P(J = j) = P(j)$ and $P(J = false) = P(J = \neg j) = P(\neg j)$. Now if we are, for example, interested in the probability that Sharon went by bicycle, and that she wasn't late while there was heavy traffic jam and no rain, and that chairman didn't wait for anyone and the boss gave the demo, we can calculate it easily as follows:

$$P(j, \neg c, \neg r, \neg l, \neg w, d) = P(j)P(\neg c)P(\neg r)P(\neg l \mid \neg c, j, \neg r)P(\neg w \mid \neg l)P(d \mid \neg l)$$
$$= 0.05 \times 0.25 \times 0.75 \times 0.95 \times 0.85 \times 0.5 = 0.0038 \qquad (6.3)$$

We can also make *diagnostic* inference, for example, the probability that Sharon went by car if the boss gave the demo and chairman didn't wait with opening the meeting, while it didn't rain and there was huge traffic jam: $P(c \mid j, \neg r, \neg w, d)$. Such probabilities can be calculated using the basic rules of probability and the given CPTs. However, for large Bayesian networks, we will need to use some efficient inference mechanism, like belief propagation that we address later, in Section 6.3.5.1.

We were dealing so far with the case where all the random variables were *discrete* and, in particular, Boolean. Often, we have a mix of discrete and *continuous* random variables, and we are then modeling the problem with a *hybrid* Bayesian network. Let us modify our example as follows: Sharon goes on foot and she can take or not take an umbrella, which is described by a Boolean random variable U. Instead of simply considering rain or not, we now model the intensity of rain by a continuous random variable R. Similarly, we replace the event of being late or not by a continuous random

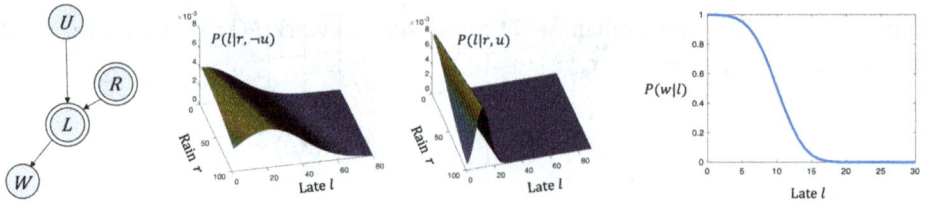

Figure 6.4: Left: A hybrid Bayesian network. Double circles denote continuous random variables. Middle: A linear Gaussian model for the probability distribution of L given its parents; Right: A probit model ($\mu = 20$; $\sigma = 3$) for the probability of W being true given the value of L.

variable, which describes how much Sharon is late. The decision to wait or not at the beginning of the meeting naturally remains Boolean. This new situation is depicted in Figure 6.4, where we use double circle to denote a continuous random variable. Now we have to specify two new types of the probability distributions, for a

- continuous random variable with a mix of continuous and discrete parents (node L), and a
- discrete random variable with continuous parents (node W).

The former one is typically described by a linear Gaussian model: $P(l|r, u) = \mathcal{N}(l; a_t r + b_t, \sigma_t^2)$; $P(l|r, \neg u) = \mathcal{N}(l; a_f r + b_f, \sigma_f^2)$, as in Figure 6.4. Observe that the mean value changes linearly with the continuous parent value r and the parameters of this linear function as well as the spread of the distribution depend on the value of the discrete parent. The probability distribution of a discrete child given a continuous parent is in fact a soft-threshold function: in our example, let's say the chairman waits with high probability if the participants are late a couple of minutes and almost never if they are late more than 15 min, with some transition in between.

This is well modeled by a probit distribution: $P(w|Late = l) = \Phi((\mu - l)/\sigma) = 1 - \Phi((l - \mu)/\sigma)$, where $\Phi(x) = \int_{-\infty}^{x} \mathcal{N}(x; 0, 1) dx$ (see Figure 6.4, right) or alternatively by inverse logit models (using the inverse of a sigmoid function).

6.3.2 Modeling noncausal problems with Markov random fields

The need for modeling *noncausal* statistical dependencies among random variables as well as allowing for *cyclic* dependencies arises in many domains, perhaps most notably in computer vision but also in sensor networks, gene expression analysis, and in general matrix completion problems. In all these types of problems, a probabilistic graphical model known as *Markov random field* (MRF) provides a convenient way to model the global statistical distribution by encoding prior knowledge about local interactions among the network nodes, which represent the random variables (see Fig. 6.5).

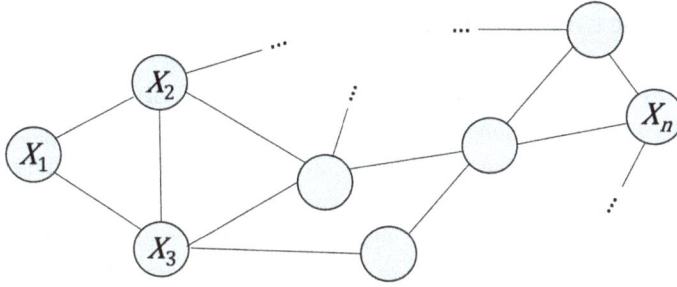

Figure 6.5: A general structure of a noncausal probabilistic reasoning problem modeled by MRFs. The involved random variables can also be continuous (we call it then a continuous MRF model).

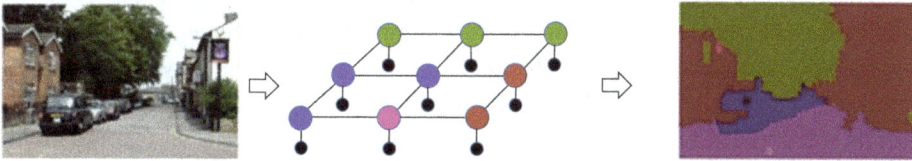

Figure 6.6: An image labeling problem, where a class label (colored circle—hidden nodes x_s) is assigned to each pixel (black circle—observable nodes y_s).

Take as an example the image labeling problem illustrated in Figure 6.6, where the goal is to segment the input image by assigning a class label $X_s = x_s$ (such as *road, cars, buildings, trees*, etc.) to each image pixel $s \in \{1, \ldots, n\}$. A classification approach that assigns a class label to each pixel independently based on the local observation y_s alone is likely to fail due to noise and because pixel color and textural appearances in a small window can be very similar for different classes. We can obtain better results by making use of our *prior knowledge* about typical image scenes, for example, that we do not expect a piece of a sky in the middle of a road.

But how are we going to use this prior knowledge to better solve our problem? One common approach is maximum a posteriori (MAP) estimation. The MAP estimate of the unknown labels $\mathbf{x} = [x_1, \ldots, x_n]$ given the available observations $\mathbf{y} = [y_1, \ldots, y_n]$ for the corresponding nodes is

$$\hat{\mathbf{x}}_{MAP} = \arg\max_{\mathbf{x} \in \mathcal{X}} P_{\mathbf{X}|\mathbf{Y}}(\mathbf{x}|\mathbf{y}) \tag{6.4}$$

where $\mathcal{X}$ denotes the set of all possible realizations $\mathbf{x}$. Using the Bayes' rule, we can rewrite the problem above in terms of the product of two factors: the likelihood and the prior distribution. Hence, our MAP estimate becomes

$$\hat{\mathbf{x}}_{MAP} = \arg\max_{\mathbf{x} \in \mathcal{X}} \underbrace{P_{\mathbf{X}|\mathbf{Y}}(\mathbf{x}|\mathbf{y})}_{\text{posterior}} = \arg\max_{\mathbf{x} \in \mathcal{X}} \underbrace{P_{\mathbf{Y}|\mathbf{X}}(\mathbf{y}|\mathbf{x})}_{\text{likelihood}} \underbrace{P_{\mathbf{X}}(\mathbf{x})}_{\text{prior}} \tag{6.5}$$

Specifying the likelihood distribution $p_{Y|X}(y|x)$, is where we "learn" from the observed data starting typically from some parametrized model. The parameters are estimated using the "Maximum likelihood estimation" method (cf. the Appendix for details) and in practice often iteratively using the expectation-maximization algorithm (see, for more details, "Solving the image labeling problem" part for the interested reader).

In the rest of this section, our focus is on specifying the prior $P_X(x)$. So, a prior model function will be shaped to encode our prior knowledge. For example, if we want to express that neighboring pixels are likely to belong to the same class we should assign a higher probability to the cases where neighboring pixel labels are of the same type than to the cases where isolated labels of one type appear in the middle of a differently labeled area.

This type of a priori knowledge is being very efficiently encoded in a Markov random field (MRF), or Markov network. It is an undirected graph model, which expresses a (generally complicated) joint probability distribution of all the network nodes as a composition of simple local terms, which describe the "interactions" within some local groups of the nodes. The joint probability of the model $P_X(x)$ becomes a "computable" function, being decomposed over the local functions with some (hyper)parameters.

In our example above, the nodes of a MRF were image pixels, thus arranged on a grid, but this is not necessarily the case (e. g., the network nodes can correspond to vertices of a deformable mesh or to moving body parts in tracking human motion). Regardless of this underlying physical arrangement of the entities that we model with our Markov networks, we simply denote each node by a one-dimensional index s, which belongs to a set of all indices $S = \{1, \ldots, n\}$. Depending on the problem description, we define some reasonable notion of a *neighborhood* system δ, which defines for each node s the set of its *neighbors* $\delta(s)$ (i. e., those that are mutually dependant or, as we say, nodes that "interact locally"). In imaging problems, $\delta(s)$ typically comprises four nearest neighbors of s (the so-called first-order neighborhood) or eight nearest neighbors of (the second-order neighborhood).

Now we can define a Markov random field more formally as a random field[1] where the probability of a label at node s given all other nodes reduces to the probability conditioned on the neighbors of s only:

$$P(X_s = x_s | \mathbf{X}_{S\backslash s} = \mathbf{x}_{S\backslash s}) = P(X_s = x_s | \mathbf{X}_{\delta(s)} = \mathbf{x}_{\delta(s)}) \tag{6.6}$$

$S \backslash s$ is the set of all nodes except s. The main strength of Markov random fields and the reason for their popularity is the ability to elegantly *express the global probability of a given configuration of nodes in terms of their local interactions*. The probability $P(\mathbf{x})$ is decomposed into contributions of simple terms that depend on relatively few nodes,

---

1 A random field means that any of its realizations is possible.

which form the so-called *cliques* (subgroups of nodes within a local neighborhood). Concretely, the probability of a MRF takes the form of a Gibbs distribution

$$P_{\mathbf{X}}(\mathbf{x}) = \frac{1}{Z} \exp\left(-H(\mathbf{x})/T\right) \tag{6.7}$$

where the *energy* $H(\mathbf{x})$ is the sum of the clique potentials. Z is the normalizing constant and T a constant called temperature (by default equal to 1), which controls the peaking of the distribution.

MRF cliques and clique potentials. Breaking down the global probability into contributions of *clique* potentials is central to MRFs. Formally, a clique c is a set of nodes that are all neighbors of each other. Figure 6.7 illustrates the cliques for the first- and second-order neighborhood in 2D. Observe that in the first-order neighborhood, the possible clique types are only a single node and two horizontally or vertically adjacent nodes; in a higher-order neighborhood we have more versatile cliques. To encode *a priori* knowledge about the entity, we are modeling (e. g., local spatial continuity in natural images or certain interactions among moving body parts), we assign an appropriate potential V_c to each clique c, which depends on the clique type and on the values of the nodes within that clique. The location of the clique is irrelevant for the potential unless the model is inhomogeneous.

The *Hammersley–Clifford theorem* establishes the equivalence between (6.6) and (6.7) with $H(\mathbf{x}) = \sum_c V_c(\mathbf{x})$, that is, the fact that the probability of a Markov random field is a Gibbs distribution where the energy is a sum of the clique potentials. For details see, e. g., Li (2009).

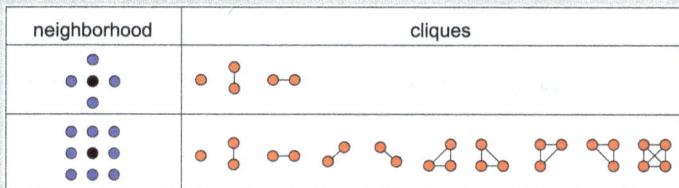

neighborhood	cliques

Figure 6.7: Two common neighborhood systems and the corresponding types of cliques.

To encode a priori knowledge that neighboring pixels are likely to belong to the same class, we shall assign a positive potential to the cliques, which contain different labels and a negative potential to the cliques that consist of equal labels. Most often, MRFs are formulated with only single- and pairwise cliques, having thus the energy of the form

$$H(\mathbf{x}) = \sum_s V_1(x_s) + \sum_{\langle s,t \rangle} V_2(x_s, x_t) \tag{6.8}$$

where V_1 and V_2 are the single-site and pair-site potentials, respectively, and $\langle s, t \rangle$ is a clique composed of two nodes s and t. Even with larger neighborhoods, the potentials of higher-order cliques are typically set to zero to limit the amount of parameters and the overall complexity. In some domains, like texture analysis, better model expressiveness with higher-order cliques justifies the increased complexity.

The celebrated *Ising* model has the energy as in (6.8), assumes the first-order neighborhood, binary labels $x_s \in \{-1, +1\}$, and the potentials $V_1(x_s) = \alpha x_s$ and $V_2(x_s, x_t) =$

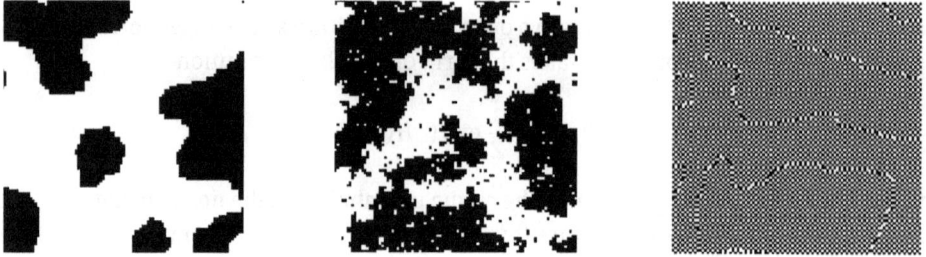

Figure 6.8: Random samples from the Ising MRF model for $\alpha = 0$ and $\beta = -2$ (left); $\beta = -0.5$ (middle) and $\beta = 2$ (right).

$\beta x_s x_t$. The coefficient α governs the preference to one type of the labels and β the strength of the interactions. Figure 6.8 illustrates samples from this random field for different values of β. Observe strong clustering of the labels of the same type when β is large negative and their strong repulsion when β takes large positive value.

Generalization to multiple labels. With more than two possible labels $x_s \in \{1 \ldots L\}$, $L \geq 2$, this model generalizes to $V_1(x_s) = \phi(x_s) = a_{x_s}$ (to enable different level of preference for each label) and the interaction potential

$$V_2(x_s, x_t) = \psi(x_s, x_t) = \begin{cases} -\gamma & \text{if } x_s = x_t \\ \gamma & \text{if } x_s \neq x_t \end{cases} \tag{6.9}$$

An extension where different γ's are defined for different orientations of the cliques results in an *anisotropic model*.

Solving the image labeling problem. We now show how to solve concretely the image labeling problem in Figure 6.6 with the MAP-MRF approach. Let $\hat{\mathbf{x}}_{MAP}$ be the MAP estimate of the class labels, and $\mathbf{y}$ the observed image. Recall that

$$\hat{\mathbf{x}}_{MAP} = \arg\max_{\mathbf{x} \in \mathcal{X}} P_{\mathbf{X}|\mathbf{Y}}(\mathbf{x}|\mathbf{y}) = \arg\max_{\mathbf{x} \in \mathcal{X}} p_{\mathbf{Y}|\mathbf{X}}(\mathbf{y}|\mathbf{x}) P_{\mathbf{X}}(\mathbf{x})$$

The prior model is given in (6.7) and we illustrate now the specification of the likelihood model (also called the *conditional model* or the *data distribution*). To make the model tractable, conditional independence is often assumed $p_{\mathbf{Y}|\mathbf{X}}(\mathbf{y}|\mathbf{x}) = \prod_s p(y_s|x_s)$. If the input image is affected by additive Gaussian noise, it is reasonable to assume that the observations y_s are conditionally Gaussian given x_s, that is, that $p(y_s|x_s)$ is a Gaussian distribution with the mean and standard deviation that depend on the class label x_s. It means that in our model noise-free values of all image pixels from a particular class have the same ideal value, and their noisy observations follow a Gaussian distribution with mean being that ideal noise-free component and the standard deviation depends on the particular class. The corresponding likelihood model is

$$p_{\mathbf{Y}|\mathbf{X}}(\mathbf{y}|\mathbf{x}) = \prod_{s=1}^{n} \frac{1}{\sqrt{2\pi\sigma_{x_s}^2}} \exp\left(-\frac{(y_s - \mu_{x_s})^2}{2\sigma_{x_s}^2}\right)$$

How do we determine the means and variances of each class? In a naive setting, if we assume that we have sufficient amount of labeled data for each class, we can determine the mean and the standard deviation for each class empirically. In a more realistic setting, we need to update the parameters in parallel to estimating the class labels and we would use an iterative algorithm like expectation-maximization (EM). With the likelihood model given above, we can compactly express the MAP estimation problem from (6.5) as

$$\hat{\mathbf{x}}_{MAP} = \arg\min_{\mathbf{x} \in \mathcal{X}} \sum_s \left(\frac{(y_s - \mu_{x_s})^2}{2\sigma_{x_s}^2} + \phi(x_s) \right) + \sum_{\langle s,t \rangle} \psi(x_s, x_t)$$

This problem is typically solved using graph cuts Boykov et al. (2001) or approximate inference algorithms that are covered in the subsequent sections.

6.3.3 Modeling noncausal problems with conditional random fields

The image labeling problems described above, as well as many other probabilistic inference problems can be modeled with a conditional random field (CRF) as well. Although the CRF model was first proposed in the context of segmenting and labeling data sequences (see Lafferty et al. (2001)) with applications in natural language processing, it has been widely adopted in computer vision, too, and in other domains including recommender systems.

While in the MRF-approach, we were modeling the prior distribution $P_{\mathbf{X}}(\mathbf{x})$ explicitly, along with the likelihood model $p_{\mathbf{Y}|\mathbf{X}}(\mathbf{y}|\mathbf{x})$, a CRF models directly the *posterior distribution* $P_{\mathbf{X}|\mathbf{Y}}(\mathbf{x}|\mathbf{y})$. Thus, we are only interested in the output structure *conditioned* on the input, hence the name *conditional* random field. We assume that a random field $\mathbf{X}$ obeys Markov property, when conditioned on the evidence variables $\mathbf{Y}$. By analogy with MRF, the joint probability distribution of a CRF is $P_{\mathbf{X}|\mathbf{Y}}(\mathbf{x}|\mathbf{y}) = (1/Z) \exp[-H(\mathbf{x}|\mathbf{y})]$. If we again restrict the clique size to single and pairwise only, the corresponding energy function is

$$H(\mathbf{x}|\mathbf{y}) = \sum_s V_1(x_s|\mathbf{y}) + \sum_{\langle s,t \rangle} V_2(x_s, x_t|\mathbf{y}) \tag{6.10}$$

Typically, the assumption is made that x_s depends only on the value of $\mathbf{y}$ at the corresponding location, and similarly that the pairwise potential between two labels is influenced only by the two observations at the corresponding locations. With this, the *unary potential* is typically denoted as $V_1(x_s|\mathbf{y}) = \phi(x_s, y_s)$ and referred to as the *association* potential (which measures the cost of assigning label x_s to the observation y_s), and the *pairwise potential* as $V_2(x_s, x_t|\mathbf{y}) = \psi(x_s, x_t, y_s, y_t)$, which is often called the *interaction potential* (as it encodes the interactions among the labels). Thus a standard form of a CRF is

$$P_{\mathbf{X}|\mathbf{Y}}(\mathbf{x}|\mathbf{y}) = \frac{1}{Z} \exp\left[-\left(\sum_s \underbrace{\phi(x_s, y_s)}_{\text{association}} + \sum_{\langle s,t \rangle} \underbrace{\psi(x_s, x_t, y_s, y_t)}_{\text{interaction}} \right) \right] \tag{6.11}$$

A common form for the interaction potential is

$$\psi(x_s, x_t, y_s, y_t) = \gamma \mu(x_s, x_t) \sigma(y_s, y_t) \tag{6.12}$$

where γ is a hyperparameter determining the weight of the pairwise term with respect to the unary term. $\mu(x_s, x_t)$ is a function that computes the agreement between the labels x_s and x_t, and is often referred to as the *compatibility function*. $\sigma(y_s, y_t)$ denotes similarity between the observations y_s and y_t. In computer vision problems, $\sigma(y_s, y_t)$ is often suppressed and the interaction potential reduced to a function of two labels only $\psi(x_s, x_t)$.

We now illustrate the application of CRF modeling in two problems: recommender systems and image inpainting. These examples enable the reader to understand better the meaning of the association and interaction terms and how are they actually computed in different practical scenarios. The case of the recommender system is illustrated in Figure 6.9. What is especially interesting here is that the observation y_s is a "user-item" pair, and the label x_s is the score assigned to that entry. Initially, only a relatively few scores are available (the matrix is sparse) and the goal is to complete the missing values for all user-item pairs. The case of image inpainting is illustrated in Figure 6.10. Now the observation y_s is a damaged (or completely missing) image patch and the label x_s is its estimated ideal counterpart. Interesting here is that the association and interaction potentials are now computed from the differences among the pixel values in image patches. Both examples are worked out in detail in the parts for the interested reader.

Example: *Recommender systems based on CRF*

Consider building a recommender system, which predicts scores that different users would give to some items (e. g., recent movies) based on relatively few available scores. Figure 6.9 (left) illustrates this problem: the available scores are sparse entries in a given relatively large matrix **M** and the goal is to complete the matrix by filling in the missing entries. This problem is known as geometric matrix completion. The observation y_s is now a particular (i, j)th pair (user, item) in the matrix and the label $x_s \in \{1, \dots L\}$ is the score $m_{i,j}$ assigned to that matrix entry $M_{i,j}$.

The initial estimates of the probabilities $P(M_{i,j} = x_s)$ can be obtained by assigning zeros to missing entries and applying a bilinear decoder with learnable weights, as explained by Nguyen et al. (2001), followed by the softmax function.[a] The unary potential (the cost of assigning x_s to y_s) can then be calculated as $\phi(x_s, y_s) = -\log(P(M_{i,j} = x_s))$. The agreement between labels $\mu(x_t, x_s)$ is typically defined such to favor smoothness unless the labels differ significantly: $\mu(x_s, x_t) = \min[(x_s - x_t)^2, \tau]$,

$$\mu(x_s, x_t) = \min[(x_s - x_t)^2, \tau]$$

where the predefined threshold τ determines the significant differences that should not be smoothed out. Observe that in the interaction function $\psi(x_s, y_s, y_s, y_t)$ defined in (6.12), the agreement between the labels x_s and x_t is modulated by the compatibility function $\sigma(y_s, y_t)$. This way, the differences between the labels can be penalized less strongly if the two (user, item) pairs y_s and y_t are more dissimilar, and vice versa. Practically, the compatibility function can be approximated by a separable function of users and items: it is easier to estimate separately the similarities between pairs of users and between pairs of items and use their product as $\sigma(y_s, y_t)$.

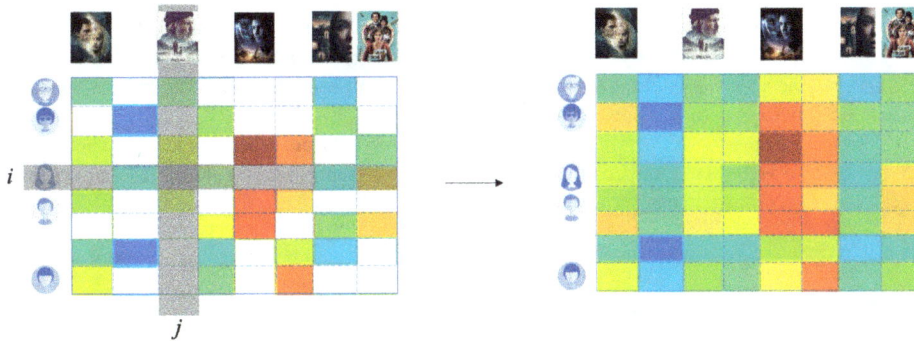

Figure 6.9: Example applications of CRFs: Matrix completion for recommender systems.

Now that the model is fully specified, the recommender system runs by maximizing the posterior probability in (6.11) applying some of the approximate inference algorithms that we discuss later. We come back to this example in Section 6.3.5.3 and show how this recommender system runs concretely with the mean field approximation approach.

a The softmax function $softmax(u) = \exp(u)/\sum_{l=1}^{L}\exp(l)$ normalizes the outputs to the range $[0,1]$.

Example: *Image inpainting based on CRF*.
Figure 6.10 illustrates another problem where CRFs have been successfully employed: patch-based image inpainting, where the goal is to fill in some missing (damaged) area in a digital image. Here, y_s is a damaged small image patch centered at spatial location s and x_s is a replacement patch taken from some undamaged image area, which should fit well with the image structure that can be seen in y_s (if any left undamaged) and in its surrounding. The replacement patches are placed such to partially overlap each other (see Figure 6.10) in order to ensure spatial continuity.

Now the unary potential $\phi(x_s, y_s)$ is some distance metric (e. g., the sum of the squared differences) between the pixel values of x_s and y_s in the undamaged part of y_s. This way, if y_s is completely damaged, $\phi(x_s, y_s)$ is zero and does not influence the selection of x_s. On the contrary, if only a small part of y_s is

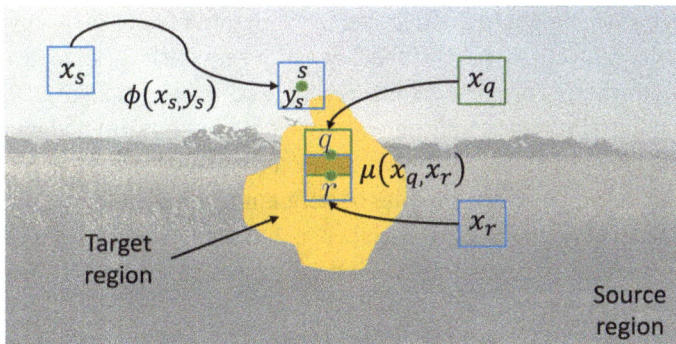

Figure 6.10: Example applications of CRFs: Image inpainting. The target region (shown in yellow) is the missing part in the image to be inpainted with image patches from the source region.

missing, $\phi(x_s, y_s)$ will enforce good agreement between the target and source patches. The interaction function is here reduced to the agreement between the labels $\psi(x_s, x_t)$, and expressed as a distance metric between the pixel values of the two adjacent patches in their overlap region. This ensures that the neighboring patches in the inpainted image mutually agree.

With this, all the components of the posterior probability in (6.11) are specified and the inpainted image is obtained as the MAP estimate $\mathbf{x} = \arg\max_{\mathbf{x}} P_{\mathbf{X}|\mathbf{Y}}(\mathbf{x}|\mathbf{y})$ with some approximate inference algorithm, like loopy belief propagation in Section 6.3.5.1.

6.3.4 A unifying representation: factor graphs

We reviewed Bayesian networks and MRFs. The question we address now is how to present them in a unified manner so that we can run the same inference algorithms in both cases. Recall that the joint probability of a Bayesian network is the product of conditional probabilities of a node given its parents: $P(\mathbf{x}) = \prod_i P(x_i | Parents(x_i))$. The joint probability of a MRF given in (6.7) can also be expressed in a product form $P(\mathbf{x}) = \frac{1}{Z} \prod_{c \in C} \psi_c(\mathbf{x})$ where $\psi_c(\mathbf{x}) = \exp(-V_c(\mathbf{x})/T)$. Thus in both cases, the joint probability distribution is a product of factors each of which is composed of a subset of the variables in a given model:

$$P(\mathbf{x}) = \frac{1}{Z} \prod_s f_s(\mathbf{x}) \tag{6.13}$$

Take as an example our Bayesian network from Figure 6.3. Its joint probability was given in (6.2) and a natural way to convert it to a factor graph is to assign a factor to each multiplier:

$$P(j, c, r, l, w, d) = \underbrace{P(j)}_{f_J} \underbrace{P(c)}_{f_C} \underbrace{P(r)}_{f_R} \underbrace{P(l|j, c, r)}_{f_L} \underbrace{P(w|l)}_{f_W} \underbrace{P(d|l)}_{f_D} \tag{6.14}$$

Such a factorization can be visualized using a *factor graph* representation as in Figure 6.11, where variable nodes are denoted by circles and factor nodes by squares. In general, a factor graph is a *bipartite graph* that expresses which variables are arguments of which local functions. For more details see, e. g., Kschischang et al. (2001). Note that in a bipartite graph the links exist only between two different types of nodes (and thus not between variable nodes and not between the factor nodes). Conventionally, a bipartite graph is drawn with two layers of nodes of different types as in the left of Figure 6.11, but another layout of the same factor graph, like the example in the right of Figure 6.11, can be more suitable for visualizing the connectivity and inference by message passing that we address in the next section.

Furthermore, observe that a factor graph representation is not unique. By this, we don't mean only visual arrangement of the same factorization as we did with two layouts in Figure 6.11, but specifying the factors themselves. There are many alternatives

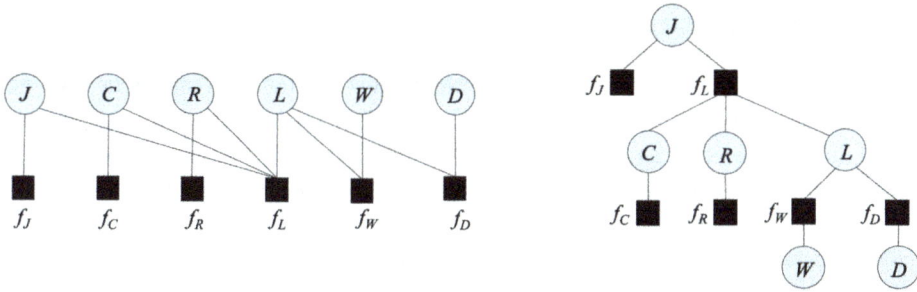

Figure 6.11: A factor graph representation, with the same factors and two equivalent layouts, for the Bayesian network *Late to Meeting* from Figure 6.3.

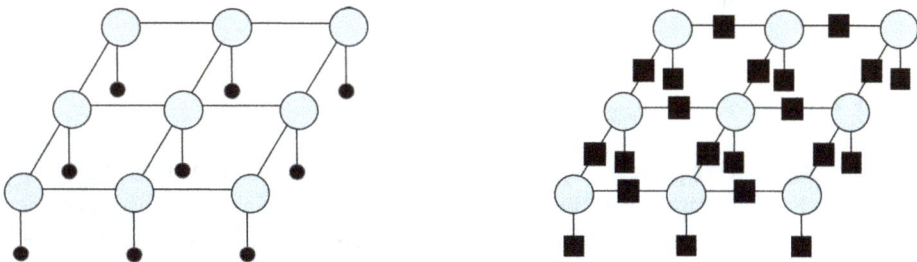

Figure 6.12: Converting a pairwise MRF with observable nodes to a factor graph.

to the choice we made in constructing the factor graph from our example above. For example, we could present the joint probability in (6.14) as a product of two factors only, f_1 and f_2, where $f_1(j, c, r, l) = P(j)P(c)P(r)P(l|j, c, r)$ and $f_2(d, l, w) = P(d|l)P(w|l)$. In that case, our factor graph representation would consist of the same six variable nodes as in Figure 6.11, but only two factor nodes: f_1 connected to the nodes J, C, R, and L, and f_2 connected to L, D, and W.

Figure 6.12 illustrates an MRF model with pairwise cliques and its factor graph. Note that the observable nodes (denoted by black circles) are replaced by equivalent factor graph functions of a single variable. The factor graph functions are here either pairwise functions $\psi(x_s, x_t)$ if they link two hidden nodes, or unary (single-node) functions $\phi(x_s, y_s)$ if they are attached to a single hidden node. Thus we represent the joint probability of the scene $\mathbf{x}$ and the observations $\mathbf{y}$ as

$$P_{\mathbf{X,Y}}(\mathbf{x}, \mathbf{y}) = \frac{1}{Z} \prod_s \phi(x_s, y_s) \prod_{\langle s,t \rangle} \psi(x_s, x_t) \tag{6.15}$$

If we are interested only in the conditional distribution of the labels given the observations (CRF formulation), we express in the same form as above the posterior distribution $P_{\mathbf{X|Y}}(\mathbf{x}|\mathbf{y})$, where ϕ and ψ are now $-\log(.)$ of those from (6.11). Observe how we model this way the image labeling problem illustrated in Figure 6.6, where x_s were the class labels and y_s the pixel intensities.

6.3.5 Approximate inference algorithms

The exact inference in probabilistic graphical models is often not feasible or even not possible (e. g., the partition function of a MRF model is in most cases an intractable function of the parameters). Hence, we need to resort to approximate inference. This is especially the case in models with cyclic dependencies (loops), hence always with MRFs and CRFs, but also with Bayesian networks in dynamic scenarios. We describe here three representative types of approximate inference algorithms. We start from belief propagation, which provides the *exact* inference in networks without loops and approximate inference in general graphical models. Then we turn to random sampling methods and variational inference, which impose numerical (sampling methods) and analytical (variational methods) approximations of the posterior probability distributions.

6.3.5.1 Belief propagation

Belief propagation is a powerful approach to solve complex inference problems by passing relatively simple "messages" between neighboring nodes of a probabilistic graphical model. This can be seen as a kind of distributed processing, where relatively simple local communications among the neighboring nodes work together to solve a desired query, which can relate to any node in the whole network or even to the joint probability of any particular values of the nodes.

Although the original belief propagation algorithm of Pearl (1988) has been derived as the exact inference algorithm for networks without loops, it is often applied as an approximate inference method in general graphical models which can contain loops (like MRFs). Running belief propagation algorithm on loopy graphs is known as *loopy belief propagation* (LBP). The algorithm became especially popular after it has been shown that turbo codes are an instance of LBP.

The message passing procedure is simple and is illustrated in Figure 6.13: a node sends a message to a neighboring node after it has received the messages from all its other neighbors. The message $m_{s,t}(x)$ from s to t has the meaning: *"I, node s think that*

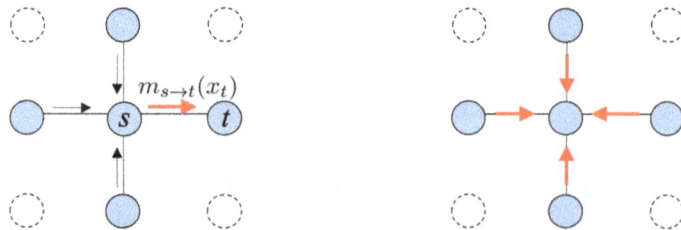

Figure 6.13: An illustration of loopy belief propagation. Left: a message from one node to another depends on the messages that the sending node received from its other neighbors. Right: each node multiplies the incoming messages from all the neighbors to calculate its belief.

you, node t, are that much likely to be in the state x." This message is based on the accumulated messages that the sending node has already received from all its other neighbors and also on its unary potential (how confident it is about its own state) and its compatibility with the receiving node.

Sum-product algorithm. Suppose the query is: *How probable is it that node n has value x_n?* Referring to our earlier examples, this can mean: "How probable is it that Sharon is late for the meeting given all we know?" or "What is the probability that the pixel (i, j) belongs to the class 'road'?" The task here is to infer the *marginal* probability of a node. The messages are then of the form:

$$m_{s \to t}(x_t) = \sum_{x_s} \left(\phi(x_s)\psi(x_s, x_t) \prod_{r \in \delta(s) \backslash t} m_{r \to s}(x_s) \right) \tag{6.16}$$

where $\phi(x_s)$ is the unary potential of the sending node (the fixed evidence variable y_s is omitted from $\phi(x_s, y_s)$ for compactness), and $\psi(x_s, x_t)$ is the compatibility between x_s and x_t. The node's belief that it is in a given state is then calculated as the product of its unary potential and all the messages that are incoming into it:

$$P(x_s) = \phi(x_s) \prod_{r \in \delta(s)} m_{r \to s}(x_s) \tag{6.17}$$

The messages are initialized to 1 (or in loopy BP sometimes to a random numbers uniformly distributed on $[0, 1]$). Because the computation of messages in (6.16) involves summation over the products of other messages, this algorithm is known as the *sum-product* algorithm. Another variant of belief propagation (*max-product*) is briefly explained at the end of this subsection.

We can also express belief propagation in a factor graph representation, where we have two types of messages: those sent from the factor nodes to the variable nodes and the messages from the variable nodes to the factors. One variable node is selected as the root and the inference proceeds in two stages: messages flowing from the leaves toward the root and then in the opposite direction. This process is illustrated with an example in Figure 6.14. The exact expressions and the detailed analysis of the selected example is in the part for the interested reader.

Sum-product in a factor graph representation. In a factor graph representation, we have two types of messages: $m_{f_i \to X_s}(x_s)$ from a variable node X_s to a factor node f_i and $m_{f_i \to X_s}(x_s)$ from a factor node f_i to variable node X_s. Let $\mathcal{X}_i$ be the set of variable nodes that are connected to f_i. The messages are then of the form:

$$m_{f_i \to X_s}(x_s) = \sum_{\mathcal{X}_i \backslash x_s} f_i(\mathcal{X}_i) \prod_{r \in \delta(i) \backslash s} m_{X_r \to f_i}(x_r)$$

$$m_{X_s \to f_i}(x_s) = \prod_{k \in \delta(s) \backslash i} m_{f_k \to X_s}(x_s) \tag{6.18}$$

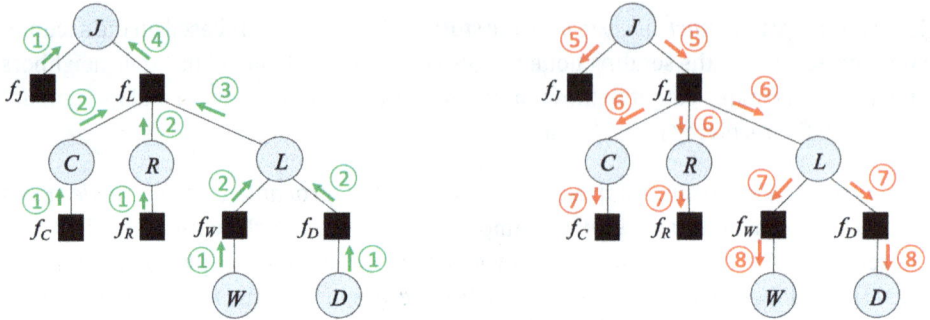

Figure 6.14: Belief propagation algorithm in action shown for the factor graph from Figure 6.11, which corresponds to the Bayesian network from Figure 6.3. The messages are shown in each step, first starting from the leaf nodes towards the root (on the left) and then from the root toward the leafs (on the right).

and the marginal probability of a node is

$$P(x_s) = \prod_{i \in \delta(s)} m_{f_i \to x_s}(x_s) \tag{6.19}$$

Observe that different potentials that appeared above in the equations for standard belief propagation in (6.16) and (6.17) are now simply "absorbed" into the factors f_i and the probability of a node is the product of the messages incoming into it from all the neighboring factor nodes.

Example of BP in a factor graph representation. Figure 6.14 illustrates the message passing process on the factor graph that was constructed previously for our Bayesian network *late to meeting*. The algorithm is initiated by choosing the root node (this can be any variable node; here we chose J). The messages flow first from the leaves toward the root. If the leaf is a variable node, its message is initialized to 1 and if it is a factor node, this initial message is the corresponding factor itself.

In our example, in step ① , five messages are simultaneously sent from the five leafs: two from the variable nodes W and D, both being 1, and three from the three-leaf factor nodes: f_J, f_C, and f_R, each being equal to the corresponding factor, for example, $m_{f_J \to J}(j) = f_J(j) = P(j)$. Each node waits until it receives messages from all its neighbors except one. At that time, it can compute the message toward the remaining neighbor (which can be regarded as its "parent" for this pass). For example, in Step ② , node L receives messages from f_W and f_D, and at the same time f_L receives messages from C and R. In Step ③ , L sends its message to f_L, which is $m_{L \to f_L}(l) = m_{f_W \to L}(l) m_{f_D \to L}(l)$. At that time, f_L can compute its message to J and send it in Step ④ : $\mu_{f_L \to J}(j) = \sum_c \sum_r \sum_l f_L(c, r, l, j) m_{C \to f_L}(c) m_{R \to f_L}(r) m_{L \to f_L}(l)$.

The computed messages are saved at each node. After all the messages reach the root, message passing continues toward the leaves, until all the nodes have received the messages on all their edges. The belief (marginal posterior probability) at each node is then simply computed using (6.19).

Max-product algorithm. A variant of belief propagation that computes the maximal *joint* probability in a given probabilistic graphical model is known as the *max-product* algorithm because the summation operator in the expressions for messages, both for standard BP (6.16) and the version on factor graphs (6.18) is replaced by the maximization. To understand this, recall that the marginal probability is $P(x) = \sum_{\mathcal{X} \setminus x} P(\mathbf{x})$, and clearly we need to replace the summation with a maximization operator if we want to determine the maximum joint probability $\max_{\mathbf{x} \in \mathcal{X}} P(\mathbf{x})$. Since the product of many small numbers can become unstable, the logarithm is often applied in this case, resulting in an equivalent algorithm called *max-sum*.[2] The max-product (or max-sum) algorithm with *backtracking* allows us to find *the most likely configuration* of the node values and is known as the *Viterbi* algorithm. For more details see, e. g., Bishop (2006).

6.3.5.2 Markov chain Monte Carlo samplers: Metropolis algorithm

Random-search inference algorithms are attractive because they are robust and universally applicable to any type of probabilistic graphical models. Markov chain Monte Carlo (MCMC) samplers construct a chain of sample configurations $\mathbf{x}^{(1)}, \mathbf{x}^{(2)} \dots$ leading to a desired equilibrium (e. g., converging to the MAP estimate $\hat{\mathbf{x}}_{MAP}$). The term Monte Carlo refers to repeated random sampling and the name Markov chain tells us that the probability of a next configuration $\mathbf{x}^{(i+1)}$ depends only on the current one $\mathbf{x}^{(i)}$ and not on earlier configurations, when the current one is known.

In a nutshell, MCMC samplers start from some initial configuration of the network nodes and apply random perturbations of one or a couple of labels at a time. The obtained "candidate" configuration is accepted or rejected based on the change in the posterior probability. Essentially, as Figure 6.15 illustrates it, we occasionally need to accept "bad moves" (i. e., decrease in posterior probability) in order to reach the global optimum. This is the main underlying idea of the *Metropolis–Hastings* algorithm. The initial configuration can be completely random but the algorithm will typically converge faster if the initial estimate is already good (e. g., obtained by a maximum likelihood approach).

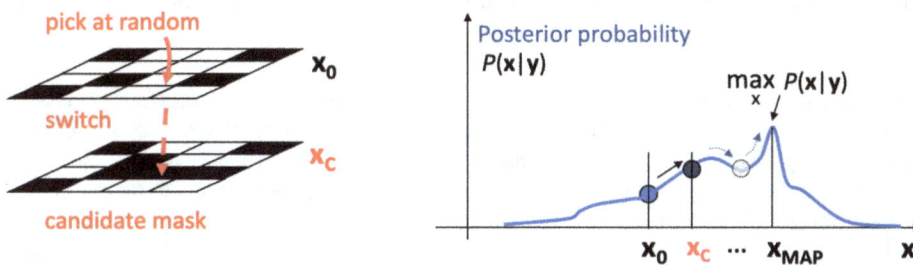

Figure 6.15: Inference by random search. **Left**: a random perturbation creates a candidate for the next configuration. **Right**: the candidate is accepted or not based on the change in the posterior probability. To reach the top, we need to accept occasionally a decrease in the posterior probability.

---

2 The product is replaced by a sum because the logarithm of a product is the sum of the logarithms of the factors.

The key question is how to decide whether or not to accept a new "candidate" configuration $\mathbf{x}^C$, obtained by randomly changing one or few labels of the current $\mathbf{x}$. We need to evaluate the change in the posterior probability, or equivalently the change in the posterior energy, which is, remember, inversely proportional to the probability: $P(\mathbf{x}|\mathbf{y}) \propto \exp[-H(\mathbf{x}|\mathbf{y})/T]$, with $T > 0$. For compactness, let $\Delta H = H(\mathbf{x}^C|\mathbf{y}) - H(\mathbf{x}|\mathbf{y})$ denote this change in the energy. Clearly, if $\Delta H \leq 0$, the change is favorable (the candidate has lower energy, i. e., larger probability), and thus we always accept it. If $\Delta H > 0$, the change is unfavorable, but we still need to accept it with some probability p in order to escape from the local optima. Practically, a random number is generated with a uniform distribution on $[0,1)$ and $\mathbf{x}^C$ is accepted if $\exp(-\Delta H/T)$ exceeds this random number.[3] When all the network nodes have been visited and their labels updated, one iteration is completed.

Let us apply this algorithm to solve the image classification problem using MRF prior, that was illustrated Figure 6.6. We assume that all the classes are a priori equally probable (the unary potential is zero for all the labels), the interaction potential is given by (6.9), and $T = 1$; hence, the prior probability is $P(\mathbf{x}) = (1/Z) \exp\left[-\sum_{\langle s,t \rangle} \psi(x_s, x_t)\right]$. We need to evaluate $\exp(-\Delta H)$, which is the ratio of the posterior probabilities of $\mathbf{x}^C$ and $\mathbf{x}$. Suppose that we always perturb only one label at a time, which is indeed the most common way of running these algorithms. In this case, $\mathbf{x}^C$ and $\mathbf{x}$ differ only in one label x_s. Assuming conditional independence as before $p(\mathbf{y}|\mathbf{x}) = \prod_s p(y_s|x_s)$, we have that

$$\frac{P(\mathbf{x}^C|\mathbf{y})}{P(\mathbf{x}|\mathbf{y})} = \frac{p(\mathbf{y}|\mathbf{x}^C)}{p(\mathbf{y}|\mathbf{x})} \frac{P(\mathbf{x}^C)}{P(\mathbf{x})} = \frac{p(y_s|x_s^C)}{p(y_s|x_s)} \exp\left(-\sum_{t \in \delta(s)} \left(\psi(x_s^C, x_t) - \psi(x_s, x_t)\right)\right) \tag{6.20}$$

Observe that the ratio of the prior probabilities reduces to evaluating the differences in the affected cliques only (which is the beauty and the power of MRF models) and this boils down to the differences in the local neighborhood $\delta(s)$ of the node s whose label has been changed. We work out this example further in detail for the case of binary classification (see parts for the interested reader).

Binary classification using the Ising model and MCMC sampler. For binary classification of image pixels, it is convenient to use the Ising MRF model, with labels $x_s \in \{-1, +1\}$, where $x_s^C = -x_s$, and $\psi(x_s, x_t) = \beta x_s x_t$, with $\beta < 0$ (to express a priori preference for spatial continuity of the two classes in the image, see Figure 6.8). Assuming again no prior preference for any label type ($a = 0$), the expression above becomes concretely

$$\frac{P(\mathbf{x}^C|\mathbf{y})}{P(\mathbf{x}|\mathbf{y})} = \frac{p(y_s|x_s^C)}{p(y_s|x_s)} \exp\left(2\beta x_s^C \sum_{t \in \delta(s)} x_t\right)$$

---

3 We can put this algorithm into an outer framework of **simulated annealing** by gradually reducing the temperature T, and allowing this way in the beginning larger bad moves and later on smaller and smaller ones.

To make it even more practical, note that it is sufficient to recompute for each visited spatial position s only $\sum_{t \in \delta(s)} x_t$ and plug it into

$$r = \frac{p(y_s|+1)}{p(y_s|-1)} \exp\left(2\beta \sum_{t \in \delta(s)} x_t \right)$$

We then compare r (when switching from -1 to $+1$) or $1/r$ (when switching from $+1$ to -1) to a random number from $(0,1]$ to decide whether or not to accept that change. This method does not guarantee the global minimum solution, but a low energy configuration is found with a large probability. After a sufficient number of iterations, the sequence of configurations will converge to the MAP solution $\hat{x}_{MAP}$ defined in (6.5). In general, a sufficient number of generated sample configurations to reach the equilibrium is 50 times the size of the image. In our experience, in binary image labeling problems as few as 10 iterations usually suffice if the initial estimate is good. **Gibbs sampler** is a special type of Metropolis–Hastings algorithm, where samples are drawn from the conditional probability distribution and the acceptance probability is one.

6.3.5.3 Mean field approximation

A class of inference algorithms known as *variational Bayes* is an alternative to MCMC samplers for solving complex inference problems. While MCMC techniques provide a numerical approximation of the exact posterior probability distribution using a set of samples, variational Bayes' methods impose an *analytical approximation* of the posterior probability distribution and deliver the exact solution under this analytical model. The posterior distribution $P(\mathbf{x}|\mathbf{y})$ is approximated by a *variational distribution* $Q(\mathbf{x})$, which is simpler than the true distribution and fully factorized:

$$Q(\mathbf{x}) = Q(x_1, \ldots, x_n) = \prod_{s=1}^{n} Q(x_s) \tag{6.21}$$

Here, $Q(x_s)$ stands for the approximating probability $Q(X_s = x_s)$ that X_s takes the value x_s, which is one of the possible labels $l \in \{1 \ldots L\}$. For brevity, let us denote $Q(X_s = l) = q_s^l$.

The approximating distribution Q is found by minimizing some dissimilarity metric with respect to the true distribution P, typically the *Kullback–Leibler divergence*[4] $D_{KL}(Q\|P)$. This yields the approximating probabilities q_s^l for all nodes $s \in \{1 \ldots n\}$ and all their possible labels $l \in \{1 \ldots L\}$ through an iterative procedure called the mean-field update equation. For details, see Koller and Friedman (2009). For the CRF model, with the posterior probability in (6.11), this mean-field update equation is

$$q_s^l = \frac{1}{Z_s} \exp\left\{ -\left(\phi(l, y_s) + \sum_{t \in \mathcal{N}_s} \sum_{k=1}^{L} q_t^k \psi(l, k, y_s, y_t) \right) \right\} \tag{6.22}$$

---

4 The Kullback–Leibler divergence between $Q(\mathbf{x})$ and $P(\mathbf{x}|\mathbf{y})$ is defined as $D_{KL}(Q\|P) = \sum_{\mathbf{x}} Q(\mathbf{x}) \log \frac{Q(\mathbf{x})}{P(\mathbf{x}|\mathbf{y})}$.

$\mathcal{N}_s$ is the neighborhood of the node s and Z_s is a normalization factor such that the resulting Q is a valid probability distribution, that is, that the probabilities of all labels sum up to 1 for each node s. The mean field algorithm repeats this update for all the nodes $s \in \{1 \dots n\}$, and all labels $l \in \{1 \dots L\}$ until a certain stopping criterion is reached, typically until the changes in all q_s^l become smaller than a given tolerance value. The MAP solution $\hat{x}_{MAP}$ defined in (6.5) is then simply found by assigning to each node s the label $l \in \{1 \dots L\}$ that gives the maximum q_s^l.

How do we apply this approach concretely to some of our earlier examples? Easily! We just need to specify the corresponding ϕ and ψ for a particular application and plug them into the update equation above. In the example with recommender systems, we would write $\psi(l, k, y_s, y_t) = \gamma\sigma(y_s, y_t)\mu(l, k)$, where $\sigma(y_s, y_t)$ models the similarity between different (user, item) pairs (e. g., with some convenient separable function over users and items) and $\mu(l, k) = \min[(l - k)^2, \tau]$ models the compatibility between labels l and k. In the inpainting example, $\psi(l, k, y_s, y_t)$ was reduced to $\psi(l, k)$ only, and both ϕ and ψ functions were sums of squared differences between the pixel values that are shared by the two image patches (the damaged and the replacement one for ϕ and the two overlapping replacement patches in the case of ψ).

6.3.6 Probabilistic reasoning over time

So far, we addressed static scenarios where the values of random variables remain fixed, at least while the agent is deliberating. In real life, most of the time we deal with *dynamic* scenarios: the world is changing, we need to track these changes and to predict the next states (values, positions, etc.) of various entities of interest. Think of traffic monitoring or robot motion, but also management of physiological processes or online social network analysis, etc., the examples are endless.

In many of these scenarios, we are dealing with temporal sequences, where the time is discrete with some step size, which depending on the dynamics of the problem can be a fraction of a second but also of the order of hours, days, or even years. We denote by $X_{1:t} = X_1, X_2, \dots X_t$ a sequence of random state variables in t consecutive time instants, and by $E_{1:t} = E_1, E_2, \dots E_t$ the sequence of evidence variables in the corresponding time instants. By our convention, $x_{1:t}$ and $e_{1:t}$ are the sequences of their respective values. $P(E_t|X_t)$ is the **sensor model** (also called the observation model), and $P(X_t|X_{t-1})$ is the **transition model**.

To make the inference tractable, we need to assume that each state variable depends only on a limited number of the previous ones, which is also well justified in practice. This is the **Markov assumption**. For the **first-order** Markov process, $P(x_t \mid x_{1:t-1}) = P(x_t \mid x_{t-1})$. Moreover, we assume that this process is **homogeneous**, that is, $P(x_t \mid x_{t-1})$ is the same for all t. Also, in practice we typically assume that the evidence variable at each time instant depends only on the state variable in that particular time instant (and not on the previous state variables and previous observations), that is, $P(e_t \mid x_{1:t}, e_{1:t-1}) =$

$P(e_t \mid x_t)$. This is called the **sensor Markov assumption**. Under these settings, we can solve efficiently various queries regarding state variables in time.

Example. *Office mates.*
Your office mate, Sam, is a software engineer and is today mainly coding but from time to time he is also chatting with his friends online. Your desks are arranged so that you are facing each other, with computer screens back-to-back. You are not particularly inspired to work right after the lunch break today and you amuse yourself by trying to guess whether your colleague is programming or chatting by taking a glance at his facial expression from time to time, say every 5 min. You know that when Sam is programming he is mostly frowning (70 % of the time), but sometimes also smiling (30 % of the time), typically because he fixes a bug. While chatting, he is almost always smiling (with probability 0.8) and rarely frowning (with probability 0.2). If he is currently programming, the chance that he will keep coding in the next 5 min is about 0.9, and in other 10 % cases he will turn to chatting. Once chatting, he tends to continue with it in the next 5 min with a probability of 0.6 and return to programming with probability 0.4. You assume that the probability that Sam starts coding immediately after the noon break is 0.7.

In the example above, the time is divided in intervals of 5 min. The state variable X_t is Sam's activity in the interval t, and it can take two values: programming ($X_t = p$) or chatting ($X_t = c$), so $x_t \in \{p, c\}$. The evidence variable is facial expression, also with two possible values: smiling ($E_t = s$) or frowning ($E_t = f$), so $e_t \in \{s, f\}$. We also know that the transition model is $P(X_t = p \mid X_{t-1} = p) = 0.9$ (and thus $P(X_t = c \mid X_{t-1} = p) = 0.1$), and $P(X_t = p \mid X_{t-1} = c) = 0.4$ (and thus $P(X_t = c \mid X_{t-1} = c) = 0.6$). This situation is depicted in Figure 6.16, and we will use it to illustrate the mechanisms of probabilistic reasoning over time, making use of a sequence of observations $\mathbf{e}_{1:t}$. For example, if we observed that Sam was frowning in the first two time instants and then smiling and frowning again, we have $\mathbf{e}_{1:4} = (f, f, s, f)$.

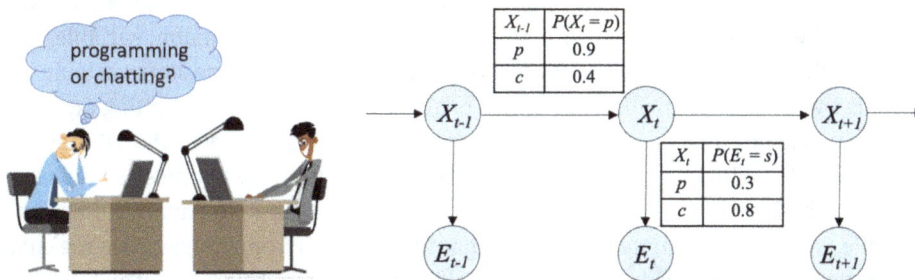

X_{t-1}	$P(X_t = p)$
p	0.9
c	0.4

X_t	$P(E_t = s)$
p	0.3
c	0.8

Figure 6.16: An illustration of the *Office mates* example, with the temporal probability model, and the corresponding transition model and the sensor model in the respective tables.

Temporal reasoning problems. Inference problems in reasoning over time are by Russel and Norvig (2021) categorized into four main types: filtering, smoothing, prediction, and most likely explanation. We can summarize them and illustrate with the example above as follows:

- **Filtering** – Determine the posterior probability distribution of the *current state* variables given all evidence received so far: $P(x_t|e_{1:t})$. In our example: *Is Sam chatting now given all his facial expressions until now?* This is the **belief state** (agent's belief about the current state), which serves as input to the decision process (which action should be taken *now*).

- **Prediction** – The posterior probability of a *future state* given all evidence received so far: $P(x_{t+k}|e_{1:t})$. For example: *Will Sam be chatting in the time instant 10 minutes from now given all his facial expressions so far?* We need prediction for evaluating the course of possible actions.

- **Smoothing** – The posterior probability of a *past state* given all evidence received so far: $P(x_k|e_{1:t})$ for $0 \le k < t$. For example: *Was Sam chatting 5 minutes ago given all his facial expressions so far?* This allows better estimation of past states, thus essential for learning.

- **Most likely explanation** – Find the most likely sequence of state variables given the sequence of observations: $\arg\max_{x_{1:t}} P(x_{1:t}|e_{1:t})$. For example: *Determine the most likely sequence of Sam's activities during four consecutive time intervals given the sequence of his facial expressions* $e_{1:4} = (f, f, s, f)$.

The filtering, prediction, and smoothing problems are solved with recursive computations, with message passing similar to the belief propagation that we covered earlier on in this section. The most likely explanation is being solved with the Viterbi algorithm (a variant of max-sum belief propagation with backtracking). All these inference mechanisms are compactly described in Russel and Norvig (2021) with nice illustrative examples.

Hidden Markov models. A widely used temporal probability model is hidden Markov model (HMM), in which the state is described by a *single* discrete random variable. The example illustrated in Figure 6.16 is an HMM because it has only one state variable being Sam's activity X_t, which can be either programming ($x_t = p$) or chatting ($x_t = c$). HMM can also be employed in cases where we have multiple state variables if they are combined into a single composite variable ('megavariable) whose values are tuples of values of the individual variables. Suppose, for example, that we have another state variable, which is the volume of messages that Sam receives from his friends in the chat, with three possible values {*none, moderate, high*}. Then the state variable of an HMM for this problem takes values that are tuples like $(p, none)$, $(p, moderate)$, . . ., $(c, high)$. HMM models are often visualized with a state transition diagram, like the one in Figure 6.17.

HMMs are also characterized by a compact description of the transition model, in matrix form. If the state variable X_t has N values, the transition model is a $N \times N$ matrix **T**,

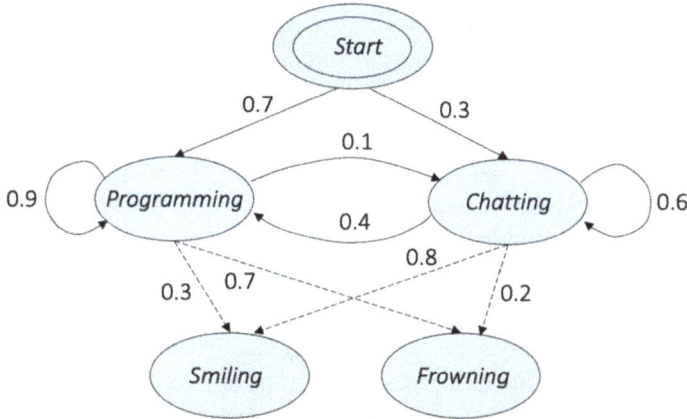

Figure 6.17: A state transition diagram for the HMM model of the *Office mates* example.

with entries $T_{i,j} = P(X_t = j | X_{t-1} = i)$. In the *office mates* example from 6.16, this transition model is

$$\mathbf{T} = \mathbf{P}(X_t | X_{t-1}) = \begin{pmatrix} 0.9 & 0.1 \\ 0.4 & 0.6 \end{pmatrix}$$

The sensor model is represented by a diagonal matrix $\mathbf{O}_t$, with diagonal elements $P(e_t | X_t = i)$. In the example from Figure 6.16, with the sequence of observations $\mathbf{e}_{1:4} = (f, f, s, f)$, we have

$$\mathbf{O}_1 = \mathbf{O}_2 = \begin{pmatrix} 0.7 & 0 \\ 0 & 0.2 \end{pmatrix}, \quad \mathbf{O}_3 = \begin{pmatrix} 0.3 & 0 \\ 0 & 0.8 \end{pmatrix}, \quad \mathbf{O}_4 = \begin{pmatrix} 0.7 & 0 \\ 0 & 0.2 \end{pmatrix}$$

because $e_t = f$ for $t = 1, 2, 4$, and $e_t = s$ for $t = 3$. This matrix form allows a compact representation of the inference procedure. For example, for the filtering problem, if we denote by $\mathbf{f}_{1:t} = \mathbf{P}(X_t | e_{1:t})$ the vector whose elements are the probabilities $P(x_t | e_{1:t})$ for all possible values x_t, then the filtering problem is easily solved by recursive computation: $\mathbf{f}_{1:t+1} = \alpha \mathbf{O}_{t+1} \mathbf{T}^\top \mathbf{f}_{1:t}$, where $\alpha > 0$ is a normalizing constant. Other inference tasks of temporal reasoning described above are with HMMs represented with similar compact formulations.

Kalman filtering. So far, we considered reasoning over time in the case of discrete-time processes. In some cases, like object tracking, control, guidance, and navigation of moving objects, we are rather dealing with continuous space variables and continuous time variables. Kalman filtering is widely used in these applications due to its mathematical elegance, which leads to relatively simple recursive computations and clear performance guarantees in cases where the assumptions of the model are met or hold up to some reasonable approximation. The assumptions that underline Kalman filtering are the linearity of state dynamics and observation process, as well as the normal distribution of noise in state dynamics and in measurements. Thus the prior model is a Gaussian distribution, and the transition model and the sensor model are linear Gaussian models. Under these models, the state prediction also follows a multivariate Gaussian distribution, and in a nutshell, its recursive estimation boils down to updating the covariance matrix and the mean.

The classical applications of Kalman filtering are in point target tracking (like radar tracking of aircrafts and missiles) but it is also used in tracking vehicles and—depending on the scale—people tracking, and in various other domains, which involve continuous state variables and noisy measurements (e. g., in signal processing and econometrics). The assumptions made (linear Gaussian transition and sensor model) are often too strong in practice and variants such as the *extended Kalman filter* and the *switching Kalman filter* exist to alleviate these limitations and to extend the application to some nonlinear systems.

Dynamic Bayesian networks. The models known as dynamic Bayesian networks (DBN) are an extension of classical Bayesian networks that we covered in the previous sections, to include random variables that are changing over the time. These modes can also be seen as generalization of both HMM and Kalman filtering. Each HMM is also a DBN, but the opposite does not hold: DBNs make use of conditional independence assertions (inherited from the classical Bayesian networks) and allow problem description with fewer parameters than HMM. Also, each Kalman filter can be represented as DBN but the opposite does not hold since DBNs allow more general transition and sensor models, beyond the linear Gaussian. Naively, the inference in a DBN could be realized by unrolling this model to its time slices and applying some of the inference methods for classical Bayesian networks. However, the complexity of such an inference approach would be prohibitive in most cases of practical interest. Thus, while DBN is a very flexible model, which allows us to represent efficiently very complex temporal processes with many sparsely connected random variables, the exact inference in such a setting is very difficult and often not feasible. Typically, approximate inference methods are being applied, among which *particle filtering* is a widely used approach relying on well-established methodology.

Particle filtering. A family of algorithms called particle filtering are sequential Monte Carlo methods with importance sampling. The key idea is that each probability distribution can be represented by a set of samples (particles) and that representative particles for the posterior distribution can be recursively formed from the initial samples taken from the prior distribution, by appropriate weighting of the samples and by resampling the sample population. In each iteration, the particles are propagated from the previous iteration based on the transition model $P(x_{t+1}|x_t)$ and weighted by the likelihood they would assign to the new evidence $P(e_{t+1}|x_{t+1})$. The particle population is resampled such that each sample is selected with a probability proportional to its weight. The new sample population is a refined approximation of the posterior probability and the process continues recursively. Numerous implementations and toolboxes for particle filtering are freely available online (in Matlab, OpenCV, Python,etc.) and typically easy to plug-in and use.

6.4 What are the limitations of probabilistic reasoning?

Let us reflect first on when do we need probabilistic reasoning in AI. Many complex reasoning and planning problems are being solved with (purely) logical approaches, searching and game playing strategies, and logic-based hierarchical planning. We need probabilistic reasoning when we are faced with uncertainties, arising from various sources such as nondeterministic aspects of the environment, uncertain outcomes of the actions, partial observability, incomplete domain knowledge, inability to run all the relevant tests, etc. Probabilistic reasoning is often the only viable or at least the most solid approach to tackle such problems. But it also faces limitations, at different levels.

We can distinguish between subjective and objective limitations of probabilistic reasoning. By subjective limitations, we mean common errors that people make in probabilistic reasoning, which are then translated to wrong selection and use of the models, wrong specification of the involved data distributions, and wrong interpretation of the involved probabilities and inference results. The examples of these errors are plentiful, and below we highlight some common causes. By objective limitations, we mean those that arise from the necessity to make various simplifications in our models and inference strategies to enable feasible solutions and the necessity to rely on the available statistical data and/or to estimate the parameters of the assumed models from the available data. This also means that probabilistic reasoning in AI inherits many of the general limitations of machine learning as well.

Subjective limitations of probabilistic reasoning are connected to difficulties in building and applying the right probabilistic approach in practice due to misconceptions and incompatibilities between how we refer to uncertainties in common life and how they need to be translated to formal models. To start with, the designers of a probabilistic reasoning model often need to translate highly ambiguous everyday expressions of probability such as "small chance," "doubtful," "highly likely," etc., into numeric values. Various common difficulties with probabilistic reasoning are known to have psychological origins, such as the gambler's fallacy (the incorrect belief that successive independent events causally influence the outcome of later events, e. g., believing that an event that occurred more frequently in the past is less likely to happen in the future and vice versa) or our tendency to characterize events that are special in some way (e. g., "royal flesh" in poker or a particular alignment of stellar object) as "low probability" events. Similarly, our judgement of the probability of occurrence of some events is often highly affected by their saliency. Common errors include also mixing up the retrospective and predictive probabilities in the problem description that is given to a system. Take the following example from Eddy (2013), which shows how statistical data from one study propagate wrongly to the next one: The original study reports: "87.5 % of the 'X-ray carcinoma' (or "positive") group had biopsy-proven malignant lesions, thus $P(ca|pos) = 87.5$ %, and $P(benign|neg) = 84$ %." A subsequent medical study builds on this but translates it wrongly to "A correct mammographic diagnosis was made in 84 % of those with benign lesions and in 87.5 % of those with carcinoma," while in fact, the true-positive rate $P(pos|ca)$ in this study was only 66 % and the true-negative rate $P(neg|benign) = 54$ %. These and similar types of misconceptions may affect in practice the specification of the data distribution models, and in extreme cases invalidate a probabilistic approach that needs to rely on them.

Objective limitations of probabilistic reasoning are those that we cannot avoid by just careful and correct implementation of the underlying theory and algorithms. Often, we need to abstract away many aspects of the real world and to construct simplified models and simplified inference engines to arrive at a solution in feasible time or to support decision making in real time. In some instances, these simplifications will be more severe than in others and potentially affecting more the accuracy and the range of the validity of the obtained results. It is therefore always important to know under which assumptions a given model has been derived and what performance guarantees does it offer. Sometimes we need to make a trade-off between the accuracy of the model that we impose and the achievable accuracy of the inference procedure under such a model. For example, MCMC samplers that we treated that we treated in Section 6.3.5.2 yield an *approximate solution* (although approaching the exact one with high probability) under the *exact probability distribution*. In contrast to this, the variational inference approaches Section 6.3.5.3 yield the *exact solution* under an approximate *simplified model* for the probability distribution. The belief propagation methods, strictly speaking, yield the exact solution only for tree-structured probabilistic graphical models, and hence not for many others that have loops. But, practice teaches us that loopy belief propagation works typically very well in general networks with loops, which was also proven by the huge success of Turbo codes. When dealing with reasoning over time, dynamic Bayesian networks with particle filtering proved as a robust approach in many practical situations. One of the important challenges in practice is how to learn the structure and the parameters of a probabilistic graphical model from the available data. This is where various data-driven strategies work hand in hand, merging often probabilistic reasoning and deep learning as it is now the case with emerging Bayesian neural networks and Bayesian graph neural networks.

6.5 Industry examples

6.5.1 Hunting for anomalous sessions using Markov chain model in Microsoft Sentinel

Karishma Dixit, Aleksandra Pižurica, Emmanuel Gillain

6.5.1.1 Business context

Large organizations store vast amounts of logs generated from the various IT products and services they use. Those logs are typically used for audit and security purposes and can be leveraged to spot anomalous activities, assuming here that malicious activity will be defined as anomalous when compared to legit activity. Many audit logs contain multiple entries that can be thought of as a sequence of related activities, or a session: a timebound sequence of activities linked to the same user or entity. A session will then be considered "anomalous" when sequences, of events or activities, aren't seen in other typical user sessions.

A lot of existing **security detections** tend to be rule-based with the aim of detecting specific known attacks such as password spray (password spraying is an attack that will usually feed a large number of usernames into a program that loops through those usernames and tries a number of passwords; it's a brute force attack). However, cybersecurity attacks on businesses are constantly evolving, making it hard to detect new types of attacks. As a consequence, more adaptive approaches should be adopted. A **Markov chain model** can bring such flexibility to adapt and surface possible activity from an adversary, whose behavior differs from that of a normal user. **Microsoft Sentinel**, a security information and event management service in the cloud, provides such a modeling approach to help customers explore and detect potential malicious activities. This modeling approach is provided via a Jupyter Notebook,[5] which can be configured to run on a schedule. Once the probability of an anomalous activity passes some threshold, it can be configured so a security incident is automatically generated and sent to security analysts or other systems to take further action. See Figure 6.18 to understand what the Microsoft Sentinel Jupyter Notebooks UI looks like. See Figure 6.19 to see an example incident that has been raised by a scheduled Jupyter notebook run in Microsoft Sentinel.

To illustrate how Markov chain models are applied to help customers detect **anomalous sessions**, let's assume a fictional company "Contoso" stores audit logs for their Microsoft Office365[6] exchange usage and their Azure Active Directory[7] sign-ins.

---

5 A Jupyter Notebook is a human-readable document, which can contain both executable python code as well as descriptions, figures, tables, etc.

6 Cloud based subscription services offered by Microsoft, including the Office desktop applications and hosted mail (exchange) and collaboration services.

7 Cloud based identity and access management service offered by Microsoft.

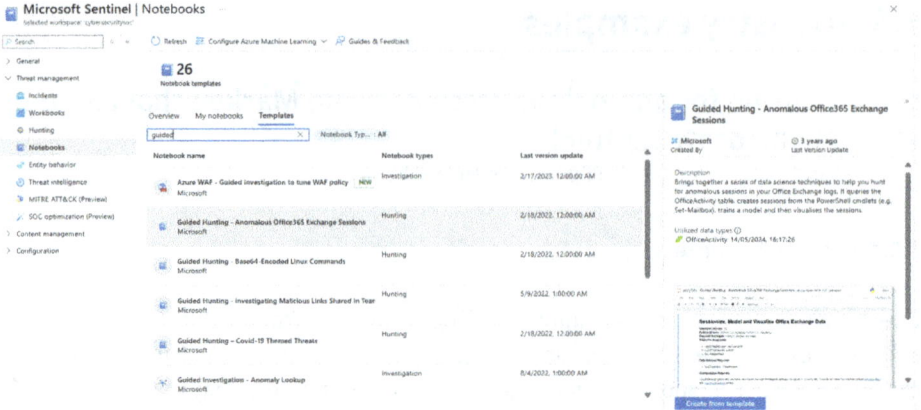

Figure 6.18: Screenshot of Microsoft Sentinel Jupyter Notebooks UI.

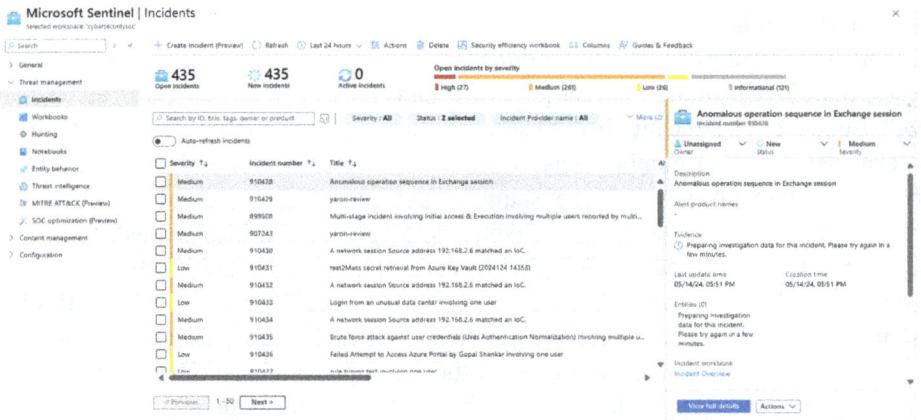

Figure 6.19: Screenshot of Microsoft Sentinel Incidents UI displaying an incident raised by a scheduled Jupyter Notebook run.

6.5.1.2 Overall approach

In a context where labeled data for security breaches is difficult to obtain and rapidly outdated, we have to deal with a shortage of labeled attack data. In such situations, it tends to be better to use unsupervised learning techniques, since they have the flexibility to discover new types of attacks. The proposed **Markov chain model** is sufficiently general, so that it can be used to model any type of sequence or session data.

Raw audit or security logs data are first organized into **"sessions,"** a **time-ordered sequence of events** typically associated to a user account, possibly with other attributes. The likelihoods of such sessions are then calculated by applying the Markov assumption, as illustrated below for a sequence of events such as A, B, C.

$$
\begin{aligned}
P(\text{session}) = P(A, B, C) \\
= P(C|A, B)P(A, B) \\
= P(C|A, B)P(B|A)P(A) \\
= P(C|B)P(B|A)P(A) \quad \text{(by applying the Markov assumption)}
\end{aligned}
$$

(6.23)

We can then calculate the **maximum likelihood estimates** for the transition probabilities (e. g., probability of transitioning from event A to event B). Once we have calculated estimates for all the transition probabilities in a given session, we can multiply them together to end up with a likelihood score for the given session. The lower the likelihood score, the more unusual the activity in the session, therefore, the more likely we would consider the session to be anomalous.

From logs to events, from events to transition of events and sessions

Let's go deeper by illustrating the concepts with some example sessions, using the fictional company Contoso's Office365 audit logs. In this case, the events are commands issued to the Office365 service through an application programming interface (API) (computing interface that defines interactions between multiple software intermediaries). For example, an event could be the usage of the "Set-User" command. We group the events into chronological sessions on a per user and IP address basis. See some example sessions in Figure 6.20.

UserId	ClientIP	begin	end	session	param_session	param_value_session
a_user_id	an_ip	2020-05-30 23:43:48+00:00	2020-05-30 23:43:49+00:00	[Set-User, Set-User]	[Cmd(name='Set-User', params={'Identity', 'SyncMailboxLocationGuids'}), Cmd(name='Set-User', par...	[Cmd(name='Set-User', params= {'Identity': '4b2462a4-bbee-495a-a0e1-f23ae524cc9c\\a2409f54-2a30-4...
a_user_id	another_ip	2020-05-31 01:21:11+00:00	2020-05-31 01:21:11+00:00	[Set-User, Set-User]	[Cmd(name='Set-User', params={'Identity', 'SyncMailboxLocationGuids'}), Cmd(name='Set-User', par...	[Cmd(name='Set-User', params= {'Identity': '4b2462a4-bbee-495a-a0e1-f23ae524cc9c\\a2409f54-2a30-4...
another_user_id	different_ip	2020-05-31 02:02:05+00:00	2020-05-31 02:02:06+00:00	[Set-ConditionalAccessPolicy, Set-ConditionalAccessPolicy, Set-ConditionalAccessPolicy, Set-Cond...	[Cmd(name='Set-ConditionalAccessPolicy', params= {'PolicyLastUpdatedTime', 'PolicyIdentifierStrin...	[Cmd(name='Set-ConditionalAccessPolicy', params= {'Identity': 'seccxpninja.onmicrosoft.com\\83a05...
a_user_id	an_ip	2020-05-31 05:13:06+00:00	2020-05-31 05:13:08+00:00	[Set-ConditionalAccessPolicy, Set-ConditionalAccessPolicy, Set-ConditionalAccessPolicy, Set-Cond...	[Cmd(name='Set-ConditionalAccessPolicy', params= {'PolicyLastUpdatedTime', 'PolicyIdentifierStrin...	[Cmd(name='Set-ConditionalAccessPolicy', params= {'Identity': 'seccxpninja.onmicrosoft.com\\235be...

Figure 6.20: Identification of a session with a user ID (column1) and IP address (column2), with a timestamp (columns3 & 4). Three different formats for each of the sessions: sequence of simple events like ["Set-User", "Set-Mailbox"] (column 5), sequence of events with parameters (Ex: Cmd(name="Set-User", params={"Identity", "Force"}), or sequence of parameters and values (Ex: Cmd(name="Set-User", params={"Identity": "test@example.com", "Force": "true"})).

Creation of the Markov chain model and training on data to estimate the probabilities

Let's now look at the modeling of such a session, with their events and parameters, using a **Markov chain** model. The **likelihood of a session** of simple events $[A \rightarrow B \rightarrow C]$ was given in (6.23) as $P(C|B)P(B|A)P(A)$.

Case 1. Consider first sessions that are made of simple sequences of events and define a transition probability as the probability of going from the previous event-state to the current event-state:

$$P(\text{event}_i|\text{event}_{i-1})$$

Once the model is in place, we estimate the transition probabilities using the maximum likelihood estimation. In this case, the **maximum likelihood estimate** simply comes down to a ratio of counts:

$$P(\text{event}_i|\text{event}_{i-1}) = \frac{\#(\text{event}_{i-1} \rightarrow \text{event}_i)}{\#\text{event}_{i-1}}$$

A subtlety to note is that we prepend and append start-tokens and end-tokens, respectively, to each session by default. The start event A would then be conditioned on the start-token, and we would have an additional transition probability in the likelihood calculation of the session terminating given the last event C in the session. Once we have computed the above estimates for all possible transitions of states, the likelihood of a session can be calculated by multiplying a **sequence of transition probability estimates** together. We can hence compute a likelihood score for each session in our data. The lower the score, the more anomalous the session activity is.

Sliding window. A point worth noting is that the likelihood calculations for longer sessions (more events) will involve multiplying more transition probabilities together. For example, a session which contains ~1000 events will involve multiplying ~1000 transition probability estimates together. Because transition probabilities are between 0 and 1, this likelihood calculation will converge to zero as the session length gets longer. This could result in sessions being flagged as anomalous simply because they are longer. To circumvent this, we can use a sliding window to compute a likelihood score per session. For example, let us consider the following session = $[A, B, C, D]$. Let us also fix the sliding window length to be 3. Then we would compute the likelihoods of the following windows:

- $[A, B, C]$
- $[B, C, D]$
- $[C, D, \#\#\text{END}\#\#]$

We would then take the likelihood of the lowest scoring window as the score for the full session. Notice that we are still using a start-token in addition to the end-token shown.

The end-token means we include an additional probability for the session terminating after the final event D, whereas the start-token appears implicitly when we condition the first event A on the start-token.

Case 2. To go a step further in the modeling of our session, we can differentiate sessions by adding the parameters of the commands. This can help us to differentiate between automated and benign usage of a command and a more unusual usage. An event can then be defined as a vector x_i:

$$x_i = \{c_i, [p_1, \dots, p_k]\}$$

where c_i is the command used in the ith event of the session and each p_j is either 1 if that parameter has been set for c_i, or 0 if not. For example, c_i could be the "Set-Mailbox" command with 3 distinct parameters: ["Identity," "DisplayName," "ForwardingSmtpAddress"]. A parameter vector for this event with value $[1, 0, 1]$ would mean "Identity" set, "DisplayName" not specified, and "ForwardingSmtpAddress" set.

If we denote the parameter vector by $\{p_k\}_i$ as a shorthand, the probability model of the current event conditional on the previous event becomes:[8]

$$
\begin{aligned}
P(x_i|x_{i-1}) &= P(\{c_i, \{p_k\}_i\}|\{c_{i-1}, \{p_m\}_{i-1}\}) \\
&= P(\{p_k\}_i|c_i, c_{i-1}, \{p_m\}_{i-1})P(c_i|c_{i-1}, \{p_m\}_{i-1}) \\
&= P(\{p_k\}_i|c_i)P(c_i|c_{i-1}) \quad \text{(by modelling assumption)} \\
&= P(c_i|c_{i-1}) \prod_{j=1}^{k} P(p_{ji}|c_i) \quad \text{(by modelling assumption)}
\end{aligned}
$$

where
- the parameters $\{p_k\}_i$ used for the current event depend only on the current command c_i and not on the previous event x_{i-1}
- The current command c_i depends only on the previous one c_{i-1} and not on the previous parameters $\{p_m\}_{i-1}$
- the presence of each parameter p_{ji} is modeled as independent **Bernoulli**[9] random variables, conditional on the current command c_i

The **learning** becomes then the estimation of the probability of a parameter being used for a given command from our sessionized data as follows:

---

8 In the real implementation, we take the geometric mean of this product: $\prod_{j=1}^{k} P(p_{ji}|c_i)$, by raising the product of probabilities to the power of $1/k$. The reason for this is because the commands can have a vastly different number of parameters set on average. By taking the geometric mean, we can have a fairer comparison of how rare sets of parameters are across the different commands.

9 A Bernoulli random variable can take on two values, 1 and 0. It takes on a 1 if an experiment with probability p resulted in a success and 0 otherwise.

$$P(p_{ji} = 1|c_i) = \frac{\#(\text{param}_j \text{ is present for cmdlet } c_i)}{\#(\text{cmdlet}_i \text{ appears})}$$

The probabilities of the parameters conditional on the command can now be calculated, as well as the transition probabilities like before. The likelihood calculation for a session now involves multiplying a sequence of probabilities $P(x_i|x_{i-1})$ together where each $P(x_i|x_{i-1})$ can be decomposed as shown above. The same sliding window approach as before is used, so we can more fairly compare the likelihoods between sessions of different lengths.

Case 3. Finally, the case where the values of the parameters are also modeled alongside the parameter names is discussed. Some of the commands can accept parameters that have higher security stakes. For example, the "Add-MailboxPermission" command has an "AccessRights" parameter, which can accept values such as "ReadPermission" and "FullAccess." Because the "FullAccess" value could be used by an attacker for privilege escalation, it could also be worth including the values of the parameters in the modeling of the sessions. However, not all the values are going to be useful in the modeling, since parameters such as "Identity" can take arbitrary strings such as email addresses as their values. In particular, parameters which accept values from only a small, predefined list such as ["high," "medium," "low"] would be useful to include in the modeling, whereas parameters which accept any string value would be too unique to be useful. Some rough heuristics are therefore used to determine which parameters take values that are categorical (e. g., high, medium, low), as opposed to arbitrary strings. In the modeling for parameters, only the values which have been deemed suitable by the heuristics are included. However, there is the option to override the heuristics in the model. This time, we denote an event as follows:

$$x_i = \{c_i, \{p_k\}_i, \{v_k\}_i\}$$

where c_i is the command used in the ith event of the session, each p_k is either 1 or 0 as above, and v_k is the value set for the parameter p_k (if the parameter was set). The same decomposition of probabilities as in the cases above can then be applied, using similar conditional independence assumptions. The learning of the probabilities follows the same approach as above:

$$P(v_{ji} = \text{some\_value}|p_{ji}) = \frac{\#(\text{param } p_j \text{ is set to some\_value})}{\#(\text{param } p_j \text{ appears across the data})}$$

Once the likelihood scores for the sessions are computed, we can then rank them in ascending order and visualize them in an interactive plot such as in Figure 6.21.

Digging deeper, a typical suspicious session coming top of the list would look like the one displayed in Figure 6.22.

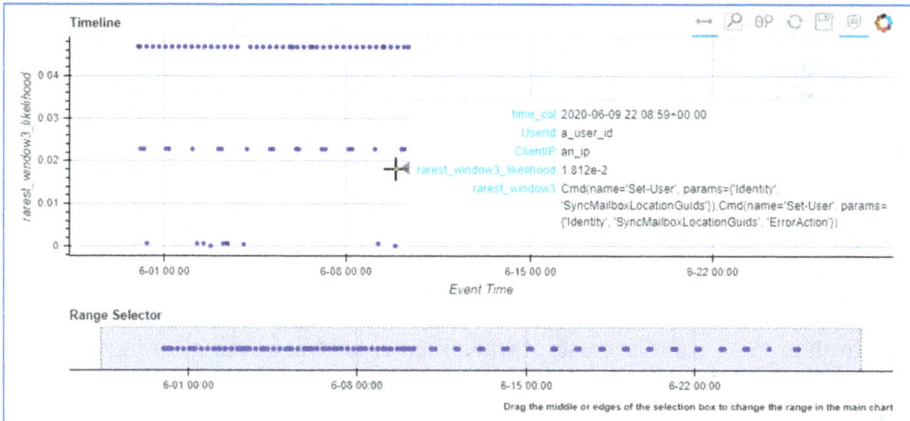

Figure 6.21: Each point on the plot represents a session. The likelihood score is on the y-axis and the time of the session is on the x-axis. The sessions with the lower likelihood scores correspond with the more anomalous sessions.

```
data.sort_values('3lik3').iloc[3]['3win3']   # Look strange!!

[Cmd(name='New-MobileDeviceMailboxPolicy', params={'Name': '              ', 'AlphanumericPasswordRequired': 'False', 'MinP
asswordLength': '', 'MaxPasswordFailedAttempts': 'Unlimited', 'AllowSimplePassword': 'False', 'IsDefault': 'False', 'Passwo
rdEnabled': 'False', 'MaxInactivityTimeLock': 'Unlimited', 'MinPasswordComplexCharacters': '3', 'PasswordHistory': '0', 'Pa
sswordExpiration': 'Unlimited', 'RequireDeviceEncryption': 'False', 'AllowNonProvisionableDevices': 'False'}),
 Cmd(name='Set-MobileDeviceMailboxPolicy', params={'Identity': '              ', 'AllowNonProvisionableDevices': 'True'}),
 Cmd(name='Set-CASMailbox', params={'Identity': '          ', 'ActiveSyncMailboxPolicy': '              '})]
```

Figure 6.22: This session shows activity where the user is creating a very weak mobile device mailbox policy with no password enabled, no device encryption required and also allows nonprovisionable devices.

6.5.1.3 Summary

By using a **Markov chain model**, we were able to model Microsoft Office365 **audit logs** to hunt for and visualize anomalous user sessions. We saw how **Microsoft Sentinel** can be used to run such analysis on an automated schedule to raise security incidents to be investigated further by security analysts. While the methods outlined here were applied to Office365 audit logs, they can of course be applied to other types of audit or security logs in a similar way.

6.5.2 Crop forecasting with Bayesian inference for agriculture producers

Matthew J. Smith, Pierre Stratonovitch, Richard Tiffin

6.5.2.1 Overview

Context

Fresh produce is a term in agriculture for crops that are consumed in their harvested form. For example, radishes, sweetcorn, and strawberries are all "fresh produce." Growers of fresh produce want forecasts of when their crops will be ready for harvesting so that they can make the best economic decisions. This case study is about the process of producing such forecasts. It represents a simplified version of crop forecast models that have been used by producers to reduce waste and gain production efficiencies. It is especially concerned with probabilistic forecasts because the probabilities of the different outcomes they provide are particularly important in grower's decision-making.

Matching the supply and demand of fresh produce can be particularly challenging because both supply and demand vary dynamically over time. On the supply side, the weather is a significant determinant of the timing of when a crop reaches maturity, and thus becomes sellable. This supply side risk is amplified by risks on the demand side where seasonality, public holidays, retailer and brand promotions, social events, and weather conditions are major determinants. A potential mismatch between supply and demand is important because there is usually a narrow time window post-harvest when they are sellable. In ambient conditions, fresh produce crops have high perishability, meaning they begin to lose their "freshness" and value rapidly. Evidently, an ability to forecast supply (and demand) can enable management actions to be taken to minimise or mitigate such mismatches. Producers want to maximize the probability that they can meet consumer demand, while minimizing the probability that they overproduce and cause waste.

The specific challenge dealt with here is that of building a model of fresh produce growth that produces probabilistic forecasts of yield and harvest timing to help growers balance the risk of overproduction with not being able to meet demand.

Key techniques

We use **Bayesian inference** to learn the parameters of a model that incorporates insights from the biology of plants in the formulation; so called "process based" models. We employ a **Markov chain Monte Carlo** (MCMC) sampler with the Metropolis–Hastings algorithm to estimate the probability distributions of the model parameters as described elsewhere in this Chapter. The probabilistic nature of the parameter estimates is carried through to the model's forecasts, which become probabilistic as a result.

6.5.2.2 Description

Process-based crop models
A process-based crop model for industrial applications is a computational algorithm that predicts properties of crop growth using expressions that contain some representation of the underlying biological processes that determine crop growth. Typical target crop properties are maturation date, yield, and quality.

One benefit of **process-based models** is that they can capture well the nonlinear response of crop growth to growing conditions without requiring extensive data. A disadvantage is that such a formulation will only ever be an approximate representation and so care must be taken to identify when the model predictions are misleading. Consequently, process-based models still need to be **calibrated** to be accurate.

A high-level representation of a process-based crop model is

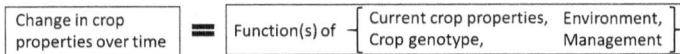

Figure 6.23: High level representation of a process-based crop model formulation.

The crop genotype (crop variety) is typically represented implicitly in the model via the model formulation and parameters. Mathematically, the crop model above can be expressed as

$$C(t + \tau) = F\big(C(t), E(t), M(t), \theta\big), \tag{6.24}$$

where $C(t)$ is a vector of crop properties at time t, τ is a small increment in time (typically one day), F is a vector of functions, $E(t)$ is a vector of environmental conditions (e. g., mean daily temperature), $M(t)$ is a vector of management conditions, and θ is a vector of parameters. Process-based crop models typically need to be solved iteratively (a crop growth simulation) to generate predictions.

Example: the radish maturation model
A simple hypothetical example of a crop model that can be trained to data is one for radishes (*Raphanus raphanistrum* subsp. *sativus*[10]). Radish is an edible root vegetable commonly used in salads that grows rapidly. Radish producers will want to forecast **likely harvesting dates** of planted crops so that they can ensure production levels that meet customer demand. Imagine a producer who has recorded **historical sowing and harvesting dates** for 3 farms between 2017 and 2018 for a single field in each farm and for crops sown approximatively every 3 weeks from the 1st of March (Figure 6.24).

---

10 https://en.wikipedia.org/wiki/Radish

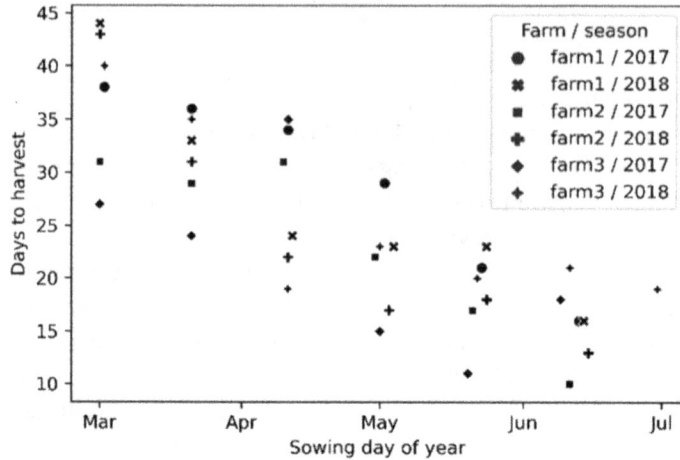

Figure 6.24: Hypothetical observations of days to harvest for radishes for three farms and over 2 years. These data are used as training data to enable us to learn the parameter distributions for our predictive crop model.

The data in Figure 6.24 show a widely observed trend. As the weather gets warmer, the duration between sowing and harvest shortens. Temperature limits the rate of biological processes and the warmer the temperature gets the faster the crops grow (provided it doesn't get too hot and other factors don't become limiting). A common approach to relate this temperature effect on crop development is to calculate the accumulated heat units from day of sowing

$$\begin{cases} A(T) = 0 & T < T_S \\ A(T+1) = A(T) + \max(0, Y(T+1)) & T \geq T_S, \end{cases} \tag{6.25}$$

where $A(T)$ is accumulated temperature, T is time in discrete days, T_S is the day of crop sowing, and Y is mean daily temperature. Here, we use T to denote time in discrete daily increments in contrast with continuous time t in (6.24) above. Figure 6.25 shows the dynamics of $A(T)$ for three radish fields in 2018 sown at different times.

For our model, we assume crops are ready to be harvested once accumulated temperatures exceed a threshold Γ

$$\begin{cases} H(T) = 0 & A(T) < \Gamma \\ H(T) = 1 & A(T) \geq \Gamma, \end{cases} \tag{6.26}$$

where $H(T)$ is a Boolean indicator of whether the crop is mature for harvesting ($H(T) = 1$) or not ($H(T) = 0$). The harvest date T_H is defined as the first occurrence when accumulated temperatures have reached the specific threshold:

$$T_H = \min_T H(T) = 1 \tag{6.27}$$

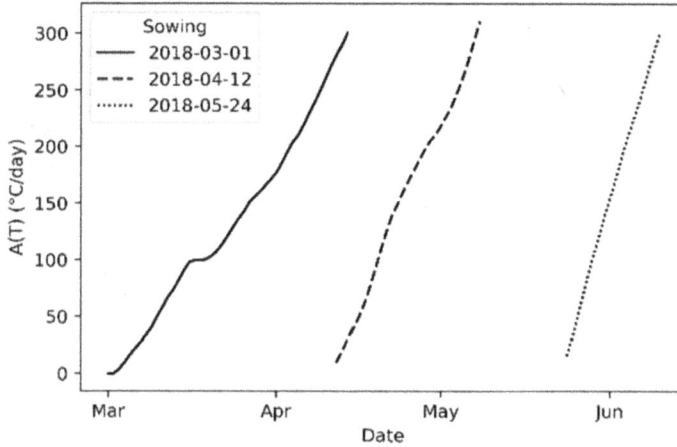

Figure 6.25: Accumulated temperature, $A(T)$, by radish crop plants planted at three different times of the year. Note how $A(T)$ increases faster later in the year.

Note that the crop model in this example is relatively simple and the vector of crop properties, $C(T)$ in this case, as defined in our general crop model (6.24), is a vector indicating the accumulated temperature by the crop and whether the crop is ready to be harvested:

$$C(T) = \begin{bmatrix} A(T) \\ H(T) \end{bmatrix}.$$ (6.28)

Driver and calibration data

Process based models can use a variety of data types as inputs. Driver data is input data that is used to represent conditions the crop is experiencing. This is the "Environment" and some of the "Management" data in Figure 6.23 and represented by $E(t)$ and $M(t)$ in (6.24) and is the temperature data experienced by the radish crops in our example.

Process-based crop models also require the specification of initial conditions, both for the crop and the environment. In our radish example, the initial conditions are the dates at which the crops are planted.

Calibration data (or training data) is used to enable the **learning of uncertain model parameters**, represented by θ in (6.24). Such calibration data typically represents properties of the crop at times in the past that can be compared against model predictions. Such data has traditionally come from on-farm measurements, manual surveys, and research projects. More recently, the increased availability of remotely sensed observations is used, typically from satellites. In our radish example, the calibration data are the farm records shown in Figure 6.24 indicating the timing of sowing and harvesting the crops.

Model calibration/parameter inference

To calibrate the process-based model, we apply **Bayesian parameter inference** to learn the most likely probability distribution of the model parameters θ, given what was observed in the past. We do this by first expressing the relationship between the model predictions and the calibration data as a probability that the data was created by the process simulated by the model (equations (6.25–6.28)). This relationship is illustrated in a factor graph in Figure 6.26: we assume that the accumulated temperatures observed by the harvested plants is drawn from a **normal distribution** centred on an inferred "critical accumulation temperature" Γ, and with an **unknown observational error** σ. These are the model parameters θ.

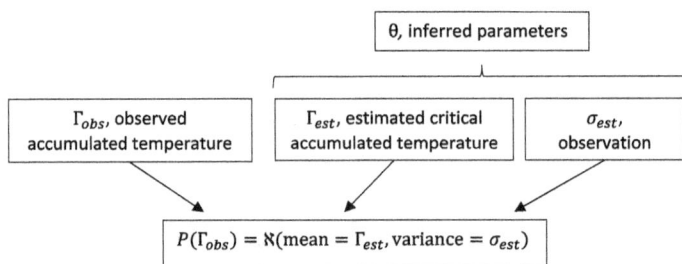

Figure 6.26: Factor graph showing the relationship between the probability of the observations and the learned parameters of the radish model (2–5).

In this simple case, which also represents a typical starting point for model calibration, we use just one model and do not make any prior assumptions about the probability distributions of θ. This has the simplifying effect of transforming our probability relationship down to one in which we approximate the probability distribution of the model parameters, given the observations as

$$\log(P(\theta|\Gamma_{obs})) \approx \sum_{i=0}^{n} \log(\aleph(\Gamma_{obsi}, \text{mean} = \Gamma_{est}, \text{variance} = \sigma_{est})), \qquad (6.29)$$

where $P(\theta|\Gamma_{obs})$ is the probability of the model parameters given the observed accumulated temperatures (worked out from the harvest timings) and $\aleph$ is a normal distribution with a mean of Γ_{est} and a variance of σ_{est}. In other words, we approximate the log probability of the parameters as the sum of probabilities of the observed accumulated temperatures to maturity. We infer these parameters using MCMC rather than sweeping through possible values of the parameters because it makes it simpler in future to extend the model to having more complexity and parameters than is practical to evaluate through basic parameter sweeps.

Implementing the **MCMC** algorithm for the radish model generates Markov chains for the threshold accumulated temperature before the crop is mature, Γ, and the observational error σ as shown in Figure 6.27.

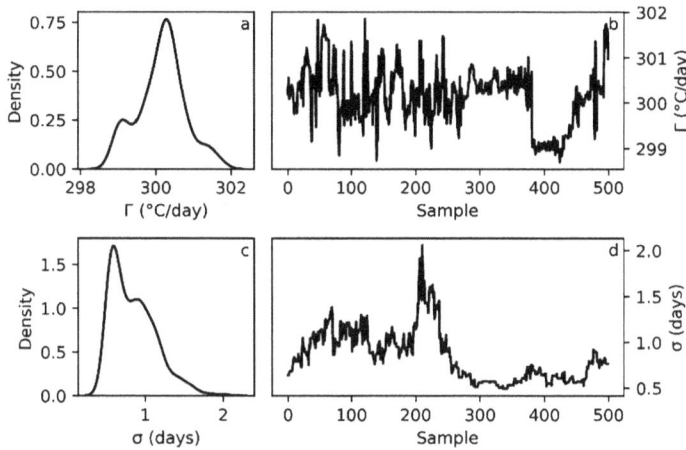

Figure 6.27: Example Markov chains of parameter estimates for the Radish model defined above (2–5). Inferred probability distributions (a) and (c) are approximated via **the sampled Markov chains** (b) and (d), respectively.

Forecasting and uncertainty propagation

Fresh produce forecasting typically involves using crop production forecasts throughout the season. Forecasts for the whole year enable experimenting with alternative planting regimes to identify the best match to expected demand throughout the year. Forecasts during the year allow growers to adjust their crop management to adapt to changes in growing conditions and demand.

For our radish example, we make forecasts of crop maturation using the **learned values of Γ and σ.** Rather than forecast the maturation of a single crop over time, we assume that the producer is planting multiple crops to grow sufficient quantities to meet retailer demand. The producer can use the trained model to forecast maturation dates for multiple crops and combine these to generate an expected **supply distribution.** For our example, here we assume that the producer would sow blocks of 10,000 plants every 3 days from March to July. The resulting supply distribution is shown in Figure 6.28.

Figure 6.28 shows the emergent dynamics of radish supply from the producer following a regular planting schedule. Those supply dynamics are notably irregular as a consequence of the interaction between crop growth and the weather. The challenge for the producer is to manage their planting, management, and harvesting so as to best match their supply to market demand. The additional probability information on the supply distribution gives the producer an indication of the relative likelihood of differ-

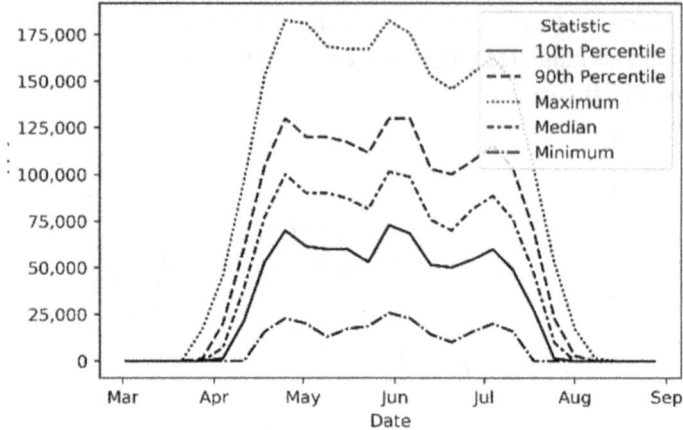

Figure 6.28: Forecasted supply over time for radishes using our calibrated radish crop growth model for a producer who sows blocks of 10,000 plants every 3 days from March to July.

ent levels of supply and, therefore, the relative risk of over or under supply given a level of demand.

Problems and alternatives

The **Bayesian parameter estimation** method described above is just one of many methods that can be used to calibrate crop models. A challenge with the approach is that parameter inference is slow, computationally expensive and requires an inference specialist. Consequently, the method is best suited for model development prior to operational model deployment. Alternative methods are better suited to **real-time inference** during the growing season such as ensemble **Kalman filter** and **particle filtering** approaches, as described elsewhere in this Chapter.

6.5.3 Estimating players' skills with XBOX TrueSkill™

Nicolas Vercheval, Aleksandra Pižurica

6.5.3.1 Business context

With millions of active gamers at any time, the success of extensive online gaming services such as Xbox Live® relies on their ability to estimate players' skills correctly and quickly match them accordingly. Playing against an opponent of comparable skill grants the user a challenging experience with fair chances of victory.

Rating points reflect a player's skill level and are updated to an increased value after a win or a decreased value after a loss, adjusting the winning (or losing) points by the level of opposition. Early rating systems rank players consistently and extract reliable winning probabilities but do not estimate the confidence that the rating is correct and require additional parameters to compensate when, for example, a new player joins the platform.

In 2005, Xbox Live introduced TrueSkill™, a light Bayesian rating system that models the game's uncertainty and human factors with probability distributions. It adapts the update according to the reliability of the ratings, shows rapid convergence, and naturally extends to scenarios that had previously required arbitrary and ad hoc modifications, such as players gathering in teams or multiple teams competing in the same game. This example demonstrates how factor graph representation and belief propagation allowed Trueskill to be up to the challenges posed by Xbox Live.

6.5.3.2 Overall approach

The outcome of a game has many elements of chance: game luck, players' concentration, untimely pizza delivery, and so on. Matching two players according to their game history involves factoring in countless possibilities that may be very specific to the game or entirely unknown. Probabilistic reasoning helps navigate this uncertain environment by describing *latent causes* in terms of their likely effect on the evidence.

Trueskill captures the randomicity of a game outcome in a random variable for each player's performance. The winner of a game is then the player with the highest performance, and the question: "Who is going to win?" becomes "Whose performance is more likely to be superior?". The assumption is that performances are mutually independent and only determined by the player's skill and luck. The skill, seen as a performance average, allows strong players to win against weaker opponents consistently. Luck explains why the weaker player sometimes wins, or a head-to-head match does not always give the same outcome.

A possible simplification assumes a player's skill is an unknown constant that does not vary over time. Nonetheless, the belief of its likely value evolves with every game they play. Inferring the distribution for the current belief is thus an example of probabilistic reasoning over time, precisely, a *filtering* problem. This illustrative example keeps this assumption for simplicity. In practice, allowing a player's skill to vary over time helps describe periods of higher or lower performances (such as an off-day where a player loses significantly more games than expected) and catch up with new players as they improve.

In Trueskill, a player's *rating* is not just a number but a compact representation of the belief about their skill. This solution allows better decision making and more precise inference; conversely, it requires an approximation to be consistent for all the players. Trueskill approximates its belief for a player's skill with a Gaussian distribution, whose mean and variance reflect the perceived skill and the confidence in estimating their skill as data are collected. Since the Gaussian distribution is well known to describe additive noise, a player's luck is also normally distributed, and their performance is the sum of the their luck and skill.

User experience requires quick inference during the pairing. The *dynamic* environment continuously registers the outcomes of games from other players, but factoring those in would endlessly delay the inference. Similarly, explicitly modeling all the past game history would have an infeasible cost in complexity. The assumption of the Markov property greatly simplifies the model by limiting the inference to a simple update: each game updates the beliefs of the player's skills only based on their previous beliefs and the game outcome. It also allows a *factor graph* representation, elegantly explaining inference with *belief propagation*.

Trueskill extends to more general setups by adding other model assumptions. A latent Gaussian variable represents a player's skill fluctuation from game to game, allowing it to vary. A team's performance is the sum of the individual ones. Draws occur when performance differences are too small to be decisive. These extensions make the factor graph more complex but maintain the same inference algorithm. We do not present these details here; instead, we refer interested readers to Winn (2023), which explains the complete matchmaking strategy and ranking criteria. A Python implementation, closely following this example, is available at Vercheval (2024).

In this example, two players, Jill and Fred, play against each other in a game where no draws are allowed. In the last section, a third player, Steve, joins in.

6.5.3.3 Predicting the outcome

When the game is between two players, their winning probabilities have a complicated analytical solution. Belief propagation helps break down their calculations into steps. We visualize these steps with a factor graph (see Figure 6.29) describing the joint distributions of Jill's and Fred's skills, luck, and performances.

We run belief propagation on the graph top-to-bottom to update the probability that Jill wins, or in other words, to infer the marginal distribution of the event that Jill wins. Since no draw is allowed, Fred's winning probability immediately results from it. The correct algorithm for this purpose is the sum-product algorithm we have seen in (6.3.5.1) for discrete variables. Even though most variables here are continuous, the algorithm is entirely equivalent.

The first step assigns the current belief to the initial variables. In particular, we use the players' current ratings to estimate their skill distributions. According to our assumptions, the skill of player P follows a normal distribution with mean μ and variance σ^2 given by their rating:

$$\text{Skill}^P \sim \mathcal{N}(\mu, \sigma^2).$$

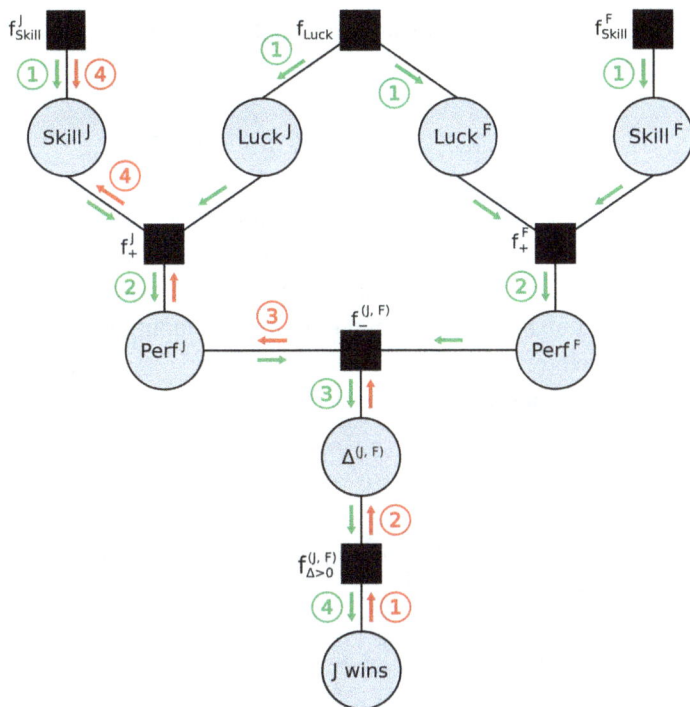

Figure 6.29: The messages in green infer the winning probabilities. The likelihood messages in red update Jill's rating.

In our example, Jill has a rating of $(\mu = 120, \sigma^2 = 1600)$ and Fred has a rating of $(\mu = 100, \sigma^2 = 25)$; Jill is likely the stronger player, but her actual ability is unclear.

The factors f_{Skill}^J and f_{Skill}^F describe Jill's and Fred's skill distributions. Given that these are continuous variables, the factors, and thus their messages, are probability density functions. For simplicity, we use the same notation for densities and relative distributions. With this notation, the messages for the nodes Skill^J and Skill^F are:

$$m_{f_{\text{Skill}}^J \to \text{Skill}^J} = \mathcal{N}(120, 1600), \quad m_{f_{\text{Skill}}^F \to \text{Skill}^F} = \mathcal{N}(100, 25).$$

The two players' luck variables, noted as Luck^J and Luck^F, receive a similar message, but the message, in this case, is the same for every player P:

$$f_{\text{Luck}} = m_{f_{\text{Luck}} \to \text{Luck}^P} = \mathcal{N}(0, 25).$$

This distribution is a fixed setting. Thus, the players' luck variables are independent.

Factor f_+^P represents player P's performance, a sum of their skill and luck. A known property of the Gaussian distribution is that the sum of two independent, normally distributed variables follows a Gaussian distribution with the sum of their means as mean and the sum of their variances as variance. We apply this formula to the performance Perf^P:

$$\text{Perf}^P = \text{Skill}^P + \text{Luck}^P \sim \mathcal{N}(\mu + 0, \sigma^2 + 25). \tag{6.30}$$

The message from factor f_+^P is the above distribution's density and has a functional form. The sum-product algorithm for continuous variables typically involves a complex integral that can only be approximated numerically. However, the properties of the Gaussian distribution enable us to avoid intensive computation and calculate the messages through exact formulas.

In the second step, we leverage the above property and send the following messages:

$$m_{f_+^J \to \text{Perf}^J} = \mathcal{N}(120, 1600 + 25), \quad m_{f_+^F \to \text{Perf}^F} = \mathcal{N}(100, 25 + 25).$$

The performance nodes then relay the message to the factor $f_-^{(J, F)}$, representing the difference between the two performances.

In the third step, we use an analogous formula for the difference of two independent Gaussian variables and get the following message:

$$m_{f_-^{(J, F)} \to \Delta^{(J, F)}} = \mathcal{N}(120 - 100, 1625 + 50) = \mathcal{N}(20, 1675).$$

We forward the message to the final factor $f_{\Delta>0}^{(J, F)}$, which converts the estimation of their performances into a winning probability.

This is possible by setting factor $f_{\Delta>0}^{(J, F)}$ equal to the following step function:

$$f_{\Delta>0}^{(J, F)}(d) = \mathbb{1}_>(d) = 0 \text{ if } d < 0 \text{ or } 1 \text{ if } d \geq 0.$$

With this factor, the sum-product message corresponds to the probability that the performance difference is more than zero, implying that Jill's performance is higher than Fred's, given what we know about their skills.

To get this probability, we need the value of the cumulative function $F_{\Delta^{(J, F)}}$ of the performance difference distribution calculated in zero. We obtain it by calculating the cumulative distribution of a standard Gaussian $\Phi(x)$ on the normalized performance differential:

$$m_{f_{\Delta>0}^{(J, F)} \to J \text{ wins}} = P(\Delta^{(J, F)} > 0) = 1 - F_\Delta^{(J, F)}(0) = 1 - \Phi\left(\frac{-20}{\sqrt{1675}}\right) \approx 68.7\% \tag{6.31}$$

Since "J wins" is the last node in the propagation, the only incoming message is its inferred probability (step 4). As expected, Jill is the predicted winner.

Updating the rating

We update the rating with the posterior distribution of a player's skills after observing the game outcome, in this case, "J Wins". According to the Bayes' rule:

$$p(\text{Skill}^P | \text{J Wins}) = \frac{p(\text{J Wins} | \text{Skill}^P) p(\text{Skill}^P)}{Z}, \tag{6.32}$$

where $\frac{1}{Z}$ is a normalization constant.

The posterior distribution of Skill^P given the outcome must again be Gaussian to respect the model assumptions, and its mean and variance become the new rating of Player P. The posterior distribution is Gaussian when both the prior distribution and the likelihood function $p(\text{J Wins} | \text{Skill})$ are.

More precisely, let $\mathcal{N}(s; \mu, \sigma^2)$ be a prior density on the variable s and s be in turn the mean of a likelihood function $\mathcal{N}(x; s, \bar{\sigma}^2)$, then:

$$\frac{\mathcal{N}(x; s, \bar{\sigma}^2) \mathcal{N}(s; \mu, \sigma^2)}{Z} = \mathcal{N}\left(s; \left(\frac{x}{\bar{\sigma}^2} + \frac{\mu}{\sigma^2}\right)\sigma', \sigma'\right), \tag{6.33}$$

where the new variance is equal to $\sigma' = (\frac{1}{\bar{\sigma}^2} + \frac{1}{\sigma^2})^{-1}$. Unfortunately, it turns out that $p(\text{J Wins} | \text{Skill})$ is not Gaussian. To show where the problem originates, we run belief propagation to node Skill^J (see again Figure 6.29), starting from the observed outcome.

Let us say that Jill wins. In the first step, the variable "J wins" signals factor $f_{\Delta>0}^{(J,F)}$ that the difference in Jill's and Fred's performance is more than zero because Jill has the higher performance. In the second step, factor $f_{\Delta>0}^{(J,F)}$ translates this information into the following message:

$$m_{f_{\Delta>0}^{(J,F)} \to \Delta^{(J,F)}} = \mathbb{1}_{>}(d). \tag{6.34}$$

This message is problematic because it is not a Gaussian density, and because of that, none of the following messages are Gaussian. Even worse, it is not the density of any distribution, and no density can approximate it numerically.

Trueskill does not approximate the message directly. Instead, it replaces the original message with a Gaussian density that impacts the probabilistic graph in a similar way. Specifically, it chooses the new message so that the resulting inference on the performance difference given the outcome is as close as possible to its exact posterior distribution. This technique, called *expectation propagation*, is more advantageous than approximating the posterior distribution of the skill directly because it ties the error to the performance node and allows exact inference in the rest of the graph, which can be much more complex than in this example.

We calculate the exact posterior distribution of the performance difference by multiplying the message containing the information that Jill has won, which is the likelihood function, with the message about the expected performance difference, which is

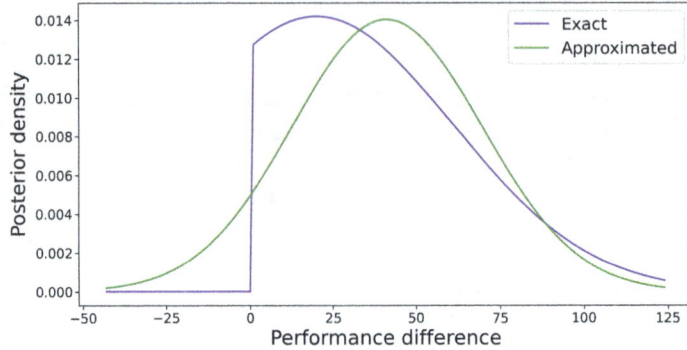

Figure 6.30: Jill has won and, therefore, the true posterior distribution of the performance difference cannot be negative. The normal approximation has a smoothing effect and leaves a small chance that Fred's performance was higher.

the density of the prior distribution we calculated in the previous section. By the Bayes' rule, we get a function proportional to the posterior density:

$$p_{\Delta^{(J,F)}|J \text{ wins}}(d) = \frac{(m_{f^{(J,F)}_{\Delta>0} \to \Delta^{(J,F)}})(m_{f^{(J,F)}_{-} \to \Delta^{(J,F)}})}{Z} = \frac{\mathbb{1}_{>}(d)\mathcal{N}(d; 20, 1675)}{Z}. \tag{6.35}$$

The Z in the normalizing constant $\frac{1}{Z}$ is the probability $P(\text{J wins}) = 0.687$ that we have calculated in the previous section.

We show the density in (6.35) in Figure 6.30. A complicated formula we omit for readability gives us its mean (≈ 41.1) and variance (≈ 809.2). We use these statistics to perform approximate inference.

We replace the likelihood message $\mathbb{1}_{>}(d)$ with $\mathcal{N}(60.8; d, 1565.6)$. Using the formula in (6.33), we get:

$$\frac{\mathcal{N}(60.8; d, 1565.6)\mathcal{N}(d; 20, 1675)}{0.687} = \mathcal{N}(d; 41.1, 809.2),$$

which approximates the exact posterior distribution of the performance difference with a Gaussian of identical mean and variance (see Figure 6.30).

Now that we know a good approximation for:

$$m_{f^{(J,F)}_{\Delta>0}} \to \Delta^{(J,F)} \approx \mathcal{N}(60.8, 1565.6),$$

we conclude the second step by forwarding the message to $f_{-}^{(J,F)}$.

In the third step, we proceed to update Jill's rating. We reuse the message about Fred's expected performance calculated in the previous section and the formula for the sum of two independent Gaussian variables to infer Jill's performance from the performance difference:

$$m_{f_{-}^{(J,F)} \to \mathrm{Perf}^J} = \mathcal{N}(60.8 + 100, 1565.6 + 50) = \mathcal{N}(160.8, 1615.6).$$

In the fourth and last step, we calculate the likelihood function for Jill's skill. Factor f_+^J receives the forwarded message of the previous step and the message from Luck^J that we have already calculated. The factor takes into account the noise from that variable and sends a message that is less confident about Jill's demonstrated ability. The formula of the previous section still applies, and we get:

$$m_{f_+^J \to \mathrm{Skill}^J} = \mathcal{N}(160.8, 1615.6 + 25) = \mathcal{N}(160.8, 1640.6).$$

The message-passing algorithm terminates with the product of the messages to the final variable, Skill^J, which is proportional to the posterior in (6.32). The first message is the density of the prior distribution from f_{Skill}^J that we have calculated at the start of this example; the second is the likelihood function approximate after seeing the outcome. Using the formula in (6.33):

$$p(\mathrm{Skill}^J | J \text{ wins}) = \frac{\mathcal{N}(120, 1600)\mathcal{N}(160.8, 1640.6)}{Z} = \mathcal{N}(140.1, 810.0).$$

The resulting density gives us a rating of ($\mu = 140.1, \sigma^2 = 810.0$). After the game, the model is more confident about Jill's skill level. Jill's rating now has a higher mean and a lower variance. The algorithm for updating Fred's rating is analogous and returns a new rating of ($\mu = 99.7, \sigma^2 = 24.8$).

6.5.3.4 Multiplayer setting

In the previous sections, belief propagation only helped explain approximated inference. In a multiplayer setting, the computation graph becomes fundamental for inference.

When Steve joins the game, its outcome becomes a ranking. Steve's rating is ($\mu = 140, \sigma^2 = 1600$). Steve is supposed to be the better player but places third, Fred second, and Jill first. We lay the players' performances in line according to their ranking (Figure 6.31) and compare the adjacent ones.

Jill places first and, therefore, must have a higher performance than Fred's. The algorithm works the same as before (steps 1, 2, and 3) and returns the likelihood message on Fred's performance:

$$m_{f_-^{(J,F)} \to \mathrm{Perf}^F} = \mathcal{N}(120 - 60.8, 1565.6 + 1625) = \mathcal{N}(59.2, 3190.6).$$

The update tells us that Fred may not be in his best form. Instead of immediately updating Fred's rating, the algorithm uses this information to contextualize its comparison to Steve's performance. Therefore, in step 4, it multiplies the likelihood message with the prior information on Fred's performance and sends the resulting message:

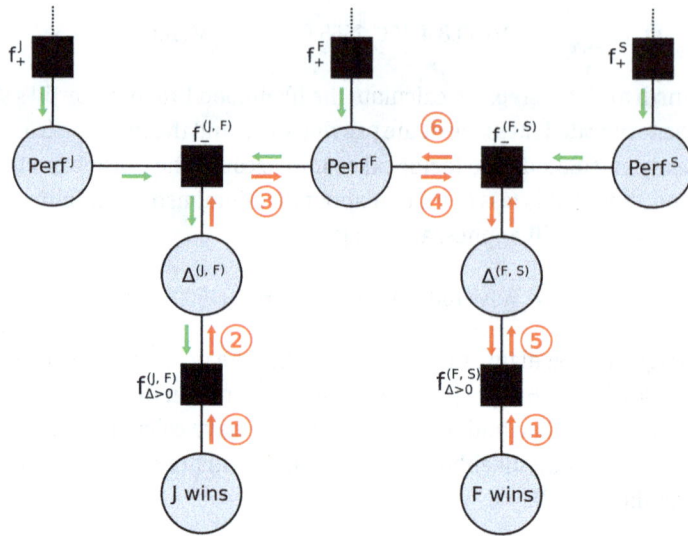

Figure 6.31: The messages containing prior information about the variables are in green. The likelihood and posterior messages in the first iteration of expectation propagation are in red.

$$m_{\mathrm{Perf}^F \to f_-^{(F,S)}} = (m_{f_+^F \to \mathrm{Perf}^F})(m_{f_-^{(J,F)} \to \mathrm{Perf}^F}) = \mathcal{N}(99.4, 49.2).$$

The message above approximates $p_{\mathrm{Perf}^F|J\text{ wins}}$ and contains posterior information about Fred's performance after losing to Jill, but no evidence that it is higher than Steve's. The algorithm treats it as prior information for what concerns the estimation of their performance difference and proceeds as in the previous section (step 5).

After seeing that Fred's performance is superior to Steve's (step 6), the model is more optimistic about it and sends a new likelihood function to Perf^F:

$$m_{f_-^{(F,S)} \to \mathrm{Perf}^F} = \mathcal{N}(177.1, 2043.4).$$

The resulting posterior density on Fred's performance distribution given the previous outcome:

$$p_{\mathrm{Perf}^F|F\text{ wins}} \approx (m_{f_+^F \to \mathrm{Perf}^F})(m_{f_-^{(F,S)} \to \mathrm{Perf}^F}) = \mathcal{N}(101.8, 48.8)$$

is a bit higher than the original prior density and could be sent back to $f_-^{(J,F)}$ to reassess the performance difference between Jill and Fred. This reassessment is necessary because expectation propagation performs an approximation whose error propagates from one performance difference to the other. We calculate the likelihood approximation one last time. In general, expectation propagation is an iterative method that stops when the messages change by a tolerated value. In our case, the current estimation of

$p_{\text{Perf}^F|\text{F wins}}$ is good enough, and we can use it to update Jill's rating as we did in the previous section. Similarly, we use the new estimation of $p_{\text{Perf}^F|\text{J wins}}$ to update Steve's rating.

Jill's new rating ($\mu = 141.1, \sigma^2 = 791.1$) is slightly higher and more definite than in the previous section because she also beat Steve. Conversely, Steve's new rating ($\mu = 80.6, \sigma^2 = 376.6$) is much lower because the model realized its initial misplaced estimation.

To update Fred's rating, we multiply the two incoming likelihood messages to his performance node to obtain aggregate information about his success, and then continue the propagation from the following message as before:

$$m_{\text{Perf}^F \to f_+^F} = (m_{f^{(J, F)} \to \text{Perf}^F})(m_{f^{(F, S)} \to \text{Perf}^F}) = \mathcal{N}(130.8, 1234.1).$$

Fred's new rating ($\mu = 100.6, \sigma^2 = 24.5$) is similar to its initial estimation. Indeed, the model used his rating, in which it was already confident, as a reference for the other two.

6.5.3.5 Conclusions

In previous sections, we illustrated Trueskill's basic functionality. Its adaptable modelling approach allows for extending the graph by introducing more assumptions and adding application-specific constraints. This flexibility is valuable for adapting to changing model requirements and ensures easy maintenance. It also enables portability across different games, providing a consistent user experience and reducing development hours. Trueskill also serves as an excellent tool for integrating into policies and decision-making processes, thanks to its ability to quantify confidence in its estimation. With theoretical properties translating into concrete advantages, Trueskill showcases how rigorous probabilistic inference can offer the flexibility and reliability that businesses seek.

6.6 Useful reminders of probability theory

6.6.1 Basic concepts in probability

In probabilistic reasoning, we are dealing with randomness, arising because the process (or a "trial") in whose outcomes we are interested *didn't happen yet* (e. g., tomorrow's match) or the *already existing outcome is uncertain* (e. g., due to various measurement imprecisions). Here, we review briefly the basic principles of probability and we explain the notation that we will use.

Random events and axioms of probability. The space of all possible outcomes of a given trial is called the *sample space*. Let Ω be this sample space and $\omega \in \Omega$ its elements.

A random **event** is any subset of the sample space $\theta \subset \Omega$ and its elements ω are *atomic events*. Take as an example rolling a die. In this case, $\Omega = \{1, 2, 3, 4, 5, 6\}$. An event "die roll > 4" is a subset $\{5, 6\} \subset \Omega$, which contains atomic events 5 and 6. The basic principles underpinning the probability theory are: *the probability of each outcome lies between 0 and 1*: $0 \leq P(\omega) \leq 1$, and *the probabilities of all the possible outcomes sum up to 1*: $\sum_{\omega \in \Omega} P(\omega) = 1$. Furthermore, *the probability of any event is the sum of the atomic events where this event is true*: $P(\theta) = \sum_{\omega \in \theta} P(\omega)$. For example, assuming a fair die, the probability of each outcome is 1/6, their sum is 1, and P(die roll > 4) = $P(5) + P(6) = 1/6 + 1/6 = 1/3$. These three principles are the **axioms of probability**. From these three axioms, all other rules that hold in the probability theory can be derived, including the one which says that the probability of a union of two events is the sum of their probabilities minus the probability of their intersection: $P(a \lor b) = P(a) + P(b) - P(a \land b)$.

Random variables versus events. What is the difference between events and random variables? An event, as was defined above, corresponds to the term *proposition* in logical reasoning, and informally it can be described as any meaningful statement about the experiment we are interested in (e. g., "die roll > 4" or "die roll is an even number", etc.) Thus, an event happens or not with some probability, while a **random variable** is a variable whose value is affected by some random phenomenon. Typically, the values of random variables are *real numbers* but in some cases also Boolean values {*true, false*}. In the die roll experiment, we can define the event "die roll > 4" as a random variable X whose possible values are *true* or *false*, or if we define X as the outcome of rolling, its possible values are $\{1, 2, 3, 4, 5, 6\}$. In general, we will denote by a capital letter a random variable, for example, X and by the corresponding small letter its particular value that we call *realization*, for example, x. We use boldface letters to denote vectors of random variables $\mathbf{X} = (X_1, \ldots X_n)$ and their realizations $\mathbf{x} = (x_1, \ldots x_n)$. We also say that $X = x$ is an event where the random variable X takes the value x.

Discrete random variables. In the examples above, we illustrated *discrete* random variables, which may take on only a countable (finite or infinite) number of distinct values. Another example of a discrete random variable is a class label, where x is one of the possible classes. The probability that X takes the value x is commonly denoted by $P(X = x)$ or by $P_X(x)$. Also, for compactness, we will use sometimes only $P(x)$ when there can be no confusion to what this refers. We apply the same convention to the vectors of random variables $P(\mathbf{X} = \mathbf{x}) = P_\mathbf{X}(\mathbf{x})$, and for short by $P(\mathbf{x})$ when no confusion is possible. Boldface $\mathbf{P}$ in $\mathbf{P}(X)$ (or in $\mathbf{P}(\mathbf{X})$) denotes a vector where each entry is the probability of a particular realization of X (or $\mathbf{X}$).

Continuous random variables. A *continuous* random variable can take infinitely many values. Examples are height or weight of a person, air temperature, distance covered, or blood sugar level. Continuous random variables are characterized by the **probability**

density function (pdf), also called *density*, $p_X(x)$. The pdf integrates to 1: $\int_{-\infty}^{\infty} p_X(x) = 1$, and the *cumulative* function is

$$F_X(x) = P(X \le x) = \int_{-\infty}^{x} p_X(x)dx \tag{6.36}$$

In the problems that we addressed in this chapter, the data distribution is typically a continuous random variable and the labels associated with each data sample are discrete random variables.

Prior and conditional probabilities. Prior probabilities express a priori belief (about the value of some random variable), without any (new) evidence (i. e., without any measurements or observations). For example, let R be the random variable expressing "it will rain tomorrow," with two possible values $r \in \{true, false\}$. Then the prior probability $P(R = true) = 0.3$ is reasonable in July in Antwerp (30 % of days are rainy), and $P(R = true) = 0.04$ in Lisbon. *Conditional* probabilities express *belief given some evidence*. For example, we know the current value of barometric air pressure B (in mbar) in Antwerp, and we refine the probability of rain as $P(R = true|B = 992.21) = 0.43$. Both for prior and conditional probabilities, we often use the more compact representation, same as we explained above, for example, $P(R = r|B = b) = P_{R|B}(r|b)$, and for a conditional probability density function $p_{B|R}(b|r)$ (since B is a continuous random variable), and when no confusion possible we also use a shorter notation, like $P(r|b)$ or $p(b|r)$. Formally, we define the conditional probability of x given y as

$$P(x|y) = \frac{P(x \wedge y)}{P(b)} \quad \text{if } P(y) \neq 0 \tag{6.37}$$

The *product rule* gives an alternative formulation: $P(x \wedge y) = P(x|y)P(y) = P(y|x)P(x)$, and successive application of the product rule gives the well-known **chain rule**:

$$P(x_1, \ldots x_n) = P(x_n|x_1, \ldots, x_{n-1})P(x_{n-1}|x_1, \ldots, x_{n-2}) \ldots P(x_1) = \prod_{i=1}^{n} P(x_i|x_1, \ldots, x_{i-1}) \tag{6.38}$$

For example, $P(x_1, x_2, x_3) = P(x_1)P(x_2|x_1)P(x_3|x_2, x_1)$. This chain rule has many practical uses and we will see later on how its special case applies to Bayesian networks.

Statistical independence and conditional independence. Random variables X and Y are statistically independent if $P(x|y) = P(x)$ or equivalently $P(y|x) = P(y)$ or equivalently $P(x \wedge y) = P(x)P(y)$. The statistical independence, when it holds, simplifies a lot the probabilistic modeling and inference: if we have n binary random variables, 2^n parameters are needed in general to describe their joint probability distribution, and if they are all statistically independent, this amount is only n. Hence, statistical independence reduces the amount of the parameters exponentially. Unfortunately, strict statistical independence among the random variables of interest may not hold in practice. However, *conditional* independence often applies to some subsets of the involved random variables. For example, low barometric pressure increases the chance of rain and headache, but the probability that it will rain is conditionally independent of whether one has headache or not, given that the air pressure is known: $P(rain|lowpressure, headache) = P(rain|lowpressure)$.

Bayes' rule. In probabilistic reasoning, the *Bayes'* rule (also known as the Bayes' theorem), named after Thomas Bayes (1701–1761), has a central place as it connects the *diagnostic* inference (inferring the cause given the effects) to the *causal* inference (the probability of effects given the cause). Formally, the Bayes' rule is

$$P(x|y) = \frac{P(y|x)P(x)}{P(y)} \tag{6.39}$$

Observe that this expression follows directly from the expression for the conditional probability (6.37) and the product rule that was written below it. The true value of the Bayes' rule is that it facilitates the inference. Think of medical diagnosis, where $P(x|y)$ is the probability of disease x given symptoms y. Inferring $P(x|y)$ directly is very difficult because the same symptoms can arise due to various diseases (e. g., symptoms like headache or fever can point to numerous diseases, from benign to very serious ones). The Bayes' rule allows us to solve this problem by inferring instead the probability of symptoms given the disease $P(y|x)$, and the prior probabilities of the disease $P(x)$ and symptoms $P(y)$ (e. g., in a population of interest). Typically, these can be readily estimated based on experience, collected data, and earlier statistical analyses.

6.6.2 Maximum likelihood estimation and Bayesian estimation

We describe three common estimators in statistical inference and probabilistic reasoning, and we comment on their use and mutual relationships.

Maximum likelihood estimation. In statistics, probabilistic reasoning, and machine learning, *maximum likelihood estimation* (MLE) is commonly employed for estimating the parameters of an assumed probability distribution, given observed data. The parameter values are found by maximizing the likelihood that the process described by the assumed model produced the observed data. The *likelihood function* $L(\theta|y) = p(y|\theta)$ measures the support provided by the data y for each possible value of the parameter θ. If $L(\theta_1|y) > L(\theta_2|y)$, we know that our observed data point y is more likely to have been generated under the model with $\theta = \theta_1$ than under the value $\theta = \theta_2$, so the observed data point tells us that θ_1 is more plausible than θ_2.

Hence, if our observations are y, and if we model their actual (unknown) distribution with a statistical model $p(y|\theta)$, with parameters θ, we find θ by maximizing the likelihood function $L(\theta|y)$:

$$\hat{\theta} = \arg\max_{\theta} L(\theta|y) = \arg\max_{\theta} p(y|\theta) \tag{6.40}$$

Suppose that we have observed measurements $y_1, \ldots y_N$, and we believe the underlying process y follows a normal distribution $\mathcal{N}(\mu, \sigma^2)$. We want to find the parameters μ and σ under which the assumed model represents the observed data in the best way possible. The likelihood function is the joint probability $p(y_1, \ldots, y_N|\theta)$, where $\theta = (\mu; \sigma)$. If our observations are statistically independent,

$$L(\mu; \sigma|y_1, \ldots, y_N) = p(y_1, \ldots, y_N|\theta) = \prod_{i=1}^{N} p(y_i|\theta) = \left(\frac{1}{\sigma\sqrt{2\pi}}\right)^N e^{-\frac{1}{2\sigma^2}\sum_{i=1}^{N}(y_i-\mu)^2} \tag{6.41}$$

Maximizing this likelihood function is equivalent to maximizing its logarithm $\ell = \ln L$, that is,

$$\ell(\mu; \sigma|y_1, \ldots, y_N) = \ln L(\mu; \sigma|y_1, \ldots, y_N) = -\frac{N}{2}\ln(2\pi) - \frac{N}{2}\ln(\sigma^2) - \frac{1}{2\sigma^2}\sum_{i=1}^{N}(y_i - \mu)^2 \tag{6.42}$$

since the logarithm is a monotonic function. The MLE estimates of the parameters μ and σ^2 are then obtained by setting the derivatives of ℓ with respect to μ and σ^2 to zero, resulting in the estimates:

$$\hat{\mu} = \frac{1}{N}\sum_{i=1}^{N} y_i, \quad \text{and} \quad \hat{\sigma}^2 = \frac{1}{N}\sum_{i=1}^{N}(y_i - \hat{\mu})^2 \tag{6.43}$$

which are exactly the empirical mean and variance. This estimation is illustrated in Figure 6.32. We can generalize this procedure to the case where our observations are vectors $\mathbf{y}_i$ (i.e., each measurement is not a single number but consists of d components $\mathbf{y}_i = [y_{i,1}, \ldots y_{i,d}]$) modeled by a *multivariate* normal distribution $\mathbf{y} \sim \mathcal{N}(\boldsymbol{\mu}, \Sigma)$, with mean $\boldsymbol{\mu}$ and the covariance matrix Σ. Applying the same procedure as above, we obtain the MLE estimates of the parameters as

$$\hat{\boldsymbol{\mu}} = \frac{1}{N}\sum_{i=1}^{N} \mathbf{y}_i, \quad \text{and} \quad \hat{\Sigma} = \frac{1}{N}\sum_{i=1}^{N}(\mathbf{y}_i - \hat{\boldsymbol{\mu}})(\mathbf{y}_i - \hat{\boldsymbol{\mu}})^{\top}. \tag{6.44}$$

Examples with the normal distribution are intuitive, but we can apply the same procedure to estimate the parameters of other, arbitrary statistical models.

In some cases, the conditional probability distribution of the observations y given some underlying unobservable data x is known. For example, x are ideal pixel intensities of a digital image and y are their measured values affected by additive white Gaussian noise of zero mean and known standard deviation. Now we don't need to estimate the parameters of a distribution, but we rather want to employ MLE estimation to find the most probable values of x given what we know. This is a different use of the MLE estimation from what we described above, although it boils down to the equivalent mathe-

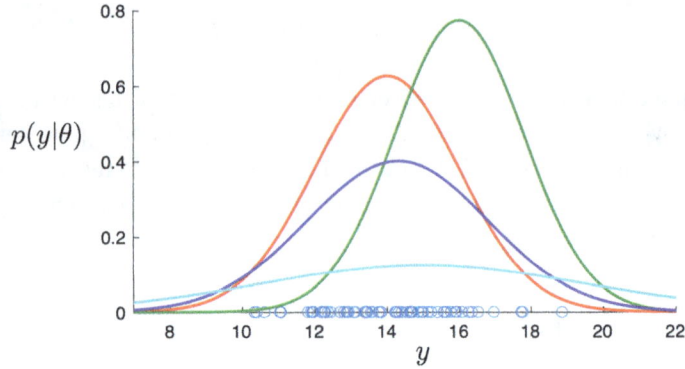

Figure 6.32: An illustration of the MLE parameter estimation. Circles on the horizontal axis are the measurements and four Gaussian distributions, with different parameters $\theta = (\mu, \sigma^2)$, are shown as candidates to fit the true distribution. (The data were actually generated from the distribution with $\mu = 14$ and $\sigma = 2$, which corresponds to the curve shown in red, so the MLE estimation should identify this distribution as the best fit.)

matical formulation. In this setting, the conditional probability $p(y|x)$ describes the likelihood of y given x under some fully specified model. The maximum likelihood estimate of the unknown values x of unobservable random variables X, given observations y is

$$\hat{x}_{ML} = \arg \max_x p(y|x) \tag{6.45}$$

This estimate is useful when we have no reliable prior knowledge about x, and is also used to *initialize* an iterative estimation procedure, which does employ some prior knowledge about x but benefits from a good initialization (like the Metropolis sampler that we describe in Section 6.3.5.2).

Maximum a posteriori (MAP) estimation. One of the most widely used Bayesian estimators is the one that maximizes the posterior probability of the unknown x given the observed y, hence the name *maximum a posteriori* estimate:

$$\hat{x}_{MAP} = \arg \max_x P(x|y) = \arg \max_x p(y|x)P(x) \tag{6.46}$$

$p(y|x)$ is called the *conditional model, data distribution,* or the *likelihood model*. $P(x)$ is the *prior model* that it describes a priori knowledge about x.

Example: Suppose that x is Laplacian distributed: $P(x) = (1/2\lambda) \exp(-|x|/\lambda)$, and that y are noisy observations of x, corrupted by zero-mean additive white Gaussian noise of standard deviation σ, which means that $p(y|x) = (1/(\sigma\sqrt{2\pi})) \exp(-(y-x)^2/(2\sigma^2))$. It can be shown that in this case the MAP estimate becomes simple soft-thresholding:

$$\hat{x}_{MAP} = \text{sgn}(y)(|y| - \sigma^2/\lambda)_+$$

where sgn(y) is the sign of y, and $(g)_+$ equals to g if $g \geq$ and 0 otherwise.

This is a realistic scenario in practice, where x are wavelet coefficients of an ideal noise-free image and y the corresponding coefficients of the observed noisy image version acquired by a digital camera.

In Section 6.3.2, we address the case where the prior model is a Markov random field. The MAP estimate then does not have a closed-form solution but we solve it by some approximate inference methods as those that we describe in Section 6.3.5.

Note that when the prior on x is uniform (i. e., when we have no prior knowledge about the actual distribution of x), the MAP estimator reduces to the maximum likelihood estimate in (6.45).

Minimum mean square error (MMSE) estimation. Another common Bayesian estimator is the *minimum mean square error estimator*, which minimizes the mathematical expectation of the square error given the observation: $E((X - \hat{x})^2|Y = y)$. One can show that this estimator is the *conditional mean*:

$$\hat{x}_{MMSE} = E(x|y) = \int_{-\infty}^{\infty} xp(x|y)dx \tag{6.47}$$

When the prior and the conditional model are both Gaussian, the MMSE estimator becomes a linear estimate (*Wiener* estimate).

Bibliography

Bishop C. M. Pattern Recognition and Machine Learning, Springer, 2006.

Boykov Y., Veksler O., and Zabih R. Fast approximate energy minimization via graph cuts. IEEE Transactions on Pattern Analysis and Machine Intelligence, 23(11):1222–1239, Nov 2001.

Eddy D. M. Probabilistic reasoning in clinical medicine: Problems and opportunities. In Judgement Under Uncertainty: Heuristics and biases, D. Kahneman, P. Slovic and A. Tversky, editors. Cambridge University Press, 2013.

Koller D. and Friedman N. Probabilistic Graphical Models: Principles and Techniques. Cambridge, MA, USA: MIT Press, 2009.

Kschischang F. R., Frey B. J., and Loeliger H.-A. Factor Graphs and the Sum-Product Algorithm. IEEE Transactions on Information Theory, 47(2):498–519, Feb 2001.

Lafferty J., McCallum A., and Pereira F. C. N. Conditional random field: probabilistic models for segmenting and labeling sequence data. In Proceedings of the 18th International Conference on Machine Learning, pp. 282–289, 2001.

Li S. Z. Markov Random Field Modeling in Image Analysis, Springer, 2009.

Nguyen D. M., R.Calderbank, and Deligiannis N. Geometric Matrix Completion With Deep Conditional Random Fields. IEEE Transactions on Information Theory, 47(2):498–519, Feb 2001.

Pearl J. Probabilistic Reasoning in Intelligent Systems: Networks of Plausible Inference, Morgan Kaufmann, 1988.

Russel S. and Norvig P. Artificial Intelligence: A Modern Approach. Pearson, 2021.

Winn, J. M. Meeting your match. In Model-Based Machine Learning, CRC Press, 2023. URL: https://www.mbmlbook.com.

Vercheval, N. Trueskill step by step. 2024. URL: https://github.com/nverchev/TrueSkill-step-by-step.

Hendrik Blockeel

7 Learning from data

7.1 Why is learning important within the broader AI domain?

The general idea behind artificial intelligence (AI) is that the computer solves new problems independently, without any humans telling it *how* to solve those problems. To do
that, the AI system must reason about the problem and about possible solutions, using a
certain body of knowledge that it has: constraints, rules, models of its environment, etc.
The domain of knowledge representation and inference studies how knowledge can be
encoded in a formal language that the AI system can reason with.

Until now, we have not discussed the question: How does the AI system *obtain* that
knowledge?

It has mostly been assumed that this knowledge is encoded by human programmers
into rules, equations, etc. However, certain types of knowledge are very hard to encode.
For instance, we all recognize Mickey Mouse when we see him on TV, because we know
what he looks like. But how can we put that knowledge in a computer? We could store
a picture of Mickey, and hope that the computer can recognize him on other pictures by
comparing them with this picture. However, we then need to program a procedure for
comparing pictures, and decades of research have shown this to be remarkably difficult.

As another example, we all know how to drive, but we cannot really explain how
we do it—at least not in a sufficiently precise manner that a student could do it herself
after simply hearing our explanation. We can share some knowledge, but the student
needs practice, too, in order to create her own "knowledge" of exactly how much the
car decelerates when you hit the brake pedal, for example.

In cases such as these, it is necessary that the AI system can build knowledge itself,
through observation of, or interaction with the world. That is what we call learning—
or, when done by an artificial system, **machine learning**. An often-quoted definition of
machine learning is that of Mitchell, in his textbook *Machine Learning* (1997):

> "Machine learning is the study of computer algorithms that allow computer programs to automat-
> ically improve through experience."

Specifically in an AI context, where computers use some body of knowledge to reason
with, machine learning implies that this knowledge is extended through experience (observation, interaction), or the way in which it is used (inference algorithms) is improved
(for instance, by learning shortcuts in the search space).

With machine learning in place, the general structure of an AI system can be described as follows (see Figure 7.1). The AI system performs tasks or solves problems independently, using powerful reasoning mechanisms. These are based on a model. That

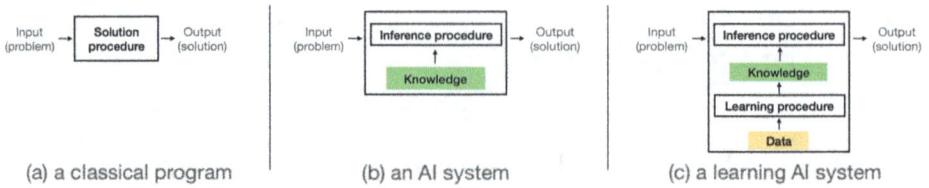

(a) a classical program | (b) an AI system | (c) a learning AI system

Figure 7.1: Software systems with increasing independence. (a) No AI: the solution is programmed. (b) AI: The solution is found by reasoning with preprogrammed knowledge. (c) Learning AI: The solution is found by reasoning with automatically obtained knowledge.

model is learned from data. The data is obtained through observation or interaction with the environment. From this point of view, learning systems could be called "AI 2.0": in a classical AI system, we do not program the solution but encode knowledge about the problem; in a learning system, this knowledge is obtained automatically, rather than hard-coded into the system.

Consider chess, for instance. A programmer could write a program that plays the game in the same way that the programmer would (option *a* in the figure). Such a program could never be a better chess player than its maker. Alternatively, the programmer could just encode the rules of the game into the program (option *b*). But knowing the rules of the game does not automatically imply you can play it well. Playing chess requires assessing the possible consequences of a move, and for that you need to know the rules *and* reason about how they might be used by you and your opponent in this particular situation and the situations it will lead to. Making computers able to perform this type of reasoning in an efficient manner has been a challenge for AI research until well into the nineties, and even in the new millennium important advances have still been made (for instance, Monte Carlo tree search).

Apart from knowing the rules themselves, human chess players quickly learn that the rook is more powerful than the bishop. We can program that into a computer but how can we quantify it? What numerical value should we assign to the rook and bishop? Furthermore, the value of a piece is really an average: depending on the position of the piece and on the current game state, a piece may be more or less valuable. It is difficult to define these numbers. However, values that work well can be learned from experience. A chess-playing program could play many games with slightly different configurations for these numbers, and eventually find out which configurations work well. This is an example of how knowledge that is difficult to provide manually can be learned from data (option *c*).

Advanced AI systems typically combine all three forms: preprogrammed solutions, domain knowledge, and a learning component. Imagine a smart travel planner for railway travel: the procedural knowledge may contain a planning algorithm, the domain knowledge expresses the railroad network, and learned knowledge might express things such as the probability that a particular train is delayed (a planner that takes this learned knowledge into account may suggest better solutions than one that does not). As another example, consider AlphaGo, the first computer program to beat the world champion at the game of Go. Go is much harder than chess for computers, for two reasons: (a) the number of moves at each possible state in the game (its "branching factor") is much larger; (b) to play Go well, one must reason not only about individual stones but about *patterns* formed by *groups* of stones. Programming these patterns is challenging. The breakthrough that AlphaGo achieved was that it *learned* complex relevant patterns using deep learning.

7.2 What category of problems does machine learning solve?

Generally speaking, machine learning tasks consist of constructing, from a given dataset, a data structure that models certain aspects of the dataset and/or the population it was drawn from. We will refer to this data structure as the **model** learned from the data. Machine learning methods vary greatly in terms of the algorithm they use, the assumptions they make, the format of the model they return, and the ways in which the resulting model can be used. Some examples of model formats are decision trees, neural networks, support vector machines, hierarchical clustering, Bayesian networks. These formats are discussed in more detail later on.

At the highest level, two types of models are often distinguished: predictive and descriptive ones. A **predictive model** represents a function that, given a partial description of an instance, can be used directly to predict the missing parts. A **descriptive model** identifies a certain structure that underlies the dataset and the population it is drawn from. In the following subsections, we discuss these categories and subdivide them into more specific tasks.

7.2.1 Learning predictive functions

Types of predictive learning

Imagine an email system that tries to sort the incoming mails into spam and nonspam for the convenience of the user. The sorting procedure can be seen as a function that maps any mail onto two possible values: spam/nonspam; or, equivalently, a function that classifies mail into one of these two categories. Such a function is called a **classifier**. In a machine learning context, the term **classification** is used both for the process of classifying objects (i. e., *using* the classifier), and for the task of *learning* such a classifier from data. The number of classes need not be two, it can be any number.

Classification can be seen as learning a function that predicts a *nominal* (a. k. a. categorical) variable. The term **regression** is commonly used for the process of learning a predictive function that predicts a *numerical* variable. An example of a regression problem is to learn (from data) a function that takes as input a picture of a person and produces as output an estimate of the person's age.

While classification and regression are the archetypical examples of predictive learning, there exist many other variants. In some cases, the output to be produced is not a single class label, but a *set* of labels. This is called **multilabel classification**. Typical examples are label a text with a set of keywords or annotate a picture with a set of objects found in the picture.

A set is a very simple structure; one may also wish to predict sequences, trees, or graphs. A generic term for this type of problem is **structured output prediction**. Within structured output prediction, we can distinguish methods that predict the structure itself from methods that use a given structure as input and simply annotate that structure. For instance, we may try to predict for a set of people which ones of them are smokers, using their social network connections as input, or to predict the social network structure itself based on properties of the individuals.

The term **multitarget prediction** is sometimes used to refer to the general case of predicting multiple variables at the same time. One could say this is a type of structured output prediction where the structure is fixed across instances.

The tasks mentioned above are all of the following nature: given a set of (x, y) pairs, called the **training set**, learn a function f that maps x values to y values in a way consistent with the data. The (x, y) pairs are called **examples** or **labeled instances**; x is called an **instance** and y its **label**.

The extent to which f is consistent with the data is typically measured by a **loss function** l. The **loss** of f for a labeled instance (x, y) is $l(f(x), y)$. Given a set of labeled instances S, the **empirical loss** of f on S is the summed loss of all its instances. The empirical loss on the training set D is called the **training set loss**. The lower the training set loss is, the better f fits the given data. Viewing D as a random sample from some population, the **expected loss** of f, also called its **risk**, is the expected value of $l(f(x), y)$ when (x, y) is drawn randomly from the population. *The goal of predictive learning is usually to find a function f with small expected loss.* Indeed, in the spam example for instance, we are not so much interested in a model that correctly classifies the mails we received earlier (such a model could simply remember all the labels by heart), but in using this model to classify future mails.

Levels of supervision in predictive learning

When learning predictive functions, the standard assumption is that for all the training instances, the desired value of the target variable y is provided. This setting is called **supervised learning**. The term supervision refers to the fact that some supervisor informs the learner what the desired outcome of f for each x is, so that the learner can compare the predictions of the function it has learned with the desired values.

A very different setting is that of **unsupervised learning**. Here, no y values are given. It may seem hopeless to try to learn a predictive function f that correctly approximates some target function t, if $t(x)$ is not provided for any instances. But under the assumption that similar cases tend to belong to the same class, it may be possible to learn a classifier f that is equivalent to t, up to a renaming of the classes. For example, if we provide pictures of fruit and ask the system to distinguish apples, pears, and bananas, without providing any labels, it may well come up with a classifier that classifies fruit correctly, except that it uses labels A, B, and C instead of apple, pear, banana.

In between supervised and unsupervised learning, there is a wide range of settings where partial or indirect supervision is available. These settings have a variety of names. In **semisupervised** learning, we typically have many instances, but only some of them are labeled. At first sight, it may seem that the unlabeled instances can just be ignored, since they offer no information about the instance-class relationship. But this is not correct, as Figure 7.2 illustrates. Adding many unlabeled instances to a small set of labeled instances may improve predictive accuracy substantially. **PU-learning** is a special case of semisupervised learning where only instances of one particular class can possibly be labeled.

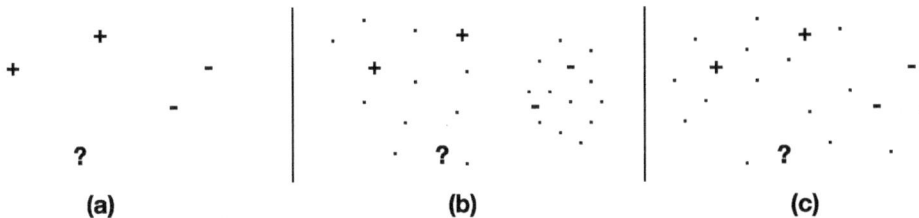

Figure 7.2: Semisupervised learning. The labeled examples alone provide insufficient information about the class of instance "?" (a). Seeing some unlabeled instances can make it more obvious what class the instance likely belongs to (b, c).

In **superset learning**, each instance is labeled with a set of labels, one of which is the correct one, but the learner does not know which one. Conversely, in **multiinstance classification**, instances are grouped into bags, and a bag is labeled with some class if and only if it contains at least one instance of that class (e. g., learn to recognize apples in pictures, when the supervisor tells you for each picture whether it contains an apple, but not which part of the picture is the apple). Both are special cases of **learning from label constraints**, where information about labels is provided in the form of constraints. **Weakly supervised learning** is the most general term for levels of supervision between supervised and unsupervised. It subsumes learning from label constraints, but also learning from probabilistic information about the labels.

In an **active learning** setting, the learner is to some extent in control of the supervision: the learner can choose which instances should be labeled. The learner will typically help minimize the labeling effort by focusing on those instances whose label is most informative to the learner (roughly, the instances it is most uncertain about).

Anomaly detection

Anomaly detection is the task of identifying situations, cases, or patterns that are somehow abnormal. A classic example is fraud detection: banks analyze credit cards payments and other transactions in the hope of quickly identifying suspicious activities, before large losses occur. As with classification, the term anomaly detection refers to both the use of a model to detect anomalies, and the learning of the model itself.

One could say that anomaly detection is an umbrella term for a variety of applications, rather than a separate task. This is only partially true. In some contexts, anomaly detectors are learned from data in a supervised manner: we provide examples of normal and anomalous cases, and the algorithm needs to learn from this a classifier that classifiers new cases into normal or anomalous. To some extent, this is a standard supervised learning task. Yet, there are typically a number of issues that make this task harder. First, there is an extreme **class skew**: anomalies are often very rare compared to normal cases. Very few classification methods can handle this well. Second, there may be a substantial amount of class noise: while instances labeled as anomalies are typically anomalies indeed, many anomalies may not have been recognized as such, and in fact these unrecognized anomalies may be the majority. A PU-learning setting may be more realistic in these cases.

In some cases, no examples of anomalies are given at all. There may also be **concept drift** in the anomalies: for instance, an entirely new type of fraud may start occurring at some point. Ideally, an anomaly detector would pick this up, even though it has never before seen this type of anomaly. Unsupervised methods are typically based on learning a model of normal behavior. When an instance deviates strongly from that normal behavior, it is flagged as an anomaly. This approach relies on having very few anomalies in the training data; indeed, if there are too many of them, the anomalies would be considered normal.

The concept of an anomaly is subjective. For instance, maintenance to a server may cause a brief disruption of service or a temporary change of behavior. If this effect is expected, it is not considered an anomaly. Hence, not every deviation from normal behavior is an anomaly. In this sense, anomaly detection is an underspecified ask. Entirely unsupervised learning is therefore often not realistic: some form of semisupervised learning is required.

Preference learning, ranking, and recommender systems

Preference learning refers to a setting where we need to learn a user's preferences from data. More specifically, given two instances x_1 and x_2, we need to learn to predict which of these two is preferred over the other. We can generalize this to **learning to rank**, where the task is to learn a model that, given k instances, ranks them from high to low value. Ranking lies at the basis of **recommender systems**, which given a set of objects (books, web pages, etc.) have to predict which ones are of most interest to the user (i. e., return the top-ranking objects).

In some sense, ranking can be reduced to regression: if we can predict the "level of interestingness" of each object, then the ranking follows by sorting them according to this prediction. This view is useful when the data consists of examples of the form (u, i, y), with y the level of interest of user u in item i. However, very often we do not have much (or even any) data of this kind. In some cases, we know whether user u preferred item i_1 over item i_2 on some occasion, and that suggests that $y_1 > y_2$, but we do not know the actual values y_1 and y_2. The typical learning setting that recommender systems face, combines aspects of semisupervised learning and weakly supervised learning.

Explanation as a secondary goal

Although the primary goal in predictive learning is making predictions, in some application domains of AI, it is crucial that the system can also *explain* its predictions. For instance, a bank that uses an algorithm to decide whether or not a certain client can get a mortgage is legally obliged to explain to the client why the loan was refused. Similarly, algorithms used for hiring personnel, deciding how much credit a customer can be given on their credit card, convicting people in court, etc., must explain their decisions. This is often a hard requirement, for example, under European legislation, EU subjects have a right to explanation regarding all decisions that involve them. As a result, **explainable AI** has quickly become a very active research subject within AI.

It is useful to distinguish two levels at which explanations can be provided: explaining the *model* as a whole, versus explaining individual *predictions*. A client who is refused a loan is entitled to an explanation of why the loan was refused in his specific case but need not understand or be granted access to the whole model used by the bank. In fact, some model formats are inherently too complex for complete models to be understood, while concrete predictions may still be explainable.

More on the need for explainability is written in Chapter 9.

7.2.2 Learning descriptive models

When, in predictive learning, we learn from (x, y) examples a function f that for any x can predict the corresponding y, we are in fact identifying a certain pattern in the data, a relationship between x and y in the given examples (x, y). But the concept of patterns, or structure in the data, is much broader than that. For instance, a system could detect, by analyzing a set of employee rosters, that there is always at least a 16-hour period of rest between 8-hour work shifts. This pattern has very low predictive value: with only this information, we cannot predict very accurately who will work when. But it can still be useful, for instance, for checking the validity of rosters. We call such patterns descriptive patterns. They identify certain structural aspects of the data. More generally, **descriptive models** are models that expose a certain global structure of the data. The ultimate goal of learning descriptive models can be the model itself (as it describes the population and as such provides insight into it), or the use of this model for a variety of tasks, including, but not limited to, prediction.

Many types of structure can be identified. Generally, structure can be defined as any kind of deviation from a uniform distribution. A simple type of structure is when data occurs in clusters: groups of similar data, separated by empty areas in the instance space. **Clustering** refers to the task of identifying the clusters in a dataset. The task can easily be illustrated in two- or three-dimensional spaces (see, e. g., Figure 7.3a), but such illustrations make it look deceptively simple: identifying clusters in high-dimensional data is not an easy task. **Hierarchical clustering** refers to building a hierarchy of clusters, or a taxonomy, where clusters get merged into larger clusters such that many clusterings, at different levels of granularity, are obtained.

Apart from occurring in clusters, it is also possible that the population under study is relatively uniformly spread, but in a relatively small volume within the entire instance space. It may lie entirely in a lower-dimensional **subspace** (such as a hyperplane), or in a narrow range around such a subspace. More generally, the data may live in or near a lower-dimensional **manifold**. For the purpose of this text, a manifold can be interpreted as a kind of nonlinear version of a subspace (imagine a two-dimensional sheet in a three-dimensional world: contrary to a plane, the sheet can be bent). Figure 7.3 illustrates these concepts. The statistical technique known as principal components analysis (explained later in this text) is an example of a method that identifies a lower-dimensional subspace.

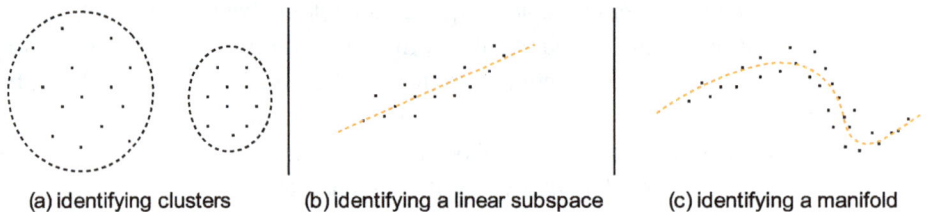

(a) identifying clusters   (b) identifying a linear subspace   (c) identifying a manifold

Figure 7.3: Identifying structure in the data: the instances may occur in clusters (a), near a subspace (b) or near a manifold (c).

Viewed from a probabilistic perspective, identifying dense clusters of subspaces can be seen as special cases of **density estimation**: identifying dense and sparse regions in the instance space. This is typically done by fitting a probability distribution to the training set.

Association rules express dependencies between different variables. Rules such as "basketball players tend to be tall" or "people who buy wine and bread are more likely (than average) to also buy cheese" provide insight in certain domains (sports, customer behavior) that may be interesting in itself but also actionable. For instance, supermarkets can exploit such knowledge in their promotional campaigns.

Some learners construct probabilistic models, which indicate stochastic dependencies between variables. A stochastic dependency links a probability distribution over the values of one variable, rather than a single value, to the values of another variable. In the context of probabilistic modeling, models that aim specifically at predicting one variable (in our terminology, predictive models) are sometimes called **conditional** or **discriminative** models. Given the value of one or more variables, they can be used to predict the value of another variable. In the same context, descriptive models are called **generative**. A descriptive model describes the joint distribution over all variables, and can be used, among other things, for sampling from this distribution, in other words, for generating data. More on this can be found later in this chapter (learning probabilistic models) and in Chapter 6.

Reinforcement learning

Reinforcement learning is a machine learning setting that differs from other settings in multiple respects. It is perhaps best illustrated by referring to a very old learning system: Donald Michie's MENACE (1960). MENACE is a system that learns to play tic-tac-toe. It is implemented not on a classical computer but using approximately 300 matchboxes that contain colored beads (see Figure 7.4 for a picture). There are 9 colors, one for each position on the board. The matchboxes are labeled: for each game state, there is a matchbox for that state. The system plays as follows: when a move is to be made, the operator takes a random bead from the matchbox associated with that state and draws a cross at the position indicated by the bead. Thus, the number of beads of each color in a matchbox determines how likely it is that the system plays the corresponding move. After a game has been played, the content of the opened matchboxes is updated as follows: if the game was lost, the drawn beads are not put back; if it is a draw, each bead is put back where it comes from; if the game was won, the beads are put back and an additional bead of the same color is added. Thus, after a won game, the moves played during that game will become a bit more likely to be chosen again at a future occasion. Michie showed experimentally that this simple system indeed exhibits learning behavior: its performance improves with the number of games it has played.

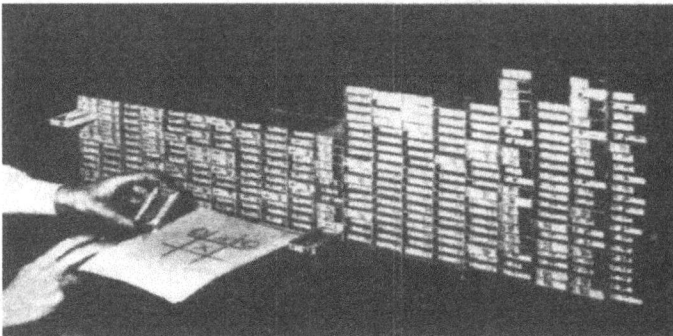

Figure 7.4: MENACE, a reinforcement learning system implemented with matchboxes. (Source: D. Michie, The Computer Journal, 6(3):232–236, 1963. https://doi.org/10.1093/comjnl/6.3.232).

The above type of learning is now known as **reinforcement learning**. The setting has in common with predictive learning that a function f mapping state x to action y (in this context called a *policy*) is learned. There is no direct supervision, however: no-one tells the learner what the right y for a given x should be. The consequences of taking a particular action are complex: it may lead to an immediate reward (e. g., winning the game), but more generally it changes the state of the learner, and the new state may have higher or lower potential for future rewards. The task, in reinforcement learning, is to learn a policy that maximizes the expected overall future reward. While performing this task, many reinforcement learning systems build a descriptive model of their environment at some level of abstraction (e. g., which action in what kind of state leads to what expectation in terms of future rewards), and in this sense they can also be seen as descriptive learners.

Reinforcement learning is relevant in any case where an agent controls a process of which the yield should be optimized in the long run, and where actions may have long-term effects. Applications range from game-playing (e. g., learning to play PacMan or Breakout) to controlling industrial processes where an operator's actions may have long-term consequences.

7.3 How are learning problems solved?

7.3.1 From data to model: an overview

It can be instructive to look at the whole process that eventually leads to learned models.

In the context of data mining, the CRISP-DM methodology (Cross-Industry Standard Process for Data Mining) has been proposed as a model of how data analysis methods (including machine learning) can be applied in a business context. Six stages are distinguished: business understanding, data understanding, data preparation, modeling, evaluation, deployment. There are multiple feedback loops between these stages (see Figure 7.5). This view mostly stresses the importance of understanding the data, which in a business context is indeed crucial. In a machine learning context, *obtaining* the data is often a challenge in itself.

7.3.2 Data collection and preparation

Data collection is by itself a nontrivial task. There are ethical concerns, related to ownership, security, privacy, and possibly undesirable biases in the data; see Chapter 9 for more on this. Data sources may contain noisy or incomplete data, so a **data cleaning** phase is often included. When merging data from multiple sources, the merging process itself (**data fusion**) can be challenging. Even when data is easy to obtain (from the web, from sensors, etc.), adding labels to the data (for supervised learning) may be costly. The

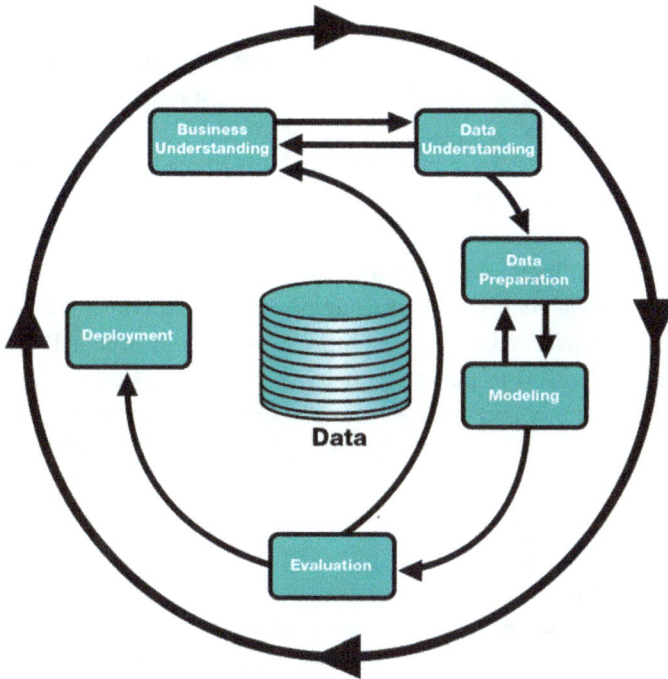

Figure 7.5: Schematic overview of the CRISP-DM methodology: the different phases and their feedback loops among them. (Source: http://crisp-dm.eu/reference-model/).

data may not be in a format that is immediately useable by machine learning systems: transforming it into the right format is called **data wrangling**. All these different steps are part of the **data preparation** phase.

7.3.3 Format of the data

In introductory courses on statistics, it is typically assumed that the data consists of a set of instances, and each instance is described by the same set of variables. These variables may be nominal (they take a symbolic value) or numerical (their value is a number).[1] As such, the whole dataset is easily represented as a table, where columns are variables and rows are instances. This format is also called the **standard format**. A row is also called a **tuple** in this context.

In machine learning, we often encounter very different types of data, which are not so easily represented in a tabular format. In the following, we provide a categorization.

---

[1] More fine-grained categorizations exist, but those are not relevant here.

Tensorial data

An image is typically represented as an array of numbers or vectors, where each number represents the brightness, or each vector the color, of a single pixel. Thus, an image with a resolution of 1000×1000 pixels, where the color of each pixel is represented using a three-dimensional RGB vector (which contains the values of the red, green, and blue components of the color) is represented by 3 million numbers. While syntactically this format may look very similar to the standard format (1 picture = 1 row containing 3 million numbers), semantically it is very different. In the standard format, each variable has a well-defined meaning and is considered relevant by itself. A single pixel in a picture, however, carries very little meaning, and insofar it does, its meaning is practically identical to that of its neighboring pixels. Groups of nearby pixels are more likely to jointly carry some meaning. But in the standard format, the relationship between different variables (namely, their relative position in the image) is lost.

A piece of video is essentially very similar to an image; one simply adds a time dimension to the image. In a similar way, one can handle 3-dimensional images, and possibly add time as a fourth dimension.

In all these examples, the data is numeric, but rather than being ordered in a single row, the numbers are put in a table with two or more dimensions. Such a table is called a **tensor**. The tensor structure follows from a notion of "nearness": cells can be near in the X, Y, or Z dimension, in the time dimension, or in other ways. For instance, for a time series that contains one number per day, the numbers could be arranged into a tensor where one dimension indicates the weekdays, and numbers are "near" in this dimension if they are exactly n weeks apart. A tensor representing video, with two space dimensions and one time dimension, is somewhat similar to a flip book, where each page is one picture and consecutive pages contain slightly different pictures.

Tensors are quite similar to the notion of vectors (for one-dimensional tensors) and matrices (2-dimensional tensors) and are in some sense a generalization of these concepts. Figure 7.6 illustrates this.

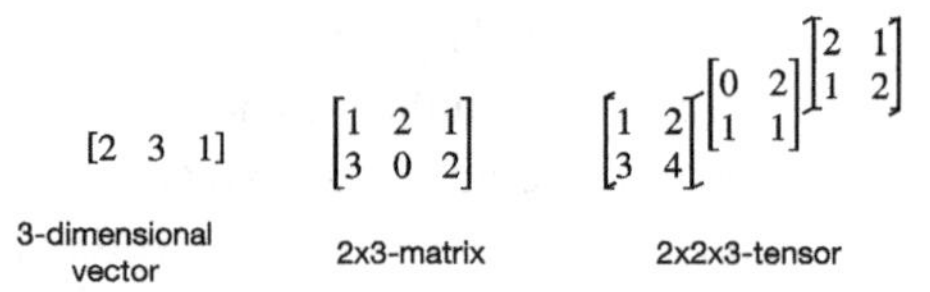

$$[2 \quad 3 \quad 1] \qquad \begin{bmatrix} 1 & 2 & 1 \\ 3 & 0 & 2 \end{bmatrix} \qquad \begin{bmatrix} 1 & 2 \\ 3 & 4 \end{bmatrix}\begin{bmatrix} 0 & 2 \\ 1 & 1 \end{bmatrix}\begin{bmatrix} 2 & 1 \\ 1 & 2 \end{bmatrix}$$

3-dimensional   2x3-matrix   2x2x3-tensor
vector

Figure 7.6: Vectors, matrices, and tensors.

Graphs, trees, and sequences

A **graph** is a mathematical structure that consists of nodes and connections between them. Graphs are typically used to represent network-like structures. For instance, a road network can be represented as a graph: edges represent roads and nodes represent junctions. In addition, molecules, which consist of atoms connected by bonds, essentially have a graph structure.

The nodes and edges in a graph can be labeled with symbols, numbers, or any other type of data. For instance, the atoms comprising a molecule would typically be labeled by their element type, the roads in a network could be labeled with their length, etc.

While graphs are relatively simple structures, they are relatively difficult to handle. A graph can be represented by a matrix, the so-called adjacency matrix: it contains one row and column per node, and the element on row i and column j is 1 if nodes i and j are connected, and 0 otherwise. Note that such a representation imposes an order on the nodes, which is not defined by the graph itself. The question of whether two adjacency matrices really represent the same graph boils down to whether a reordering of the rows and columns of one matrix exists so that one becomes equal to the other. No truly efficient method is known for answering this question, which is also known as determining graph isomorphism.

A path is a sequence of edges that lead from one node to another (or the same; in the latter case, the path is called a cycle). A **tree** is a graph that is connected (there is a path between any two nodes) and has no cycles. Sometimes one node in a tree is singled out and called the root of the tree. Rooted trees are very common as a data structure. For instance, web pages (and generally all HTML documents) internally have a rooted tree structure.

A **sequence** is a (somewhat degenerate) rooted tree that never splits into multiple branches; that is, for any two paths starting from the root, one path is completely contained in the other. Edges become irrelevant at this stage: abc can only mean the graph a-b-c, no other structure is possible. A finite sequence in which each node is labeled with a symbol from some finite alphabet is also called a **string**. A **time series** is a sequence in which each consecutive element is annotated with a point in time. Figure 7.7 shows a few examples of graphs, trees, and sequence data.

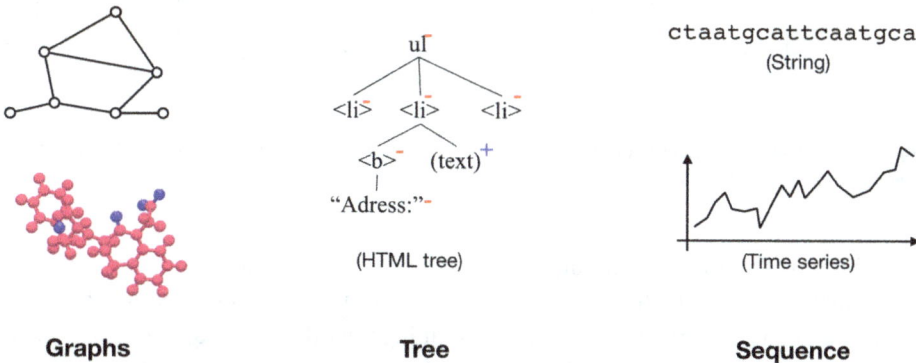

Graphs	**Tree**	**Sequence**

Figure 7.7: Graphs, trees, and sequences.

Like tensors, graphs have an internal structure. Typically, however, the internal structure of tensors does not vary among different instances, while that of graphs does. For instance, tensors representing 2-dimensional pictures with constant resolution all have the same structure; apart from the border pixels, each pixel has one left, right, upper and lower neighbor. In a social network, not every person has the same number of friends, and there is no concept such as "your left friend."

Relational data

In practice, data is often stored not in a single table or graph format, but in a relational or object-oriented database. We will here use terminology from relational databases, but statements apply equally to object-oriented or other advanced formalisms.

In a relational database, information is stored not in a single table, but in multiple tables (see Figure 7.8 for an example). The tuples in these tables are connected with each other using foreign key references. These connections essentially form a kind of graph. The main difference between relational database representations and graph representations is that in graphs, the graph structure is often the primary information that is being conveyed, whereas in relational databases, the primary information is in the tuples.

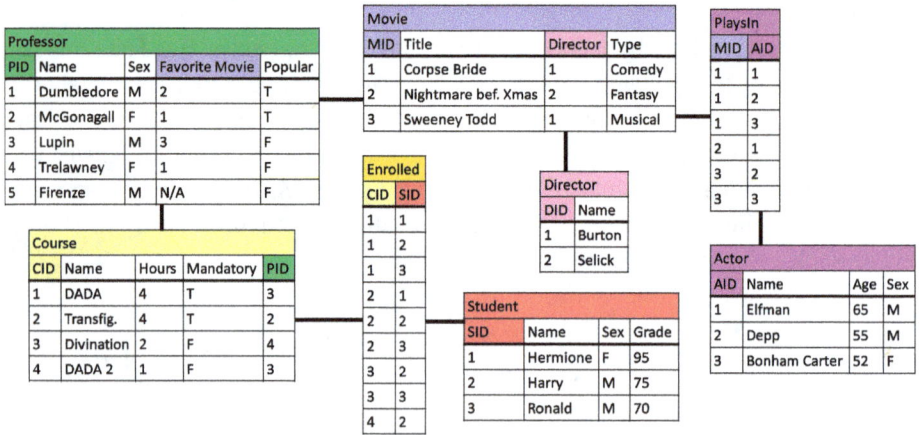

Professor

PID	Name	Sex	Favorite Movie	Popular
1	Dumbledore	M	2	T
2	McGonagall	F	1	T
3	Lupin	M	3	F
4	Trelawney	F	1	F
5	Firenze	M	N/A	F

Course

CID	Name	Hours	Mandatory	PID
1	DADA	4	T	3
2	Transfig.	4	T	2
3	Divination	2	F	4
4	DADA 2	1	F	3

Movie

MID	Title	Director	Type
1	Corpse Bride	1	Comedy
2	Nightmare bef. Xmas	2	Fantasy
3	Sweeney Todd	1	Musical

Enrolled

CID	SID
1	1
1	2
1	3
2	1
2	2
2	3
3	2
3	3
4	2

Director

DID	Name
1	Burton
2	Selick

Student

SID	Name	Sex	Grade
1	Hermione	F	95
2	Harry	M	75
3	Ronald	M	70

PlaysIn

MID	AID
1	1
1	2
1	3
2	1
3	2
3	3

Actor

AID	Name	Age	Sex
1	Elfman	65	M
2	Depp	55	M
3	Bonham Carter	52	F

Figure 7.8: A relational database. Information related to, for example, Professor Dumbledore, is spread over multiple tables. (Illustration by Jonas Schouterden).

Consider, for example, a recommender system for recommending movies. There is data about users (one table, containing, e. g., each user's age, gender, native language, etc.), data about movies (another table, containing the title, genre, director, year of release, etc.), and data about how users rate movies (a third table containing in each tuple a reference to a user and a movie, and information pertaining to this combination, such as how this user rated that movie). The movie table might also be connected to an actor table through a relation "... plays in ...," which might contain information about what role the actor played in that movie, etc.

We say that data is **relational** when there are relationships between instances, or each instance has itself an internal structure that is relational. For example, a molecule contains atoms and bonds between them: this is an internal relational structure. In a social network, if we are analyzing persons (one person is one instance), the network describes relations between instances.

Knowledge

The broadest type of information is knowledge. Knowledge can include factual data, such as the age of a particular person, but also general knowledge about the domain. That knowledge can consist of logical constraints, mathematical formulas, etc. A knowledge base allows us to represent all these kinds of knowledge. Representation languages for knowledge bases are often based on subsets of first-order predicate logic. Working with those subsets allows for efficient computations. Description logics are the most typical example of this. Languages based on description logics allow us to define ontologies and reason with them. One could, for instance, indicate in a medical database that the flu is caused by a viral infection. Data analysis methods can take this into account when detecting patterns.

Description logics are useful for conceptually rich contexts, but less so for the description of physical systems, where equations over real numbers rather than categorical data are used. Such knowledge bases could, for instance, represent knowledge in the form of differential equations. We refer to Chapter 5 for a more in-depth discussion.

Converting data to the standard format: feature selection and construction

Most machine learning methods assume a specific type of data: standard format, tensorial data, graph data, logic-based knowledge bases, etc. Many advanced methods, however, have been developed specifically for the standard format. Therefore, in many cases, the success of machine learning lies in converting the data into the standard format in such a way that the most relevant aspects of the data are preserved, while irrelevant aspects are removed.

When converting data from its raw form into a meaningful representation in the standard format, the variables in this standard format are usually called **features**. The process of finding the right features is called **feature engineering**; it involves both **feature selection** (selecting the most relevant features from a large set of candidate features) and **feature construction** (creation of new features by combining existing ones).

In many cases, *feature construction is by far the most time-consuming part of the machine learning process*. Given the right features, many learning methods achieve similar performance in terms of predictive accuracy, and the choice of method matters less than the choice of features. Unfortunately, this task is very hard to automate. If deep learning has forced a breakthrough in computer vision, this is largely due to its capacity to automatically define useful features (see the section on deep learning for more information on this).

When data is available from multiple sources, the integration of information available from these different sources is called **fusion**. This integration can happen on the level of the raw data (**early fusion**), the features, the models, or the actual predictions (**late fusion**).

A specific type of data conversion is needed for methods that assume all inputs to be numerical. Nominal data are then encoded using so-called **dummy variables** or a **one-hot encoding**. One-hot encodes a nominal variable with k possible values as a vector of k binary variables, one of which is 1 and all others 0. Dummy variables as used in linear regressions, for example, work in a similar way, except that only $k - 1$ variables are used, and the kth value is indicated by setting all variables to 0.

7.3.4 Paradigms for learning

Having discussed the preparatory steps of data collection and data preparation, we now come to the main topic: methods for learning from data.

The field of machine learning is very broad, and different views on the area as a whole exist. To name but a few examples: Pedro Domingos, in his best-selling book *The Master Algorithm*, distinguishes five different "tribes" of machine learners (called symbolists, connectionists, evolutionaries, Bayesians, analogizers). Peter Flach, in his textbook *Machine Learning: the Art and Science of Algorithms that Make Sense of Data*, considers three paradigms: geometric, logical, and probabilistic. Chris Bishop's *Machine Learning and Pattern Recognition* tries to provide a uniform view on this broad area and to this aim takes a Bayesian perspective. Leo Breiman, in his 2001 article *Statistical Modeling: The Two Cultures*, distinguish methods that assume a particular stochastic model of data generation from those that do not,[2] and views this as the major difference between classical statistics (which assumes such a model) and machine learning (which often does not).

Machine learning approaches can be categorized in many dimensions. We can distinguish them according to the *type of problem* they try to solve, the *type of data* they can handle, the *format of the output* they produce, certain *issues* they can deal with, etc. To illustrate some of these things, consider the following example task.

An email program wants to classify mails as spam or nonspam, in order to make it easier for the user to sift through them. To do that, it needs to build a model of what

---

2 For instance, classical least-squares linear regression assumes noise-free measurements of the independent variables, Gaussian noise on the dependent variables, and a linear relationship between the independent variables and the dependent variable's conditional expected value. To a statistician, understanding an approach implies understanding this model.

spam mails (or regular mails) look like. That model can be in the form of a predictive function: a function that takes as input a description of the mail and produces as output an answer to the question whether the mail is spam. Its output could be one of two categorical values (spam/nonspam), but it could also be a number that indicates how likely it is that the mail is indeed spam. This number could be a probability: a number between 0 and 1, with the property that, for instance, among all mails that score 0.9, one expects 90 % of them to be spam.[3] But it does not have to be: it might also be any positive real number, with the property that the higher the number is, the more indications there are that the mail is indeed spam.

The input to the predictive function is a description of the mail. This might be a relatively simple description, such as the number of times particular words occur, the number of typos in the mail, etc. It may also be a more complex type of description, one that does not simply count words or typos, but tries to "understand" the text, which among other things requires taking into account the sentence structure. Finally, the description used may simply be the actual text: just a string. In the latter case, we say the learner takes as input the "raw data,"un-preprocessed, as opposed to the earlier cases, where certain descriptive features are defined, and the mail is preprocessed by computing values for these features before a prediction is made about it.

The predictive function can differ both in terms of the output it gives, and the input it takes. However, even among functions that use exactly the same format for inputs and outputs, there is great variety in the representation of the function itself. Functions with numerical inputs and outputs may describe the link between outputs and inputs using a mathematical formula, as for a linear regression. But many other formats are possible. For instance, certain types of functions can be described as a set of if-then rules. One such rule might be: if the mail sender occurs in the user's address book, the mail is not spam. Rules of this kind are usually very easy to understand, compared to mathematical formulas. We will discuss these and many other formats in the remainder of this chapter.

The mentioned learning paradigms constitute only the top level of the taxonomy of machine learning methods. Within a particular type of method, different algorithms exist, each with a different behavior. Individual algorithms typically have a number of so-called **hyperparameters**[4] that affect the behavior of the algorithm.

There are many more ways in which machine learning approaches can be described. OntoDM[5] is an ontology for data mining that attempts to bring order to the landscape of machine learning algorithms, among other things.

---

3 If this link between the predicted number and some well-defined proportion is not present, it is not very meaningful to call the produced number a probability.

4 Hyperparameters are just settings for the learning algorithm. The term is used to avoid confusion with numerical variables that are part of the learned model, which are then simply called parameters.

5 http://www.ontodm.com/

In the following sections, we discuss different approaches to machine learning, categorized according to the format of the model they produce. We discuss their hyperparameters as we go along.

7.3.5 Instance-based methods

Instance-based methods are machine learning methods that simply remember the instances they have seen (i. e., their "model" is the training set itself). At the time a prediction needs to be made, they compare the case for which a prediction needs to be made to the stored cases and make a prediction by analogy. The assumption that these methods rely on is: if two cases are similar with respect to many observed variables, they are probably also similar with respect to the variable to be predicted. *If it looks like a duck, walks like a duck, and quacks like a duck, it's probably a duck.*

Methods in this class are also called nearest neighbor methods. The most basic of these is called "k-nearest neighbors," or k-NN. The prediction algorithm is simple: to classify a new case, find the k most similar cases in the training set, see which class is most common among those, and use that class for the prediction. In a regression context, one would typically predict the mean of the target values of the k nearest neighbors. k-NN like methods can also be used in the broader predictive context, such as for multilabel prediction, and even for descriptive learning, such as density estimation. For instance, in a Euclidean space, if the k'th nearest neighbor is at distance r, this means there are $k + 1$ instances in a hyperball with radius r, so $\frac{k+1}{V}$ (with V the volume of the hyperball) is a rough estimate of the local data density.

Methods like these strongly rely on having a suitable notion of similarity. In many cases, when inputs are numerical, a distance measure such as Euclidean distance is used: two cases are considered similar if they are closer together in the input space. This also allows for intuitive visualizations, as in Figure 7.9, but such a simple definition of similarity leads to problems. For instance, assume person A is 1.80 m tall and weighs 80 kg, and is represented using the vector A = (1.80, 80). Assume, similarly, B = (1.60, 79) and C = (1.83, 82). One could argue A is more similar to C than to B, given that A and C are approximately equally tall and heavy. However, the Euclidean distance between A to B is smaller than between A and C. This is because a large difference in length maps to a small number. If we chose to express length in centimeters rather than meters, the Euclidean distances would be very different, and A would be more similar to C. Clearly, any classification method that gives a different outcome depending on the unit of measurement used in the training data (which is an entirely random choice) is not a good method.

As illustrated above, nearest neighbor methods suffer from high sensitivity to rescaling of individual dimensions. They also tend to underperform in high-dimensional input spaces, especially when many of the input attributes are not relevant. The Euclidean distance between two instances is then easily dominated by irrelevant dimensions, which makes the classification highly random.

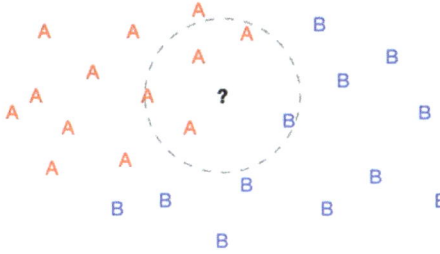

Figure 7.9: k-NN for $i = 5$ classifies the instance marked "?" with the majority of its 5 nearest neighbors (within the circle). As 4 of these are A and 1 is B, the predicted class is A.

Instance based methods can work very well, if an appropriate similarity measure is used. Unfortunately, as illustrated above, it can be very hard to find such a measure. Another weakness of nearest neighbor methods is that they can be slow at prediction time. When a large training set is used, finding the k nearest neighbors among possibly millions of cases can take a lot of time. The solution to this is to construct an appropriate index structure on the dataset, one that makes it possible to find the nearest neighbors of an instance without searching the whole dataset.

7.3.6 Decision trees

What are decision trees?

Decision trees underly many of today's most successful learning approaches. A decision tree represents a very simple type of decision procedure. It starts with one test, and depending on the outcome of that test, it is decided what the next test will be. This continues until a decision is reached. Figure 7.10 shows an example of a decision tree.

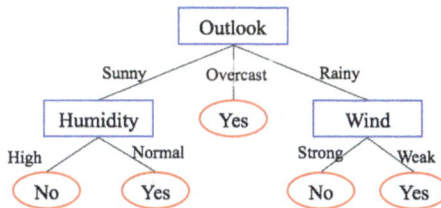

Figure 7.10: A decision tree that a person might use to decide whether to play tennis today. The person first checks the outlook; if that is sunny, then the humidity is relevant: if the humidity is normal, it's a good day for playing tennis. (Example after T. Mitchell, Machine Learning).

The decision can be a prediction, and in that case the decision tree represents a predictive model. The tree is called a **classification tree** when a nominal variable is predicted, a **regression tree** when a numerical variable is predicted. In principle, the tree can also predict multiple variables at once (or, equivalently, a tuple-valued variable); such trees are sometimes called **multitarget trees**. Classification trees that do not merely predict a class but define a conditional probability for each class given the input, are called **probability estimation trees**.

As it repeatedly defines a dataset into subsets, a tree implicitly defines a hierarchical clustering. Trees learned for this purpose are called **clustering trees**. The difference between a hierarchical clustering defined by a clustering tree, and one defined by other clustering algorithms, is that each cluster in a clustering tree is defined precisely by a set of test outcomes.

Density estimation trees partition the dataset into regions of high and low density and, as such, can be used to describe the joint probability distribution of the data.

How can we learn them?

Given a dataset, how can we learn a decision tree that accurately models it? The procedure for learning such a tree is remarkably simple. It works as follows: given the whole dataset and a target variable, find the test whose outcome is most informative about the target variable (in other words, the test that correlates best with the target). Once that test is found, split the dataset into subsets based on the outcome of the test. Repeat this procedure for each subset thus created, and for their subsets, and so on, until further splitting is no longer useful.

This procedure requires a few details to be filled in. First, what is a "test"? In the standard framework, a test is typically based on a single attribute. Nominal variables can simply be tested by asking for their value; this test has as many outcomes as there are values in the variable's domain. For nominal variables with many values, it may be better to use a binary test: the domain of the variable is partitioned into two subsets, and the outcome of the test equals the subset to which the variable's value belongs. For numerical attributes, the value of the attribute is typically compared with some threshold, and there are two possible outcomes: the value is less than or equal to, or greater than the threshold.

A second question is: how do we measure how "informative" a test is for the target? For classification trees, the concept of "information gain" can be used: this is the average reduction of class entropy caused by the test. Entropy is a measure from information theory that can be interpreted as diversity. A set of 50 dogs and 50 cats is more diverse than a set of 90 dogs and 10 cats, and a set of 30 cats, 30 dogs, and 40 sheep is even more diverse. The less diverse a set is, the easier it is to predict the class of a randomly chosen element.

For numerical targets, diversity is typically measured as variance, and the best test is then the one that reduces variance the most (i. e., the average variance of the target within each subset must be as small as possible). Variance can also be used for higher-dimensional target variables, and for nominal variables if they are encoded using a one-hot encoding (in this case, the variance of this encoding is called Gini impurity).

A third question is: when is it no longer useful to split a subset into smaller subsets? For classification trees, it is clear that when a subset has zero class-entropy (i. e., all cases in the subset have the same class), further splitting is no longer useful. For regression trees, the equivalent would be zero variance, but that is almost never achievable. Some learners stop splitting when the best test does not lead to a significant reduction of entropy or variance. When the subset to be split is very small, further reductions are almost certainly not significant; for that reason, many learners only split subsets whose size is above some threshold value.

It is known that too large trees tend to **overfit** the data: they fit the training data well but tend to perform worse on other data. Figure 7.11 illustrates this problem. Ideally, a tree learner stops splitting just before such overfitting occurs. However, it turns out it is very hard to determine the right moment. Statistical significance tests do not work well in this context, and it is perfectly possible that even if no single test leads to a substantial improvement, a combination of tests will. For this reason, many learners grow the tree beyond its optimal size, and prune away useless branches afterwards. This pruning process typically makes use of a so-called **validation set**: a set of data not used for learning the tree but used to estimate the quality of the full tree and its pruned variants during the pruning process. Since the validation set was not used while growing the tree, it provides an unbiased view of the actual predictive accuracy of the tree. The pruning process then consists of pruning away branches that did not lead to a higher accuracy on the validation set (i. e., the improvement they gave on the training set was most likely due to overfitting).

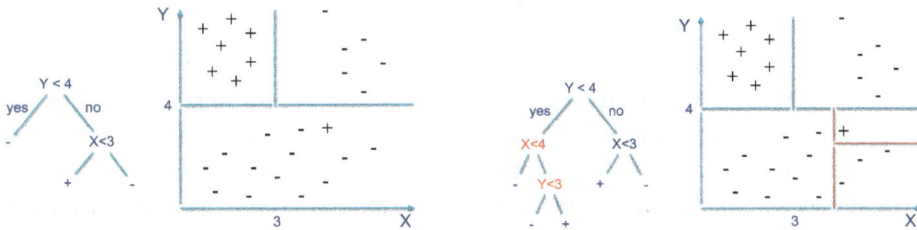

Figure 7.11: Left: a small tree fits the training data almost perfectly. It can be grown to fit perfectly (right), but a relatively large area to the right will then be predicted positive, while the data contains very little evidence for this.

Finally, there is the question of what prediction to make in a tree leaf. The most straightforward choice is to predict the class that occurs most frequently in the leaf (the "mode"), or the mean of the numerical target values in it. For probability estimation trees, the probability distribution of the classes in the leaf is typically returned. For density estimation trees, a local density model is learned and stored in the leaf. For regression trees, one can consider including in the leaf, instead of a constant value, a linear model that predicts the target from the other variables. This linear model is then learned using linear regression on only the data in this leaf. Such trees are called **model trees**.

From a machine learning point of view, decision trees have a number of important advantages. The learning algorithm is very fast: trees are often learned in under a second, even from large datasets. The prediction procedure is even faster: making a prediction using a single tree typically involves less than a dozen comparison operations, which means one can easily make millions of predictions per second. It is easy to understand how trees are built and how they make their predictions. It is easy to explain a prediction: "the provided instance has properties X, Y, Z, and among all instances with those properties, 90 % belong to class A, therefore it is likely that this instance also belongs to class A." The main disadvantage of decision trees is that their predictive accuracy, while good, typically lags slightly behind the accuracy of other models. In cases where accuracy has priority over the rest, this is often a reason for preferring more complex models. However, some of those more complex models internally still rely on trees, as the next section illustrates.

7.3.7 Ensembles of decision trees

Bagging

In machine learning, an **ensemble** is a predictive model that simply consists of multiple simpler models, combined in some way. Just like a committee of experts can be wiser than a single expert, especially when those experts have complementary expertise, a committee of models (an ensemble) can be more accurate than any of its components.

To learn an ensemble, the first question we face is: how do we get multiple component models? Let us say we have one decision tree learner, and a single dataset. We can use the learner to learn a tree from this dataset. If we use the same learner a second time, on the same dataset, we will get the same tree. But there is no point in learning multiple trees, if they are all identical. A simple way to get different trees then is to learn them from different datasets. If we have only one dataset, we can create multiple variants of it through random sampling. One specific procedure for this, called **bagging**, works as follows: given a dataset containing N instances, choose N times a random instance from it. The same instance can be chosen multiple times. The result is a "bag" containing N instances, in which each instance from the original set occurs zero, one, or multiple times. This bag is a first variant of the original dataset. Because the procedure is randomized, it can be used repeatedly, creating a new variant with each run. Thus, from a single dataset, k datasets are created with the same size, and whose elements follow roughly the same distribution. Running our learner once on each of them yields k models. These likely differ from each other, as they were learned from different datasets. Still, they are all expected to generalize equally well toward the population, as each was learned from an equally large sample from the same distribution. Any differences between these models are necessarily due to coincidence in how the particular datasets were formed and, therefore, are related to overfitting.

The next question is: how do we combine the models at prediction time? The choice made in the technique called bagging is simple: we consider the predictions of individual models as votes. For classification, we use the majority of these votes for the final prediction; for regression, the mean.

Thus, the procedure called bagging can be summarized as follows: create k variants of the given dataset by sampling with replacement, learn one model from each variant (always using the same learner), and have those models vote.

Because the bagged model is less dependent on coincidental patterns in the dataset, it tends to have better accuracy than the individual models. It has been confirmed experimentally that the difference is often substantial.

Random forests

An obvious disadvantage of bagging is that learning takes longer: instead of learning one model, we need to learn k models. Furthermore, for bagging to work well, it is essential that the different models are sufficiently different. Both of these issues are quite minor, in the context of decision trees: tree learning is very fast, and it is relatively unstable, meaning that small differences in the training set suffice to get significantly different models. Yet, the bagging procedure can be improved to a procedure that is known as random forests.

Random forests add just one small thing to the bagging procedure. When learning an individual tree, instead of considering at each node all possible tests, and selecting the best one, random forests pick the best from a randomly chosen *subset* of possible tests. Given a total of m attributes, the number of attributes actually considered for testing is typically $\sqrt{m}$, or even less. Especially in high-dimensional datasets, which contain many attributes, this speeds up the learning process considerably. Obviously, the actual best test may not be considered in this way, but this is not a big deal, because optimality is to some extent coincidental (had we learned from a different sample, a different test would have seemed optimal), and even with suboptimal tests, trees tend to work well in practice. Moreover, the randomness thus introduced increases the variation among the trees in the ensemble. This positively affects the predictive performance of the ensemble, and often overcompensates a possible loss of accuracy of the individual trees.

For well over a decade since its publication by Breiman in 2001, random forests were widely considered the best way of building tree ensembles, in terms of predictive accuracy. Only rather recently, a technique that is about equally old has led to a decision-tree based method that pushes accuracy still a bit higher: gradient boosting.

Gradient boosting, gradient boosted trees

Boosting is, like bagging, an ensemble method. It differs from bagging in how it obtains different models, and how these models are combined. There have been multiple interpretations of how and why boosting works. In his text, we take a relatively recent interpretation: that of gradient boosting.

The basic idea behind boosting is the following. Assume we have learned a predictive model from a given dataset. Even on the training set, the predictions by the model may differ from the observed values (models do not necessarily have 100 % accuracy on the training set).

Observing this, we can try to learn a second model that tries to fix the errors that the first model makes. In the context of regression, a natural approach would be: once a model f_1 is learned, replace each target value y_i in the dataset by $y_i - f_1(x_i)$ and learn a model f_2 from this new dataset. f_2 basically learns how much we have to add to f_1's prediction to make it equal to y_i. Once we have learned f_2, we should combine f_1 and f_2 by predicting for any $x, f_1(x) + f_2(x)$. If there is still a difference between the observed y_i and the predictions $f_1(x_i) + f_2(x_i)$, this procedure can of course be repeated.

Though intuitive, the above procedure is a bit too naïve to work well. However, a principle has been developed that has sound theoretical support, and of which the above is a special case: **gradient boosting**. The principle is motivated by statistical approaches to learning, which are discussed later in this text. It essentially formulates the simple idea above as a gradient descent approach, where an approximation to the target function is iteratively improved by "moving downhill" in a search space, in such a way that the loss is maximally reduced for a given step size. Basically, the procedure learns f_1 and f_2 as above, but determines a factor a such that $f_1(x_i) + af_2(x_i)$ minimizes the loss, and on top of that, typically using a learning rate such that a smaller step is set in this direction than the one that theoretically minimizes the loss. These adaptations to the simple intuition above are motivated by the gradient descent viewpoint. This viewpoint also makes it possible to apply the principle to other loss functions and in, for example, a classification context. An older boosting method called AdaBoost turns out to be derivable in this manner.

The principle of gradient boosting can be applied to any type of learner that makes numerical predictions and uses a differentiable loss function. In the context of decision trees, the method is generally known as gradient boosted trees. A well-known implementation of this method is XGBoost. XGBoost has been shown to yield very accurate predictions under broad circumstances, and is at the time of writing, considered by many the most effective learner for a wide range of learning problems.

Isolation forests

While all the above methods are set in a predictive learning context, tree-based methods can also be used in other contexts. One of the most popular methods for unsupervised anomaly detection is called isolation forests (IF). The basic idea behind IF is to recursively split the data based on random tests, until each instance is in a separate leaf (is "isolated" from all others). Instances that are very different from other instances tend to get isolated more quickly. Due to this, the depth of an instance's leaf is an indication of how anomalous the instance is. Isolation forests learn multiple trees in this way, and eventually assign an anomaly score to instances on the basis of the average depth of their leaf in these trees.

7.3.8 If-then-rules

If-then-rules have been popular representation formats in AI from its very beginning. An **if-then-rule** is simply an expression of the form:

```
IF conditions THEN decision
```

Depending on the context, the "decision" can be a predicted value for some target variable, a recommended action, a diagnosis, or just about anything else. The condition part typically contains a set of individual conditions, all of which have to be satisfied in order for the rule to apply.

A **rule set** is a set of such rules. A **rule list** (also called **decision list**) is an ordered set of such rules. In a rule list, rules are typically evaluated from first to last, and the first rule whose condition is fulfilled is the one considered relevant.

The difference between rule sets and rule lists is easily illustrated on a small example that everyone is familiar with: the definition of leap years (which have 366 days). A year is a leap year if it is a multiple of 4, except for years ending in 00: those are only leap years when they are a multiple of 400. This definition can be written using if-then-rules as follows:

– As a rule set:

```
if multiple of 4 and not multiple of 100 then leap year
if multiple of 400 then leap year
if not multiple of 4 then regular year
if multiple of 100 and not multiple of 400 then regular
    year
```

– As a rule list:

```
if multiple of 400 then leap year
else if multiple of 100 then regular year
else if multiple of 4 then leap year
else regular year
```

The rule list is written using the "else if" format to remind the reader that a rule only applies when none of the earlier rules apply.

Rule lists tend to be more compact than rule sets. When a simple rule exists that is almost correct but has some exceptions, the exceptions can be handled using separate rules that precede this rule. In a rule set, the exceptions have to be excluded explicitly from this rule.

Rule sets, on the other hand, are easier to interpret. An individual if-then-rule is easy to understand, but the rules in a rule list are only accurate when considered in the context of the rules preceding it.

Just like decision trees, decision rules can be used for both classification and regression; they are then called classification rules and regression rules.

Learning classification and regression rules

Any decision tree can trivially be turned into a rule set: with each leaf of the tree, we simply associate one rule that has all the test outcomes on the path from the root to that leaf as its conditions, and the leaf's prediction as its decision. However, there are also algorithms that learn rule sets or lists straight from the data, without the detour via decision trees. They typically learn one rule at a time. A typical approach, sometimes called the *covering* approach, is to repeatedly try to learn one rule that makes only correct predictions, and apart from that is as widely applicable as possible. Each consecutive rule focuses on cases that have not yet been handled correctly by previous rules. A single rule is typically learned by starting without any conditions, and repeatedly adding a single condition in such a way that the condition maximally improves the rule's accuracy. Though relatively old, Cohen's RIPPER algorithm (Cohen, 1995) is still state-of-the-art for classification rules. An implementation of it can be found in the WEKA toolbox for data mining.

Learning association rules

In the above discussion, the rule set was interpreted as a predictive function. However, due to their interpretability, rule sets are also very useful for descriptive learning, where the goal is to understand the data. When used for that particular purpose, the standard rule learning algorithms are no longer effective.

A *descriptive* rule is useful when it adds information that was not provided by other rules. A *discriminative* rule is useful when it contributes to a better discrimination between different classes.

To illustrate the difference: if we want a rule set that allows us to distinguish dogs and cats, then the rule "if it barks, it's a dog" suffices. The rule "if the animal buries bones, it is a dog" may be equally accurate, but we have no need for it if the previous rule already works well enough. A predictive rule learner will therefore not add the second rule to the set. A descriptive rule learner, however, will typically return both rules, as both may be interesting to a user who wants to gain insight into the behavior of animals.

Association rules are one type of descriptive rules. Syntactically, they are very similar to classification rules. They have the following format: $P \rightarrow P'$, where P and P' are sets of properties that an instance may have. The rule expresses that when an instance has all properties in the set P, it is more likely to also have all properties in the set P'. Two numbers are typically associated with an association rule, called the support and confidence of the rule. The **confidence** is the fraction of instances with properties P that also have properties P'. The **support** is the fraction of instances that have both P and P'. Support expresses how broadly applicable the rule is, confidence expresses how accurate it would be if it were interpreted as an if-then rule: "if P then P'."

An association rule describes a pattern in the data, and the goal of association rule mining is typically to find all patterns in the data that meet certain quality criteria. The APRIORI system was among the first to solve the following problem: return all association rules whose support and confidence are above given thresholds s and c.

The prototypical application of association rules is market basket analysis: what associations exist between products bought by customers in a supermarket? For instance, assume that among all customers buying mozzarella and ham, 30 % also buy tomatoes. If the overall percentage of customers buying tomatoes is only 20 %, this rule indicates a clear positive association between buying mozzarella and ham and buying tomatoes. Such rules provide insight into customers' buying behavior, which can be invaluable to the supermarket.

Although association rules look very similar to classification rules, there are important differences. Classification rules predict the value of just a single variable (the target variable), they aim at making this prediction with 100 % accuracy, and a rule set typically contains as few rules as possible. In a set of association rules, a single rule may predict multiple properties, the predicted properties vary from one rule to another, the rules need not have near 100 % accuracy in order to be useful, and a rule set may contain many rules that would be redundant when used for the sole purpose of prediction.

Inductive logic programming

As explained in Chapters 4 and 5, many formalisms for knowledge representation and reasoning rely on some form of logic. Mathematicians distinguish *propositional logic*, which have propositions (atomic statements) and logical connectives such as "and," "or," "not," and *predicate logic*, which uses a richer vocabulary where one can refer to objects in some universe of discourse, to properties of these objects or relationships among them (using variables where needed). In predicate logic, the fact that a country can have only one capital could for instance be expressed as $\forall x, y, z : \text{Capital}(x,y) \wedge \text{Capital}(x,z) \rightarrow y = z$, which is to be read as: for *all* x, y, z, if y is the capital of x and so is z, then y and z must be equal. This rule cannot be expressed using propositional logic, it simply does not have the vocabulary for that.

While the field of knowledge representation has moved beyond propositional logic decades ago, most of machine learning (and related fields such as probability theory and statistics) still employs propositional representations. All the example rules shown above can be expressed in propositional logic. **Inductive logic programming (ILP)** is an exception: it is a learning paradigm in which predicate logic is used to represent both the data we learn from, and the results of the learning process.

Suppose we have a geographic knowledge base, containing, among other things, which city is the capital of which country. Such a rule, stating that cities can be the capital of only one country, could be discovered automatically by an ILP system by analyzing the data. In terms of application potential, the discovery of such patterns is not essentially different from standard rule discovery: we can learn predictive rules that allow us to fill in missing values or guess unknown information, we can learn descriptive rules that explain properties of the data, we can detect errors or anomalies by looking for patterns that are almost always satisfied (the exceptions then being considered anomalies), etc. The main difference is that the increased expressiveness of predicate logic allows the system to identify patterns that standard rule-learners could never find, because they simply cannot express them.

ILP has been applied to a wide variety of concrete problems. Maervoet et al. (Maervoet et al., 2012) show how it can be used for anomaly detection in geographical information systems (GIS). In this application, an ILP system was used to find common patterns in a road network, which were next used to identify errors in the data. For instance, the system could discover the pattern that road crossings of one-way streets always have both incoming and outgoing streets (to us this is obvious: a crossing with only incoming streets could never be left without violating traffic regulations), that road segments adjacent to primary schools always have a 30 km/h speed limit, etc. This led to the identification of errors in the database. The ability to automatically discover unusual situations in such a database, without first having to define what is unusual, is a major advantage in the maintenance of complex knowledge bases, especially when these evolve. For instance, a change in traffic law stating that road segments near primary schools must have a 30 km/h speed limit, would be automatically picked up by such a system, and exceptions to the rule reported. Clearly, this can be a major help for keeping GIS up to date.

A road network is essentially a graph, and indeed analysis of such networks could also be done with graph-based learning methods. What makes first-order predicate logic interesting is that it can express any knowledge that can be expressed using graphs and, moreover, it is a well-studied mathematical formalism.

First-order predicate logic is expressive enough to allow the construction of complete computer programs in it. The programming language Prolog exploits this fact. To illustrate this: a Prolog implementation of the well-known quicksort algorithm (which efficiently sorts an array from low to high) would typically contain the following rule:

$$\text{Sorted}(x, y) \leftarrow \text{Split}(x, x_l, x_r) \wedge \text{Sorted}(x_l, y_l) \wedge \text{Sorted}(x_r, y_r) \wedge \text{Concat}(y_l, y_r, y)$$

which can be read as follows: if you Split an array x into two parts x_l and x_r such that all elements of x_l are smaller than all elements of x_r (this information would be in the definition of the Split predicate, not shown here), and y_l and y_r are the Sorted versions of x_l and x_r, then Concatenating y_l and y_r gives a Sorted version of x. This can be read as a simple if-then-rule, but at the same time, it is operational: it allows Prolog to actually sort arrays by splitting, sorting the sublists separately, and concatenating. Prolog programs consist entirely of such sets of rules.

Interestingly, *simple programs of this kind can be learned from examples*. The idea is to show some input/output pairs for a program, and have the computer discover a program that is consistent with these pairs. This kind of task is called **program synthesis**. ILP is one approach to program synthesis. With the current state of the art, relatively complex programs such as quicksort can indeed be learned from examples, but only when the right auxiliary predicates are already given. For instance, the above sorting algorithm could be learned, if the predicates Split and Concatenate are provided. The automatic discovery of useful auxiliary predicates is called **predicate invention**. This is essentially an unsupervised learning task and, therefore, much harder. The lack of effective methods for predicate invention has stalled ILP for decades, but recent developments in (among other areas) deep learning have created new opportunities.

An example of what is currently realistic, in terms of program synthesis, is the automatic discovery of spreadsheet equations. Excel's Flashfill tool can automatically fill in a column that nontrivially depends on other data, using a program synthesized from very few examples. In a similar vein, the Tackle tool (Kolb et al., 2020), illustrated in Figure 7.12, is able to discover patterns that involve multiple tables and predict the value of cells on the basis of the discovered patterns.

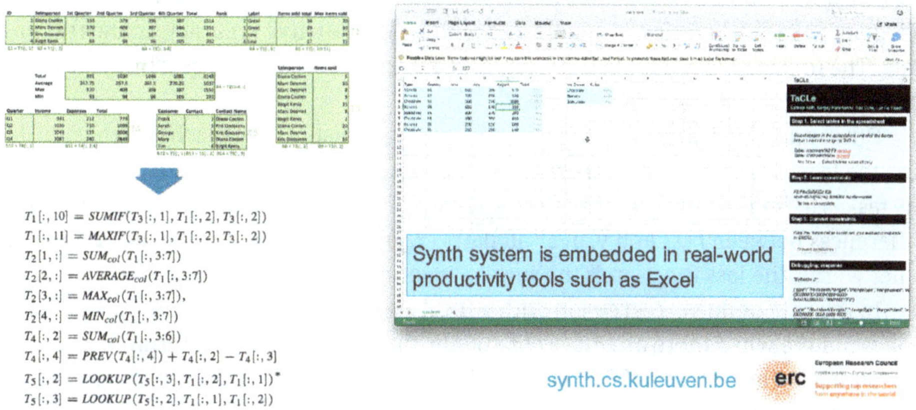

$$T_1[:, 10] = SUMIF(T_3[:, 1], T_1[:, 2], T_3[:, 2])$$
$$T_1[:, 11] = MAXIF(T_3[:, 1], T_1[:, 2], T_3[:, 2])$$
$$T_2[1, :] = SUM_{col}(T_1[:, 3:7])$$
$$T_2[2, :] = AVERAGE_{col}(T_1[:, 3:7])$$
$$T_2[3, :] = MAX_{col}(T_1[:, 3:7]),$$
$$T_2[4, :] = MIN_{col}(T_1[:, 3:7])$$
$$T_4[:, 2] = SUM_{col}(T_1[:, 3:6])$$
$$T_4[:, 4] = PREV(T_4[:, 4]) + T_4[:, 2] - T_4[:, 3]$$
$$T_5[:, 2] = LOOKUP(T_5[:, 3], T_1[:, 2], T_1[:, 1])^*$$
$$T_5[:, 3] = LOOKUP(T_5[:, 2], T_1[:, 1], T_1[:, 2])$$

Synth system is embedded in real-world productivity tools such as Excel

synth.cs.kuleuven.be   erc

Figure 7.12: The Tackle tool, developed as part of Luc De Raedt's SYNTH project, analyzes spreadsheets and invents equations that show how some cells can be computed automatically from other cells.

7.3.9 Statistical model fitting

Minimizing a loss function
Many machine learning methods essentially try to fit a function to data, and in this sense, machine learning has a clear connection with statistics.

A machine learning task is often described as follows: a model class (sometimes also called the hypothesis space) $\mathcal{F}$ is given, and the task is to find, among all models in that class, the "best" model. The quality of a model is typically expressed using a so-called **loss function** l; the "best" model is the one with the lowest loss. The loss function typically evaluates how well the function fits the observed data D, but may also take other criteria into account, such as the complexity of the model, or some preexisting preference for certain types of models. In the tree and rule learning examples we saw earlier, $\mathcal{F}$ would be the set of all possible trees, rule sets, or rule lists.

Formally, machine learning is thus an optimization problem: find $f^* = \arg\min_{f \in \mathcal{F}} l(f, D)$, that is, a function f^* such that $l(f^*, D)$ is smallest among all $l(f, D)$ where $f \in \mathcal{F}$.

In a purely predictive setting, where we do not care about the complexity of the model but only want to find the model with the best predictive performance, we typically want the most accurate model on the *population*, which may differ from the most accurate model on the training set D (otherwise there would be no such thing as overfitting). That means that the loss function l minimized during training is not necessarily identical to the loss function whose expectation (over the population) we want to minimize.

In some machine learning approaches, the optimization problem is implicit. Decision tree learners, for instance, try to find a model with good predictive accuracy on the population, by doing a heuristic search through the space of all possible trees. Its heuristics point is in the direction of small trees with high predictive accuracy on the training data. The underlying motivation is that large trees can more easily overfit the training data.

Other machine learning approaches, including statistical approaches, explicitly define the loss function to be minimized. In least squares linear regression, for instance, the task is to find a function f that fits the data best in the sense that the sum of squared differences between $f(x)$ and y, over all pairs (x, y) observed in the data, is minimal. In other words, the loss function is the sum-of-squared-errors $\sum_{(x,y) \in D} (f(x) - y)^2$.

In the case of linear regression with a fixed set of input variables, the optimal solution is unique when there are more data points than variables, and a formula exists for computing it. For instance, if $f(x)$ is of the form $ax + b$, formulas exist for computing the values of a and b that minimize $\sum_{(x,y) \in D} ((ax + b) - y)^2$.

There are also situations where such a formula does not exist or is too complicated to calculate. A step-by-step procedure, or *algorithm*, must then be used instead. An example of such a procedure is **stepwise linear regression.** This procedure is used when a tradeoff needs to be made between the complexity of the model (as measured by the number of variables included in it) and how well it fits the data. Stepwise linear regression starts with building a model with a single input variable. To find the best such model, it considers each potential input variable in turn, finds the last squares solution when only that variable is used, and among all models thus obtained, keeps the one with lowest loss. It then adds a second variable by trying each other input variable in turn, evaluating how much it improves the model when added to the already chosen variable, and keeping the one that gives the biggest improvement. It continues doing this for a third, fourth, ... variable, until the additional improvement is no longer compensated by a sufficient decrease in loss.

Regularization

Stepwise linear regression is biased toward simple solutions in an algorithmic manner. That is, the procedure is defined in such a way that it tends to yield simple models. But we can also include a bias toward simple models in the loss function. Often, the loss function contains two terms: one that measures the fit with the training data, and a second one that penalizes models that have undesirable properties, for example, they are complex or likely to overfit. This second term is often called a **regularization term**. Regularization is a very generally applicable technique. In linear regression with many variables, overfitting is often avoided by trying to keep the regression coefficients small. The regularization term typically sums the squares of these coefficients, or their absolute values. Thus, if we have, say, two input variables and one output variable, the linear function to be constructed is of the form $y = a_1 x_1 + a_2 x_2 + b$, and the optimization ask is to find the values for a_1, a_2, b that minimize $\sum_{(x,y) \in D} ((a_1 x_1 + a_2 x_2 + b) - y)^2 + C(a_1^2 + a_2^2)$.[6] The first term ensures that the function fits the data well, the second term ensures that the coefficients do not grow too large. The method's hyperparameter C allows the user to influence the tradeoff. The statistical methods known as **ridge regression** and **lasso** are variants of this type of regularization. They impose an upper bound on, respectively, the sum of squares and the sum of absolute values of the coefficients. Using the sum of squares or sum of absolute values each has its own advantages. The sum of absolute values criterion, as used by lasso, is that it tends to make many coefficients exactly zero, which is not the case for the sum of squares. **Elastic nets** combine the advantages of both by using both types of regularization terms.

Although these linear methods are most naturally used for regression with numerical input variables, they can easily be used with categorical input variables by encoding these using groups of binary variables called dummy variables. For instance, a nominal variable with values a, b, c can be replaced by two dummy variables that jointly take the values $0, 0$ for a, $0, 1$ for b, and $1, 0$ for c. Less trivially, these regression techniques can also be used for classification. **Logistic regression** is an archetypical example of this approach. In this approach, we try to predict the probability that an instance belongs to a particular class. In the binary case where the classes are represented as 0 and 1, we learn a model that predicts $P(y = 1 \mid x)$ from x. A problem with this, however, is that probabilities are restricted between 0 and 1, whereas linear regression, by nature, learns a linear function that is not restricted to any interval. This mismatch is solved by actually

---

6 The bias term b is usually not included in the regularization term, because a large b does not increase the model's fitting capacity.

predicting not $p = P(y = 1 | x)$, but the so-called **log-odds ratio**, $l = \log(\frac{p}{1-p})$. From this formula follows $p = \frac{e^l}{1+e^l}$, so $P(y = 1 | x)$ can easily be derived from l. The loss function used in logistic regression is the so-called cross entropy: $y \log(p) + (1 - y) \log(1 - p)$.

While all the above cases illustrate the use of statistical model fitting for learning predictive functions, exactly the same principle can be used for clustering, density estimation, probabilistic modeling, etc.

The principle of minimizing some loss function is present, either implicitly or explicitly, in almost all machine learning approaches. It may seem that explicit loss minimization, as done in statistical model fitting, is in general a better approach, especially when an optimization method is used that provably returns the optimum. However, one should keep in mind that when learning predictive functions, the loss function that we *actually* minimize (a combination of fitting and regularization) is not the one that we ultimately *want* to minimize (expected error on unseen examples). The practical benefits of using (computationally expensive) exact optimization methods may therefore be small.

The bias-variance tradeoff

The concepts of bias and variance, as known from statistical inference, are frequently used in machine learning to describe properties of learning methods. To explain these concepts, we need the concept of an **estimator**, a formula that estimates some population parameter from a sample. (For instance, the mean of a randomly drawn sample is often used to estimate the mean of a population.) Given one sample, the estimator yields one value, but for different randomly drawn samples typically yield different values. We can therefore view the estimator's value as a random variable with a certain mean and variance. The **statistical bias** of an estimator is the difference between its mean and the parameter value it estimates. It indicates whether the estimator has a tendency, on average, to overestimate or understimate. For instance, the sample mean is an unbiased estimator for the population mean: indeed, there is no reason to believe that the sample mean would, on average, be higher than the population mean (nor that it is lower, on average). In contrast, if we want to estimate the *range* of some population (the difference between highest and lowest value), the range of a random sample can at most be equal to that of the population (this happens when the sample happens to include both the highest and lowest values in the population) and will often be less; hence, on average, taken over all random samples, the sample range will be smaller than the population range. Thus, this estimator has a negative bias.

The **variance** of an estimator is simply the variance of its probability distribution. It indicates how strongly the estimated value may differ from one random sample to another. Specifically, it is the expected value of the squared difference between the estimator's value and its mean.

The concepts of bias and variance can be generalized to the setting of machine learning in different ways. Consider the case where a predictive function $\hat{f}$ is learned from a sample from the population; is supposed to approximate the "real" function f. Given an instance x for which a prediction is made, the learner's bias at x is the difference between the expected value of $\hat{f}(x)$, taken over all possible samples from which $\hat{f}$ might be learned, and $f(x)$. The variance at x indicates how strongly $\hat{f}(x)$ may vary around its mean.

Generally, there tends to be a tradeoff between bias and variance, and this tradeoff is related to the tradeoff between underfitting and overfitting. Indeed, high variance essentially means that we might have obtained very different predictions if we had learned from a different dataset (drawn randomly from the same population). This shows that the learned model expresses accidental properties of the dataset, not actual properties of the population. Conversely, a strong bias implies that the model will systematically underestimate or overestimate the target value at a particular point, no matter what dataset it was fed.

Regularization can be seen as a way of controlling the bias-variance tradeoff: by insisting on small values for certain parameters, it reduces the variance of the learner, at the cost of an increased bias.

7.3.10 Support vector machines

Support vector machines (SVMs) became popular during the 1990s, and for over a decade they were considered the method of choice for learning classifiers. The intuition behind them is simple and appealing, and while the technicalities are somewhat involved, they exploit standard methods from convex optimization, a well-developed field. Together with SVMs, the concept of *kernel-based methods* was introduced in machine learning. Below, we first discuss the basic principles behind linear SVMs. Next, we will discuss the role of kernels.

Linear support vector machines

Imagine a plane in which two types of instances ("positive" and "negative" instances) are positioned, and imagine you are looking for a linear separator, that is, a straight line that separates the positives from the negatives. Obviously, such a line represents a decision criterion and, therefore, a model for predicting the class of unseen cases.

Any straight line that separates the positive training instances from the negatives has 100 % training accuracy. Still, not all these lines are equally natural. Consider Figure 7.13. Lines A and B are both perfect separators. But line A line implies that the instance marked "?" is positive, even though it is very close to a negative and there is absolutely no evidence that this instance should be positive. Line B is only slightly better. Generally, a separator that comes unnecessarily close to any training instance implies an unintuitive conclusion for unseen instances near that instance.

This leads to the following intuition: when creating a linear separator, we prefer a line that is as far away from all training instances as possible. In other words, we look for

Figure 7.13: Left: two linear separators. Both get unnecessarily close to labeled instances, yielding unintuitive predictions. Right: the max-margin separator keeps the largest possible distance from all instances.

a line with **maximal margin**, where the "margin" around the separator is the smallest distance from the separator to any training instance.

Learning a support vector machine is nothing more than *finding the linear separator with maximal margin*. When the two classes are linearly separable, this max-margin separator exists and is unique. Moreover, it can be found by solving a well-understood type of optimization problem, namely finding the minimum of a quadratic function over a convex domain. Whereas finding the minimum of a function is hard in general, it is relatively easy in this particular case, and efficient off-the-shelf solvers exist.

Assume a dataset with elements (x, y) is given, with x a vector in the input space and y its label, which is 1 for positive cases and -1 for negative cases. We will look for a function $f(x) = ax + b$ with the property that $ax + b \leq -1$ for all negative instances, and $ax + b \geq 1$ for all positive instances. This function will be used for classification as follows: x is positive if $f(x) > 0$, negative otherwise. Clearly, with this classification rule, the separator is entirely consistent with the training data, and the separating line is defined by $f(x) = 0$.

There are many such f. All of them are consistent with the data. The minimization problem now states that we want that f for which $\|a\| = \sum_{i=1}^{m} a_i^2$ is minimal. One can view this as a kind of regularization, but here it has a very concrete interpretation: the f with this property turns out to be the max-margin separator.

Constrained optimization problems are often solved by constructing a so-called Lagrangian function. We leave out the technicalities here; what's important is that the Lagrangian associates one variable (called a Lagrange multiplier) with each constraint and returns, together with the solution, a value for each of these variables. This indicates to what extent the constraint affects the optimal solution. If it is 0, it means this constraint does not actually constrain the solution: dropping the constraint would have yielded the same solution. If it differs from 0, it indicates that a better solution could have been found if this constraint were absent.

In the context of an SVM, we have one constraint per training instance. Training instances with a nonzero value for their Lagrange multiplier are the so-called **support vectors**: they are the ones closest to the separator. If such an instance were dropped, we might find a better solution with a wider margin. The other instances do not affect the solution at all. In Figure 7.13, on the right-hand side, it is easy to see that there are three support vectors.

This leads to an important insight: if we know the support vectors and their values, we know the separator. It turns out that the max-margin separator $f(x) = ax + b$ can be rewritten using only the support vectors s_i and the corresponding Lagrange multipliers λ_i:

$$f(x) = \sum_{s_i} \lambda_i y_i x_i x + b$$

Not only do the Lagrange multipliers suffice for making predictions, the optimization problem itself can also be rewritten directly in terms of them. More specifically, instead of finding the a that minimizes $\|a\|$, we find the values for all λ_i that maximize

$$\sum_i \lambda_i - \frac{1}{2} \sum_i \sum_j \lambda_i \lambda_j y_i y_j x_i x_j$$

under the constraints $\lambda_i \geq 0$. In optimization terminology, this is called the **dual problem**.

We present this formula here because we want to draw attention to its structure. The training set influences the formula only through the dot products of all training instances (multiplied by $y_i y_j = -1$ for instances with opposite labels). If these dot products are computed in advance, no further access to the training instances is needed at this point.

Thus, for a dataset with N m-dimensional instances, the separator is uniquely defined by a and b (together $m + 1$ values), but also by the nonzero λ-values and the corresponding support vectors. The number of these support vectors cannot exceed N. The smallest representation of these depends on the dataset. When learning from a small dataset in a high-dimensional input space, the separator is more efficiently represented using the support vectors.

All the above insights are essential for understanding the main strength of SVMs, which we will describe next.

Kernels and nonlinear SVMs

We have seen that learning a linear max-margin separator boils down to an optimization problem that requires as inputs, apart from the labels, only the dot products between training instances.

Now, how can we learn nonlinear separators? The SVM approach is as follows. Imagine that we transform the input space to some higher-dimensional space, learn a linear separator there, and transform the result back to the original space. For instance, the transformation $\Phi(x_1, x_2) = (x_1, x_2, x_1 x_2, x_1^2, x_2^2)$ transforms a 2-dimensional input space to a 5-dimensional one. A linear separator in this transformed space has the form $a_1x_1 + a_2x_2 + a_3x_1x_2 + a_4x_1^2 + a_5x_2^2 + b = 0$. This is at the same time a quadratic form in the original space. Hence, we can learn quadratic separators by following this procedure. These separators do not have the max-margin property in the original space, but they do in the transformed space. Figure 7.14 illustrates the process for a simpler version of the transformation.

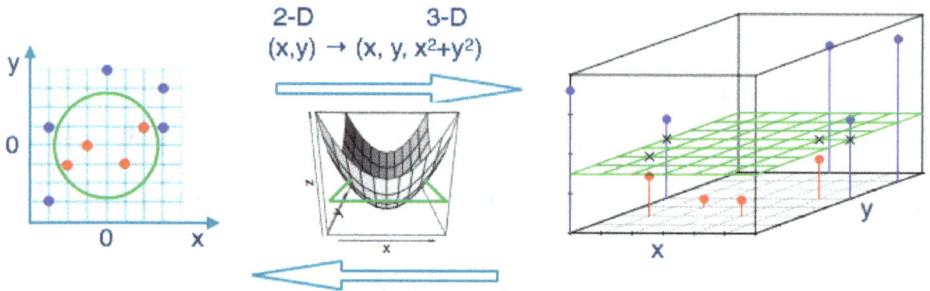

Figure 7.14: A set of instances of two classes (red/blue) that is not linearly separable is transformed to a 3-dimensional space. In that space, they can be separated by a plane. Finding the max-margin plane and transforming it back to the original space gives a circular separator in the original space. (In practice, the transformed space is usually of much higher dimension than shown here.)

Now, because learning an SVM requires only dot products, learning an SVM in the transformed space requires only computing the dot products $\Phi(x_i)\Phi(x_j)$ for all training instances. The computation of these dot products is done by a **kernel function** $K(x_i, x_j)$.

In practice, when learning SVMs, we directly choose a kernel function, without making the detour of choosing Φ. Not each function is suitable as a kernel function; the function must be such that some transformation Φ exists such that $K(x_i, x_j) = \Phi(x_i)\,\Phi(x_j)$. Technically, a kernel function must satisfy *Mercer's condition*: the kernel matrix must be positive semi-definite for any dataset. Predefined families of suitable kernel functions exist. Well-known examples are the **polynomial kernels**, $K(x_i, x_j) = (1 + x_i x_j)^p$, and **the radial basis function kernels**, $K(x_1, x_2) = \exp(-\|x_1 - x_2\|^2/(2\sigma^2))$, with σ a parameter.

From the point of view of feature representations, the above can also be interpreted as follows: the kernel implicitly defines a set of relevant features, and the SVM constructs a linear separator in this feature space.

The above kernels are defined on Euclidean vector spaces, but kernels can be defined for different types of inputs too. For instance, to learn a support vector machine that classifies graphs, all that is needed is a kernel function that works on graphs. Intuitively, such a kernel function gives an indication of how similar two graphs are.[7] We

---

7 For example, if we consider graphs similar if they have a similar number of edges and nodes, and only that matters, then we could decide to use $K(G_1, G_2) = (n_1, e_1, 1) \cdot (n_2, e_2, 1)$, where n_i and e_i are the number of nodes and edges of graph G_i, respectively. In practice, more complex kernels are typically used.

use the kernel function to create the kernel matrix (containing the pairwise similarity between each pair of graphs in the input data), we feed this to the optimizer, and out comes an SVM that can classify the graphs.

The strength of support vector machines is that they have a very strong regularization criterion. They can learn from small datasets in very high-dimensional spaces, a task that is very challenging for most other approaches. On the other hand, the necessity to construct an N-by-N kernel matrix (with N the number of instances) makes the approach computationally complex for large datasets.

We have discussed SVMs in the context of binary classification, but similar techniques have also been developed for regression and other variants of predictive learning. SVMs are often used for data that is not in the standard format, for example, classification of text, sequences, graphs, etc., or where the number of dimensions is very high, relative to the number of instances.

7.3.11 Neural networks

Neural networks are a relatively old approach to artificial intelligence that has gone in and out of fashion multiple times. A neural network is a specific type of highly parametrized model that can be fit to very diverse types of input data. Neural networks are to some extent inspired by biological nervous systems, which include the human brain, but in the interest of conciseness we will not discuss this aspect and go straight to a description of the format.

Simple neural networks

A neuron is a simple unit that takes a number of numerical inputs and aggregates them into a single numerical value, which is then output. The aggregation operator typically consists of two parts: first, a weighted sum of the inputs is computed; next, a so-called **activation function** is applied to the weighted sum. If we represent the inputs as a vector x and the corresponding weights as a vector a, then the weighted sum equals the dot product of these vectors: $ax = \sum_i a_i x_i$. Thus, a neuron outputs $\sigma(ax)$, with σ the activation function. Activation functions often "binarize" the output in the sense that they map their input to two clearly distinguished levels, "low" or "high." Figure 7.15 shows a few commonly used activation functions.

Many descriptions of neurons describe the first part as a function of the form $ax + b$, with b called the bias parameter, rather than simply ax. Both formulations are equivalent if we implicitly extend each input vector with one additional component that is always 1; the weight of this additional component equals the bias. For simplicity of notation, here we use the simpler notation ax.

A single neuron can be used as a predictive model in itself. When training such a neuron, the activation function is assumed fixed, so training boils down to finding a weight vector that minimizes some loss function. For instance, given a training set D, we

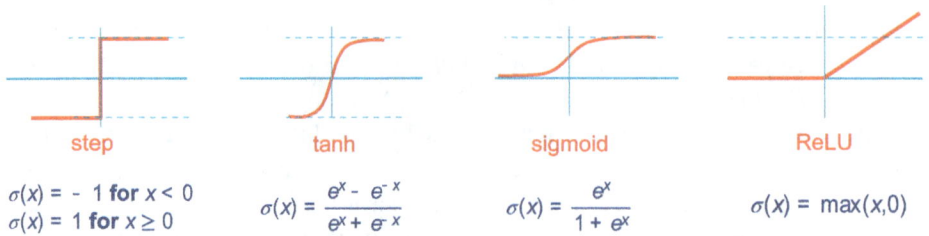

step	tanh	sigmoid	ReLU

$\sigma(x) = -1$ **for** $x < 0$
$\sigma(x) = 1$ **for** $x \geq 0$

$\sigma(x) = \dfrac{e^x - e^{-x}}{e^x + e^{-x}}$

$\sigma(x) = \dfrac{e^x}{1 + e^x}$

$\sigma(x) = \max(x, 0)$

Figure 7.15: Four different activation functions commonly used in neural networks.

try to find the a that minimizes the squared loss $\sum_{(x,y)\in D}(\sigma(ax) - y)^2$. If we assume for σ the identity function, then this actually corresponds to least-squares linear regression. Whereas least-squares linear regression has a closed-form solution, this is not generally the case when σ is not the identity function.

The true strength of neural networks comes from combining these simple neurons into a network. Here, the output of one neuron is fed as an input into another neuron. The layout of the neurons (how many are there, how are they connected) is called the **architecture** of the neural network. This architecture is typically given in advance, so that the learning process remains limited to optimizing the parameter values.

Figure 7.16 shows a *fully connected, 2-layered, feedforward neural network*. It is called *2-layered* because there are two layers of neurons, *feedforward* because the output of a neuron is fed into the neurons of the next layer, and *fully connected* because each neuron is connected to *all* neurons of the next layer.

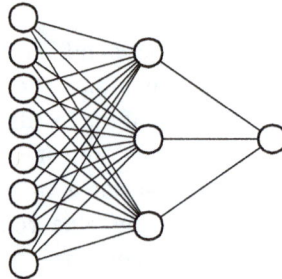

Figure 7.16: A fully-connected two-layered feedforward neural network.

Whereas a single neuron is strongly limited in terms of the kind of functions it can represent, a 2-layered neural network can represent a wide variety of predictive functions. For instance, in the classification context, a single neuron can only construct linear separators, but a 2-layered neural network can construct any kind of separator, as long as it has enough neurons in its first layer (see Figure 7.17). It is essential here that the neurons have a nonlinear activation function. If linear activation functions are used, their combination is still a linear function and there is no increased expressiveness.

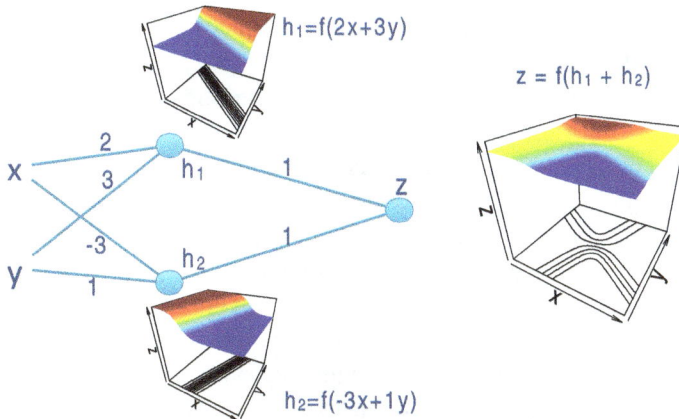

Figure 7.17: Activation regions of the neurons in a small neural network (for sigmoid activation function). For the layer-1 neurons, the regions of high and low activation are linearly separable. By combining the outputs of the layer-one neurons, the layer-2 neuron creates a high-activation region that is no longer linearly separable from the rest (see the contour lines).

Training a neural network

In the context of predictive learning, training a neural network involves finding values for its parameter such that the network fits the data well. This training process typically takes use of a general optimization procedure called **gradient descent**. To explain this procedure, we need some basic mathematical concepts.

The **derivative** of a function $f(x)$ is a new function that indicates the slope of $f(x)$ at point x. For instance, the function $f(x) = x^2$ has a slope of 0 at $x = 0$ and 2 at $x = 1$. The function that indicates for each point x what the slope of f is at that point is, in this case, $g(x) = 2x$.

For a function of multiple variables $f(x, y, z)$, the **partial derivative** of f toward x at point (x, y, z), denoted $\frac{\partial f(x,y,z)}{\partial x}$, indicates the slope of $f(x, y, z)$ in the direction of the X-axis at point (x, y, z). The **gradient** of f at point (x, y, z), denoted $\nabla f(x, y, z)$, is a vector that contains the partial derivative of f toward each input:

$$\nabla f(x,y,z) = \left[\frac{\partial f(x,y,z)}{\partial x}, \frac{\partial f(x,y,z)}{\partial y}, \frac{\partial f(x,y,z)}{\partial z} \right].$$

The gradient has the property that *it indicates the direction in which f grows fastest.* Imagine standing on a hillside: the steepest direction uphill is the gradient. If you put a ball on the floor and release it, it starts rolling in exactly the opposite direction of the gradient. The length of the gradient vector (technically, its 2-norm) indicates how steep the function is. Where it is 0, f is tangential to a horizontal plane.

The basic intuition behind gradient descent is simple: to find a minimum of f, start at a random position and take a small step in the opposite direction of the gradient. Repeat this process at your new position. Keep repeating this until the gradient is zero.

This basically means you go down until you cannot go down any further; at that point, you have reached a **local minimum**. The term "local" stresses that you have reached a lowest point in the immediate neighborhood of your location, but not necessarily a globally lowest point. Starting at a different position may yield a different local minimum. Among all local minima, the smallest one is called a **global minimum**.

Now, let us apply all of this to curve fitting. There is some loss function, and we are trying to determine the parameter values that minimize this loss. To do that, we need to consecutively compute the gradient of the loss function for the current parameter values (and then adapt the parameters in the direction of that gradient). The size of our step in the opposite direction of the gradient determines how fast we move down; a large step size moves faster but increases the risk of "jumping over" the minimum.

Consider the neural network drawn in Figure 7.18. Intuitively, the partial derivative of the loss function toward one parameter can be found by imagining we increase that parameter by a very small value ε. How does this tiny change affect the output, and hence, the loss? In the shown network, the propagation of this change for a single (x_1, x_2) input is shown. In practice, we have to repeat this for all (x_1, x_2) pairs in the training set, and add up all the effects, to find partial derivative for that parameter. To compute the gradient, this procedure needs to be repeated for all parameters.

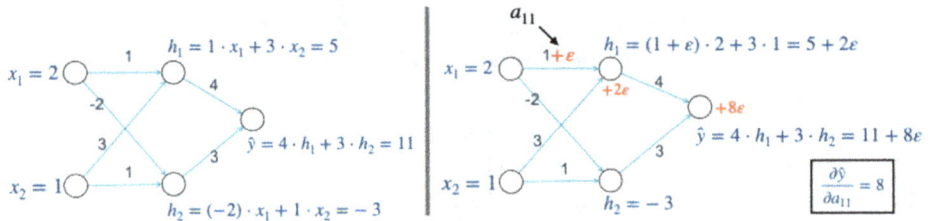

Figure 7.18: A given neural network with inputs $(2, 1)$ gives 11 as output. How does the output change if we add a small number ε to parameter a_{11}? We can recompute the output for the changed value (for simplicity, linear activation functions are used here). The ratio between the output change and ε is the partial derivative.

A single computation of the gradient clearly requires a lot of work. There are two ways in which this can be reduced. First, instead of computing the gradient of the loss function on the entire dataset, we can compute it on a random subsample. This is acceptable because moving in the opposite direction of the gradient is essentially a heuristic procedure: there is no guarantee that this is the best direction to move in, we just choose the gradient direction for lack of something better.[8] This procedure is called **stochastic gradient descent**.

---

8 Actually, better versions do exist—their motivation is too technical for this text—but that only reinforces the argument that we need not invest too much in computing the gradient exactly.

Second, when computing the partial derivatives for many parameters, a large part of the computations overlap. The change in a parameter value propagates toward the loss function through multiple nodes; the change in the parameter affects some node n, which in turn affects the output, which in turn affects the loss. Other parameters affect the same node n and, instead of recomputing the effect of the change in n on the loss, we can reuse outcomes computed earlier. This idea materializes in the concept of **backpropagation**. Backpropagation essentially associates with each internal node the partial derivative of the loss toward that node, then determines the effect of each parameter on the loss by determining its effect on the nodes it is directly connected to.

The concept of backpropagation is applicable to any function that can be expressed as a so-called computational graph. In a computational graph, an edge $x \rightarrow y$ indicates that the output of node x is an input for node y, and each node represents an operator that computes the node's output from its inputs. Very complex models can be defined and trained in this manner. This leads to a machine learning approach called **differentiable programming**, where the "program" defines the structure of the computational graph while the parameter values are learned from data.

Deep learning
Consider a network with n layers. The output of the network (the n'th layer) is computed entirely from the $n - 1$'th layer; there are no shortcuts to earlier layers. Therefore, the $n - 1$'th layer must have all the information needed to make an accurate prediction. Similarly, the state of the $n - 1$'th layer must be computable entirely from the $n - 2$'th layer, and so on, up to the first layer, which is computed straight from the input x. In such a network, *each layer forms a different representation of the input.* A layered neural network can be seen as performing a series of transformations from one representation to another. These transformations are learned in a largely unsupervised manner: we do not tell the network what kind of representation it should have in its first, second, etc. layer. The network discovers suitable representations by itself. In a sense, the backpropagation algorithm does not only yield suitable settings for the parameters, it also yields a suitable sequence of internal representations. The fact that these representations are learned in an unsupervised manner is interesting. We have seen earlier that it is easier to make accurate predictions if the right features are available. In a sense, multilayered neural networks invent the right features in such a way that the penultimate layer defines a feature space from which the target is very easy to predict.

The more layers a neural network has, the more "unsupervised" the learning becomes. The term **deep learning** is used for learning networks with many layers. The further away a layer is from both the (observed) input and output spaces, the more it defines an underlying, "deep" representation that is learned without supervision. The learning of useful deep representations is sometimes considered a goal in itself; this task is called **representation learning**. Deep neural networks turn out to be very good at this task, and one of the best ways to illustrate this is to look at how they work in the context of computer vision. That is the subject of the next section.

Convolutional neural networks

Convolutional neural networks are neural networks with a specific architecture that is adapted to a particular type of processing. They are frequently used in the context of computer vision (for instance, for recognizing objects in pictures or classifying images), but also for processing audio signals, time series, and other sequences. In this text, we describe them and their objectives from the point of view of visual object recognition.

For ease of discussion, we assume here that the pictures fed to the neural network are black-and-white, c pixels wide, and r pixels high. Such a picture can be represented as a matrix with r rows and c columns, where each cell has a number between 0 and 1 that indicates the brightness of the corresponding pixel.

A convolutional neural network typically consists of many layers. The first layers in this network are typically special-purpose layers: so-called *convolutional layers* and *pooling layers* occur intermittently. The layers closer to the output are typically general-purpose layers. The idea behind this architecture is that the first layers construct a feature representation that is useful for object recognition, whereas the later layers get trained for specific recognition tasks using these features as inputs.

A **convolutional layer** consists of multiple "convolutional nodes." A convolutional node is an array of neurons, all with the same input weights, but connected to a different part of the input image. More specifically, each neuron is connected to a different k-by-l "window" on the matrix and gets activated when a particular pattern is present in that window. As all neurons share the same weights, they recognize the same pattern. The neurons are arranged in such a way that nearby neurons are connected to nearby (overlapping) windows. As such, the convolution node forms a "map" that indicates in which part of the matrix a particular pattern occurs (see Figure 7.19). Note that, although the node is an array of neurons, we can conceptually think of it as a single neuron with a matrix-valued output (the map).

36 inputs

16 neurons with *same* activation pattern, connected to different inputs

Figure 7.19: A convolutional node is a matrix-valued node that indicates where in the input a certain pattern occurs. Left: a single node in the map is connected to a particular subimage and gets activated when a certain pattern occurs there. Each node in the map has the same activation pattern. Right: illustration of what a map for a particular pattern might look like.

As the map returned by a convolutional node has a similar format as the input (a matrix), we could in principle define a second layer of convolutional nodes on top of it. However, before doing that, a **pooling layer** is typically used to lower the resolution of the maps. The pooling layer groups multiple nearby "map pixels" into a single map pixel, thus reducing the size of the map by, for example, 4 or 9, if 2-by-2 or 3-by-3 pixel squares are pooled. The pooling typically happens by computing the average or maximum of the pooled pixels. The motivation for this operation is that, first, the exact position (up to a single pixel) where a low-level visual feature was found typically does not matter, and second, the number of neurons in later layers is strongly reduced in his manner.

After the first pooling layer, a second convolutional layer is typically included. This layer learns medium-level features that are formed by combining low-level features. It is typically followed by a pooling layer, and possibly more pairs of convolution and pooling layers that construct higher-level features. When it is no longer necessary to define even higher-level features, a general-purpose neural network (typically with few layers) maps the features onto a prediction. Figure 7.20 illustrates how increasingly higher-level features can be learned in this way.

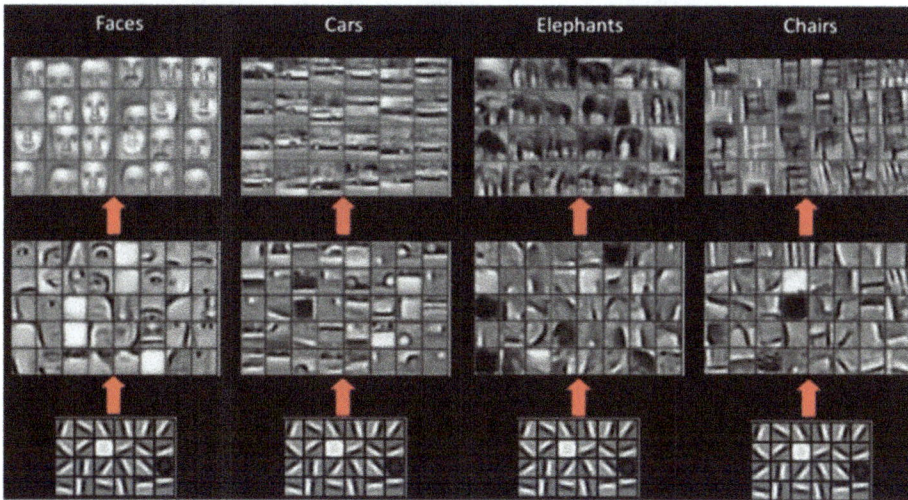

Figure 7.20: Convolutional networks are good at recombining lower-level features (lower part of the image) into medium and high-level features. The same low-level features can be recombined into different patterns. (Picture after Lee et al., "Convolutional Deep Belief Networks for Scalable Unsupervised Learning of Hierarchical Representations").

Note that the local patterns that the network looks for in a picture are defined by the weights of the convolutional nodes. In principle, we could make it look for predefined features by fixing the feature weights appropriately. However, by letting the network learn these weights, it creates by itself sets of features that are relevant for image processing. This ability is perhaps the most important leap that deep learning caused in computer vision: until the early 2000s, computer vision relied on cleverly engineered features, but with deep learning, even better features were discovered fully automatically. This came with a price: training deep networks requires huge amounts of data and enormous computing power. However, the lower-level features turn out to be relatively stable. What happens in practice is that people use a pre-trained network in which the early layers (i. e., the visual features) are fixed, and only the final layers of the network are tuned for the task at hand. This often provides excellent accuracy for a reasonable training cost.

Although deep learning has revolutionized computer vision, it is not without its challenges. It is possible to fool vision systems based on deep learning in ways that would never fool humans. Figure 7.21 illustrates how humans can be made "invisible" to computer vision, when we have no problem at all seeing them. Fooling the computer in this way requires inserting carefully constructed signals in the image.

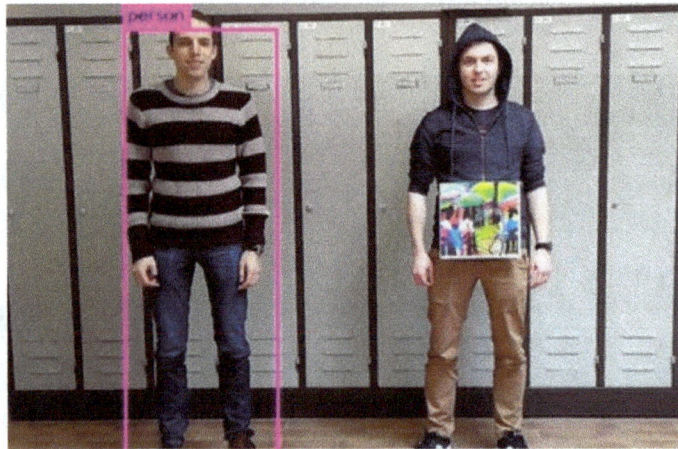

Figure 7.21: Automatic person detectors can easily be fooled by inserting carefully crafted subimages into a picture. (Source of picture: https://nieuws.kuleuven.be/en/content/2019/ku-leuven-researchers-make-themselves-invisible-to-ai-cameras).

Long short-term memory networks (LSTMs)

Up until now, we considered only feed-forward networks, where the output of a layer is fed into the next layer. Other architectures are possible. A **recurrent network** is a network where the output of a neuron is fed back into the same layer, or even an earlier layer (Figure 7.22). Such networks are useful for learning from sequences. Suppose we have a sequence of inputs x_1, x_2, x_3, and a single label y for the whole sequence. In a recurrent network, we feed first x_1 to it, then x_2, then x_3, and at that point observe y. In a feed-forward network, y would be determined by x_3 alone but, in a recurrent network,

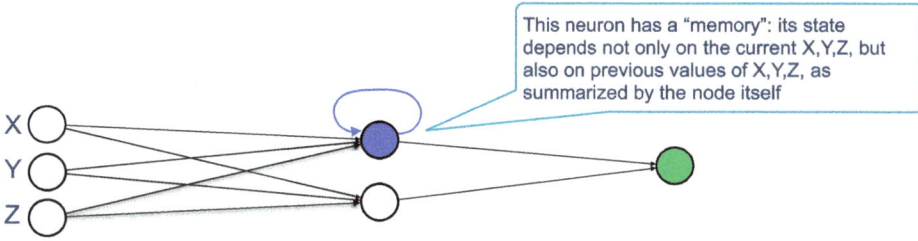

This neuron has a "memory": its state depends not only on the current X,Y,Z, but also on previous values of X,Y,Z, as summarized by the node itself

Figure 7.22: A neural network with one node whose output is fed back into it.

part of the input for y is information that was computed for x_2 (and fed back into the input together with x_3), and the computations for x_2 relied on x_1 for the same reason. The prediction of label y thus depends on the whole input sequence, not just on the last input.

Recurrent networks are usually trained with a procedure called backpropagation-through-time. This is the standard backpropagation algorithm, applied to an "unfolded" version of the recurrent network (where the same node at different points in time is represented as different nodes).

There is a significant problem with recurrent networks, which is inherent to the way neural networks compute their outputs. Signals get multiplied by the weight of the connection they travel through. When a signal is fed back into an input multiple times, it also gets multiplied by the weight of that connection multiple times, so the strength of that signal either increases or decreases exponentially (comparable to what happens with microphone feedback). In the context of training, this leads to the vanishing gradient or exploding gradient problem. To control this effect, more advanced architectures for recurrent networks have been proposed. The **long short-term memory network (LSTM)** is one of the best-known examples.

An LSTM (Figure 7.23) has nodes that serve as a kind of memory. A feedback loop from a memory node to itself ensures that the node's value can be remembered from one time step to the next (it can simply be copied). The memory node has trainable "gates": an input gate, which determines when a signal should be stored in the memory, a forget gate, which determines when the memory should be cleared, and an output gate, which determines when the memorized information should be used. Each gate is a small neural network in itself.

LSTMs have been very successful for natural language processing, among other things. We refer to Chapter 8 for more information on that topic.

Auto-encoders

Consider a feedforward neural network with as many output nodes as there are inputs, but (much) fewer nodes in each intermediate layer. Assume this network is trained so that its output matches the input. This may seem useless, since we simply learn the iden-

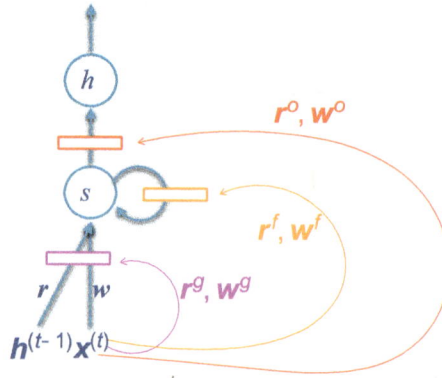

Figure 7.23: Schematic overview of the structure of an LSTM. The state node feeds back into itself. The purple, orange, and red "gates" decide are controlled by the current input and most recent output and decide when the memory is updated, emptied, or used.

tity function. However, since each layer in the feedforward network is computed from solely the previous layer, each layer must have complete information about the input (i. e., sufficient information to accurately restore the input from this). For a layer with few nodes, this implies the input is mapped losslessly to a lower-dimensional space. The network thus defines a lossless compression mechanism. The part of the network that maps the network to the lower-dimensional representation is called an **encoder**, the part that restores the input from that is called the **decoder**. The network itself is called an **auto-encoder**: it learns to efficiently encode the inputs in an unsupervised manner.

In the context of deep learning, such auto-encoders are sometimes stacked onto each other: the encoding constructed by an auto-encoder is itself encoded by a new auto-encoder, and so on. This leads to the concept of stacked auto-encoders.

Auto-encoders as described above are deterministic: an input is mapped to a single point in the lower-dimensional space (also called latent space), and it can be reconstructed from that point. There also exist probabilistic auto-encoders. Here, the encoder maps an input to a distribution in the latent space, rather than to a single point; and similarly, the decoder maps a single point to a distribution over the output space. This might seem like a strange idea, but it is not essentially different from what probabilistic graphical models (Section 7.3.14) do. For example, for a feedforward neural network with sigmoid activation functions, it suffices to define output nodes as binary-valued (0/1) and interpret the output of the activation function as the probability that the output node is 1, rather than as the value of that node itself, to have a network that maps an input to a distribution. Restricted Boltzmann machines (RBMs) use exactly this principle. Variational auto-encoders (VAEs) are another example of probabilistic auto-encoders. VAEs are trained in such a way that a certain level of continuity is guaranteed: nearby points in the latent space should represent inputs that are somehow similar; interpolation between two points in latent space yields an input that is in some way "in between" the corresponding input points, etc.

While deterministic auto-encoders can be used to reconstruct inputs, probabilistic auto-encoders are useful to generate new data points that are similar, but not identical, to a given input. They are very useful in the context of generative AI, for instance, for generating pictures. One can generate a picture that is similar to a given picture, change certain aspects of a picture, mix aspects of different pictures into one (through interpolation in the latent space), or simply generate a picture from scratch.

The encoder–decoder structure is also used by transformer models, though the input and output spaces may differ there: for instance, a piece of text (input) may be mapped onto a point or distribution in the latent space, from which, through sampling, an output picture could be generated. We refer to Chapter 8 for more information on transformers and their use in natural language processing.

Further considerations about neural networks

Roughly since 2010, deep learning has been dominating machine learning. It is relevant for a broader class of models than just neural networks (e. g., probabilistic models), but it has revived the area of neural networks in a spectacular manner. Deep learning is indispensable when the information to be learned from contains raw data: data for which no (or not enough) meaningful features are predefined. That includes images, video, audio, and written text.

7.3.12 Dimensionality reduction

The dimensionality of a dataset is simply the number of variables, or features, with which each instance is described. High dimensionality poses problems for just about all learners, to the extent that it is often referred to as the "curse of dimensionality." In linear regression, too, high dimensionality causes overfitting and makes the solution nonunique. In nearest neighbor methods, high dimensionality makes the similarity metric highly uninformative. Density estimation is very hard, as even large datasets are usually distributed extremely sparsely in a high-dimensional space. Support vector machines are more or less the only approach that does not suffer from high dimensionality, but to the contrary exploits it.

Given this situation, much research has gone into **dimensionality reduction**: mapping the data into a lower-dimensional space before any further analysis is performed.

Many successful systems implicitly perform a kind of dimensionality reduction. For instance, image data are very high-dimensional. Part of the reason that convolutional networks work well on such data, is that the convolutional layer is essentially a way of performing dimensionality reduction. It defines a relatively small (compared to the number of pixels in an image) number of features that make it possible to interpret the image. Part of the dimensionality reduction method is hardcoded (the convolution principle), part of it is learned (the parameters, which define the patterns the system should look for).

For other approaches, the dimensionality reduction aspect is more explicit, or can even be considered the goal of the approach. For instance, as described above, an auto-encoder maps a dataset to a lower-dimensional space in such a way that the dataset can be reconstructed from this lower-dimensional representation. Hence, an auto-encoder can be seen as a method for dimensionality reduction.

In statistics, arguably the most basic method for dimensionality reduction is principle components analysis (PCA). Intuitively, PCA works as follows: assume a given dataset with n numerical features. PCA defines new features that are linear combinations of the original features. In a scatterplot, these linear combinations can be interpreted as directions. The first new feature is then the linear combination for which the variance of the data maximal; in other words, the direction in which the data varies the most. We call this the first principal component. Each consecutive feature (the second, third, etc. principal component), is orthogonal to all previous components and is, among all features with that property, the one with highest variance. Figure 7.24 illustrates this for a two-dimensional dataset.

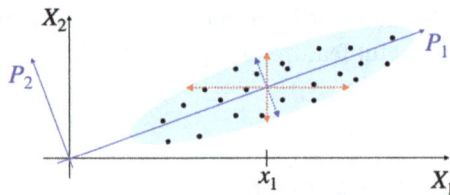

Figure 7.24: A two-dimensional point cloud. P_1 and P_2 are its principal components. The variance of the dataset is maximal along P_1, minimal along P_2. The red arrows indicate the uncertainty on the location of the point given one of its coordinates. If its coordinate along P_1 is known, the uncertainty about its location is minimal (blue arrow).

From the point of view of dimensionality reduction, this construction has the following advantage: consider a 2-dimensional dataset with features x_1, x_2, as in Figure 7.24, and assume we want to reduce it to 1 dimension. We could simply keep x_1 and drop x_2: if we know the x_1 value of a point, we know the approximate location of the point. The remaining uncertainty about the location is given by the variance of x_2: the more x_2 varies around its mean for x_1, the more uncertainty remains. We could also choose to drop x_1 and keep x_2. This would be a less good solution, as the variance of x_1 given some value of x_2 is high. However, instead of reducing the dimensionality by projecting points on one of the original axes, we can also project them onto its first principal component, p_1. The uncertainty about the point's location is then smaller, as the variance in the direction of p_2 is smaller than the variance in the direction of either x_1 or x_2. Because the data vary maximally in the direction of p_1, they vary minimally in directions orthogonal to p_1. As such, if we want to indicate a point's location with maximal precision using only one number, that number should be its position on the p_1 axis, which is the first principle component.

Mathematically, computing the principle components boils down to computing the eigenvectors of a matrix derived from the dataset. More precisely, if D is a matrix with n rows and m columns that represents the original table with m features and n instances, then $D^\top D$ is an $m \times m$ matrix. The eigenvectors of $D^\top D$ are the principal components, and the corresponding eigenvalues are the square-root of the data's variance along this component.

PCA can be seen as identifying features that are linear combinations of the original features and that are suitable for defining a lower-dimensional representation that allows to reconstruct the data with maximal precision. In that sense, an auto-encoder can be seen as a nonlinear version of PCA, as the identified feature are not merely linear combinations of the original features, but linear combinations to which a nonlinear threshold function has been applied. This makes them somewhat more expressive. The price to pay for this is that training an auto-encoder is a more complex procedure than performing PCA.

7.3.13 Matrix factorization and tensor decomposition

To motivate the concept of matrix factorization, consider data that might be used to train a recommender system for books. We have a matrix where rows are users, columns are books, and the cells indicate to what extent a user likes a book. In practice, many of the cells are empty, and we want to predict their values (which means we will predict, for a user who has not read some book, to what extent that user will like the book).

Matrices are studied in the very well-developed field of linear algebra. One particular property that is relevant here is the question to what extent the rows of a matrix are linearly independent. A vector is linearly dependent of other vectors if it can be written as a sum of rescaled versions (a. k. a. a *linear combination*) of them. For instance, $(2, 4, 6)$ is linearly dependent of $(1, 2, 3)$ (it suffices to multiply the latter by two), but not of $(1, 2, 4)$. $(2, 4, 6)$ is linearly dependent of the *pair* $\{(1, 2, 4), (0, 0, 1)\}$, as $(2, 4, 6) = 2 \cdot (1, 2, 4) - 1 \cdot (0, 0, 1)$.

In the remainder of this section, we drop the word "linearly" and simply talk about (in)dependence. The *rank* of a matrix indicates how many independent rows it contains. In a rank 1 matrix, each row is a rescaled version of the first (nonzero) row. For instance, the matrix on the left-hand side of the following equation contains 3 rows, where the second and third are rescaled versions of the first (resp., by a factor 3 and 4). We can write that matrix as a matrix product of a single row and column as follows:

$$\begin{bmatrix} 1 & 2 & 3 & 4 \\ 3 & 6 & 9 & 12 \\ 4 & 8 & 12 & 16 \end{bmatrix} = \begin{bmatrix} 1 \\ 3 \\ 4 \end{bmatrix} \begin{bmatrix} 1 & 2 & 3 & 4 \end{bmatrix}$$

Indeed, following the standard definition of matrix multiplication, the right-hand side yields a matrix with three rows where the first row equals 1 time [1 2 3 4], the second row 3 times the same, and so on.

A rank-k matrix can be written as a sum of k rank-1 matrices, or equivalently, into a product of a k-column matrix and a k-row matrix, where row i of the first matrix contains the coefficients with which each row of the second has to be multiplied such that the multiplied rows sum up to row i of the rank-k matrix. For instance,

$$
\begin{bmatrix}
1 & 1 & 0 & 0 & 0 & 0 & 0 \\
3 & 3 & 0 & 0 & 0 & 0 & 0 \\
0 & 0 & 0 & 0 & 0 & 0 & 0 \\
0 & 0 & 0 & 2 & 2 & 2 & 0 \\
0 & 0 & 0 & 2 & 2 & 2 & 0
\end{bmatrix}
=
\begin{bmatrix} 1 \\ 3 \\ 0 \\ 0 \\ 0 \end{bmatrix}
\cdot \begin{bmatrix} 1 & 1 & 0 & 0 & 0 & 0 & 0 \end{bmatrix}
$$

$$
+ \begin{bmatrix} 0 \\ 0 \\ 0 \\ 2 \\ 2 \end{bmatrix}
\cdot \begin{bmatrix} 0 & 0 & 0 & 1 & 1 & 1 & 0 \end{bmatrix}
=
\begin{bmatrix}
1 & 0 \\
3 & 0 \\
0 & 0 \\
0 & 2 \\
0 & 2
\end{bmatrix}
\cdot
\begin{matrix}
1 & 1 & 0 & 0 & 0 & 0 & 0 \\
[0 & 0 & 0 & 1 & 1 & 1 & 0]
\end{matrix}
$$

How is all that matrix algebra useful in the context of recommender systems? Let us use an example to illustrate this. Suppose some users like horror novels, others like sci-fi novels, and still others like both (and love books that combine both). Assume the first type of users gives 1 to horror novels and 0 to all others; the second gives 1 to sci-fi novels and 0 elsewhere; and the third gives 1 to horror and sci-fi novels, 2 to novels that combine both, and 0 elsewhere. In this context, if we have a row per user, we find that there are two independent rows in the matrix: one representing "horror" and one representing "sci-fi." Some rows are a combination of both. In this particular case, the scores matrix has rank 2.

This example illustrates how the rank of a scores matrix can be interpreted as the number of different "genres" of books, where genre is to be interpreted in the following manner: a user's score for a novel depends on which genres it belongs to, and on the user's preference for each genre. Matrix factorization is basically a way of analyzing a matrix that lays bare its underlying structure, in terms of how many different (independent) genres *exist*, to what extent these genres are *present in each book*, and to what extent they are *liked by each user*. See Figure 7.25 for an illustration. In matrix algebra, **singular value decomposition (SVD)** is one basic approach to matrix factorization. SVD identifies the genres, their relative importance for each user and relative presence in each movie, and the overall importance of each genre for the scores. The scores matrix may technically have high rank but be well approximated by a low-rank matrix. Using SVD, an optimal approximation for a given low rank can be found by simply ignoring the genres with low overall importance.

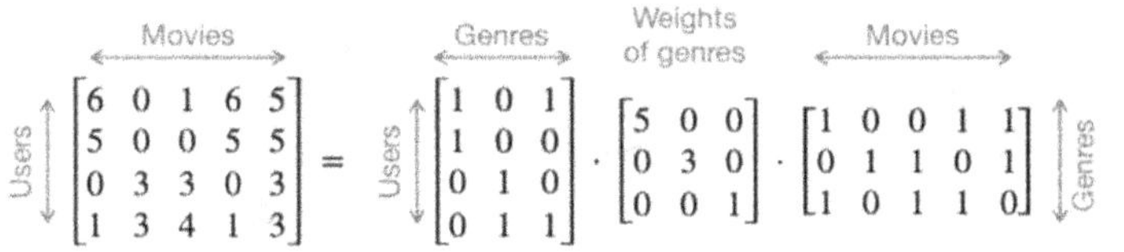

$$
\begin{bmatrix} 6 & 0 & 1 & 6 & 5 \\ 5 & 0 & 0 & 5 & 5 \\ 0 & 3 & 3 & 0 & 3 \\ 1 & 3 & 4 & 1 & 3 \end{bmatrix} = \begin{bmatrix} 1 & 0 & 1 \\ 1 & 0 & 0 \\ 0 & 1 & 0 \\ 0 & 1 & 1 \end{bmatrix} \cdot \begin{bmatrix} 5 & 0 & 0 \\ 0 & 3 & 0 \\ 0 & 0 & 1 \end{bmatrix} \cdot \begin{bmatrix} 1 & 0 & 0 & 1 & 1 \\ 0 & 1 & 1 & 0 & 1 \\ 1 & 0 & 1 & 1 & 0 \end{bmatrix}
$$

Figure 7.25: A matrix of user-movie scores is "explained" by decomposing into three matrices, identifying 3 genres, their weights, and how users and movies connect with them. (This decomposition is not an SVD, it merely illustrates the principle of how matrix decompositions identify structure in a matrix.)

The idea of dropping some variables is of course related to what we did in PCA, and indeed there is a link between SVD and PCA. Whereas PCA tries to represent the original data in a lower-dimensional space, SVD tries to represent the connection between two types of entities (in this case, users and movies) using a lower-dimensional space.

Identifying the structure in the user-movie matrix is useful because once this structure has been found, it can be used to make all kinds of inference. For instance, given a new movie, if we know some user's scores for this movie, we can infer its combination of genres, and given that, we can predict the scores of other users for the same movie. That is obviously useful in the context of movie recommendation, or more generally, for recommender systems.

Though we illustrated the principle using "genres" of movies, it is clear that the underlying structure does not have to correspond to identifiable genres such as horror, SF, etc. More generally, the dimensions of the lower-dimensional space that defines the overall structure are called **latent variables**.

The principle of matrix decomposition can be extended to tensors. Matrix factorization and tensor decomposition are a way to identify underlying structure in matrix- or tensor-shaped data. This includes recommender systems, but also many other types of data.

7.3.14 Probabilistic graphical models

Probabilistic graphical models (PGMs) are used to model a population by means of its joint probability distribution (JPD). Once the JPD is known, all kinds of reasoning can be performed with it: see Chapter 6 for details on that. Here, we focus on how JPDs can be learned from data.

Joint probability distributions

Assume the instances in a dataset are described using a fixed set of variables, to which values are assigned. For ease of discussion, we assume discrete variables for now.

The **joint probability distribution** $p(x)$ is a function that describes the population in the following way: for any instance x, $p(x)$ gives the probability that an instance drawn randomly from the population equals x. For instance, if the population consists of

10 yellow, 5 red, and 5 green marbles in a jar, for any particular marble m, $p(m) = 1/20$. If we do not distinguish marbles of the same color and consider x to be simply the color of the drawn marble, then $p(\text{red}) = 5/20$. If a marble is described by multiple variables, say *size* (big/medium/small), *material* (glass/agate), and *color* (red/blue/yellow/green/white), then $p(s, m, c)$ gives for any size s, material m, and color c, the probability that a randomly drawn marble has the specified size, material, and color.

Factorizations of JPDs

In principle, the joint probability distribution p could be defined using a table, listing for each combination of values for the variables the probability of drawing an instance that has exactly this combination. In the above marble example, we would need 30 rows but when there are many variables and possible values, the number of combinations quickly becomes astronomical. It is then impossible to even store such a table. Even when that is possible, estimating the probability of each individual combination from data will require all existing combinations to occur at least once in the data, and probably multiple times if we are to estimate the probabilities accurately; so, for a table with billions of rows, we would also need billions of data points to learn from, before we can effectively fill it in.

In such cases, a probability distribution is typically written as a product of functions with fewer variables. We call this a factorization. The simplest example of a factorization is $p(x_1, x_2) = p_1(x_1)p_2(x_2)$: the JPD is simply written as a product of two independent functions (called **factors**). Note that not each JPD over x_1, x_2 can be written as such a product. For instance, if (x_1, x_2) denotes the outcomes of a roll of two dice, then sure, $p(2, 3) = 1/6 \cdot 1/6 = 1/36$. With 6 equally likely outcomes for a single die, there are 36 combinations, and all are equally likely. But if our jar contains 2 marbles, one (green, glass, small) and one (red, glass, large), then $p(\text{green}) = 1/2$ (if I draw a random marble, there is a 50 % chance it is green) and $p(\text{small}) = 1/2$, but $p(\text{green}, \text{small}) = 0$ as there are no small green marbles in the jar, and obviously $0 \neq 1/2 \cdot 1/2$. In statistics terminology, when $p(x, y) = p(x)p(y)$, variables x and y are said to be **stochastically independent**. In the remainder of this section, (in)dependent means stochastically (in)dependent.

When we write a JPD as a product of multiple simple functions (with fewer variables), we are making assumptions about a certain independence among variables. The advantage is that the factors may be much easier to estimate from data, as they contain fewer variables. The disadvantage is that if the assumptions are violated, we get a poor approximation of the JPD.

The following example illustrates this.

Suppose someone owns a loaded die; the sides 1–6 do not all have an equal probability, but you do not know the probabilities. Now if you throw this die 5 times, what are the chances you obtain 5 sixes? You could find out by doing the experiment (throwing 5 times) a million times, counting how often you got 5 sixes, and dividing this number by 1 million. If you repeat the experiment only 1000 times, it is likely that you never get 66666, which gives an estimated probability of 0—a poor approximation. But you could also estimate $p(6)$ (the probability of obtained a six if you throw once) by throwing 1000 times, counting the number of sixes, and dividing that by 1000. This gives a good approximation of $p(6)$, and if we assume the different throws are independent (i. e., the outcome of a new throw does not depend on what the previous outcome was), then $p(66666) = p(6)^5$. Making the assumption of independence allows us to obtain the answer from much less data.

Naïve Bayes

Perhaps the simplest kind of factorization, in the context of predictive learning, is used by a method called **naïve Bayes**. Consider a target variable Y and input variables $X_1, \ldots, X_m$. We write the variables X_i jointly as $\boldsymbol{X}$. According to the definition of conditional probability, given values for $\boldsymbol{X}$, the probability that Y takes a certain value is

$$P(Y \mid \boldsymbol{X}) = P(\boldsymbol{X}, Y)/P(\boldsymbol{X})$$

Assume we want to find the most likely value for Y, given $\boldsymbol{X}$. Since $P(\boldsymbol{X})$ does not depend on Y, the value of Y that maximizes $P(\boldsymbol{X}, Y)$ also maximizes $P(Y \mid \boldsymbol{X})$, so it suffices to estimate $P(\boldsymbol{X}, Y)$, the joint distribution of the inputs and outputs. Now, naïve Bayes estimates this as follows:

$$P(X_1, X_2, \ldots, X_m, Y) = P(X_1 \mid Y)P(X_2 \mid Y) \ldots P(X_m \mid Y)P(Y).$$

This equality implies a condition called **class-conditional independence**: within a certain class Y, the probability of X_i taking a particular value is not influenced by the value of X_j for $j \neq i$. To illustrate what this means: suppose we want to classify people into two classes, say, football players and basketball players. Both populations are characterized by variables such as height, weight, speed, etc. Naïve Bayes essentially assumes that both height and weight depend on the class (e. g., the height of basketball players is distributed differently from that of football players) but given that a certain person is a football player, knowing their height does not provide any information on their weight (an unrealistic assumption, in this particular case).

Under the assumptions made by the model, the JPD can be estimated efficiently: all the factors have at most 2 variables, which means counting combinations of 2 values suffices.

For instance, assume someone has a large bag containing hundreds of marbles with size, material and color as explained before (3 sizes, 2 materials, 5 colors, yielding 50 possible combinations). Suppose you randomly draw 20 marbles from it, and writing down their properties, you get the table shown in Figure 7.26.

Now, the fact that no big white glass marble was drawn does not mean there are no such marbles in the bag: there are 50 possible combinations, of which at most 20 can

Size	Color	Material
big	red	glass
small	blue	glass
small	white	glass
medium	yellow	glass
big	blue	glass
medium	blue	glass
medium	yellow	glass
medium	red	glass
big	green	glass
medium	red	glass
big	white	glass
small	blue	glass
medium	green	glass
small	red	glass
medium	green	glass
medium	blue	agate
big	white	agate
medium	yellow	agate
medium	green	agate
small	white	agate

$P(\text{big} \mid \text{glass}) = 4/15$

$P(\text{white} \mid \text{glass}) = 2/15$

$P(\text{glass}) = 15/20$

$P(\text{big} \mid \text{agate}) = 1/5$

$P(\text{white} \mid \text{agate}) = 2/5$

$P(\text{agate}) = 15/20$

Figure 7.26: A table listing 20 randomly drawn marbles. Some count-based estimates of (conditional) probabilities are shown next to it.

occur in the table even if all 50 occur in the bag. So, to estimate the probability of a particular size/material/color combination, counting combinations of three values does not work: we easily find an estimated probability of 0. However, if we assume that size, material and color are all independent, we can estimate from the table the distribution of material (15/20 are glass), size (5/20 are big), and color (4/20 are white), and estimate the probability of drawing a big white glass marble as $P(\text{big})P(\text{white})P(\text{glass}) = \frac{15}{20}\frac{5}{20}\frac{4}{20} = \frac{3}{80}$. If we relax this assumption, and instead state that the size and color distributions may be different for different materials (but size and color are still independent given the material), we estimate $P(\text{big, white, glass})$ as $P(\text{big} \mid \text{glass})P(\text{white} \mid \text{glass})P(\text{glass}) = \frac{4}{15}\frac{2}{15}\frac{15}{20} = 0.027$.

The **naïve Bayes classifier** makes the assumption of class-conditional dependence to predict the class of an object given its other attributes. In the above example, the material variable represents the class. Let us assume that we observe that a marble is big and white. The probability that it is made of glass, respectively agate, is then

$$P(\text{glass} \mid \text{big, white}) = \frac{P(\text{big, white, glass})}{P(\text{big, white, glass}) + P(\text{big, white, agate})} = \frac{0.027}{0.027 + 0.02} = 0.57$$

$$P(\text{agate} \mid \text{big, white}) = \frac{P(\text{big, white, agate})}{P(\text{big, white, glass}) + P(\text{big, white, agate})} = \frac{0.02}{0.027 + 0.02} = 0.43$$

Naïve Bayes will predict the class with the highest probability, which is *glass*.

It is clear that the independence assumptions made by naïve Bayes are quite strong, and likely violated in many practical applications. Despite this, naïve Bayes turns out to work relatively well for classification. The reason is that the class that maximizes the estimated probability often coincides with the class that maximizes the real probability, even if the estimates themselves are not very accurate.

Probabilistic graphical models

Probabilistic graphical models impose a specific independence structure of the variables; equivalently, they imply that the JPD can be factorized in a specific manner. The structure of the factorization is defined by a graph. Different formalisms exist for defining the structure using a graph.

Bayesian networks are directed graphs where each node represents a variable, and with each node is associated a probability distribution that is conditional on the parents of that node. The JPD corresponding to a Bayesian network is obtained by simply multiplying all the (un)conditional distributions. Bayesian networks are easiest to interpret in a causal framework: an edge from x to y, "y depends on x", is then interpreted as a causal dependency. However, while it is possible to expose causal structure using a Bayesian network, there is *no guarantee* that the edges in any given Bayesian network indeed indicate a causal relationship; it is perfectly possible to define Bayesian networks where this is not the case.

Markov random fields, or **Markov networks**, are yet another format. They are undirected graphs where, with each clique (set of nodes that are fully interconnected), a so-called potential function is associated. The JPD is obtained by multiplying the potentials. Yet another format is the **factor graph**. This is a graph with two types of nodes: factor nodes and variable nodes. When the factorization contains a factor $p(x,y)$, a factor node p is connected to two variable nodes x and y.

For all these formats, it holds that the JPD is a product of functions, where each function can easily be estimated from data. For instance, for a Bayesian network, the conditional probability tables shown in Figure 7.27 can easily be obtained by counting in the data how often a burglary occurs, how often an earthquake occurs, how often John called when the alarm was (not) going off, etc.

The ability to impose an independence structure on the variables in a flexible manner gives the specialist user much modeling power. This becomes even more clear in more complex forms of PGMs, such as **dynamic Bayesian networks** (**DBNs**). A DBN models a sequential process where the values of the variables at step t may depend on other variables at step t, but also on the values of the variables at step $t-1$. Similar kinds of probabilistic models are used in natural language processing (see Chapter 8 for more on this).

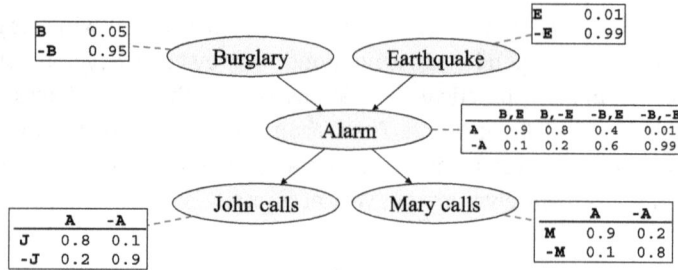

Figure 7.27: A Bayesian network defining a JPD over 5 variables. It indicates an underlying causal structure: a burglary or earthquake may cause the alarm to go off, the alarm may cause the neighbors to call. The joint probability of any value combination can be found by multiplying the conditional and unconditional probabilities found in the table.

A note on continuous variables

The above discussion assumed discrete variables. When a variable x is continuous rather than discrete, $p(x)$ is typically zero, for example, the probability that a randomly drawn person from the population of all humans is *exactly* 178.32000000000... cm tall is zero. If we round to 1 cm (i. e., we discretize the continuous variable), we get meaningful probabilities, but we could also round to 1 mm, 0.1 mm, etc. Naturally, the smaller the intervals, the smaller the probability of drawing an element in that interval. For example, if 3 % of the population is between 178 and 179 cm tall, then likely about 0.3 % is between 178.0 and 178.1 cm tall. The probability itself is not very meaningful unless you know the granularity, but the "probability per mm" remains pretty much the same. The **probability density** is the probability divided by the interval width, for very small intervals (mathematically, the limit as the interval width approaches 0). Conversely, the probability of an interval is the integral of the density over the interval.

A probability density is quite different from a probability (e. g., it can be greater than 1), and to stress this difference, $p(x)$ is often called a **probability mass function** when x is discrete, and a **probability density function** when x is continuous. When x is a tuple of multiple variables, some of which are discrete and other continuous, the function is a hybrid between the two, and it is convenient to simply use the term joint probability distribution.

While PGMS are a very powerful and flexible formalism, they are mostly used for discrete variables. A difficulty with numerical variables is that the factors can no longer be estimated and stored in tabular format. Instead, they are continuous functions that must somehow be fit to the data. This is often done using parametric methods (e. g., fitting a Gaussian). Care must be taken to use an appropriate family of functions in this case, as choosing the wrong family may give arbitrarily bad results. For example, fitting a bimodal distribution with a Gaussian will assign high probabilities to values in the middle, when in fact these values may be very improbable. This would be like concluding from the 52–48 result of the Brexit referendum that most Brits were indifferent to Brexit, when in fact opinions were strongly polarized.

7.3.15 Clustering and density estimation

Like matrix factorization, **clustering** is an unsupervised learning task where we try to find structure in the data. In its most basic form, clustering simply implies finding "clusters," groups of instances that are similar. In some variants, the clusters themselves are grouped into larger clusters; this is called **hierarchical clustering**. In other variants, the clustering system does not only group the instances into clusters but also describes these clusters; this is sometimes called **conceptual clustering**.

Flat clustering

A flat clustering is simply a partitioning of the instances in the dataset, into groups of similar instances. When we visualize the data space, clusters typically look like "islands" of instances, or dense regions amid sparse or empty area. Any method that can reliably estimate the density of the population in the data space implicitly indicates where there are clusters. Conversely, clustering methods differentiate high-density from low-density regions and in that sense can be seen as a rudimentary form of density estimation. An important difference, however, is that clustering actually assigns individual cases to subgroups, whereas density estimation does not necessarily do so.

Perhaps the best-known clustering algorithm, though by no means the only useful one, is called **k-means**. It has one hyperparameter: k, the number of clusters we want to find. It finds a partitioning of the data into k clusters, using a very simple procedure: start with k random instances called "seeds," assign each instance in the dataset to the closest seed, recompute seeds as the mean of all instances assigned to them, repeat until convergence.

A probabilistic counterpart of k-means is the **expectation maximization** procedure, **EM** for short. It is an iterative procedure where a mixture of probabilistic models is learned. For instance, it may be assumed that the population consists of several subpopulations, each of which follows a Gaussian distribution—a "mixture of Gaussians." Each iteration consists of a so-called *expectation* step, where for each instance we guess which subpopulation it belongs to, and a *maximization* step, where we update the parameters of the distributions to optimally fit the data, given the assumption about which instance belongs to which subpopulation. These steps are continued until the process converges.

Because k-means assigns points to the cluster whose center is closest, it tends to find spherical clusters. That is not always what we want. **Spectral clustering** methods work very differently: they define a graph in which nodes are instances, and edges exist between instances that are similar. Spectral clustering tends to find clusters in which instances are near other instances of the same cluster, but not necessarily near the center of that cluster; in fact, the center of the cluster (defined as the mean of all its instances) may not even belong to the cluster.

Hierarchical clustering

A hierarchical clustering roughly corresponds to the concept of a taxonomy. Constructing it is similar to what humans have been doing for centuries with, for instance, animals. We observe that there are certain types of animals that look alike: we call them dogs. While all dogs have certain similarities (which is why we recognize them as one species), we can also subdivide them into subtypes: German shepherds, border collies, great Danes, huskies, etc. At the same time, dogs, elephants, and squirrels have in common that they are mammals, which are vertebrates, etc. The taxonomy of animals that we have today is the result of a hierarchical clustering process.

Computers typically form cluster hierarchies in either a top-down way (starting with one cluster that contains the whole dataset, and repeatedly splitting clusters into subclusters) or bottom up (merging similar individual into a cluster, then merging clusters into larger clusters, etc.). The result of the hierarchical clustering process is called a **dendrogram** (Figure 7.28). It is basically like a taxonomy, except that at its lowest level it contains the actual individuals.

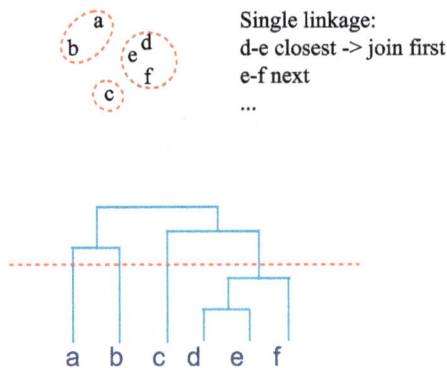

Figure 7.28: A dendrogram, and a flat clustering derived from it by cutting the dendrogram at a certain level. Here, the "single linkage" principle for merging clusters was used: the distance between two clusters is the distance between their closest members.

Semisupervised clustering

A major challenge when performing clustering is that there must be a well-defined notion of similarity. This notion may be subjective, for example, when we cluster pictures of people, do we want to group them according to clothing style, pose, physical appearance, etc.? A computer cannot know what the user means by "similar," except through a formal definition of similarity provided by the user. However, it may be hard for the user to provide such a formal definition. For instance, to cluster pictures according to the clothing style of the person on the picture, we need to define a mathematical function that, given two-pixel matrices, tells us how similar the clothing styles are—a practically impossible task.

Semisupervised clustering methods allow the user to provide some guidance by showing examples of instances that should (not) be in the same cluster. From these hints, they derive a similarity criterion that seems consistent with the user's opinion and use that to form clusters. For instance, in **constraint-based clustering**, the user states for a small number of pairs of instances whether they belong to the same cluster or not. Some systems allow the user to do this interactively: see Figure 7.29 for an example.

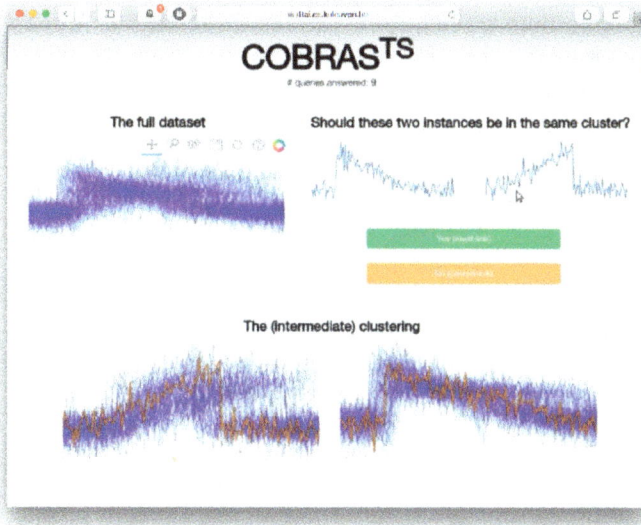

Figure 7.29: The interactive semisupervised clustering system COBRAS-TS (https://dtai.cs.kuleuven.be/software/cobras/) clusters time series based on limited interaction with the user. (Picture by Wannes Meert, KU Leuven).

7.3.16 Automata and hidden Markov models

Finite automata

We earlier mentioned Flashfill, an Excel plug-in that can learn data transformations on strings from examples. Such transformations require finding a common pattern in strings. There are many ways in which string patterns can be described. Among them are **regular expressions**. A regular expression or **regex** is a pattern that can "match," and to some extent parse, strings. A simple example is u[0-9]+, which stands for "the letter u, followed by one or more digits" (the [0-9] construct matches a digit, the + means there can be one or more such matches). The regex gr[ea]y matches the strings "grey" and "gray." Such expressions are commonly used to find and substitute pieces of text. When Flashfill learns, say, to turn the value pair (Clarke, Arthur, Charles) into the string "Arthur C. Clarke," it is learning a substitution rule that can be expressed using regexes.

Checking whether a substring matches a given regular expression can be done using a device known as a **finite automaton**. A finite automaton can be thought of as a machine that reads a string from left to right. It is always in one of a finite number of states and moves from one state to another upon reading a symbol. Depending on the state it is in after reading the entire string, it accepts or rejects the string. Now, for each regular expression, a finite automaton exists that accepts a string if and only if the string matches the regular expression. As such, finite automata and regular expressions are equivalent. Learning a regular expression from examples can hence be done by learning finite automata from examples, a task for which multiple algorithms exist. For instance, the RPNI algorithm learns, from example strings labeled as positive or negative, an automaton that accepts all positive example, rejects all negative examples, and can be used to predict the label of new strings. To do this, it first constructs an automaton that accepts exactly the positive strings and no other, then merges states in this automaton in such a way that the automaton will accept more strings, but never a negatively labeled string. See Figure 7.30 for an illustration.

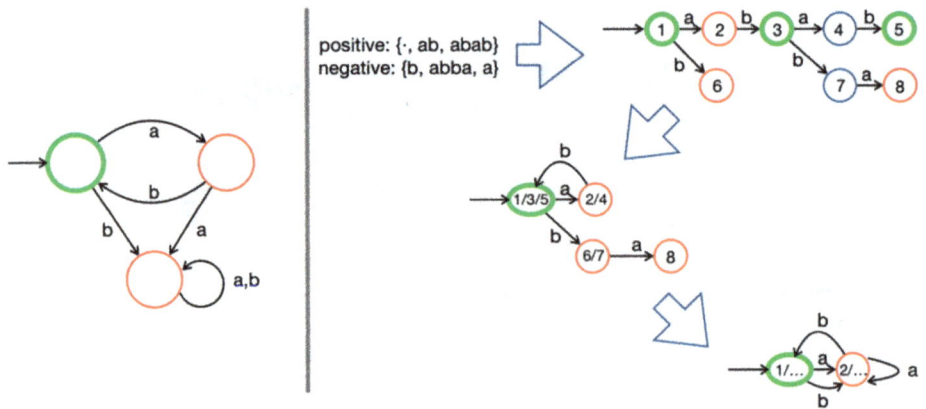

Figure 7.30: Left: a finite automaton that accepts all strings of the form ·, ab, abab, ababab, etc. and no other. Right: such a automaton can be learned by constructing an automaton that accepts exactly the positively labeled strings, and simplifying it by merging states as much as possible, subject to the constraint that when two states are merged, states reached from them by the same symbol must also be merged, and we cannot merge accepting and rejecting states. The end result, in this case, is an automaton that differs from the intended one as it also accepts the strings bb, bbbb, etc. More data would be needed to prevent that from happening.

Finite automata have a broad application potential: they are used by compilers to parse computer programs, but also for verifying the correctness of software, hardware, protocols, etc.

Hidden Markov models

A **hidden Markov model** or HMM is somewhat similar to a finite state machine, but it has a probabilistic component. Instead of reading symbols and moving states according to that, it spontaneously moves from one state to another and with each move produces an output symbol. The state it moves to, from a given state, is determined probabilistically: for each pair of states, there is a constant probability of moving from the first to the second. Put differently, a conditional probability distribution $P(s' \mid s)$ defines the probability that the automaton moves to state s' if it is currently in state s. This distribution is summarized in the so-called transition matrix. The output it produces on each move is also stochastic.

Given some strings of symbols, we can then consider the following learning task: determine the most likely transition matrix and output probability distribution of the underlying HMM. Typically, a fixed number of states is assumed in this learning process.

7.3.17 Reinforcement learning

Reinforcement learning is a relatively general setting for learning. It can be approached in different manners. We first describe the general setting. Next, we describe a concrete learning algorithm for deterministic environments, Q-learning, and discuss the challenges of exploration and generalization.

The problem of reinforcement learning can be stated as follows. An agent can be in a number of states and can take actions in these states. Each action can result in an immediate reward, and in a change of state. For now, let us assume a deterministic environment, where an action always has the same result. The goal is to find a **policy**, prescribing what action to perform in each state, that leads to maximal accumulation of rewards. Formally, the task is to find a function $\pi: S \to A$ (this is the policy) such that, if we start in any state s_0 and in each encountered state s_t take action $\pi(s_t)$, $\sum_t \gamma^t r(s_t, \pi(s_t))$ is as high as possible. Here, $r(s, a)$ is the reward that we get for doing action a in state s, and $\gamma < 1$ is a so-called **discount factor**: it reduces the value of rewards far in the future.

As an example of this problem setting, consider playing chess: the board positions are states, actions are moves, the reward for an action could for instance be +1 if you win the game with that action, –1 if you lose, and 0 otherwise. The policy tells you what move to make and finding a policy that maximizes the accumulation of rewards means, in this case, learning to play the game well. In this particular example, the outcome of an action is nondeterministic, in the sense that the next state in which you'll need to take action also depends on your opponent's response to the action. Other examples of such a setting are, for instance, industrial processes where an operator has to take certain actions to keep the process under control; the yield of the process can be used as a reward, but if something goes seriously wrong, this may result in a large negative reward (or cost).

Note that maximizing the accumulation of rewards is not the same as maximizing immediate reward. For instance, consider a car racer: breaking too early in anticipation of the next sharp turn makes him lose time (negative reward), but going at maximum speed for too long may get him into a situation where he will not be able to take the turn and crash. The initial reward is higher, but the final outcome is worse. Reinforcement learners need to plan ahead.

We can consider increasingly hard versions of the reinforcement learning problem. The simplest one is the case where the agent has a model of its environment that includes the state transition and reward functions. That is, it knows which new state it will end up in when doing an action in a given state, and it knows where the rewards are. In this case, it is possible to compute an optimal policy by simply searching the whole state space. This setting is comparable to playing chess against a computer while knowing how the computer will respond to any move, for instance. There is also a second instance of the same program where you can ask what it will do in a given state. Even though this makes the problem deterministic (you could in principle keep trying out sequences of moves until you find one where you win), the set of all possible sequences of moves is so large that this problem is not practically solvable.

A harder version of the problem is one where the agent does not know the transition or reward function. It can find out by simply trying. Such an agent could first learn all there is to know about the environment by exploring (trying each action in each possible state and seeing what state it leads to and what its immediate reward is) and then learn the optimal policy as described before. In practice, reinforcement learning algorithms merge these two phases (learning about the environment and finding the optimal policy) into one.

A well-known algorithm for solving such problems is the **Q-learning** algorithm. Q-learning tries to find an optimal policy without even constructing a full model of the environment. It only constructs a model of the quality of actions taken in any state, where quality is to be understood as the accumulated reward that will result from taking this action, assuming that you will behave optimally later on.

More specifically, Q-learning learns a function $Q(s, a)$ that represents the "quality" of performing action a in state s, where quality is defined as the accumulated reward we will get if we take action a in state s, and after that, consistently choose the action with highest Q-value. Because of this definition, the following equation holds (sometimes called a **Bellman equation**):

$$Q(s, a) = r(s, a) + \gamma \max_{a'} Q(\delta(s, a), a')$$

where $\delta(s, a)$ is the state we end up in after taking action a in state s. If these Q-values are known, the optimal policy simply consists of always choosing the action with highest Q-value.

The simplest version of the Q-learning algorithm works as follows. The algorithm keeps a table with for all s, a-combinations its current estimate of the corresponding

$Q(s, a)$ combination; this table can be initialized with 0's everywhere. The algorithm repeatedly does the following: from a starting state s_0, random actions are taken, forever or until some end point is reached (in which case we go back to the starting point). After each action, the Q table is updated by calculating the right-hand side of the Bellman equation and assigning the result to the left-hand side. This means that the $Q(s, a)$ entry in the table is updated using the immediate reward and the Q entries of the next state. Thus, Q-values in some sense get back-propagated from the end to the beginning. It is easy to see why this works when we think of games such as chess, this can work: for states near the end, it is easier to see which moves are good and which are bad, than for states at the beginning of the game. It can be proven that, if this procedure is repeated infinitely, the Q table converges to the real Q-function.

Figure 7.31 illustrates how an agent without any knowledge about the environment might explore by taking random actions and construct a model of which action is best in which states. Blue/red arrows indicate actions with positive/negative rewards, and the thickness of the arrow indicates how strongly positive/negative the Q-value is. Compare the lower left and right figure to see how an action gets assigned a positive Q-value despite having led to disaster in the last exploration run (upper red arrow leading to crocodile), because the model already knows it discovered that taking a different action after it would have led to a very positive outcome (treasure). Note that the model constructed by Q-learning (the Q-table) does not include the transition function; looking at the Q-table we do not know which new state action a in state s led to. Indeed, to determine an optimal policy it is not necessary to know the exact outcome of action a in state s, we only need to know which action is best. Thus, Q-learning gains some efficiency

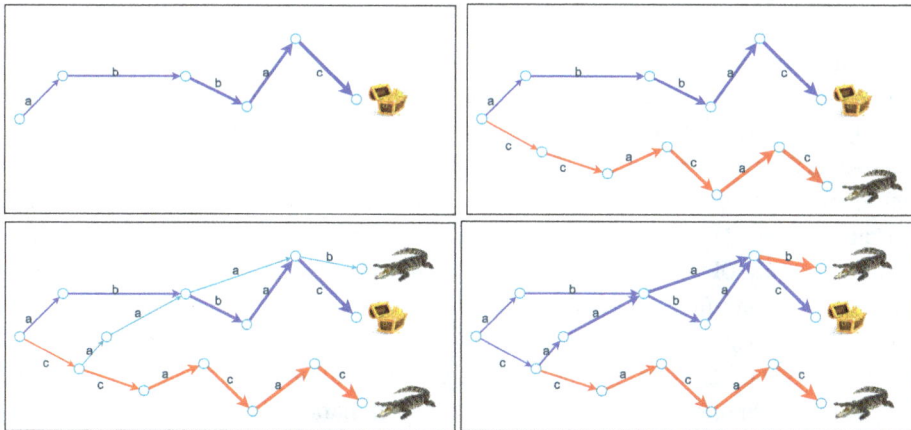

Figure 7.31: Illustration of the gradual building of an environment model using Q-learning. Blue/red arrows indicate positive/negative Q-values, thicker arrows indicate larger (absolute) values. Top left: situation after first Q-update (backpropagated). Top right: after second Q-update. Bottom left: before third update (path followed s indicated, Q-values not yet updated). Bottom right: after third update. Note that Q-learning explores the graph structure but does not actually learn (store) it; it only stores the Q-values.

by not attempting to build a full environment model but constructing an abstraction of it with just enough detail to be able to learn an optimal policy.

In a nondeterministic environment, the immediate reward as well as the next state that results from an action may vary. Typically, this variation is modeled using a probability distribution for both the next state and the reward. The decision process that the agent faces in this setting is often called a Markov decision process (MDP). Methods for learning an optimal policy in this setting resemble the ones for deterministic settings; the main difference is that, given that actual outcomes and rewards may vary, they optimize *expected* values of these variables. A crucial property of MDPs is that the agent knows at all times what state it is in. A harder version of this type of problems is when the agent cannot observe the state it is in; such problems are called POMDPs (partially observable MDPs).

Exploration strategies

In the above description, we assumed that the agent, while learning, acts totally randomly. If this goes on for long enough, the Q table converges to the real Q function. However, we are really interested in finding a good policy, not in finding an accurate Q function. To have an optimal policy, it suffices that the Q-value of the optimal action is accurate, we do not need all Q-values. To refer back to the chess example: once we know which action is best in a certain situation, we are not interested in exactly how bad the other actions are.

For this reason, it makes sense to have an exploration strategy that is not entirely random but gradually focuses more and more on the better actions, once it finds out what those are. Often, a probability distribution over actions is associated with each state, which is initially uniform, but as some actions are starting to appear better than others, their probability is increased.

Exploration strategies for reinforcement learning are a research topic in itself. This topic also relates to **multiarmed bandits**. The multiarmed bandit problem is formulated as follows. Assume you have one-armed bandits (the casino machines). With each bandit is associated a probability distribution for the payout it will give you if you pull its arm. This payout distribution may be different for each machine and is unknown to you. Obviously, as a player, you want to maximize your expected payout. If you can play infinitely many times, what is the best strategy for choosing the arms to pull? In the beginning, you want to try all of them. After playing for a while, you notice that certain bandits tend to give a higher payout on average, so you may start using those more often. Once you are certain which bandit gives the best payout, you want to always use that machine. If you decide too soon which bandit is best, there is a risk you miss the best machine. If you wait too long, you miss an opportunity to maximize your profit. The **contextual multiarmed bandit** problem is similar, but now the payout distribution of a bandit is not constant but depends on the state it is currently in.

The (contextual) multiarmed bandit problems are simplified versions of the exploration problem in reinforcement learning. They are simplified in the sense that the payout distribution describes the immediate reward, while in reinforcement learning, we try to maximize the long-term accumulated reward. The latter depends also on what state the current action will bring us to, and this, too, is unknown in the beginning. An often-used exploration strategy is the so-called **epsilon-greedy** strategy: here, at any time, the action that currently seems best (highest Q-value, in Q-learning) is selected with a probability of $1 - \varepsilon$ (for some hyperparameter ε), and with a probability ε, a random other action is chosen.

While exploration is important during reinforcement learning, there comes of course a time when we want to start using what has been learned: there is no point in learning an optimal policy if we are not going to use it. The **exploration** phase is therefore followed by an **exploitation** phase. Whereas in most learning settings, the learning phase and the operational phase are clearly separate (e. g., we first train a neural network, and then we use it; training does not continue during its use), in reinforcement learning they tend to overlap. As the learner learns what actions are valuable, it gradually starts exploiting what it has learned by choosing the better actions more often than the other. This is not only useful from the point of view of exploiting the learned knowledge, but also from the point of view of learning itself: choosing the better actions more often allows the agent to explore parts of the search space that look promising in more detail.

A good illustration of this principle is **Monte Carlo tree search (MCTS)**, an algorithm for game playing that is similar to the traditional minimax algorithm but works more probabilistically. To make the discussion more concrete, consider again chess. The minimax algorithm (call it player M) essentially evaluates a possible move by considering all possible responses by the opponent (player O), all possible responses by M to each of these, all responses by O to each of *these*, etc. Basically, it tries all possible sequences of moves M-O-M-O-M-O-M-O (leaving out some that are provably nonoptimal) up to a certain depth. Obviously, the number of such moves increases very fast: even for small depths, it easily runs in the billions. MCTS differs in that it does not try all sequences, but a random sample of them. Initially, this sample is entirely random, but when after a while it seems that some initial moves lead more often to an advantageous position than others, it will increase the probability of those moves, exploring this more promising part of the search space in more detail.

Generalization

Until now, we have assumed that the Q-values are stored in a table that has one row for each state and one column for each action. This works if there are not too many states and actions. But if we consider a case where the state is defined by the values of a number of variables, we can easily get an enormous number of states. It can even be infinite, if some variables are continuous. Q-learning only converges if each state-action-pair has been encountered many times. Clearly, this condition is unrealistic in very large state spaces.

In such case, it makes sense to store the current approximation of the Q-function not as a table, but using a different format: a neural network, decision tree ensemble, linear model, etc. Of course, the Q-updates then can no longer be made by simply updating a value in a table. Instead, the following procedure then follows: generate a large number of $Q(s, a)$ values in the regular; once there are enough, learn a model that given some s, a-pair predicts Q; use these predictions in the right-hand side of the Bellman equation shown before; store the result of the right-hand side calculation as the target value for a new training instance for $Q(s, a)$. Once we have enough new training instances, we retrain the model. Thus, generation of data using an old version of the Q-model and training a new model on the basis of this data are interleaved, and this continues until a good approximation of the Q-function is obtained. This approach, sometimes referred to as **model-based reinforcement learning**, combines the generalization power of standard machine learning algorithms with the strengths of reinforcement learning.

7.3.18 Other aspects of learning

Hyperparameter tuning

Many machine learning algorithms have hyperparameters that affect their behavior. For instance, when using a tree learner, we may wish to limit the depth of the trees, or impose a minimum on the number of instances that gets sorted into each branch; in k-nearest neighbors, the choice of k is relevant; the backpropagation algorithm for neural networks uses a so-called learning rate, which affects the size of the steps that the gradient descent procedure makes; the choice of the kernel function in an SVM affects its performance, etc. It is important to choose the right values for these hyperparameters. But how do we find them?

A straightforward way of tuning the hyperparameters is to simply try many different values and see which gives the best results. This means more work for training (to try 100 different hyperparameter settings, we need to train and evaluate 100 models). An important aspect is that the evaluations used for tuning the hyperparameters should be independent of the final evaluation (e. g., they cannot have access to the test set that will be used for the final evaluation), otherwise there is a risk of overfitting the hyperparameter settings to the test set.

When there are few parameters, it may be feasible to systematically try all combinations of their settings; this is called **grid search**. In many cases, this is not feasible and more advanced methods are used. The field of automated machine learning (**AutoML**) focuses on methods for determining what algorithm is optimal for your learning task, with what parameter settings; techniques from Bayesian optimization are often used for this. Advanced AutoML methods optimize not only the learning algorithm itself but also aspects such as data preprocessing, feature selection, etc.

Missing data and noise

When describing a method or algorithm, we explain the calculations it performs on its inputs. In practice, it happens that data are incomplete. It is important to distinguish two types of situations: missing data *during training*, and missing data *at prediction time*.

For instance, if we have a function $y = 3x_1 + 5x_2 - 2x_3$, and we want to use it to predict y for a new case where $x_1 = 1$ and $x_2 = 2$ but x_3 is unknown, how do we do that? A straightforward approach is to fill in for x_3 some representative value, for example, the mean. Somewhat more advanced, we can try to predict x_3 from x_1 and x_2 (using a regression function that was previously learned for this purpose) and fill in that value.

PGMs naturally deal with missing values *at prediction time*: the advanced inference procedures they have are developed for the general case where some variables are known (evidence) while others are not. Some other methods have special ways of handling missing values. For instance, in a decision tree, when the outcome of a test is unknown, both paths can be followed; the instance thus ends up in multiple leaves, and we can combine the predictions of these leaves into a single prediction.

Handling missing values *at training time* is a quite different task. When missing values are rare, the instances containing them can simply be ignored. When there are many attributes, and missing values are frequent enough that most instances have at least one missing value, this obviously does not work. Some learning methods can internally deal with missing values. A decision tree learner, for instance, when evaluating individual tests, can simply ignore the instances whose outcome is unknown for that specific test (while still using these instances when evaluating other tests). Some decision tree learners give a penalty to tests with many unknown outcomes: if an attribute's value is often unknown in the training set, it may also frequently be unknown at prediction time, reducing its usefulness for prediction.

For methods that do not explicitly deal with missing values, an often-used approach is to **impute** the missing values in the training dataset: replace all missing values by some reasonable value (e. g., their mean or a random value drawn from their distribution), then run the learning system. This obviously inserts some noise into the data. Methods that are robust to noise will handle this approach better than methods that are not.

Depending on the context, missing values may not occur randomly. Exit polls are a well-known example: the probability of not answering the question "who did you vote for" may depend on who you voted for, making exit polls notoriously unreliable. In statistics, different mechanisms behind missing values have been recognized. The **missing at random** (MAR) condition states that the probability that a value is missing does not depend on the real value of this or any other variables, except possibly the class. **Missing completely at random** (MCAR) is stronger and also excludes dependence of this probability on the class value. **Not missing at random** (NMAR) is the most general setting: no conditions on randomness are imposed. While there has been a fair amount of work that make machine learning methods robust against missing values, this robustness does not necessarily hold for all settings.

Apart from having missing values, there is also the problem of incorrect values, or noise.

Some methods are not affected much by incorrect values as long as there are few (even if the values themselves are very different from the right value), others can be thrown off badly by even a single noisy value. Robustness against noise is strongly related to overfitting avoidance. Indeed, a method prone to overfitting is more likely to overfit the noise. However, this connection is not perfect. Linear regression in low-dimensional input spaces, for instance, tends to underfit rather than overfit, and still, it is very sensitive to outliers, to the extent that even a single noisy value can completely change the outcome if it is extreme enough.

Integrating background knowledge

The best results with machine learning are obtained when the system exploits both the *data* and the *domain knowledge* that the user may have. Some approaches (e. g., inductive logic programming) allow the user to provide domain knowledge as part of the input to the system. In other cases, the architecture of the model allows one to express certain background knowledge (for instance, the causal or independence structure indicated by a PGM, the network architecture of a neural network). Some methods allow the user to interactively provide guidance to the system, for example, in constraint-based clustering.

For most machine learning approaches, the incorporation of background knowledge into a system requires an advanced understanding of the machine learning methods. It therefore typically requires a close collaboration between machine learning experts and domain experts. This is often challenging and requires a substantial investment but, in many cases, it pays off.

Transfer learning

Consider the following task: we want to learn a model that can serve as an early warning system for runners. A number of sensors measure all kinds of data of the runner (heart rate, etc.). When there is an increased risk of injury, an app tells the runner to take some rest. Each person is different, so ideally the model that the app uses is trained on data from this user, but it may take a while before we have enough data about this user to train an accurate model. On the other hand, there are also similarities among people, which can be captured by models trained on a larger population of people.

Transfer learning refers to a setting where a model is learned on one dataset, then transferred to a different context. In this second context, we also have data and we want to use it to train a model, but we want to prime the model with the model learned from the first dataset. Transfer learning refers to techniques for doing exactly that.

Transfer learning is useful in many circumstances. For instance, different hospitals may have different procedures, different patient populations, etc. A predictive model learned in one hospital may not fit well in another hospital but may be useful as a starting point for developing their own model. As another example, in computer vision, pre-trained networks are often used. They recognize features that are useful for a wide range of vision tasks. Special-purpose networks are often trained by starting from a general-purpose pretrained networking, and merely adapting the weights in a few higher-level layers. This, too, is a form of transfer learning.

7.4 Evaluating the results

When many different machine learning methods can in principle be used for a given task, a relevant question is: which of all these methods should I use? The answer to that depends on many different things. In some contexts, time or computational resources are limited, so the learning process must be very efficient. In other contexts, computational resources for learning are practically unlimited and our goal is simply to find the "best" model. Next comes the question: what is "best"? Many different criteria may be important: interpretability of the model, runtime efficiency (the computational effort needed to make a single prediction), noise robustness, etc. For some of these criteria, no single evaluation measure exists (e. g., interpretability is highly application-dependent and often subjective). In the context of predictive models, quality is often measured by how accurate the predictions are, but even that can be defined and measured in different ways.

Accuracy of classification

In classification contexts, the term **accuracy** is typically used for one specific thing: the probability of making a correct prediction, given an instance drawn randomly from the population. The term **error** is used for the probability of making an incorrect prediction. The sum of accuracy and error is 1.

Accuracy seems like a very natural criterion, yet it can be tricky to interpret. When some predictive model A has an accuracy of 0.999, is that a high accuracy? If the task is to predict who is a terrorist and who is not, then a model B that always predicts "no" may well achieve an accuracy of 0.999999 (misclassifying one in a million)—much better than A. At the same time, B is completely useless for finding terrorists, whereas if A flags 0.1 % of the population as potential terrorists but this turns out to include all terrorists, it may be useful, if only as a filtering tool. So, accuracy as defined above does not tell everything—a less accurate model can be more useful than a more accurate one.

In many settings, there is a certain cost associated with making an incorrect prediction, and that cost is not necessarily symmetric. For example, the cost of the fire alarm going off when there is no fire is very different from the cost of the opposite happening. If we assign a cost C_{FP} to each false positive (false alarm), and C_{FN} to each false negative (undetected fire), then the expected cost of a single prediction is

$$C_{FP} \cdot P(FP) + C_{FN} \cdot P(FN) = C_{FP} \cdot FPR \cdot P(\text{Neg}) + C_{FN} \cdot FNR \cdot P(\text{Pos}),$$

where $P(FP)$ and $P(FN)$ are the probability of having a false positive/negative. A false positive occurs when a negative instance is drawn and predicted positive; the probability of this happening is $P(\text{Neg}) \cdot FPR$, with FPR, the false positive rate, the probability of a random negative instance being predicted positive by the model. Similarly, $P(FN)$ equals $P(\text{Pos}) \cdot FNR$. Note that predictive models with lower FPR tend to have higher FNR and vice versa: an alarm that sets off very easily will tend to have a low FNR at the cost of a higher FPR. Thus, there is typically a tradeoff between these two.

Maximizing accuracy minimizes the expected cost only when $CFP = CFN$. In all other cases, a model with lower accuracy may be better in terms of cost. The optimal model depends on the FPR and FNR of the models, but also on the ratio of positives versus negatives in the population. If that ratio changes, another model may become optimal. A so-called **ROC diagram** (ROC stands for "Receiver Operating Characteristics") is often constructed to visualize the performance of models in this context. The ROC diagram plots models in a two-dimensional space, where the horizontal axis indicates the false positive rate, and the vertical axis the true positive rate TPR (which equals one minus the false negative rate). A perfect model is located in the upper left corner: it has $FPR = 0$ and $TPR = 1$. Models with the same cost are situated on a straight line (called an iso-cost line), the slope of which is determined by CFP, CFN, $P(\text{Pos})$, and $P(\text{Neg})$ (see Figure 7.32). Models with lower cost are closer to the upper left. Given a number of models, plotting them in a ROC diagram allows us to select the model that is optimal under certain operating conditions (misclassification cost ratio and class distribution). When a single model has a parameter that can push it toward predicting positive or negative (e. g., we could have a neural networks predict "positive" only if its numerical output is above some threshold, which by default is 0.5 but could be changed to, say, 0.8, so that it only predicts positive in those cases where it is most certain), changing this parameter makes the system move on a curve in the ROC diagram. Depending on the operating conditions, the system can be tuned to behave optimally under those conditions. When these operating conditions are not known in advance, the area under the curve is often used as a quality criterion for the system; it is referred to as **AUROC** or **AUC**, for "area under the (ROC) curve."

In some cases of binary classification, the positive class is of more interest to the user than the negative class, and evaluation measures take that into account. Imagine you are interested in finding a set of terrorists, and an AI system provides a number of suspects. The **precision** of the AI system is the fraction of suspects that turn out to be terrorists indeed, and the **recall** of the system is the fraction of terrorists that are suspects. (Thus, recall is actually a synonym for TPR; precision, however, is not equal to any of the above-mentioned rates.) There is typically a tradeoff between precision and recall: the more eagerly the AI system treats people as suspect, the higher its recall tends to be, but the lower its precision. Many systems have a tunable parameter that makes it possible to make them more or less sensitive. Varying the value of this parameter yields a so-called **precision-recall curve** or **PR-curve**, plotted in a **PR-diagram** shows how

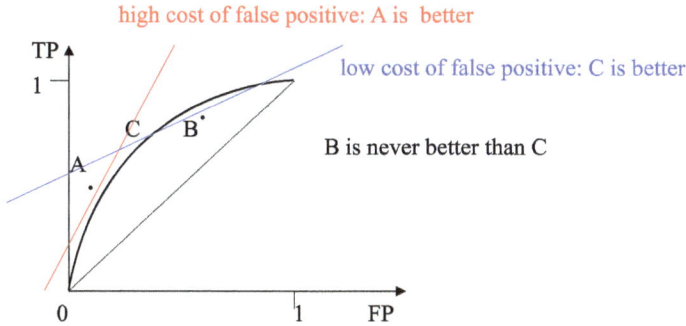

Figure 7.32: A ROC diagram allows us to choose the most suitable classifier among a set, depending on the class frequencies and cost ratios. The red and blue lines are lines of equal cost, under different circumstances. The ideal classifier is located in the upper left, at coordinates $(0, 1)$.

this parameter influences both precision and recall. Systems are sometimes compared on the basis of the area under the PR-curve, a measure often abbreviated as **AUPRC**.

Quality of regression

In a regression context, predictions are rarely entirely correct. When estimating the age of someone who's actually 51, 52 is pretty good, and certainly more accurate than 12. It is not meaningful to simply categorize predictions as right or wrong. In this context, the error is defined as the difference between the predicted and the actual value, and the **mean absolute error (MAE)**, the **mean squared error (MSE)**, and the latter's square root (**root mean squared error, RMSE**) are more meaningful measures. Given a test set T on which a model f is evaluated, the MSE of f on T is

$$\text{MSE}(f, T) = \frac{1}{|T|} \sum_{(x,y) \in T} \left(f(x) - y\right)^2$$

For RMSE a square root is added, for MAE the square is replaced by the absolute value.

Whereas in a classification context, error is always between 0 and 1, in a regression context, the MAE and (R)MSE are measured on a scale that is application-dependent. Is an RMSE of 2 good or bad? When the task is to estimate someone's age in years, it is pretty good; when estimating someone's height in meters, it is extremely bad. Sometimes, the MSE of a model is divided by the variance of the target (which equals the MSE of a model that always predicts the mean). This measure is called relative MSE, and it is typically between 0 (perfect prediction) and 1 (no better than always predicting the mean). Similarly, the RMSE can be divided by the standard deviation in the population, yielding relative root mean squared error (**RRMSE**).

Quality of clustering

As argued elsewhere, clustering is an inherently subjective task. As such, it is hard to evaluate clustering. Some clustering systems try to find clusters with minimal intracluster variance; in those cases, the average intracluster variance of the clusters is obviously a suitable criterion. However, a clustering with many clusters will naturally have lower intracluster variance, so this criterion can only be used to compare clustering with the same number of clusters. Further, the criterion favors spherical clusters, which in some contexts may not be desired.

In some cases, a reference clustering R is available with which the constructed clustering C can be compared. The **Rand index** can then be used: it expresses the probability that for a randomly drawn pair of instances, R and C agree on whether they are in the same cluster or not. When $R = C$, the Rand index is 1. For a random clustering, the index is not zero. For that reason, the "adjusted Rand index" (**ARI**) is often preferred: it rescales the Rand index so that the expected ARI of a random clustering is 0.

Efficiency, interpretability, fairness, safety

The above-mentioned measures all relate to some kind of accuracy: how well does the learned model approximate reality? But, as said, there may be other criteria.

Efficiency is often important. We can measure the efficiency of the *learning* process, in terms of computational effort (often less relevant) or how much data needs to be collected (data collection can be costly, especially when it requires manual labeling of examples). Then there is the *operational* efficiency of the learned model: how fast can it make predictions, how costly is each prediction. It is important to distinguish these two types of efficiency; they are largely uncorrelated.

Under some circumstances, **interpretability** is a requirement for AI systems (hence the large interest in "explainable AI" or XAI). When these AI systems make use of learned models, this often means that the learned model and/or its predictions must be interpretable. There is no general way of determining interpretability; often it is measured by simply asking users how satisfactory they find some explanation.

Finally, models may have to fulfill additional criteria, such as **safety** and **fairness**. These are typically expressed using constraints that the model or its predictions must fulfill. For instance, if a self-driving car uses a learned model, one may want to have guarantees that it will never decide to hit a wall at full speed because of some quirk in the learned model. Similarly, one may demand guarantees that algorithms used for scanning job applications or algorithms used in law enforcement will not behave in a sexist or racist manner, even when the data they are trained on reflect some unconscious bias. It is currently very much an open problem how to ensure fairness in machine learning and more generally in AI. We refer to Chapter 9 for more on this.

7.5 What are the limitations of learning?

Generally, machine learning should only be used when a principled solution is not available. For instance, given that a mathematical solution for least-squares linear regression exists, there is no point in using neural network methods to solve this particular type of problem. Similarly, when theory about a domain is available, hard-coding it into a knowledge base may be better than learning it from data. In some cases, a combination is possible: one can start from a model or knowledge base that represents current knowledge, and improve it, for instance by tuning certain parameters to a particular dataset using machine learning.

As an example, consider the scheduling of thesis defenses at a university. Assume we provide a computer with info on which theses have to be evaluated, which professors evaluate which theses, the exam schedule of students and evaluators, the availability of rooms, etc. Some constraints are obvious: a professor cannot be in two places at the same time, we cannot have two defenses at the same time in the same room, etc. But there may also be less obvious constraints: two rooms may be too far apart for a person to be able to attend a presentation at 10 am in room A and at 11 am in room B; some professors may by default not be available after 6 pm while others are, etc. The person who uses the program may not know all these constraints, yet ideally the program should take them into account. It could do so by observing past schedules and learning generalizable patterns from them.

Apart from deciding *whether* to use machine learning at all, there is the problem of deciding *what* machine learning approach to use. This depends on the task (classification, regression, etc.) and on certain operational constraints (such as requirements on interpretability or energy-efficiency of the models), but also on properties of the dataset: some methods are good at handling high-dimensional data, some are not; some are computationally efficient, others less so, etc.

All machine learning methods have a so-called **inductive bias**, a set of implicit assumptions they make about the target model. An approach will work well if its inductive bias matches well the task at hand. Unfortunately, inductive bias is a relatively elusive concept; we do not understand the inductive bias of many machine learning approaches very well, and also for many datasets it is unclear what inductive bias suits them well. As a relatively simple illustration, consider the following problem: we want to model how an academic's salary depends on their job, age, gender, and the country they live in. We could do this with linear regression or regression trees. Linear regression assumes the effects of different variables are additive: if, say, increasing input variable x_1 by 1 increases the outcome y by v_1, and increasing x_2 by 1 increases y by v_2, then increasing both by 1 increases y by $v_1 + v_2$. In our example, this implies, for instance, that the difference in salary between men and women does not depend on the country they live in (the technical term in statistics is that there is no **interaction**). That may be unrealistic. Decision trees do not make that assumption: if a tree notices a difference between countries, it may split the dataset based on country and learn a completely different model for each group. A disadvantage of this approach is that the amount of data available within each group quickly becomes small. As a result, the different sub-models may be highly inaccurate, as they are learned from small datasets.

Given these concerns, one might think we should strive to develop machine learning methods that have little or no bias, so they perform well under all circumstances. Unfortunately, such methods do not exist. The so-called **no-free-lunch theorems** by Wolpert and others show that learning without any inductive bias at all is theoretically impossible. Also, aiming for a kind of "minimal bias" is not optimal: methods with a strong bias will always perform better on datasets that fit their bias.

The concept of inductive bias is related to the statistical concept of bias, but not identical.

The term bias is used in many different ways in the context of artificial intelligence. Apart from the two ways in which it was used here (both reflecting an inherent preference of learners toward models with particular properties), the term may also refer to datasets being non-representative for the population (selection bias), reflecting misconceptions or prejudices among humans, etc. More on these types of biases is mentioned in Chapter 9.

7.6 Industry examples

7.6.1 Predicting the metallization rate in iron production using ensemble methods at ArcelorMittal

Oussama Chelly, Abrao Aqueri, Emmanuel Gillain

Many industrial processes that require feedback-loop regulation can benefit from supervised machine learning algorithms. In fact, the state and parameters of the process can be fed to regression models in order to predict the future state and, therefore, proactively apply the parameter adjsutments required to optimize the outcome. One such process is the direct reduction of iron in steel factories: impurities in the reagents add a degree of variability that requires periodic adjustments to the conditions of the chemical process in order to increase the metallization rate and, therefore, obtain better quality steel.

Arcelor Mittal Acindar is the leading producer of long carbon steel in Argentina with a 60 % market share. The company operates a direct reduction plant in Villa Constitución. The objective of a direct reduction is to drive off the oxygen contained in various forms of iron ore in order to purify the material and convert the ore to metallic iron without melting it. The process uses a mixture of natural gases in order to reduce the iron ore as illustrated in the diagram 7.33. Iron ore physical, chemical, and meturgical characteristics have a direct impact on the process efficiency.

The most important performance indicator of a direct reduction plant is the "metallization rate," that is, the purity of iron, which in turn determines the downstream steel quality and efficiency. Ideallly, operators must keep the metallization at its highest level, with minimum deviation in all input variables, at maximum productivity, and at minimum energy consumption. In order to determine the metallization rate, production samples are regularly taken to the lab and undergo a 2-hour-long analysis process.

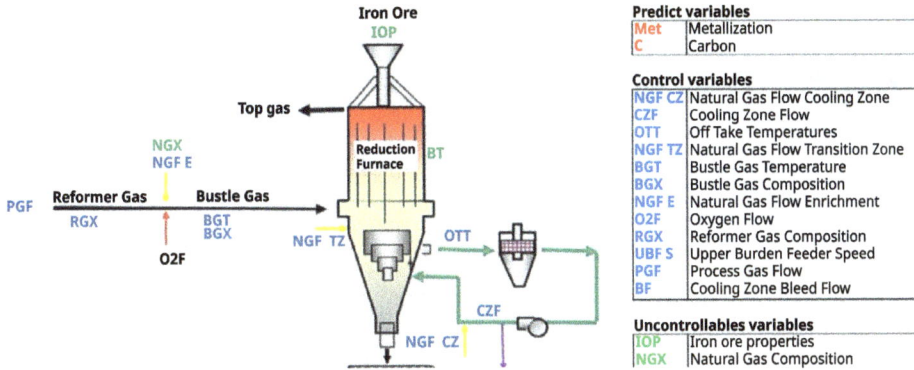

Figure 7.33: The process of direct reduction.

Hence, the obtained results are 2 hours old and the whole process could have deviated in the wrong direction during all of that time. After lab measurements are ready, the operators use a deterministic approach, combined with some human expert knowledge to reach an average of about 94.75 % metallization rate, with an estimated 0.3 mean absolute error.

In the past, the metallization rate used to be only estimated based on the aforementioned lab measurements and without any anticipation. Operators used to take reactive measures based on the measurements of the metallization rate, which are obtained with a 2-hour lattency. Anticipating these measurements and taking action based on the predicted values has a very positive impact on the process. In fact, a small variation in iron quality can have a major effect on the end product quality and on the energy costs in further downstream processes. May the reaction diverge, a high energy consumption would be needed to redirect it to the desired levels. In a quest to improve the process and product quality, Acindar launched a project to assess whether statistical methods can do better than the current deterministic approach in predicting the metallization rate with the smallest error possible, now and in the future.

One of the goals was to reduce the latency of the lab measurement process. Ideally, the panel operator would not wait for 2 hours while lab measurements are being processed but rather react instantly based on predictions made by a trained machine learning model. The lab results would be used later to retrain and improve the prediction model.

Thanks to the use of different supervised machine learning techniques using multiple predictors coming from telemetry and lab measurement data, operators can now anticipate the metallization rate within a 0.1 error and ensure they adapt the process parameters to maintain the highest possible quality of iron since they can regulate the process without the necessity to wait for lab outputs. This has led to better quality endproducts and less energy consumption thus leading to higher profits for the company.

Creating the machine learning model used in the process starts with data collection (cf. Figure 7.34). The collected data is composed of two types of features that describe ev-

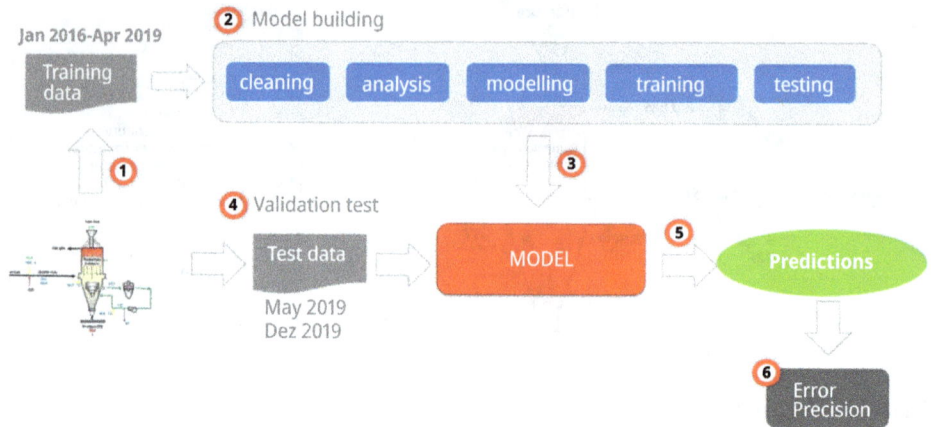

Figure 7.34: The pipeline for creating a regression model for metalization rate prediction.

ery sample. First, sensor data sampled every 30 seconds is available through a myriad of sensors measuring temperatures, pressures, gas compositions and flows, the porosity of the iron pellet, all at different steps of the process. Second, data on the current metallization rate is manually collected every 2 hours based on laboratory measures. The latter is the real label data associated with the samples. Date and time are also taken into account in the prediction model to associate the sensor data with the appropriate lab measurements, which become available 2 hours later.

The collected data is then preprocessed. First, a single entry is created from each set of 30 seconds worth of sensor data. Then the metallization rate measurements available 2 hours later are smoothed out by connecting the values at the two ends of each 2-hour interval and linearly inferring the values for every 30-second time stamp inside the interval. The inferred values—which correspond to the labels—are added to each entry with a 2-hour buffer since the goal is not to predict the current rate, but rather the rate 2 hours in the future. Additionally, this label data is transformed to make the time series stationary. This means that instead of the actual metallization rate value, the label is replaced by the difference between the "future" metallization rate and the current one. The task is no longer to predict the next metallization rate but rather to predict how much the current rate is going to increase or decrease. At this point, one data entry is created every 30 seconds and is fully available 2 hours later because of the lab process. Second, the average of the readings for every sensor is computed, which has the effect of smoothing the readings and reducing the effect of outliers. In the third step, statistics such as the slope and the skewness are also computed for the original readings. These statistics are closely related to the derivatives of the sampled sensor values, and are theoretically involved in chemical kinetics. In order to improve the training data quality, Acindar's data scientists were assisted by chemical engineers in order to enrich the data by designing additional features. This was achieved by looking into the correlations between predictors and label data, and by taking into account theoretical knowledge about physics and chemistry brought by the chemical engineers. One such additional predictor is energy, which can be measured by multiplying the temperature by the production rate. Finally, every predictor is scaled so that predictors with large values do not outweight those with small values in the subsequent process. In total, more than 60 predictors are used, and 2 years worth of data was available for the training of machine learning models.

Once data is preprocessed, predictors were assessed based on the correlation between each pair of predictors as seen in Figure 7.35. As a result, pools of highly correlated predictors were identified. Taking few predictors from each pool has the advantage of simplifying the model and reducing overfitting with very limited quality loss in the prediction performance.

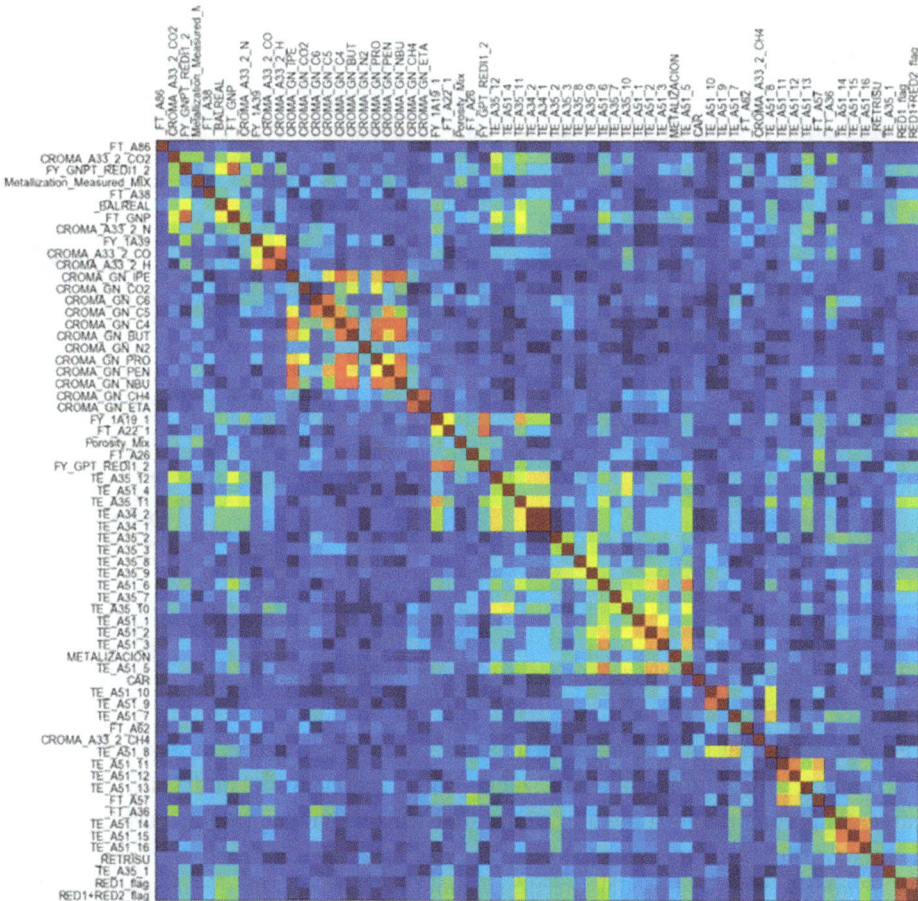

Figure 7.35: Correlation map between predictors available after preprocessing.

Many models were trained using the features selected from the preprocessed data. In particular, Linear Lasso, LGBM, Gradient Boosting, and Random Forests were trained to predict the evolution of the metallization rate. All of these models were trained in order to minimize the mean absolute error of the metalization rate. Acindar opted for a more stable model by combining the outcome of all 4 trained models in a single model. This technique is one form of "ensemble methods," where the outcome of several prediction techniques are combined into one model. This has the advantage of reducing the variability of the outcome. By taking the average of the 4 predictions, if a prediction happens to have an outcome which is much higher or much lower than the other 3, averaging would reduce the difference.

Several iterations were necessary before a final model was created. For every iteration, a model is trained and the importance of each predictor is assessed by its coefficients in the prediction process. The higher the absolute value of the coefficient, the more important the predictor (cf. Figure 7.36). Then the least important features are removed and a new model is trained with less features. This feature selection process has the advantage of reducing the model complexity and thus further reducing overfitting.

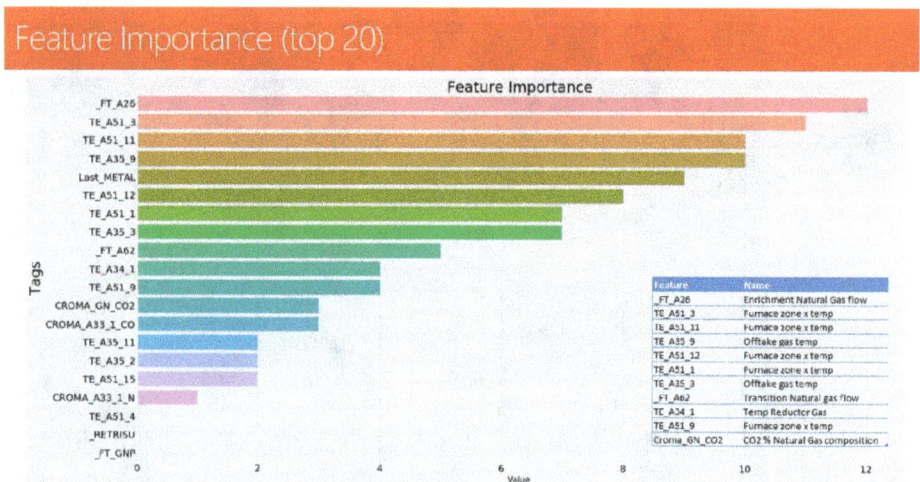

Figure 7.36: Top 20 features ranked by decreasing order of importance.

The trained model takes the set of measurements taken every 30 seconds and predicts the metallization rate—that is, label—to be achieved in 2 hours. The model can currently predict the next metallization rate with a 0.1 mean absolute error, and Acindar are working on improving the model.

Ironically, the final model's accuracy could be "altered" by the operators. In fact, as they use the prediction from the last iteration in order to improve the metallization rate, they change the outcome predicted by the iteration itself (cf. Figure 7.37). When

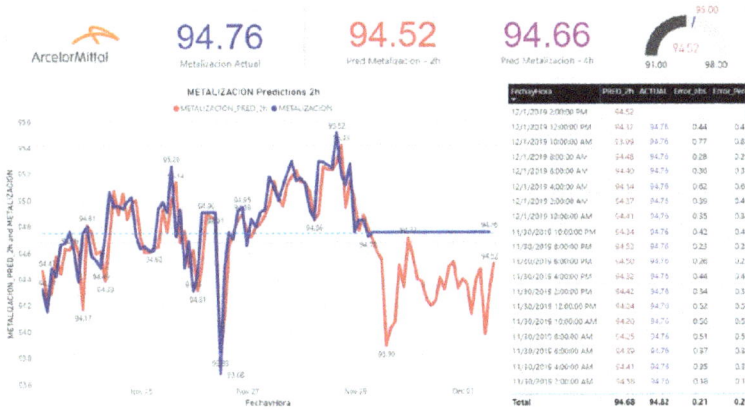

Figure 7.37: Actual vs. predicted metallization rate for a duration of 4 hours.

the operator takes the action to change control variables to improve the iron quality, he changes the "course of events" and the actual output. This is indeed the goal of making the prediction. To be fair with the model, a proper comparison of its predictive quality requires a comparison of the predicted value when there were no changes made by the operator, that is, a situation without the ML-based prediction.

It is possible to create a model that takes this phenomenon into account by acquiring new data after operators start to account for the predictions as they tune the parameters. With such a new model, quality is expected to improve, and a smaller mean absolute error could potentially be achieved. In the future, Acindar will also be looking at both simulating and predicting the impact of controlled variables to find the most optimal settings.

7.6.2 Estimating the value of real estate with supervised regression models, at KBC Group

Oussama Chelly, Michaël Mariën

Employing some 42,000 staff worldwide, KBC Group (7.5 billions EUR revenue, 280 billions EUR assets, 19.5 billions EUR equity, at the time of writing) is an integrated bank-insurance group, catering mainly for retail, private banking, SME, and mid-cap clients. The company serves approximately 11 million clients with an omnichannel approach including branches, insurance channels, as well as online and mobile channels in Belgium, Ireland, and Central Europe. KBC customers, millenials in particular, are increasingly using digital channels to access information and purchase new products or services. To fulfill the needs of digital-minded customers, KBC is continuously looking at digitalizing its sales process, aiming to be the reference as the most accessible and solution-oriented bank insurer, ensuring a full role for each distribution channel. Home insurance is one of KBC core "nonlife" core insurance product. The process of issuing home policy used to be serviced by physical channels. The calculation of the premium is based on the value of a home, which is itself derived from a questionnaire of about 20 to 30 questions. Such process and approach were internally challenged as they didn't meet the needs of digital-minded customers looking for online, accessible, and easy-to-use services.

In mid-2016, the company launched a project to reinvent and simplify the process with the goal to cover 75 % of those digital-native customers needs through a simple and digital process without losing quality and service. The classical evaluation system had to be reworked into a lean and data-driven evaluation system, which can calculate a quote based on solely the address of the property. From the address, a plethora of information can be inferred such as the type of the building, the number of floors, or the surface in square meters. At this point, machine learning **regression** techniques are used to combine these factors and infer the property value, which in turn determines the market-based premium for the client. Once the premium is calculated, the policy can then be closed immediately and online. KBC launched their new data-driven evaluation system through web and mobile application in November 2017. It is estimated that the new system accounts for more than 30 % of new policy production. By using their new intelligent evaluation system, their intermediaries can also better focus on personal advice and tailored cases.

In the current KBC home policy insurance app (cf. Figure 7.38), the customer no longer answers a 20-to-30question questionnaire. Instead, the current implementation asks the user for their address, the type of home, and the number of floors. Based on that, KBC can automatically gather hundreds of features related to the property. These features gathered from both internal and external sources include the coordinates, therefore the location, the neighborhood, the city, or town, the distance to public transportation, the price per square meter in the area, the construction date, the building surface and number of floors, the flood risk, etc.

Not all external features are freely available to KBC. Hence, with some of these features, having a substantial price tag, there is a tradeoff between costs and model quality. KBC engineers therefore created several models using different subsets of the available features, and a business decision had to be made to choose the model with the best quality versus complexity and price balance.

Various regression models were implemented using different combinations of the afore-mentioned features. The models were trained and tested using historical data from KBC's database, then compared based on both their financial cost and Mean Absolute Error (MAE). In this context, the cost refers to the financial cost associated with acquiring the data used in the model, while MAE is the average difference between a model's estimate and the asset's true value indicated in KBC's data.

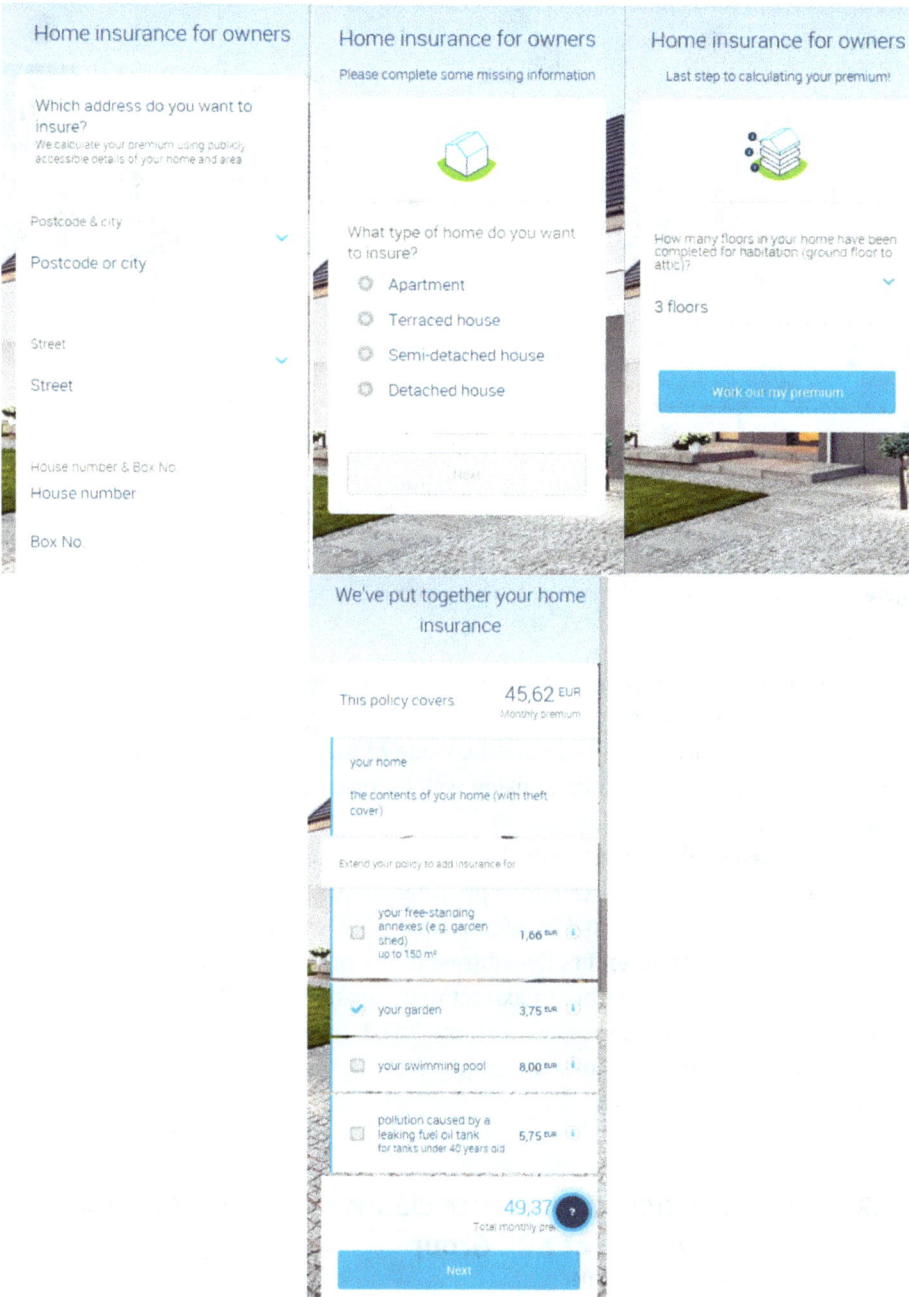

Figure 7.38: Screenshots from the new mobile app.

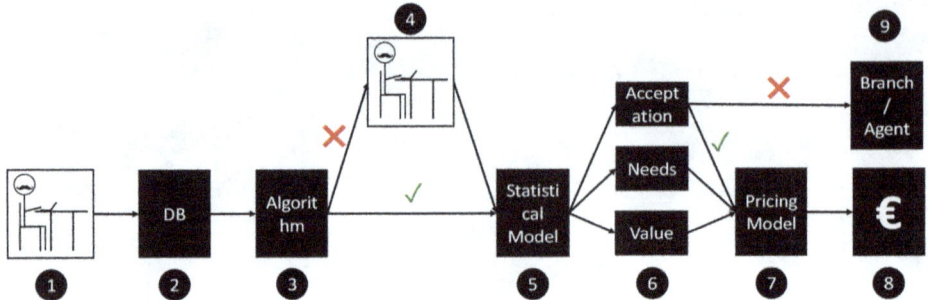

1. Prospect enters the address of property
2. Address is enriched by additional information that are stored in internal or external DB
3. Based on the quality of enriched data, algorithm decides whether there is sufficient information for the model or not:
   - If yes, we continue straight to step 5
   - If no, we continue to step 4
4. If needed, additional information are asked to the prospect
5. Statistical Model(s) run on all information from prospect together with information from said DBs
6. Acceptation, needs detection and value of a property is calculated by said statistical model(s)
   - If we are not willing to accept said prospect and/or said property, we refer said prospect to standard offline process (9)
   - If we are willing to accept said client and said property, we continue to step 7
7. Pricing model run on the outputs from said statistical model(s)
8. Premium is calculated

Figure 7.39: Architecture of the solution for insurance premium pricing.

The model that provided the best tradeoff between cost and quality measured by the MAE uses LASSO regression on key features including surface area of the asset and prices in the neighborhood to estimate the value of the property. From that point, the home insurance policy is calculated using well-defined formulas and an offer is made to the customer.

The deployed model can estimate the insurance premium based on the address. To conclude and illustrate the descriptions provided above, we hereby provide a calculation example for a house located in Nieuwstraat (Rue Neuve) in Brussels, Belgium. In the first step, the customer enters the address of the property as shown in Figure 7.40. The address allows the algorithm to extract many features that describe the property. Some of these features are described in Figure 7.41. Finally, the algorithm estimates the insurance premium and provides almost instantly an insurance offer to the customer that can be seen in Figure 7.42.

7.6.3 Fraud detection in insurance claims with unsupervised outlier detection, at KBC Group

Oussama Chelly, Michaël Mariën

Fraud in insurance claims is unfortunately a wide-spread practice. It is extremely difficult to provide accurate estimates for the financial cost of this malicious endeavor given that many, if not most, frauds go undetected. However, it is believed that in 2019 alone, fraud set European insurers around 13 bn EUR back. Obviously, the financial costs of in-

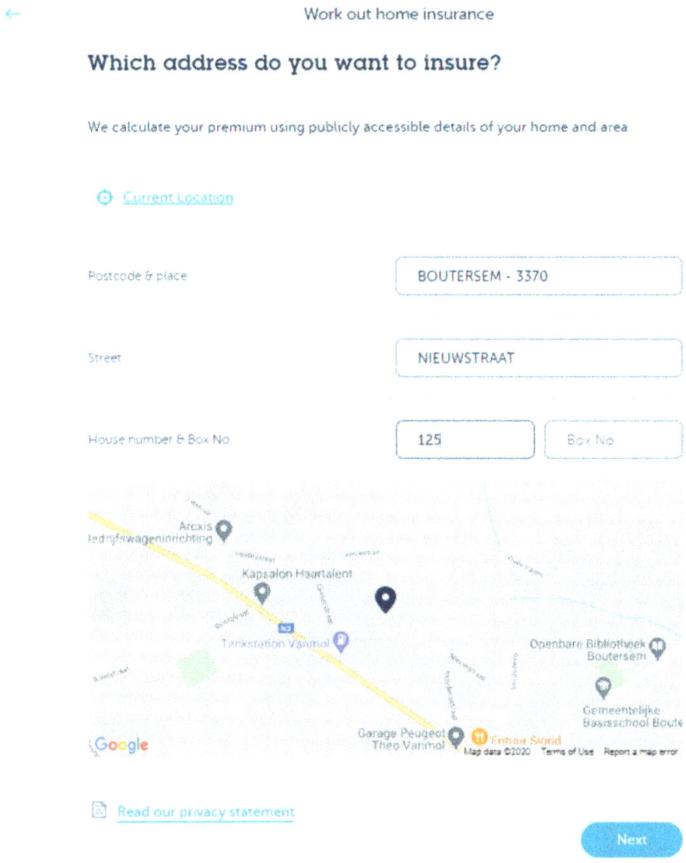

Figure 7.40: The customer enters the address.

Figure 7.41: The algorithm extracts the features associated with a given address.

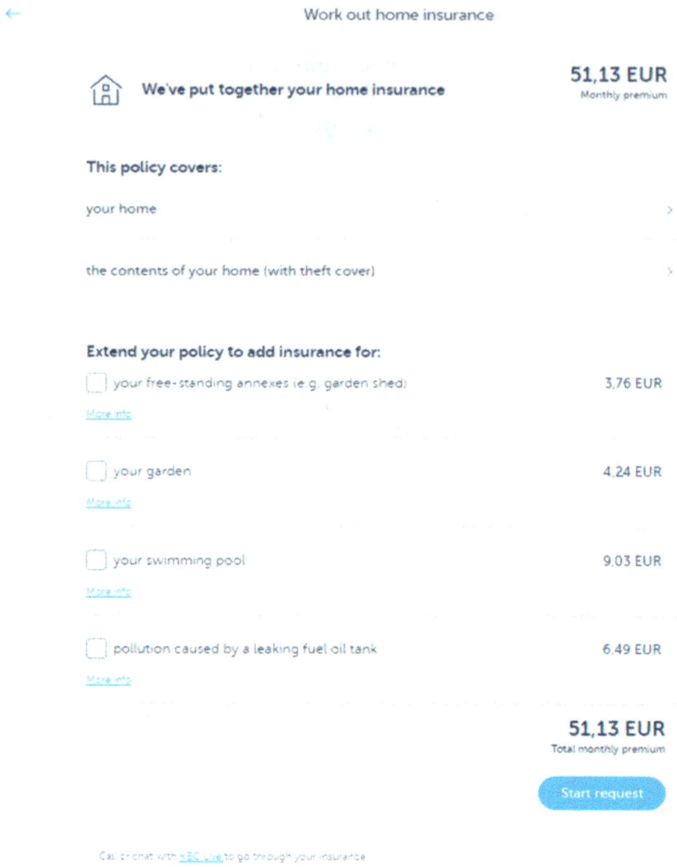

Figure 7.42: Insurance offer.

surance fraud provide more than enough motivation for insurers to invest heavily in fraud detection. Many of them, such as KBC Group (an integrated bank-insurance group with core markets in Belgium, the Czech Republic, Slovakia, Hungary, and Bulgaria), decided to design and apply machine learning models in the detection of fraud. To assess the validity of an insurance claim, and thus be able to distinguish between a legitimate claim and fraudulent one, insurance companies organize their efforts and personnel in three lines of defense. The first line consists of customer-facing agents who carry out business objectives. First-line agents are only responsible for detecting the most flagrant fraud and do not carry out a thorough assessment of each claim. The second line of investigators review and challenge the claims that make it past the first line. Meanwhile, the third line focuses on preventive measures as well as evaluation of the efforts carried out at the first and second lines. Since the early nineties, many insurance companies around the world have started to empower their second-line investigators with machine learning tools in order to help them more accurately and efficiently identify nontrivial fraudulent insurance claims.

Fraud detection is an inherently complex machine-learning task because of several factors. First, whenever labeled data is available, the labels are **noisy** at best. In fact, available data cannot be labeled with total certainty as either fraudulent (positive class) or legitimate (negative class). Many are the cases where frauders may have successfully obtained compensation. These cases are referred to as false negatives or Type II error. Meanwhile, in a few other cases—referred to as false positives or Type I error—a legitimate claim may have been refused. Consequently, semisupervized or, in this case, unsupervised learning models need to be used to overcome the absence of reliable labels.

Second, there is a severe **class imbalance** since fraud is—despite its significant volume—still rare when compared with the total volume of transactions in the insurance industry. In other words, having orders of magnitude fewer examples in the "fraudulent claim" class than in the "legitimate claim" class makes it harder for machine learning models to accurately identify the boundaries between the two classes.

Third, the two types of error have **imbalanced costs** because Type I errors are "more costly" than Type II errors both from the moral and from the financial perspectives. From the moral standpoint, tolerating some fraud is more acceptable than refusing compensation to a legitimate claim. The situation is comparable to that of a judge who would set a guilty suspect free rather than taking the risk of condemning an innocent. While the moral aspect is obvious, the business reasoning is less trivial. At first glance, rejecting a legitimate claim or accepting a fraudulent one may seem to yield identical financial costs from the insurer's perspective: both decisions would require the insurer to pay a compensation to the customer. However, error costs should not only include the compensation paid to the customer but also the costs of potential legal battles should a customer with a rejected legitimate claim decide to take legal action, the loss of this customer in the future and, more importantly, the image of the insurance company, which would be tarnished in such circumstances. With the complication of having different costs for different errors, reducing the overall error rate is not optimal from a cost perspective, and a machine learning model has to find the best tradeoff between the two error types to minimize the combined error costs rather than simply minimize the error rates.

Finally, the data often suffers from **class overlap** in that two claim files could be entirely identical with one considered positive and the other negative. In the simplest scenario, a frauder could copy a legitimate claim and ask for compensation. With more data collection, the two claims could conceivably be separated. For instance, information on the financial situation or legal history of the customer could help in distinguishing fraud from legitimate claims. However, high data collection costs in addition to data privacy legislations could limit the accessibility or the usability of such information.

To add a layer of complexity, it is not possible to conceive an all-around fraud detection algorithm that would operate on all products of an insurance portfolio. The uniqueness of each type of claim makes it hard to create a general-purpose model. In fact, the inputs involved in those claims can greatly differ from one context to the other. For example, information such as the type of road where a car accident took place are irrelevant in claims related to home insurance. Nonetheless, it is possible to design specific tools for each product in the insurance portfolio. As explained, these algorithms are domain-specific and cannot be generalized to all types of claims.

Despite these complications, investment in machine learning tools for fraud detection is still profitable. In fact, it is not necessary to achieve an impeccable error rate since any improvement on existing methods has a significant financial impact. Consequently, insurance companies developed fraud detection tools for a variety of their products. **KBC Group** has developed machine learning tools for a variety of tasks for fraud detection in job accident-related claims. We will focus on this example to illustrate how unsupervised machine learning can be leveraged in the insurance industry.

For clarification, a high-level, generic overview of the business process and the interaction with machine learning models can be found in Figure 7.43.

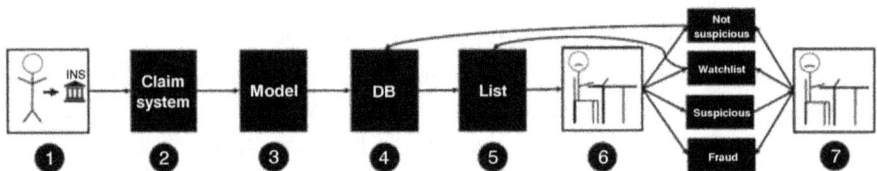

1. Client fill in the claim and deliver it to the insurance company
2. Claim is processed and stored in the system
3. Model runs on the claim data and score each claim
4. Based on the score the claims are ranked and stored in database
5. The claims are compared with the actually investigated claims and the top N not yet investigated are picked up and pushed to the list that is later presented to the claim handlers and private investigators in the webapp web app
6. Claim handlers assess the suspiciousness of the claims that are presented to them in the webapp
    a) Not suspicious (it would be showed again only in case the suspiciousness increases)
    b) Watchlist (Will stay in the list and, therefore, will be still presented to the claim handlers)
    c) Suspicious (Go to private investigators for further investigation)
    d) Fraud (Fraud procedure)
7. Suspicious claims are picked and investigated by private investigators in order to get more detailed assessment, they can choose the same actions as claim handlers

Figure 7.43: Pipeline for evaluating an insurance claim.

In a job accident claim, inputs include information on the employee and on the employer, the date, location, and type of accident, as well as the expected duration of incapacity. The latter is known to investigators to be positively correlated with the likelihood of fraud. In other words, the longer a person is claiming to be unfit for work due to the medical consequences of the job accident, the more suspicious the insurance claim.

Taking these inputs into account, KBC created a two-step approach to create a fraud detection model. In the first step (cf. Figure 7.43, steps 3–4), they used the **isolation forest** algorithm to perform unsupervised outlier detection. The algorithm associates an "outlierness score" with each observation, that is, claim. In this algorithm, anomalies are assumed to be "few and different," hence the outlierness score indicates the rarity of a given claim and its divergence from more common claims. Since the algorithm can only cope with numerical data, a preprocessing step is necessary. Indeed, categorical inputs such as the sector of activity is transformed into binary inputs. For example, the input corresponding to the sector of activity is removed and replaced with a column for each sector that appears in the data, such as transportation and construction. Then a claim that used to indicate "transportation" under the input "sector of activity" would instead indicate a "1" under "transportation" and "0" for the rest of artificially created inputs that indicate the rest of sectors (cf. Figure 7.44).

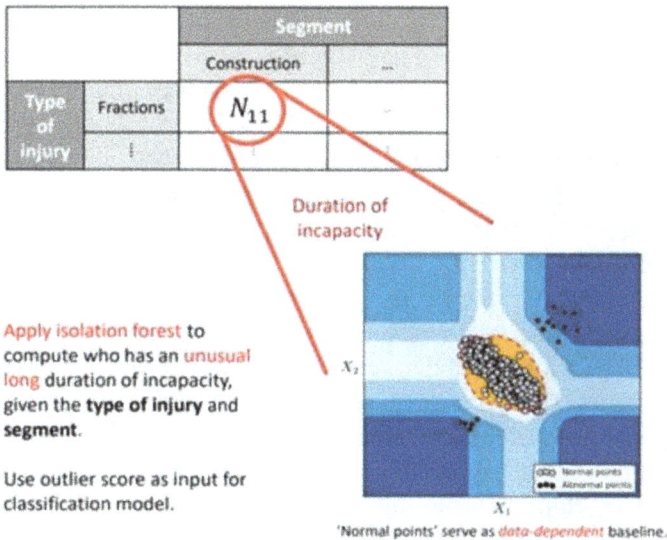

Figure 7.44: Isolation score as a measure of outlierness.

In a second step (cf. Figure 7.43, step 5), the outlierness score as well as available labels are added to the rest of the inputs and fed to a supervised binary classification algorithm. It is true that the labels in the training data carry a proportion of error, which inhibits the quality of the final model. But in the absence of clean and undisputable labels, this error can be tolerated since the final model yields much better results compared to the human investigators. The algorithm would then be trained to separate fraudulent claims from legitimate ones. Data was—as usual with supervised algorithms—separated into a training set used to train several algorithms and a testing set used to evaluate them and select the best algorithm. The different algorithms were evaluated for **lift**. This evaluation metric is the ratio of percentage of true positives identified by the model and the percentage of positives in the test data. Lift is a more appropriate metric than accuracy when working with highly imbalanced classes. Under these circumstances, the small class would be barely contributing to the accuracy score leading to bias toward the larger class.

Several models were tested including Random Forests, Linear Support Vector Machines, and Logistic Regression. They all confirm the tendency that the longer the incapacity, the higher the likelihood of a fraud. Other models such as auto-encoders do have a potential to improve current results, but logistic regression has the best lift results and was the model deployed as the final solution that assists second-line investigators of KBC with the task of identifying fraud in job-related insurance claims.

To illustrate how well the model works, two claims that were labeled by the algorithm as fraudulent are provided. In the first example, a young employee claimed that he had back injuries from lifting heavy furniture while performing his work duties. The incident falls under the category "simple fall" and the injured employee received first medical attention no earlier than 4 days after the claimed incident. Additional information from KBC's data shows that this is not the first claim made by this customer. The claim estimated to 190.000 € was flagged by the algorithm as a potential fraud. The insurance company has therefore decided to appoint a private investigator who later found elements proving that the claimed accident did not occur during work. The client has not accepted the refusal and started a legal procedure. As of the time of writing, the court has not provided a verdict.

In the second example, the victim "slipped away in the hallway on the way to the kitchen" during work times. The accident is categorized as "slip and fall." No witness was present during the accident, and the victim was taken to hospital immediately despite no apparent injuries. The claim was submitted 12 days after the accident without any mention of injuries and with limited data about the incident. The extended incapacity duration in relation with the type of accident was suspicious. These factors contributed toward all four algorithms labeling this case as fraud. A dedicated claim handler from the insurance company found no proof that accident happened the way it was described by the victim. Furthermore, no doctor could confirm the relation between the injury and the accident circumstances. However, doctors found proof of preexisting injuries prior to the claimed accident. The claim—which has been estimated at 80.000 €—was refused, and neither the employee nor his employer decided to react to the refusal by taking judicial action.

Being assisted with machine learning tools, KBC investigators can process a larger volume of claims in a shorter amount of time and with higher accuracy. In fact, the model requires a fraction of a second to assess a claim, compared to a few minutes in the first line and a few hours of work in the second line. Additionally, in the pilot phase 45 % of the top fraudulent claims were identified as suspicious by the model, compared to only about 10 % usually detected by human operators.

7.6.4 Defect detection in textile using CNN, *k*-means clustering, and unsupervised anomaly detection, at Veranneman Technical Textiles
Jonathan Kesteloot, Oussama Chelly

Veranneman Technical Textiles produces textiles for use in industrial applications. In other words, the end consumer of their product is not the retail industry and, therefore, aesthetic defects are generally of no concern during the manufacturing process. However, while aesthetic defects have, by definition, no impact on the structural integrity of the end-product, they can give the appearance of low quality textile. To give customers a sense of quality about the delivered product, each produced roll of textile is subject to a manual inspection over the complete length of the textile roll. This easily sums up to multiple kilometers of textile for a single production run.

Preemptively flagging the location and severity of defects that occur during production would relieve the human inspector of having to manually verify every single meter of textile. The focus could then be shifted to other parts of the production process. The scale of defects can range from $1\,cm^2$ for the smallest aesthetic defects to significant structural defects spanning over several meters of textile.

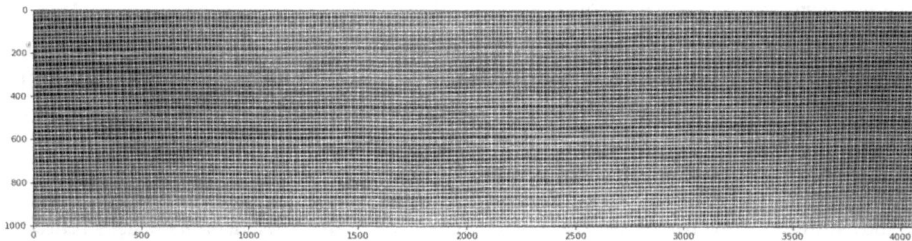

Figure 7.45: Microscope view of a portion of defect-free textile.

The solution provided by Robovision and Viu More allows for the automatic detection of defects and scoring of the quality of textile per meter. The techniques used allow for real time monitoring of 2-meter-wide textile using three 12-megapixel cameras running during the entirety of the production process. Five classes of defects shown below are being detected, and a quality score is assigned to each part of the textile.

The problem separates into two unsupervised tasks where defects are first detected and then quality scores are given to each part of the roll, and a supervised task where the highlighted defects are classified in one of the five aforementioned defects. The pivotal part of the complete solution is the unsupervised anomaly detection algorithm trained on images of good textile. Obtaining an image of a good textile is considerably easier since it is known that defects are rare.

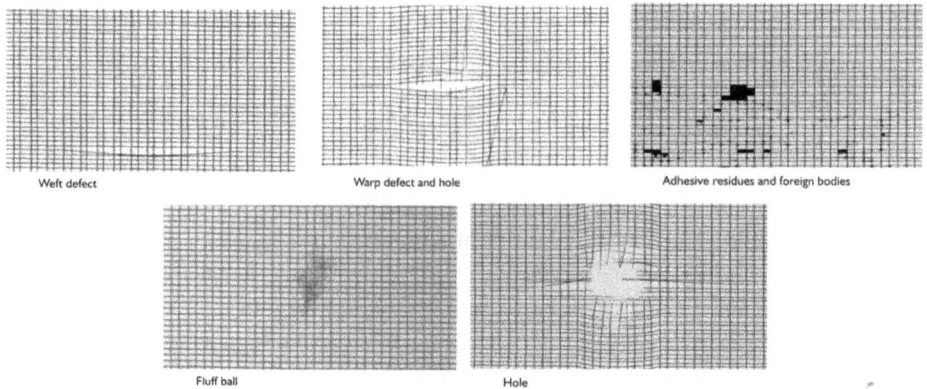

Figure 7.46: Various types of textile defects. Source: http://osif.de/Download/Flyer_GridInspector_Eng.pdf.

The task of assigning a quality score to an image of textile and highlighting defects within the image is considered unsupervised for two reasons. First, there is no consensus by textile producers on the how quality scores should be assigned; and second, defects are rare. Manually labeling all defects from scratch would require traversing through terabytes of data per roll.

The supervised task can then be stated as classifying highlighted defects into one of five possible classes. This is a common problem in computer vision and many out of the box solutions are available. For this use case, a basic ResNet50 was used.

The technique used for the anomaly detection is called a "backbone-based feature dictionary." It uses the resulting feature vector from a pretrained model as a measure of "visual similarity" of image patches. This technique is unsupervised as it uses features from an existing network trained with "good/perfect" images. It uses a combination of PCA and k-means clustering to make it fit on the problem at hand. To accommodate the images to a standard image size, it uses a sliding-window approach where patches are cropped from the image to be processed. A summary of the solution is provided in Figure 7.47 where the training pipeline and the inference pipeline are explained.

First, the initial image seen in Figure 7.48 is divided into patches as shown in Figure 7.49). From these patches, the original pixel values are transformed into a set of features. This feature transformation known as transfer learning consists of transforming the set of pixels into information about the presence or absence of certain shapes within the patch. Simple shapes usually correspond to a horizontal, vertical, or oblique lines. Then more complex shapes correspond to combinations of basic shapes, which could be used to draw any pixelated curve. This is a generic operation that is often performed in machine learning tasks on images and the extracted shape features are not specific to the use case. In this model, a ResNet18 pretrained on ImageNet was used. No further training on this backbone network was performed.

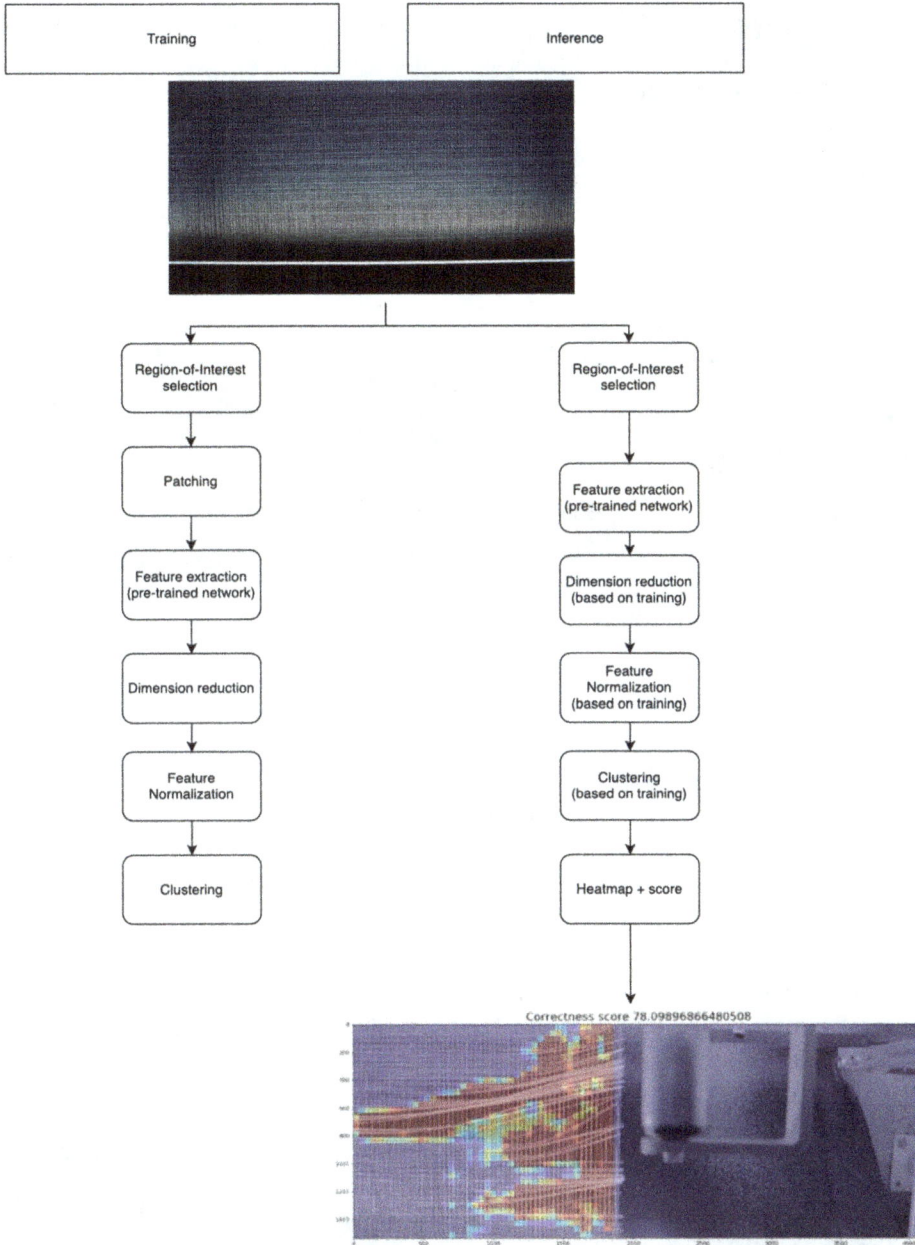

Figure 7.47: Training and inference pipelines.

Second, we perform extract features, which are specific to our use case by performing PCA on the shape features. With this operation, features that are most relevant to our images are kept while shape features that are irrelevant would be cleaned out. In

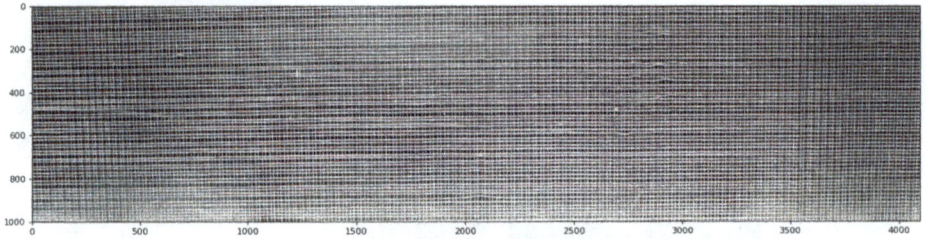

Figure 7.48: Input image of textile.

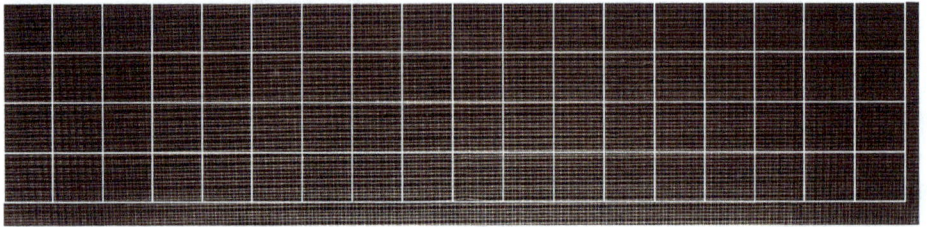

Figure 7.49: Cutting the image into patches.

other words, the use of PCA transform forces the algorithm to distill the information about what makes good textile.

In the third step, the features are normalized and used to cluster the images using the k-means algorithm. Since the input data is assumed to be good, these clusters form the representation of normal textile.

Finally, to detect defects, which are the outliers of our dataset, a distance metric is defined. Each image patch whose features are far from the existing clusters is then counted as an outlier. The definition of what distance is too far can then be set by the customer. Example output of the algorithm at inference is shown below. A heatmap with the local distance is the result. These local distances then get converted to a quality score of the entire image (cf. Figure 7.51).

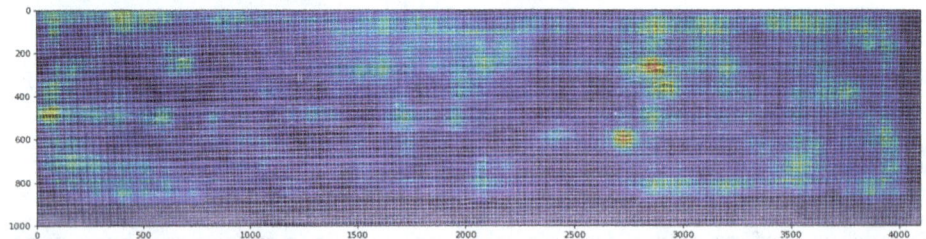

Figure 7.50: Output of the outlier detection algorithm as a heatmap where red colors indicate textile defects.

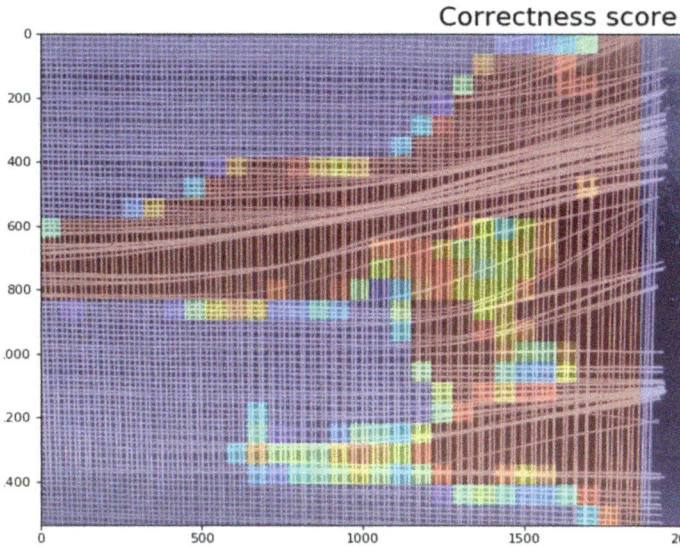

Figure 7.51: Correctness scores for the example image in Figure 7.50.

7.6.5 Extracting information from forms using clustering techniques with Microsoft Form Recognizer

Nicolae Duta, Oussama Chelly

Note. *The service and version referred to in this section is Microsoft Form Recognizer based on unsupervised learning techniques. Form recognizer now moved to a new service called Azure AI Document Intelligence, which also added supervised learning and generative pre-trained transformer (GPT) based solutions. Each of them has advantages and disadvantages, so the customers can use whatever works best for their problem.*

Companies and people gather large amounts of data in various paper and digital documents such as invoices, receipts, purchase requests, tax forms, technical and scientific literature, etc. Forms are documents that contain structured data such as key-value pairs and/or tables and can include typed text or handwriting. Unfortunately, both the format as well as the layout of the data (key-value pairs embedded in cell structures, complex tables, etc.) make automated ingestion, information extraction, and use of any semantic information very challenging. Figure 7.52 shows an invoice sample on left and the structured information to be extracted on right.

Manually consolidating data from multiple forms can be a tedious and time-consuming operation. In addition, the data layouts and formats may not be consistent across different forms, especially ones from different sources. A system that looks for specific keywords or locations will only be able to process a limited number of forms or require multiple different configurations. In contrast, newer technologies identify keys and values within documents without interaction and without being provided any

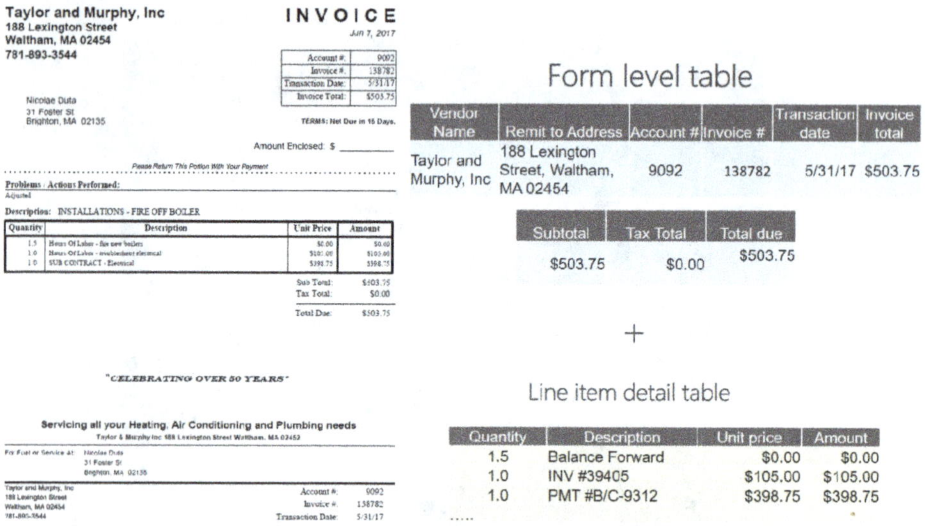

Figure 7.52: An invoice form sample (left) and the structured information to be extracted (right).

key or value locations. These systems learn the keys and the table headers from different types of forms and can then extract values for them without requiring user input regarding keys, values, or layout of the documents.

Microsoft Form Recognizer uses unsupervised learning to understand the layout and relationships between fields and entries in forms. When a set of forms is submitted, the system clusters them by type, discovers what keys and tables are present, and associates values with keys and entries with tables. This does not require manual data labeling or intensive coding and maintenance. The architecture of the unsupervised Form Recognizer is shown in Figure 7.53. There are two main processing pipelines: one for **Training** and one for **Recognition**.

Each pipeline consists of several modules starting with the same **preprocessing** module, which takes as input an image file or PDF document, splits it into pages, and then extracts all characters identified on each page and the coordinates of each of those characters. To extract the characters, this module first uses a third-party page decomposition software for digital pdf documents, then uses Microsoft's Optical Character Recognition (OCR) to translate images into text. The extracted characters are aligned on their vertical coordinates such that after sorting by vertical and then by horizontal coordinates, they appear in reading (left-to-right and top-to-bottom) order. The characters are then grouped into tokens based on multiple cues: horizontal spacing, presence of punctuation or vertical lines, common prefix, etc.

In addition to extracting characters, the preprocessing module also extracts the line structure (horizontal and vertical lines) present on each page either directly from a digital pdf or using computer vision algorithms such as the Hough transform whenever the document is in image format. The extracted lines are subsequently grouped into

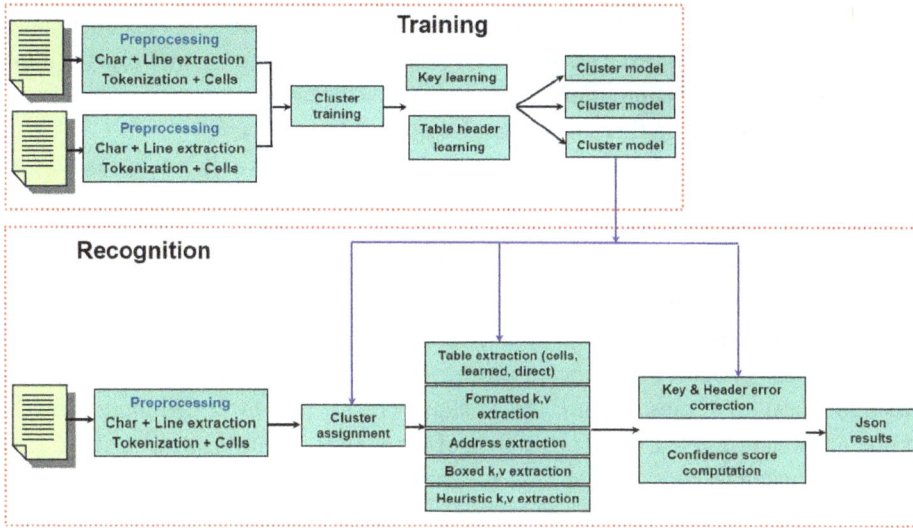

Figure 7.53: Architecture of the Microsoft unsupervised Form Recognizer.

rectangular cells and the cells into connected components. The character tokens and rectangular cells constitute the main objects used by system to compute features used for classification. These features are usually statistics (counts) on various token and cell alignment occurrences and may be quite complex. For example, the *average percentage of* (*aligned*) *numeric tokens per row* (cf. Figure 7.54) can distinguish well between

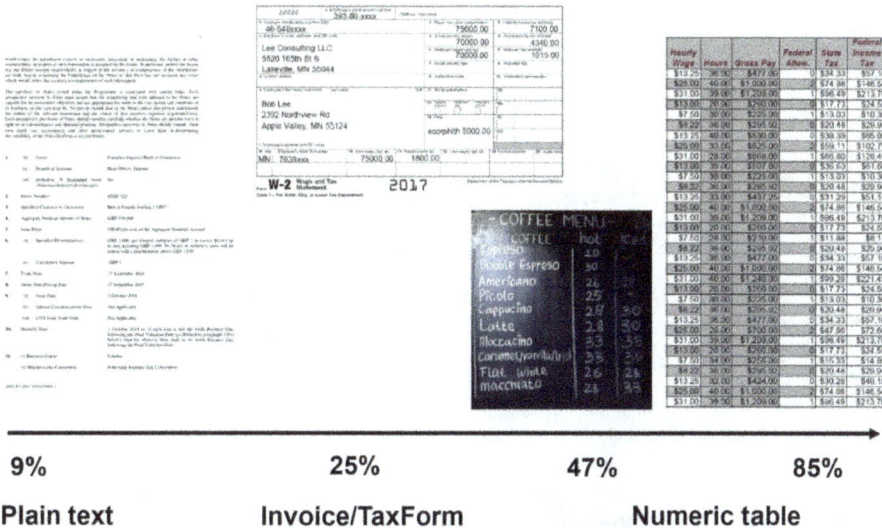

9%	25%	47%	85%
Plain text	**Invoice/TaxForm**		**Numeric table**

Figure 7.54: Example of a feature used by the Microsoft unsupervised Form Recognizer to assess page content.

various page contents: from numeric tables (high value) to invoices and tax documents (medium value) to plain text (low value). In contrast to the complex and computationally demanding feature computation, the classifiers used by the system are very simple and fast, usually based on thresholding and hashing.

The second training module is **page clustering**, which takes as input the tokens produced by the preprocessing module and identifies sets of similar pages (e. g., invoices coming from the same vendor with the same layout) and groups them in multiple clusters (cf. Figure 7.55). As the case with clustering algorithms in general, a measure of similarity is defined. Then different items are grouped in different clusters based on how similar they are. Similarity in the case of the Form Recognizer is measured by the number of tokens that are common between two documents, as well as the distance between the location of these tokens. This similarity measure is by design indicative of a common visual layout of a set of pages than of a common semantic interpretation. That is, if one just changes the provider's name in some of the documents in one cluster, they would still be clustered together while if one rearranges the location of tokens on the page while keeping the same tokens, the page would be placed in a new cluster. Clustering is implemented as a hashing of the token text and approximate location. Hence, if two pages have a certain percentage of common token text and location, they are placed in the same cluster.

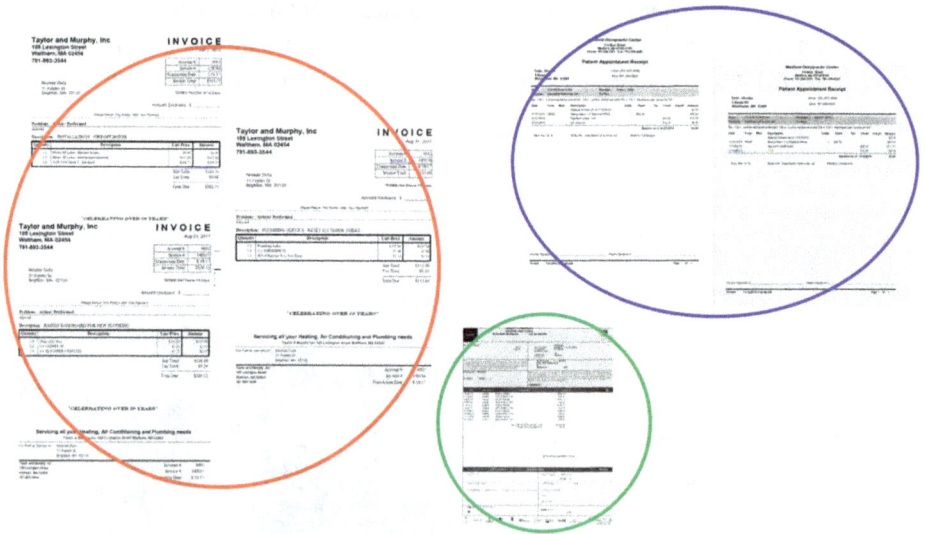

Figure 7.55: A set of 3 document clusters produced by the page clustering module.

Clustering complexity is linear in the number of pages clustered and constant in the number of clusters, that is, it will roughly take the same time whether the data contains 10,000 clusters or 10 clusters. That allows Form Recognizer to mine large amounts of data simultaneously and be limited more by the amount of physical memory than by the computation time.

Since pages in a cluster contain common tokens, it makes sense to consider some of them as being "*keys*" (semantic entities whose values are to be extracted, e. g., "Account #," "Transaction date," "Total due" in Figure 7.56) and some as being "*table header*" (*e. g.*, "*Quantity*," "*Description*," *Unit price*," "*Amount*"). One can even think of a table as a set of multirow key-value pairs. Of course, not every repeated token can be a key or a header; numeric entities (numbers, monetary amounts, dates) are usually excluded. For each cluster, Form Recognizer identifies the set of likely "*keys*" and "*table headers*" and stores them in a cluster model, which concludes the training pipeline.

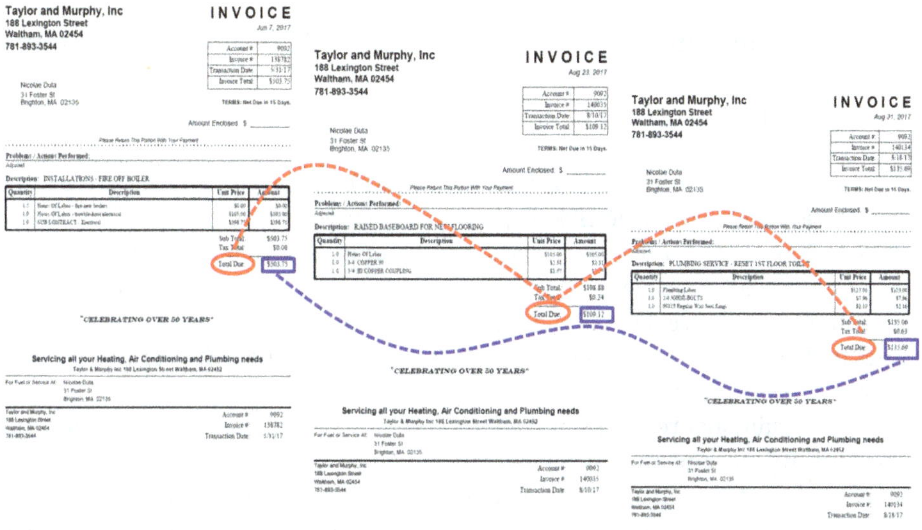

Figure 7.56: Learning keys from a page cluster.

The **recognition** (i. e., inference) pipeline takes one document, splits it into pages, and applies the same **preprocessing** module to extract the tokens and line structure. The tokens are used to assign each page into one of the clusters seen at the training stage. The key and table models for that cluster are retrieved and used to extract key-associated values and table bodies. For all documents in a single cluster, the field names will be the same while the associated values will be different. For one-of-a-kind pages, the system can still extract some key-value entities and tables based on semantics but may miss information that can only be extracted by seeing other similar documents.

Training on multiple documents of the same type can also allow for correcting typos or OCR. For example, a token "*Net price*" might be extracted by the OCR as "*Nit price*" on some document and as "*Not price*" on another. However, seeing more of "*Net price*" at some locations than versions containing random errors will prompt the system to output the more common therefore likely correct version.

The final module of the **recognition** pipeline is the **confidence score** computation. It assigns a confidence score for each extracted key-value pair and table entry. The confidence is based on the agreement of the semantic entities extracted for the same key from multiple documents in the same cluster. For example, if the key *"Net price"* has numeric values in 9 out of 10 documents in a cluster, then the 9 key-value pairs will get a confidence of 1.0 while the 10th nonnumeric one will get a 0.0.

The output of **recognition** pipeline is a file containing identified cluster IDs, keys, values, tables, and their locations, as well as confidence scores. Returning the cluster ID back to the user allows him to also perform document classification and better management of the document database.

7.6.6 Recommending pages in Microsoft News with reinforcement learning

Saheli Datta, Oussama Chelly

As the web has expanded, reading online news has become very popular around the world. A key challenge for news websites is to help their users find news articles that would be appealing. This comes under the realm of news recommendation. Microsoft News, formerly MSN, delivers high-quality news from popular and trusted publishers across the globe. A mix of human and algorithmic curation decides the most appropriate stories to show and how to position them on the page. In addition, the stories need to be relevant to users' interests.

Unlike traditional recommendation systems, news recommendation is inherently non-stationary since news become less relevant once they are outdated. In addition, the cold-start problem that is common to recommender systems is exacerbated in the news scenario. In other words, starting without prior knowledge about the user preferences leads to a starting page with news potentially unrelated to his interests. To deal with cold-start scenarios and the inherent nonstationarity of news, personalized recommendations have previously been formulated as a contextual bandit problem where the system recommends content to users based on contextual information of both users and content while also adapting its content selection strategy by incorporating user click feedback to maximize total user clicks.

When a user requests the homepage, MSN's front-end servers need to decide which content to choose from a readily available pool of content, then how to organize this content on the user's homepage. This content needs to be presented to the user in a way that is relevant to him along with optimizing overall performance or engagement metrics. Websites usually measure performance through a variety of click-based measurements such as click-through rate (CTR).

Users who come into MSN will generally fall into one of two buckets: users with history who have previously visited and interacted with content on MSN, and new users with no history. A standard approach of all recommender systems is to build up an understanding of users based on their past interactions with the system. This understanding is often referred to as user preference. Additionally, it is also common to have metadata about the content available to the system.

Contextual bandits (CBs) were introduced as a variant of the multiarmed bandit problem, while considering additional context. As in traditional bandits in casinos, the player starts with a fixed number of coins and can spend a proportion of his coins to evaluate the chances of winning in each machine. This step is known as "exploration" since the player is exploring different machines to evaluate the probability of winning in each one. Then the player typically spends the rest of his coins in the machine or set of machines with the highest winning probability in order to maximize his gains. This step is called "exploitation." The tradeoff between exploration and exploitation is key in bandit problems. In fact, the more a player invests in exploration the less risk he takes during exploitation but also the less funds are left to exploit. In some cases, the funds left for exploitation are not sufficient to cover for the exploration investment. Inversely, when a player overlooks the exploration phase by not investing enough coins, he would be taking much higher risks during the exploitation phase. In summary, spending more money in the exploration minimizes risk but comes at the price of a portion of funds otherwise available for exploitation. This tradeoff connects casino bandits to the more general full reinforcement learning (RL) problems. In particular, CBs can be considered as a 1-state RL problem.

In a contextual bandit setting, a dataset has four components, which are the context, the action, the probability of choosing the action, and the reward for the chosen action. In our scenario, the goal is to maximize user engagement using the CTR as a proxy. The content pool that is available to be shown to the user are the set of **actions** available as action choices. The probability of choosing a particular action is dictated by the **exploration** policy that is fixed a priori. The reward in our case is the user **clicks** on a piece of content.

In a first scenario, a user with previous history comes to Microsoft News. For such user with previous user history, the user preference is available along with potential demographic information to cater content. The content pool and all metadata about the content are also available. These are the information that are passed as **context** to the contextual bandit algorithm.

In a second scenario, a user with no previous history comes to Microsoft News. For such user with no user history, some demographic information such as location can be available and is used to help cater content. The content pool and all metadata about the content is also provided. These are the information that are passed as **context** to the contextual bandit algorithm in this case. Compared to the previous scenario, user preference is not part of this information. In either case, apart from the presence or absence of the user history information, there is no difference to the input to the CB learner.

Microsoft News currently uses a customized implementation of personalizer summarized in Figure 7.57. Personalizer is based on cutting-edge science and research in

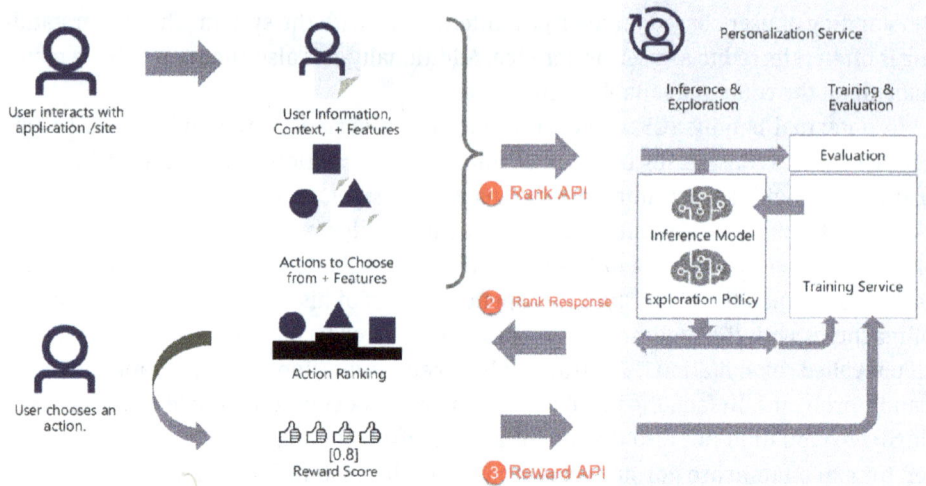

Figure 7.57: Figure illustrating how Azure Personalizer works (Picture Credit: Azure Personalizer).

reinforcement learning including papers, research activities, and ongoing areas of exploration in Microsoft research.

The primary learning loop uses machine learning to build the model that predicts the top news article for users. Actions with features and context features are sent to a ranking application programming interface (API). If a user is logged in, there is user-specific context features: the topics of news stories he clicked on in the past; otherwise only location is available. The action choices are the current set of news articles selected for curation. These articles are typically less than 50 pieces of content. Each piece has features that describe its topic. The **ranking** API decides to either **exploit** by choosing the best action based on current learning from past data, or else **explore** by randomly selecting a different action. If the API always picks the best action, the user will no longer be able to shift away from his initial choices. For example, a user interested in sports and technology could have clicked on a sports article on his first visit to MSN. Without the API exploration, the user would always be provided with sports articles to choose from, and there would be no chance for the API to discover the user's interest in technology. The current implementation uses the default **Epsilon-Greedy** exploration policy with epsilon in the range of 10–33 % in different markets. This strategy consists of always spending an epsilon-fraction of the available resources—that is, articles presented to the user—in exploration, and the rest of resources in exploitation.

Once the ranked response is shown to the user, the system collects feedback from the user—that is, whether there is a click or not—and sends it back to the training service within the API to continue learning. Models are continuously updated based on configuration settings, usually every 10 to 15 minutes. Continuous optimization is important, especially in the context of news, where breaking events are very common.

Over the course of the next few examples and for the sake of simplicity, only the following **action** choices will be considered: politics, sports, travel, and food. The system will choose a ranked order of these actions for users coming in, based on its current understanding of the world.

In a first scenario summarized in Figure 7.58, a user who has a history of reading content on MSN lands on the site. She primarily likes reading politics and health content. The system is starting from a cold start, so it starts with **exploring** available action choices by randomly sampling from the content pool with **probability** ε_n where n is the number of choices available to show to the user at the top slot. The user does not like what was provided in the top slot and does not click on it. This is sent as a negative reward signal to the learner.

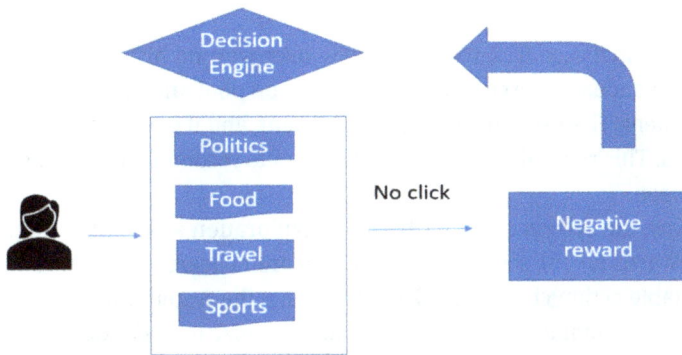

Figure 7.58: Illustration of user content interaction with negative reward.

Over the course of the next few iterations, the system continues to record signals on whether its recommendations generated positive or negative rewards. Based on this information, the system would learn a policy based on tuples of user and document context, the probability with which an action was chosen, and the reward signal associated with each iteration. By default, personalizer uses a **linear** model on the **features (context)** as the training policy although other representations are available.

In a second scenario, a different user who likes to read sports content lands on Microsoft News as shown in Figure 7.59. The user context heavily reflects his preference for sports-related content. If the learned policy exploits the user preference, it presents sport news to this user. As he clicks on the content, a positive reward signal is generated and sent back to the learner to continue online training.

In a third scenario, a new user with no history on Microsoft News opens the web page. The system chooses to employ its best guess (**exploit**) and may predict sports for this user. Assuming the user is not actually interested in sports, he would not click on any content. This is sent back as a negative reward to the learner.

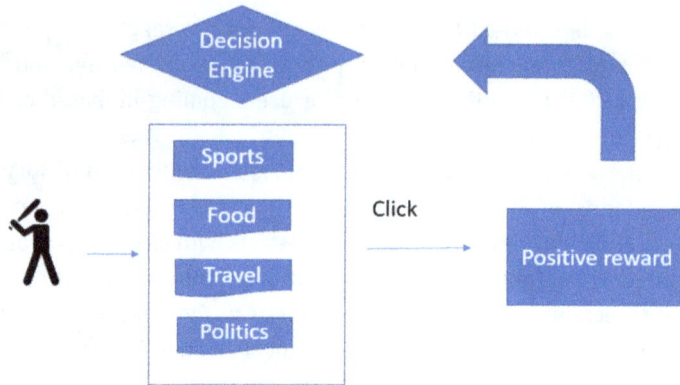

Figure 7.59: A sport-loving user opens Microsoft News.

In a fourth and final scenario, we have a new user with no history on Microsoft News and the system chooses to **explore** among its action pool and presents the user with a politics document. The user clicks on this document, and it generates a positive reward for the learner. This reward encourages the system to recommend more politics-related news to the user.

In this way, the system switches between exploration and exploitation in order to maximize rewards for the scenario. As this is an online learning system and both users and the available action choices can keep changing, the system continues to balance its exploration with exploitation in an attempt to maximize total clicks.

Bibliography

Bishop Chris. Machine Learning and Pattern Recognition. Springer, 2006.

Breiman Leo. Statistical Modeling: The Two Cultures. Statistical Science 16(3): 199–215, 2001.

Cohen William. Fast effective rule induction. In Proceedings of 12th International Conference on Machine Learning, 115–123, 1995.

Domingos Pedro. The Master Algorithm. Penguin Books Ltd., 2017.

Flach Peter. Machine Learning: The Art and Science of Algorithms that Make Sense of Data. Cambridge University Press, 2012.

Kolb Samuel, Teso Stefano, Dries Anton, and De Raedt Luc. Predictive spreadsheet autocompletion with constraints. Mach. Learn., 109(2): 307–325, 2020, https://dblp.org/db/journals/ml/ml109.html#KolbTDR20.

Maervoet Joris, Vens Celine, Vanden Berghe Greet, Blockeel Hendrik, and De Causmaecker Patrick. Outlier detection in relational data: A case study in geographical information systems. Expert Syst. Appl., 39(5): 4718–4728, 2012, https://dblp.org/db/journals/eswa/eswa39.html#MaervoetVBBC12.

Mitchell Tom. Machine Learning. McGraw Hill, 1996.

Walter Daelemans

8 Between language and knowledge

8.1 Why is natural language processing important within the broader AI domain?

Language is closely linked to intelligence and thinking. We use language to describe and explain facts and events, to express opinions and emotions, to tell stories, and for many other reasons. Sometimes what we say or write is objective and factual, other times it is an opinion or an expression of an emotion. We also use irony and humor. But most importantly, we want to *communicate*—we have something to say. An essential property of language is that we count on our listeners and readers to share sufficient contextual and cultural knowledge with us. It is because of this shared *common sense* that we can understand each other, even though what we say is often ambiguous and incomplete.

In this chapter, we will provide an overview of where the field of natural language processing (NLP) currently is, with pointers to the literature and possibly fertile directions for further research and development. NLP encompasses most of the artificial intelligence (AI) techniques described in the other chapters in this book: search and problem solving, logic and knowledge representation, statistics and machine learning (ML), and in addition information theory, linguistics, epistemology, cognitive science, mathematics, etc. It would be possible to write a 1000-page textbook about NLP, and in fact this has already been done. We gladly refer the reader to Jurafsky and Martin (2009 and the forthcoming third edition) for a more in-depth explanation of NLP algorithms and approaches.[1]

In this section, we provide a brief overview of the history of NLP leading up to the current situation in which large language models (LLMs) have become central. Section 8.2 describes the challenges of the field, and its various subtasks. Rather than striving for completeness, we focus in Section 8.3 on the current state of the art: **pretrained language models** and sequence-to-sequence models, both currently based on the **transformer** architecture in deep neural networks. Section 8.3.5 provides an overview of the more **classical modular approach** to analyzing language into meaning, and Section 8.4 addresses the limitations of the state-of-the-art in NLP: the lack of true natural language understanding.

To make the material more tangible, three examples of NLP applications have been added to the end of this chapter (Section 8.5). The first example is a cloud-based bot service called **Azure QnAMaker** that lets you create a conversational question-and-answer layer over your existing data. Hence, it is a good example of information retrieval and

---

1 Draft chapters of third edition available at https://web.stanford.edu/~jurafsky/slp3/

conversational AI. The second example is the application **LexagorIA** as used by Legal Village, an AXA business unit. LexagorIA analyzes, labels, classifies, and groups legal data in order to make it more searchable, to build logical bridges between the different documents, and to add indicators of completeness and relevance to track the evolution of law. Finally, there is **Icertis' enterprise contract management system** (ECMS) as used by Daimler. The application extracts and encodes both the content and the dynamic structure of contracts into a structured schema and a procedural code formulation so that processes, such as onboarding new suppliers, can be automated, clear actions can be defined, and compliance checks can be performed.

8.1.1 A short history of NLP

Throughout the history of natural language processing (NLP), we have seen an evolution back and forth between data-oriented and knowledge-based methods for building the models needed for developing language processing applications. Data-oriented methods combine language data with statistical pattern recognition techniques to learn models, knowledge-based methods start from formal linguistic knowledge about some task and handcraft models using this knowledge. NLP originated in machine translation (MT) in the 1940s within a data-oriented (cryptographic, information-theoretic, neural network) framework (see, e. g., Warren Weaver's 1949 'Translation' memorandum).[2] Given the hardware restrictions and data poverty of the time, however, this program remained an idea only. The engineered systems of that period did not achieve much more than word-to-word translation with limited use of context.

Because of the then-perceived limitations of this statistical approach, from the 1970s onward, models based on the implementation of linguistic rules and representations were built. These models were applied not only for MT but also for many other applications, like question-answering, dialogue systems, information extraction, summarization, etc. The familiar modular NLP pipeline (Section 8.3.5) originated here, with modules for analysis and generation at word level (morphological analysis, lemmatization, stemming, word sense disambiguation), at sentence level (syntactic analysis, sentence semantics), and at text level (coreference resolution, discourse coherence, and structure analysis). For each module in each language, linguistic data and rules had to be collected and designed. Consequently, the 1980s were the heyday of computational linguistics, where linguistic theory was driving research and development. However, the limitations of this approach soon became apparent: the models did not scale, had to be developed for each language, task, and domain from scratch, were not accurate enough for deployment in applications, and the pipeline architecture caused error percolation to downstream modules.

---

2 Reproduced in (Locke and Booth, 1955).

By the mid-1990s, most of the field had switched back to statistical and machine learning approaches. This shift was inspired by successful application in information retrieval and speech recognition and exploiting the availability of the first sizable machine-readable text collections (corpora of only one or a few million words were considered large before that time) and more computing power. They led to better accuracy and more useful applications. This was a period when many statistical and machine learning methods were championed and compared, ranging from decision tree learning and rule induction over Bayesian learning, maximum entropy approaches, transformation-based error-driven learning, and memory-based learning to different types of neural networks. By the start of the century, the clear winner within the field of NLP seemed to be **statistical learning theory** and its "optimal" machine learning method SVMs: support vector machine with kernel methods (see Chapter 7). The ML methods were applied both to individual pipeline modules (training a syntactic analyzer using statistical methods, for example) as well as directly to applications (e.g., a machine learning approach to text categorization tasks like spam filtering). Although the adage "there's no data like more data" was commonly accepted, a more specific result was often neglected, namely that as more data becomes available, apparently large differences between different machine learning methods disappear and they become interchangeable, suggesting that in NLP, as in many other fields, it is the data that matters most, not the ML method.

In the machine learning approach, designing input representations (feature engineering) was still a crucial and highly specialized skill involving linguistic and application insight. This was because of the inherent complexity: most machine learning methods are sensitive to the "curse of dimensionality" and its impact on training complexity and overfitting. Developers had to come up with a compact and relevant feature set. Apart from providing state of the art accuracy and a strong theoretical framework, SVMs allowed the use of tens of thousands of input features and automatically weighing their relevance or transforming them with kernel methods to feature spaces suited for learning nonlinear problems. Still, careful feature engineering could make a large difference.

The **deep neural network** revolution (called **deep learning**, DL from here on) from around 2010 onward, was not so much a revolution (multilayer perceptrons have been around from the start of NLP) as it was a choice for a method even better than SVMs at making feature engineering unnecessary thanks to the availability of even more data and computational resources. Current state of the art models for NLP like BERT and GPT (generative pre-trained transformer), allow for self-supervised learning on huge corpora and then **fine-tuning** the resulting language model to models for specific NLP tasks. These large background language models implicitly contain morphological, syntactic, and semantic patterns and arguably even world knowledge. Gone is the necessity for both model design and feature engineering. In essence, the focus of machine learning based NLP switched from learning models of transformations between linguistic input and output representations to learning rich reusable representations and from modular systems to end-to-end architectures.

A sobering thought is that, regardless of the technical algorithmic progress made in deep learning the last decade, most of the techniques used throughout the history of NLP were already well-established in the 20th century. What drove the evolution was the exponentially increasing availability of data and computing power. This also allowed for the end-to-end training of complex language processing tasks. The necessity to factorize a complex task into modules, each with their own input features, model, and representations disappeared.

With the benefit of hindsight, the history of NLP followed a logical route. When little data and processing power are available, linguistic knowledge-based methods compensate for that deficiency, resulting in modular and heavily engineered models. As soon as sufficient data and processing power are available, statistical and machine learning methods become possible. This modifies the role of the developer to that of a feature engineer and pipeline designer, assisting in pattern matching, or serving as an annotator of data for supervised learning. With even more data and computing power, deep neural network architectures become possible, making feature engineering unnecessary and pretrained models and end-to-end learning possible and preferable. The main recent innovation is the shift in focus from processing to representation learning, the result of **self-supervised pretraining** of language models on huge amounts of language data. These models can then be fine-tuned to specific NLP tasks. This has led to the current situation where models exist such as Open AI's GPT-3 with 175 billion parameters trained on 500 billion words (Brown et al. 2020). Such models perform a wide range of NLP tasks with an accuracy that is comparable to that of several different special-purpose models each trained for a separate task.[3]

This brief history of NLP of course does not do justice to the complexity of the field. Classical statistical and machine learning methods like SVMs are still used, especially for tasks where little training data is available (e. g., for languages and dialects for which few resources have been developed). And even linguistic knowledge-based methods, often in combination with machine learning methods, still have a role to play in many applications where explainability is crucial. Figure 8.1 provides a summary of the main evolutions in the history of NLP discussed in this section.

8.2 What category of problems does natural language processing solve?

It is not by accident that language understanding has been tightly coupled with artificial intelligence from the very start (Turing, 1950). Despite progress in artificial intelligence on many aspects of natural language processing, automatic natural language

---

3 This chapter was written just before the introduction of ChatGPT and GPT-4, which only made even more clear the superiority of pretrained large language models.

Knowledge-Based	Statistical and ML	Deep Neural Networks
Handcrafting	Feature Engineering	End-to-end
Modularity	Modular and end-to-end	

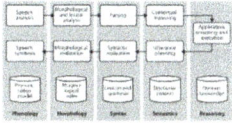

$$\mathrm{argmax}_T P(S \mid T) \times P(T)$$

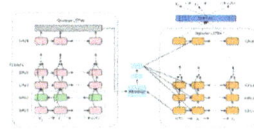

Data size and computing power
exponential growth over time

1970	1980	1990	2000	2010	2020

Figure 8.1: NLP Approaches and properties as a function of exponential data and processing availability growth over time.

understanding and generation still is partially an unsolved problem. NLP focuses on the analysis, production, translation, and transformation of language (both as speech and as text). Engineering has led to significant progress in accuracy on many subtasks. However, it may not have provided a solution for the basic research questions involved in *understanding* rather than processing language yet. Nevertheless, this improvement in accuracy does make an increasing number of applications possible. In a world where a large part of knowledge is encoded in thousands of different natural languages and the amount of available text is estimated to double every year, NLP technology plays an increasingly important role. Text and speech are *unstructured* data, like images, video, and audio data, and must be interpreted before they can be handled with computational methods. The main challenge of NLP is that the symbolic representations of language (characters, words, sentences, documents) must be made numeric while preserving meaning.

This challenge is carved up into the tasks of understanding, producing, and translating language, as well as communication in language. These diverse tasks can be reduced to a single computational problem, namely implementing **transformations** between **representations** using **models**. For example, to solve the problem of speech recognition, a model can be made that takes an acoustic signal as input representation and produces a transcription in text as its output representation. Likewise, the problem of machine translation can be modeled as a transformation between a text representation in the source language to a text representation in the target language. As there is a natural temporal order in language utterances, these are named sequence-to-sequence transformations (seq2seq). These transformations are not trivial and models implementing them must address three types of difficulties: (i) the same input can have different outputs in different contexts (**ambiguity**), (ii) the same output can be associated with many different inputs (**paraphrase**), and (iii) often the input does not even contain sufficient information to compute the output, and external or contextual knowledge is needed in the model (**inference**).

8.2.1 NLP tasks

Figure 8.2 provides a complete overview of what can currently be achieved with different transformations between text, speech, and image. In this figure, we see mappings between sequences for different NLP tasks, also between language and images. It takes some flexibility to see images as (just) sequences of pixels, though. As we will explain in the next section, all these transformations can be learned with encoder–decoder sequence-to-sequence transformer models or variants of that approach. In fact, it is difficult to find a technology other than NLP that has been so thoroughly and completely transformed and taken over by a single methodology. Transformer architectures are recently also taking over many AI fields beyond NLP, where they were developed.

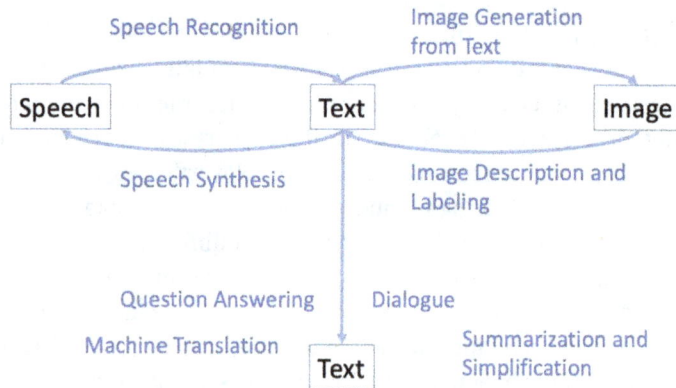

Figure 8.2: Example transformations between representations that are currently investigated in NLP.

An essential consequence of this evolution is that the NLP tasks are not split up into modules but learned end-to-end, bypassing an important aspect of NLP: constructing the meaning of language fragments (be they text or speech or gestures). What is missing, in other words, is the transformation between language and **explicit meaning**. We will return to this below in Section 8.3.5.

8.2.2 Speech to text processing

Speech output (**speech synthesis** from text) can be regarded as a largely solved problem. Indeed, its quality is so good that it has become a matter of concern recently that it is possible to clone a voice and have it say whatever you want in any language you want. Interesting remaining research questions concern the synthesis of intonation patterns.

For example, a sentence like "He bought her flowers" can have sentence accent on *he, her, flowers* or even *bought,* each intonation pattern having a different meaning, appropriate in different contexts. However, it could be argued that this problem should be solved in the text generation module. The addition of appropriate emotional load to speech output also remains a research problem.

Speech recognition is a harder challenge: despite enormous progress using deep neural network approaches, several problems remain, including modeling regional and dialectical speech, speech in languages for which little data is available, speech recognition in noisy environments, capturing emotion from speech, producing punctuation for recognized speech, etc. Nevertheless, the accuracy with which YouTube movies, for instance, are automatically transcribed (subtitled) using speech recognition would have seemed impossible less than a decade ago. It is no exaggeration to state that off-the-shelf text to speech and speech to text have become usable post- and preprocessing models to text generation and text understanding.

8.2.3 Image to text processing

A recent addition to the toolbox of NLP applications is image to text and text to image processing, routinely defined as seq2seq tasks these days. Similar to how speech synthesis and recognition are useful preprocessors to NLP and have many practical applications, image recognition and synthesis could potentially have an analogous impact (think, e. g., of environment description for the visually impaired, reporting in text or speech from camera streams, helping designers to visualize their ideas, etc.). In addition, the multimodal combination of text or speech and images can lead to further progress and additional applications.

Thanks to deep learning progress, the two necessary technologies, image processing (CNNs and transformers) and language processing (recurrent language models and transformers) can be combined in innovative ways. Going from image to text is called **image captioning**. The task is to generate a textual description of an image (or a video fragment represented as a sequence of snapshots). Given **multimodal** input, several DL training techniques have been proposed to learn how to align image and text sections for later use in text generation from images. In some architectures, the image information is injected into the **language model** while it is processing text, in others it is merged with the output of the language model in multimodal layers. Companies have played an important part in the development of this technology, thanks to their access to large training data. Similar architectures have also been developed for image generation from text. The most impressive to date is OpenAI's DALL-E, a 12-billion parameter GPT transformer model that interprets natural language inputs and generates relevant corresponding images (Ramesh et al. 2021). Figure 8.3 shows the infamous avocado armchair example. These early examples of multimodal "text-to-anything" AI applications have resulted in a boom in **generative AI (GenAI)** applications.

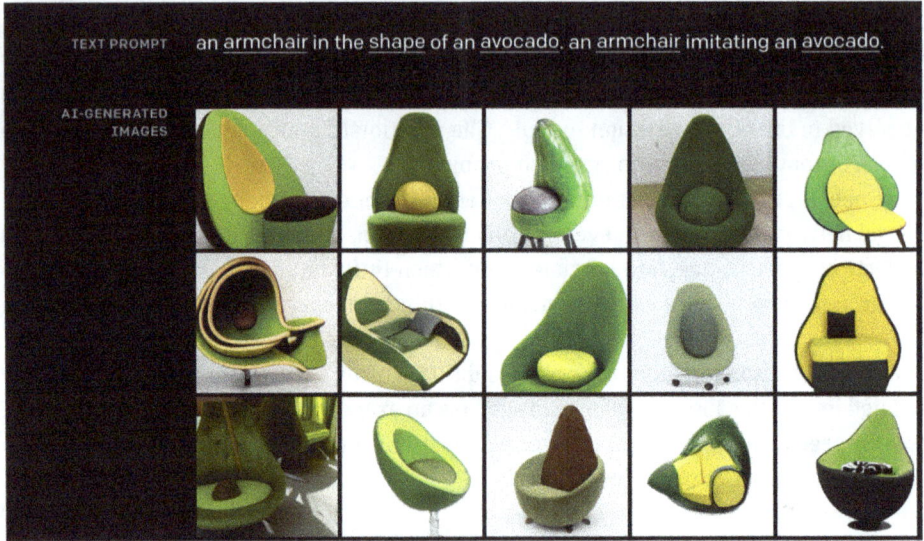

Figure 8.3: Images generated from text by DALL-E (from https://openai.com/blog/dall-e/).

8.2.4 Text to text processing

Given a text as input, we can train models to translate it, simplify it, summarize it, normalize it, answer questions about it, initiate a conversation about it, and many other transformations. All these tasks fall under the label of text-to-text processing.

The most visible NLP application in this category is **machine translation** (MT). It motivated and initiated the discipline of computational linguistics in the USA in the Cold War period because of the interest in the Russian-English language pair. Its evolution followed the historical sketch in Section 8.1 to the letter, from initial statistical ideas that were not workable with the computing power of the day, over linguistic knowledge-based approaches in the 1970s and 1980s, to statistics from the 1990s onward, and finally to deep-learning-based sequence-to-sequence models using transformers (as used, e. g., by Google Translate or DeepL). For some language pairs, accuracy of translation is considered human level or close. It is difficult to ascertain this as automatic evaluation methods that are routinely used have repeatedly been shown to correlate poorly with human translation quality assessment. On the other hand, it is clear that current quality levels for the more popular language pairs have increased to usable levels in routine translation. Of course, the models are only as good as the data on which they are trained. They contain **bias** and sometimes produce unexpected mistakes. As an example of bias, consider the following example from an MT system translation.[4]

---

4 Example from Stanowski et al., 2019; See their proceedings for a collection of recent work in this area https://aclanthology.org/P19-1164/

E.   The doctor asked the nurse to help **her** in the procedure
SP. **El** doctor le pidió a **la** enfermera que **le** ayudara con el procedimiento

In this example, the doctor is assigned male gender and the nurse female in the translation while the doctor is female are the nurse gender-neutral in the original (more about **bias** in Chapter 9).

As translated texts are needed to train these models, the methodology should be considered a supervised learning approach. However, some self-supervised models (e. g., GPT-3), trained on general internet corpora for the task of word prediction, have also been shown to be able to achieve machine translation close in performance to supervised approaches. This even eliminates the need for explicitly aligned translations.

Text summarization also fits the sequence-to-sequence learning approach. Summarization reduces the size of a text by producing a shorter alternative text that contains the most important information of the input text. It can be abstractive or extractive. In the first case, a summary is generated based on some "understanding" of (parts of) the complete text, potentially rephrasing the original. Such a system should combine extracting, integrating, and generating information into one single model. In the case of extractive summarization, where the task is relatively simpler, informative sentences in the original are selected and combined into a summary. The problem here is that a set of extracted sentences is not yet a coherent and cohesive summary (e. g., in avoiding repetition, handling pronouns, and their antecedents, etc.) and must be post-processed. In a seq2seq approach, the abstractive approach has advanced to high quality levels, and also the more complex abstractive approach is tackled. **Text simplification** can also be regarded as a translation problem (in this case into a different style of the same language). Using data such as a simplified version of Wikipedia,[5] seq2seq models can be trained to this end. One such application would be the simplification of legal and governmental texts to a level, which is understandable by people with only lower secondary school education.

Even **question answering (QA)** systems and the related task of **machine reading comprehension** can fit in the template of sequence-to-sequence learning. The huge pre-trained language models, described below, such as GPT-3, already encode a lot of factual information. We can query these language models directly to answer questions. Other QA research, however, still uses a classical knowledge-based or information retrieval-based approach. In the former, questions are linguistically analyzed and matched to similarly analyzed knowledge bases, in the latter, sentence similarity is used to search and rank potential answers to questions.

The same model also fits **conversational agents** (chatbots, dialogue systems). Most companies using conversational agents prefer to have control over the output of the

---

5 https://simple.wikipedia.org/wiki/Main_Page

system, which still is hard to achieve in deep learning-based systems. An unexpected or biased answer produced by a chatbot can do a lot of harm to an organization. Also, the dialogue management needed to keep the conversation consistent and relevant in task-based conversational agents is at this point hard to achieve with neural models, so frame-based or decision tree-based approaches to task-based dialogue still dominate. For chit-chat conversational agents, however, deep learning approaches have considerably advanced in recent years.[6] With the introduction of ChatGPT, an LLM finetuned for conversation, long dialogues can be produced that are consistent.

Considering the NLP tasks described in this section as directly trainable high-accuracy end-to-end systems, rather than as needing complex modular pipelines, has completely changed the field. Now, these complex tasks can be used as applications in their own right or used as pre or post-processing modules in more complex systems. In the next section, we go into the NLP research state of the art that has made this revolutionary progress possible.

8.3 How are natural language processing problems solved?

What are the current methods in NLP? In the previous section, we have seen how most NLP tasks fit the sequence-to-sequence, end-to-end, transformation pattern. In this section, we focus on the current workhorse methods in NLP that solve this mapping. Current approaches in NLP are based on the idea of **transfer learning** (see Chapter 7). In the case of NLP this means that models (called **language models** these days) are **pretrained** in an unsupervised or self-supervised way on large amounts of text data and are **finetuned** to the task at hand with a typically not very large amount of supervised task data. Some tasks and language models may work with just a few examples (few-shot learning) or even just a prompt (zero-shot learning). As context is important in sequence learning tasks, **attention mechanisms** are used, mostly as a component of transformer models. We briefly track the origin of this approach and describe the state-of-the-art methods.

8.3.1 Static word embeddings

One influential idea from linguistics in NLP is the theory of **distributional semantics** of the nineteen fifties, associated with linguists like Zellig Harris and John R. Firth. The idea was quite simply that words occurring in similar contexts have similar meaning. It was first introduced in NLP in the early 1990s by Hinrich Schütze. Variants of this idea,

---

6 For an overview: McTear (2020).

applied to large datasets with sufficient computing power, revolutionized NLP from 2013 onward as the concept of **word embeddings**.

Until then, for representing the meaning of words, NLP researchers often made use of **hand-crafted resources** such as WordNet[7] and similar dictionaries and ontologies that provide information about definitions and semantic relations like synonymy, meronymy (part-whole relations), and hyponymy (instance-subtype-type relations). Their advantage is that these resources are generally precise and accurate; the disadvantages are that they are expensive to construct and keep up to date, that their coverage of the complete vocabulary is usually low, and that the sense distinctions they make are often either too fine-grained or too coarse-grained for specific applications.

Distributional semantics opened the door to corpus-based, unsupervised, extraction of word semantics. By assuming that linguistic items that occur in the same contexts have similar meanings, it follows that a representation of the contexts in which a word occurs is a good meaning representation for that word. One count-based way of operationalizing this idea is to construct a word-word **matrix**. For each word (row in the matrix), count the number of times that word occurs in the context of other words (columns in the matrix). The context size could, for example, be two words to the left and two words to the right. By representing a word as the vector of these (normalized) counts of the number of times they appear with the context words, we have effectively vectorized (made numeric) the meaning of that word, and we can use **similarity functions** like cosine similarity to compute the semantic distance between words and by simple vector operations also other linguistic entities like sentences, paragraphs, and documents. However, these vectors are large and sparse, and while **dimension reduction** techniques like principal components analysis (PCA) may alleviate this problem, they may also lose information about relations between words.

Word2vec (Mikolov et al. 2013) was a milestone evolution in this approach because it created a fast and efficient way of generating dense (low-dimensional) vectors with usually better generalization than previous count-based methods (see Figure 8.4). Two algorithms were used in this approach: in the skip-gram version, given a word, the model learns to predict the left and right context of that word; in the **continuous bag-of-words** (CBOW) approach, the left and right context are used to predict the word in between.

The model is important because it showed that semantic representations that are dense (i. e., vectors without many zeroes) can be learned efficiently using unsupervised learning on very large corpora, and that these semantic representations incorporate linguistic and world knowledge. This was demonstrated with the famous word vector analogies showing that vector operations (addition, subtraction, averaging, etc.) could be used to do semantic computations: man – king + woman = queen, or Beijing – China + Russia = Moscow, etc. But syntactic analogies like singular-plural and comparatives are

---

7 https://wordnet.princeton.edu/

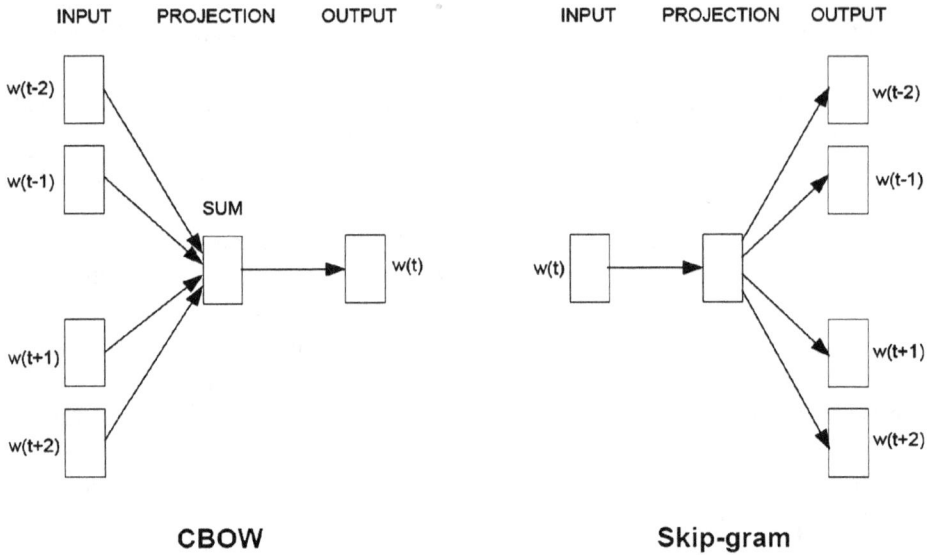

Figure 8.4: The 2 architectures in Word2vec (from Mikolov et al., 2013, arXiv:1301.3781 [cs.CL]).

implicitly present in the vector space as well. In addition, the similarity between words as operationalized in, for example, cosine distance between word vectors, seems to match well with human intuitions about word similarity. These advantageous properties of pre-trained models will return in the discussion of language models below.

What will also return is a problematic aspect of these pretrained semantic representations: they also incorporate the **bias** implicit in the corpora on which they were trained, coming up, for example, with unwanted analogies like father – doctor + mother = nurse.

8.3.2 Language models

Language is inherently sequential and has an implicit order. Language data comes in various lengths, including short and long phrases, sentences, and texts. In addition, most language processing tasks are sequence-to-sequence learning problems (transforming a sequential input representation into a sequential output representation) as explained earlier. Machine translation is a typical example of this.

Language models address the problem of sequence prediction given previous context by predicting the *conditional probability* of a next word given the previous words (the history):

$$P(w^{(t+1)} \mid w^{(1)}, w^{(2)}, \ldots, w^{(t)})$$

For example, in a count-based approach using a language corpus and a vocabulary, the probability of "cheese" given "cats like" is estimated. This is done by dividing the number of times "cats like cheese" occurs in the corpus by the number of times 'cheese' occurs. In this case the language model is an ***n*-gram** language model with $n = 3$. Because the probabilities based on complete histories are too sparse, this simplifying "Markov property assumption" is made (the complete history is not taken into account, but only the immediately preceding part of the history).

A language model also allows us to compute the probability of a sequence (e. g., a sentence or a text) as the product of the probability of each word given its (simplified) history, and to compute the *probability* of a (new) sentence or text given the language model. These language models have played an important role in NLP in areas as diverse as spelling and grammar correction, speech recognition, authorship attribution, machine translation, summarization, etc.

However, this approach runs into difficulties of **sparsity** (some probabilities cannot be computed because counts are zero, necessitating smoothing techniques such as adding 1 to all counts) and **memory requirements** for storing all the *n*-grams are high. The higher the size of *n*, the better the performance of the models in applications, but also the higher the sparsity and memory problems become. There is an analogy with the problem of count-based methods in word semantics described in the previous subsection. And also in this case, **neural models** solve the sparsity problem.

Fixed-window neural language models, which take a fixed window size as input, and the next word as output, solve these sparsity and memory problems in an efficient way by implicitly representing the *n*-gram patterns of the training data in their weights. Figure 8.5 gives an overview of the fixed-window neural language model architecture. The embeddings of the three input words are fed-forward through hidden layers to an output vector, which is associated with the following word.

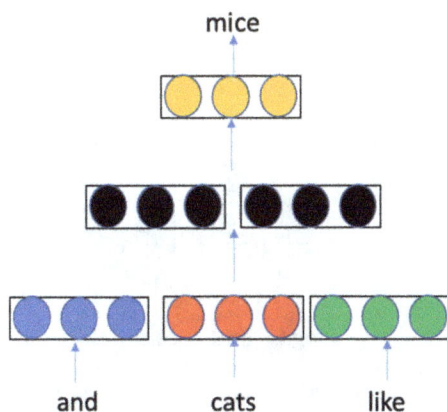

Figure 8.5: Fixed-width window neural language model. Colored dots represent neurons, arrows weight matrices. The first layer are word embeddings, the second layer is a hidden layer, and the output layer the embedding of the output word.

These models are still limited as they can only handle fixed length input and output, that is, for a particular model the input and output layer size are fixed, so the size of the input is limited, as is the size of the output. In practice, these models also don't scale very well because adding more context blows up the size of the models.

This is unfortunate because the mapping to be made in most language tasks is dependent on **context**. Sometimes local context is sufficient (e. g., to decide whether "work" is a noun or a verb only a few words of context are needed), but it is often the case that long-distance context is needed, for example, to solve problems like coreference resolution (see Section 8.3.5) and for end-c-end problems in general.

In conclusion, while fixed-window neural language models offer efficient solutions to sparsity and memory challenges, they come with certain limitations. As a solution for this, simple **recurrent neural network** (RNN) were proposed. For any variable length input sequence (and corresponding output sequence during training), these networks split up the sequence into tokens and take input token by input token, processing them by the same network. Starting from the second token, the model combines the input (the current/ second token) with the hidden state from the preceding token, creating a "recurrent" connection. This process continues until the end of the sequence. This way, information earlier in the sequence can help solve the mapping of later parts of the sequence. Figure 8.6 illustrates the approach by describing an "unfolded" recurrent network, providing a visual representation of how the network processes input sequences

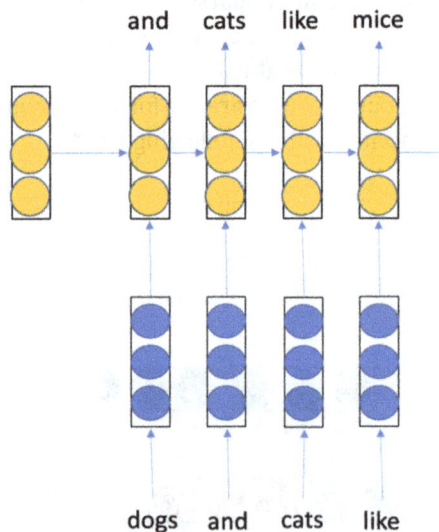

Figure 8.6: Simple recurrent neural network language model. The figure is to be read from left to right and starts with inputting "dogs" and an initialization of the internal state. The second step, 'and' is input, and the internal state when processing """dogs" previously. This goes on for the complete input sequence. Note that the figure represents an "unrolled" representation of what is happening in the same network over time.

step by step. Notice that the same network is used each time step, integrating the input embedding of a word with the hidden state of the previous word processed. As the hidden state is continuously updated, it contains information from all input that came before. Also, in this neural language model, usually **word embeddings** are used in the input.

While the network remains the same for each input token and its previous internal state, during training, the effect is akin to using a network as deep as the number of input items in the sequence. This is due to the need to compute gradients for the entire input string. This is highly problematic because of the **"vanishing gradient"** problem that this causes. The deeper a network, the smaller or bigger the weights become layer after layer and, therefore, also the gradients. The gradients go to zero or infinity depending on whether they start smaller or larger than one, leading to networks that are hard to train. In addition, although in principle information from all previous context is passed from input to next input, there is no control exactly which information is kept and which is overwritten. This problem motivated the development of **long short term memory networks** (LSTMs) in the 1990s (Hochreiter and Schmidhuber, 1997), a specific type of GRU (gated recurrent unit) network. In this architecture, a gate is simply a vector containing weights that represent the relevance of the components of another vector (in this case the hidden state passed on to the next time step in the recurrent network). Simply put, it determines what information to pass on and what to retain for later.

Long short-term memory networks have extra learned parameters to control memory and forgetting. They follow the same approach as described for GRUs but for each step they have apart from the short term state (previous hidden state) also a long term state that consist of an input gate (as in GRUs) that decides what to add to the long-term state from the previous hidden state, a forget gate that controls what to forget in the long-term state, and an output gate that decides what to use as output. This way the effect of vanishing gradients is counter-balanced and more relevant context information is kept until later in the sequence. Figure 8.7 shows the architecture of an LSTM cell.

GRU networks and LSTMs counteract the vanishing gradient problem, but like the simple RNN, they are computationally slow because they must process input sequentially making it impossible to parallelize the computation of the processing of complete inputs.

8.3.3 Transformers

The **transformer** network (Vaswani et al. 2017) (Figure 8.8) when used as a language model is a neural network design that can be easily parallelized because it does not implement recurrence of input (see Figure 8.6) but looks simultaneously at long stretches of input, which would have to be done in separate steps in an RNN. However, this does come at the cost of needing more memory. Figure 8.8 shows this approach: contrary to the RNN, the recurrence arrows that are still present in Figure 8.6 are now missing and

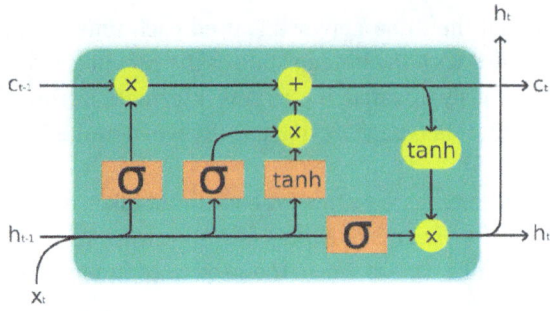

Legend:　Layer　Pointwize op　Copy

Figure 8.7: An LSTM cell with gates for input, forgetting, and output (Illustration from Wikipedia).

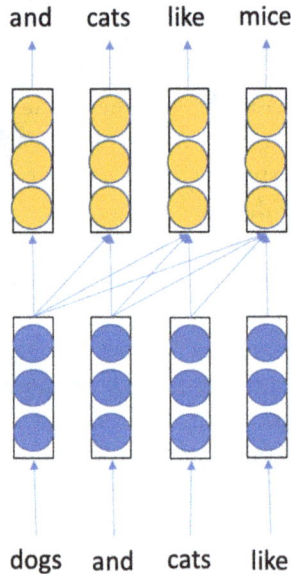

Figure 8.8: Transformer architecture for a language model (word prediction). Here, the complete input sequence is input at the same time and the hidden layers have access to all input words at the same time. The output sequence is also produced in parallel.

each hidden state has access to all previous inputs. As usual in language model neural networks, the input will be **word embeddings**.

A crucial element in this architecture is the concept of **attention**, which was also already present in some GRU recurrent models. Called **self-attention** in transformers, it allows during the processing of input words (queries), within a single transformer self-attention layer, to access all other input words (keys) weighing their learned relevance. This means that the input word can be related to every other word in its context. **Multiheaded self-attention** means that different aspects of the context can be learned within the same self-attention layer. This way different heads can pay attention to different aspects of the context.

Self-attention layers, hidden layers, and shortcut connections are combined into transformer blocks. In addition, positional embeddings, which are representations of word order, are added to the input word embeddings to allow the transformer to provide the transformer with information about the location of each word in the input.

We have now explained all the building blocks for sequence-to-sequence transformers. As shown in Figure 8.9, they contain an **encoder** part and a **decoder** part. The encoder translates the input representation into an internal representation and has access to all input, both left and right. The decoder part predicts the next output token given the previous inputs and the internal representation of the encoder. This turned out to be a successful architecture for NLP as so many NLP tasks can be described as sequence-to-sequence tasks, as we have shown above. The main strength of the architecture is its solution to the problem of long-distance dependencies thanks to the attention mechanism.

8.3.4 Contextual word embeddings

Now that we have introduced the transformer neural network architecture, we can explain another crucial element in current NLP by focusing on the left-hand side of Figure 8.9, the encoder part of a full sequence-to-sequence transformer. As a reminder, the left-hand side of the transformer network represents the encoding of text into internal representations. The **word embeddings** described earlier (Section 8.3.1) are "static" word representations because they represent the meaning of a word regardless of its context. But words have different meanings and behavior depending on their context. So, building on ideas from ELMo's contextualized word representations (Peters et al. 2018) and ULM's proposals on fine-tuning (Howard and Ruder, 2018), BERT (Devlin et al. 2018), the **bidirectional encoder representations from transformers**, and its successors, quickly became the state-of-the-art approach for **contextual word embeddings**. The way to train BERT is to mask a percentage (15 %) of the input words (while still looking at left and right context) and learn to predict the masked words. In addition, BERT also proposed to learn pairs of sentences, determining which is more likely to precede the other, but it is unclear whether this plays an important role in many tasks. BERT contextual word embeddings can be used as is, but they can also be fine-tuned to specific classification or sequence prediction tasks.

Just like we could explain contextual word embeddings by referring to the left-hand side of Figure 8.9, we can refer to text generation, that is, **generative language models** like GPT (generative pre-trained transformer), by referring to the right-hand side, that

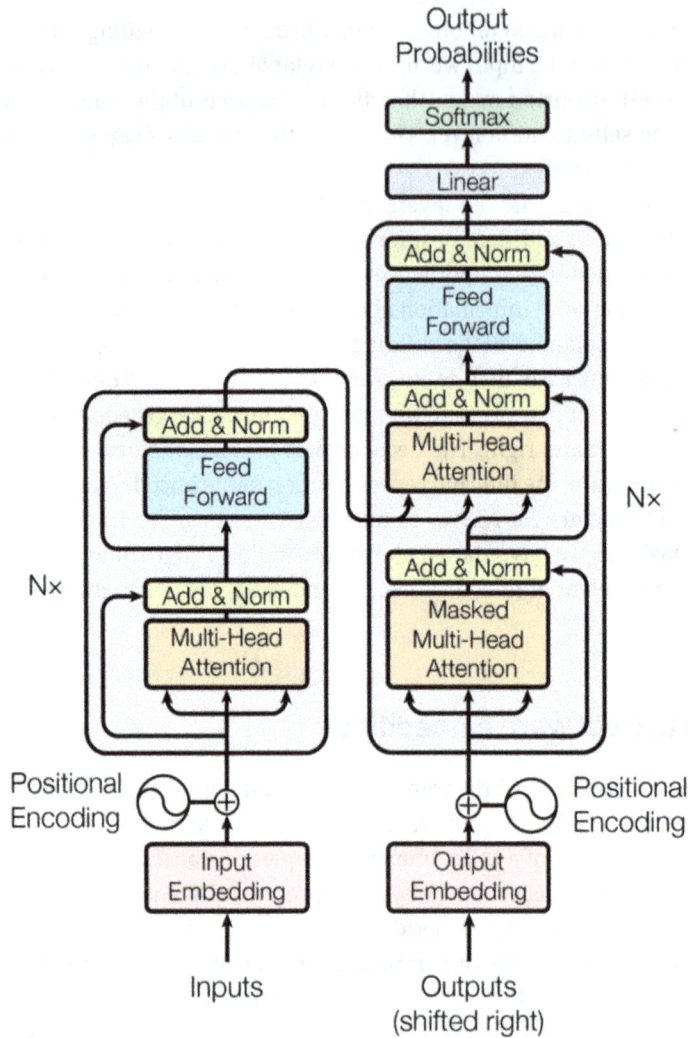

Figure 8.9: The transformer architecture (from Vaswani et al., 2017). It consists of an encoder (like BERT-type models) and a decoder (like GPT-type models), and as a whole, implements a sequence-to-sequence model (like BART-type models).

is, the decoder. In that case, only the left-hand input context is input, and the model predicts the next word like in the language models described earlier in this section.

While more classical ML approaches, such as those described in Chapter 7, are also still used in NLP, at this time, almost all work has converged on transformer-related architectures. Two concepts, implemented in an efficient way in these models, stem from this. The first is the concept of **pretraining**: by training language models (like GPT) and **contextual word embeddings** (like BERT) on large amounts of language data, these models learn the general patterns and aspects of language in a self-supervised way. The second concept is **fine-tuning**: by reusing these pretrained models in specific domains and tasks by fine-tuning them on data from those domains or tasks, we have faster convergence and better results as the linguistic and other knowledge is transferred to the new domain or task (transfer learning). Even more interesting, the pretrained models not only seem to learn linguistic patterns, but also semantic patterns and world knowledge. We already discussed the semantic analogies implicit in (static) word embeddings, but language models like GPT-3 also show competence in a great diversity of tasks not directly related to language modeling, for example, translation, question answering, conversation, etc. The introduction of ChatGPT, GPT-4, and the alternative LLMs from Google, Meta, Anthrop\c, and other companies has strongly corroborated this result.

Of course, as mentioned before, pretrained models inherit any **bias** that exists in the data that was used to train them. If training corpora scraped from the internet contain hate speech, and biased language use, the model may generate or use it at unexpected moments, even as a reaction to innocent input. The training data may also contain privacy-sensitive information that might be extracted from the model. These **ethical issues** are further discussed in Chapter 9. Fine-tuning the LLMs with human supervised input or RLHF (reinforcement learning by human feedback) alleviates this problem but does not completely solve it.

The impressive results of transformer models on sequence-to-sequence learning tasks, sometimes reaching human performance levels, should not blind us to their limitations: they only work with large amounts of training data, make unpredictable errors, and more generally, they are limited to the information used to train the language models. For example, in language models trained on data before 2020, no information will be found on COVID-19 and, therefore, no useful inferences will be made about the disease in language processing tasks. The different reactions to GPT-3, both by experts and by the general public, are informative in this respect. Hailed by some as an important step toward artificial general intelligence, it is put away by others as an expensive parlor trick. Assigning their real value to these models is probably more complex. These models are excellent pattern analyzers and generators, and their internal representations reflect knowledge in some way, but they don't *understand* anything, they have no opinions, and no reflection and their capacity for reasoning is rudimentary at best. Most importantly, they don't have anything to say, and they only want to continue a prompt. Their knowledge is second-hand, based not on direct perception of and interaction with reality, but on texts written about reality by many different people. Nevertheless, the patterns they manage to extract from that second-hand information are impressive and useful.

The good old-fashioned NLP approach with its modular pipeline approach to transforming text or speech to knowledge (semantics) therefore remains an interesting approach and provides components of commercial value. While it may eventually be overshadowed, these tools remain essential for the NLP developer. We investigate this approach in the following section.

8.3.5 From text to knowledge and back

When defining NLP tasks in an end-to-end fashion, often we find no need to concern ourselves with meaning. The difference between an MT system and a human translator is that the human translator will have understood and be able to think about what they translated. This will not be the case, at least not for now, for the MT system. In many applications, this explicit representation of meaning is nevertheless needed to allow reasoning and problem solving. A good example is the task-based conversational agents discussed earlier. They require semantic representations of both previous dialogue and of the context. Many basic research questions remain about **causal and temporal reasoning**, handling **implicit and figurative** language, understanding **situational context**, using **common sense knowledge**, etc. It is not clear yet whether it will be possible to extend pretrained language models used in sequence-to-sequence tasks to handle this. Until that time, we will need modular pipelines that extract *explicit* representations of knowledge and meaning (see Figure 8.10). These meaning representations (mostly based on **symbolic logic** or **semantic graphs**) allow inference and reasoning and are understandable by humans.

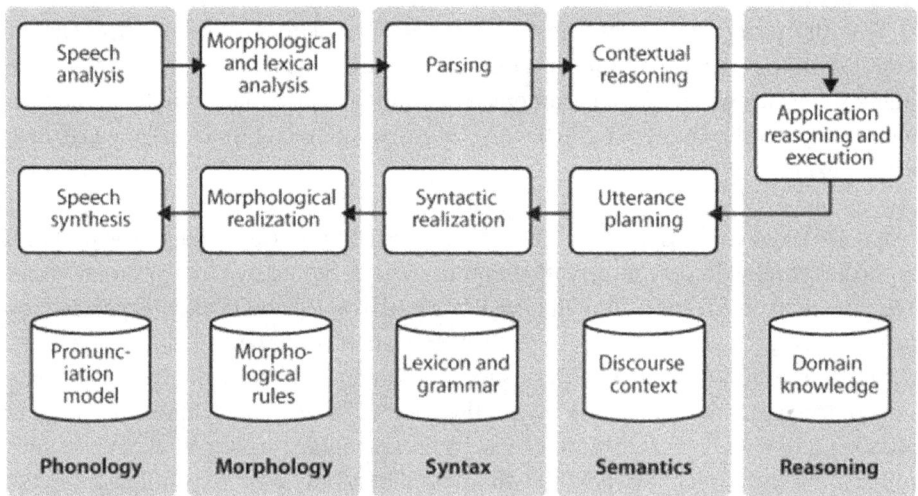

Figure 8.10: Modules in a classical NLP pipeline (from the NLTK book, Bird, Klein, and Loper, 2009).

8.3.5.1 Preprocessing modules, the word level

The initial step in an NLP pipeline inevitably remains preprocessing the text data (which may also be an automatic transcription of speech to text) into basic building blocks like words and sentences.

Tokenization finds the entities that most people would call a word: a string of alphabetic characters (e. g., "worked" is a word token). **Lemmatization** is a process in which different word tokens that are part of the same word family are mapped to the same "lemma" (e. g., work, works, worked, and working are all mapped to the lemma "work-Verb"), Lemmas are a first step toward semantics: all work tokens belonging to the same lemma partially share the same meaning. A good lemmatizer would assign two lemmas to the word token "work": work-Verb and work-Noun. In **sentence splitting**, a text is split into sentences based on capitalization, punctuation, and spacing information. For most commercially interesting languages, good sentence splitters, tokenizers, and lemmatizers are easily available. That is less the case for smaller or commercially less prominent languages. There are quite a few complexities to be solved for many languages: tokens may contain punctuation (as in "we've" as a short form of "we have": the tokenizer should decide whether to split this into we and have or to keep it together). Sentences may contain abbreviations ending in a full stop creating a spurious sentence split: abbreviation lists are a solution here. Some languages don't have spaces between word tokens and need a dedicated approach. Methods used range from regular expressions to machine learned models.

Also, in preprocessing, the neural network approach has changed the game considerably. Some models are character-based and don't bother about splitting input text into words or sentences. This approach solves the tokenization and sentence splitting problem by ignoring it as a separate problem. Other neural models may work subword-based. This approach implicitly creates its own definition of what a token is. A good example is byte pair encoding, a method that combines characters into subwords based on frequency of cooccurrence in a corpus.

Morphological analysis is a segmentation problem. A complex word like "reworking" consists of a prefix "re" followed by a stem "work" and a suffix "ing." Knowing the decomposition of a word into morphemes like prefixes, suffixes, and stems allows compositional computation of the word meaning. Just like they implicitly solve the tokenization and sentence splitting problem, recent neural network models don't delve into morphological analysis (the subwords discovered often correspond to something between syllables and morphemes, however). In a more traditional approach, methods such as finite-state morphology can be used to segment and analyze complex words. These methods are based on hand-made finite state automata that transduce an input word to morpheme structures. There are also unsupervised learning methods for morphological decomposition, but these are basically analogous to byte pair encoding.

In the final stage of the word level module in the pipeline, we have isolated the word tokens and related them to their respective types and lemmas, segmented them into morphemes (the smallest meaningful elements in language), and split off punctuation marks (which, incidentally, are also regarded as meaningful tokens). This is an important first step toward meaning representations.

8.3.5.2 Syntactic modules, the sentence level

At a level between words and sentences, **part of speech (PoS) tagging** assigns grammatical classes to word tokens in context. Tokens can be ambiguous between different classes (e. g., "bank" can be a Noun or a Verb). A PoS tagger utilizes both the context of the word and a lexicon, listing the possible PoS tags for each word. Of course, the complete arsenal of classical and neural ML methods has been applied to this classification (actually, seq2seq) subtask and PoS-tagged texts are used as separate representations in applications like stylometry, spelling correction, and information retrieval.

Classical methods of sentence analysis (parsing) are **grammar-based**. Consider a grammar fragment (a set of rewrite rules): "VP → V NP" saying that a Verb Phrase can be rewritten as a sequence of a Verb and a Noun Phrase and "NP → Det N" saying that an NP can consist of a determiner (e. g., "a" or "the") followed by a Noun. Determiner, Noun, and Verb are lexical categories, NP and VP are nonlexical categories (phrases). Lexical categories represent all words of that category listed in a lexicon (dictionary). Given these two (context-free) grammar rules, a sentence fragment like "eat a pizza" can be analyzed as an application of "VP → V NP" and "NP → Det N" giving rise to the tree structure (a "parse tree") in Figure 8.11. The right-hand side demonstrates a dependency formalism. In dependency grammar relations are defined directly between words, rather than indirectly as configurations in parse trees. As an example, the link between eat and pizza can be labeled directly as "object" in the dependency tree, whereas it must be defined indirectly as "the NP dominated by the VP" in the constituent-based approach to the left of the figure.

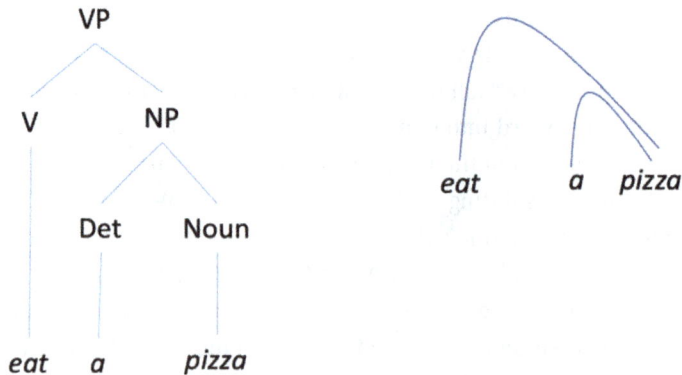

Figure 8.11: A syntactic tree.

Parsing is a search procedure using a **grammar** and a lexicon that can be either bottom up (starting with the words of the sentence to be parsed and rewriting them to syntactic categories using lexicon lookup and applying the rules right to left) or top down (starting with the top rule and testing the rules left to right until a match is found with the sentence). A grammar can also be used to generate sentences by top-down application of the rules and lexicon lookups. Grammars constructed by hand were often incomplete (missing rules or lexical items) and at the same time generated too many possible parses most of which were possible according to the grammar but unlikely for people (this is called spurious ambiguity). Rather than handcrafting a grammar, the **statistical NLP** approach introduced the idea of a treebank, a corpus of sentences that was syntactically analyzed by hand (a correct parse tree was provided for each sentence), and hence provided a sample of real language use with the associated syntactic analysis. Such a treebank can then be used to automatically learn probabilistic grammars from the data. Rather than giving each grammar rule the same probability, frequencies of occurrence in a treebank could be used to weigh the probability of each grammar rule, leading to a way to rank parse trees found according to their probability.

8.3.5.3 Semantic modules and the discourse level

We already discussed word level semantics in the previous section. **Contextual or static word embeddings** are currently the state-of-the-art for representing word meaning, but semantic lexicons like WordNet provide more explicit representations in terms of semantic relations between words, which makes it easier to integrate word meaning in symbolic knowledge representations, such as ontologies, supporting reasoning, and inference. Of course, a word can have different meanings (senses) in different contexts; so, in that approach, word sense disambiguation becomes a necessary module, for which both handcrafted knowledge-based and machine learning based solutions have been proposed.

A special case of word meaning analysis is **named-entity recognition** (NER). The task here is to find instances of a specific type of concept, for example, person names, organization names, addresses, URLs, locations, etc. Also, domain-specific concepts can be detected, for example, gene and protein names in a biomedical domain. Techniques used range from regular expressions over lists of names (*gazetteers*) and handwritten rules to machine learning methods (especially conditional random field methods and more recently sequence-to-sequence learning methods).

Given syntactic representations and word and concept meanings, semantic rules can be defined to assign a semantic representation to sentences in a compositional way. Syntactic and semantic analysis at the sentence level will be much more intertwined: semantic constraints will guide the selection of the most appropriate of possible syntactic analyses (parse trees), and the parse trees will help in detecting semantic structures such as semantic frames.[8] The target symbolic representation for this process of semantic interpretation is mostly some form of predicate logic or semantic graphs. Such a symbolic

---

8 https://framenet.icsi.berkeley.edu/

semantic representation language should support inference, be compatible with symbolic knowledge representation formalisms, and be unambiguous and canonical (same meanings should have the same representation, whatever the form). At this point, we have a literal sentence meaning, but we are only halfway computing the intended meaning of a sentence. The meaning of the sentence may change depending on the context, a process that may involve inference from background knowledge and from the meaning of other sentences of the text. One difficult task that must be solved is **coreference resolution**. In a short fictitious text like the following, we must infer from background and common-sense knowledge and from the sentence meaning that the first "his" refers to the president, that "Peterson" refers to the secretary of state, that the second "his" refers to the president and so on.

> "The president was quick to fire his secretary of state. In reaction, Peterson went straight to the press and revealed his crime."

Coreference resolution is essential in linking together the semantic representations of the sentences into cohesive units. Furthermore, we must break down larger discourses and deduce their structure, such as identifying causal relationships between sentences. It is only at this point that the real complexity of language becomes clear. Deep learning approaches swipe all this complexity under the rug of end-to-end systems, doing so efficiently and with growing accuracy.

8.4 What are the limitations of natural language processing?

We sketched an overview of the recent evolution of the field of natural language processing and the never-seen-before convergence of the field to a single paradigm: a **deep neural network (transformer)** approach based on **self-supervised pretrained** models and **supervised fine-tuning** to specific sequence-to-sequence and classification tasks. This approach fits almost all NLP tasks. Even more remarkably, this approach has also achieved a convergence between different subfields of artificial intelligence, for example, image and speech processing. If nothing else, at least the transformer deep neural network approach has brought back a useful and long-lost synergy between these subfields.

Made possible by an exponential growth in data availability and processing power, this approach has led to significant improvements in almost all areas of NLP, and has made possible new applications, for example, those translating between images and text. The success of the approach has led to end-to-end approaches for tasks that used to be carved up in separately designed or learned modules combined in special-purpose pipelines. Three examples of this improved state-of-the-art in applications are described in the following section.

For all its strengths, the approach also has limitations: pretrained models contain **bias** picked up from their training material, which may lead to undesirable effects in fine-tuned tasks. Debiasing pretrained models is currently a hot area of research and engineering. An even more serious challenge is that it is unclear which knowledge is present in these pretrained models. Many morphological, syntactic, and semantic patterns are picked up in this language model learning approach, and to some extent also world knowledge that can be used in **common-sense reasoning**. It is also clear that **transfer learning** by fine-tuning these models to specific tasks, even without or with little supervised training data (zero-, one-, and few-shot learning), makes it possible to put the learned patterns to work in a wide range of end-to-end tasks that used to require intricately engineered pipelines. However, for some tasks, especially when not a lot of appropriate data for pretraining models is available, accuracy is not that much better than the state of the art before deep learning.

While there is still a considerable journey ahead, particularly in tasks requiring reasoning, it is crucial to acknowledge that labeling anything as impossible for large language models is dangerous, given the potential advancements in subsequent versions. It is unclear for now whether remaining limitations are inherent to the deep learning approach or if, in principle, solutions are possible within this framework.

For that reason, we also described the now almost old-fashioned modular approach in NLP, which defines different modules (either based on knowledge-based handcrafting or on machine learning) needed for a pipeline that translates language into **explicit semantic representations**. These representations make it easier to connect language to symbolic background knowledge in knowledge bases and to reason with them. The downside is that systems developed this way are not scalable. And of course, the classical approach also doesn't have good solutions for many semantic problems yet.

Hybrid systems at first sight seem to be the most straightforward way ahead, and some interesting **neuro-symbolic approaches** have emerged. But for now, they have not been very influential in NLP.

With the exponential growth of data and computing power, research interest in the field of NLP and AI in general has exploded as well, so there is a good chance we may soon know whether in reaching for the moon we are just climbing a higher tree or whether NLP has really found its escape velocity.

8.5 Industry examples

This chapter features three industry examples to demonstrate the theory in a practical way. The first example discusses the main techniques involved in answering questions with chatbot technology, while the other two examples provide solutions to deal with the common problem of extracting and finding valuable information buried in countless

documents. The second example, applied to contract processing, provides a less technical overview, emphasizing the logical reasoning process behind document classification.
It serves as an introduction to the procedures of both document processing and document modeling. In contrast, the last example, supporting lawyers in their tasks, delves
more deeply into the employed techniques and articulates the reasons behind their selection.

These use cases emphasize that technology, often driven by a pursuit of efficiency
and innovation, perpetually evolves to optimize current solutions and meet expanding
needs. However, human intervention often remains crucial for approving outcomes or
handling cases beyond the software's capabilities. These advancements not only ensure
the optimization of processes, but also allow employees to allocate extra time to specific
tasks, such as intricate chatbot interactions or complex legal cases. A focused approach
in these instances yields greater benefits, including increased productivity for employees.

The AI techniques discussed are drawn from industry examples that were used
around the years 2019–2020, when they were first chosen for the book. GPT-2 was just
launched. They might not represent the state of the art anymore. They still illustrate,
however, important features and concepts of AI, such as embeddings and transformers,
that remain fundamental for the reader to understand newer developments.

8.5.1 Automate question answering using NLP with Microsoft QnAMaker

Parag Agrawal, Achraf Chalabi, Anneleen Artois

Nowadays, intelligent **chatbots** are much desired software applications that companies
seek to implement into their websites and mobile apps to hold seamless conversations
with their clients and visitors. These bots enrich the user experience by providing alternative means of communication and information retrieval. Additionally, they also
significantly reduce the need for human intervention (i. e., support teams) by handling
frequently asked questions (FAQ), anticipated queries, and other inquiries in real-time
whenever feasible. Moreover, as many websites have huge reference documents such
as FAQ pages, making it hard for users to browse through, adding a conversational layer
over raw data can speed up the process of finding relevant information.

However, many end-users express frustration with malfunctioning chatbots, highlighting a pressing need to humanize their interactions. QnAMaker, part of Azure AI Language, improves the precision and relevance of responses, leading to a more natural and
effective interaction and making the chatbot more adept at understanding user intent.

The technology serves as a conversational interface for semistructured data such
as FAQ pages, product manuals, and support documents. It stands as a viable option for
extraction and question-answering as a service. More specifically, QnAMaker aims to
simplify the process of information retrieval by extracting **question-answer** (QA) pairs

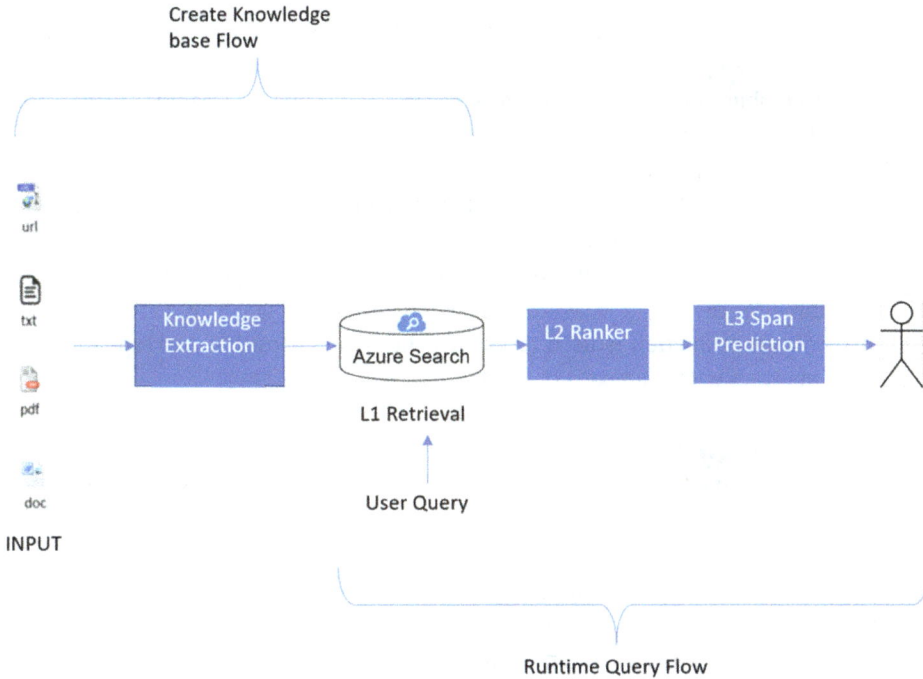

Figure 8.12: Logical process. The steps of 1. Knowledge extraction, 2. Retrieval, 3. Ranking, 4. Span prediction.

from data previously stored into a **knowledge base** (KB) and by designing a conversational layer on top (see Figure 8.12). When a developer utilizes QnAMaker to create a knowledge base, he automatically gains access to a set of NLP capabilities, which we will delve into shortly.

QnAMaker includes a persona-based chit-chat layer, enhancing the bot's capability to handle small-talk queries with personalized and human-like responses. The tone and (in)formality of the replies depends on the selected personality. Additionally, bot developers get automatic feedback from the system based on end-user traffic and interaction. This helps them in turn to enrich the knowledge base. This feature is called **active learning**. Finally, QnAMaker also allows users to add **multiturn questions and answers** to a knowledge base, particularly valuable for refining answers through follow-up questions. QnAMaker currently supports over 50 languages.

Note that technology evolves quickly and that this illustration, rooted in a 2020 snapshot, may not represent the current state of the art for QnAMaker. Newer and more advanced versions for **question answering** are now available on the market, including options from Microsoft's Azure AI Language services. The application still shows, however, some of the building blocks and concepts that will help the reader understand newer developments.

Figure 8.12 provides a comprehensive overview of the QnAMaker process, consisting of four major steps:

1. The initial phase involves knowledge extraction, whereby the system is responsible for understanding the layout of a given document and extracting potential QA pairs.
2. These QA pairs then serve as the foundation of the Azure Search knowledge base, providing information to the QnA-Maker Ranker based on the matching of user queries with existing QA pairs.
3. The Ranker system is responsible for reordering and refining the initially retrieved *top* number of results.
4. Finally, span prediction is used to predict the boundaries that correspond to the target answers within a text.

Logical Block 1: Knowledge extraction

As Figure 8.13 displays the initial phase, knowledge extraction consists of 4 phases, which we will explore individually:

1. document parsing
2. layout understanding
3. structure understanding
4. augmenting the output by adding entities

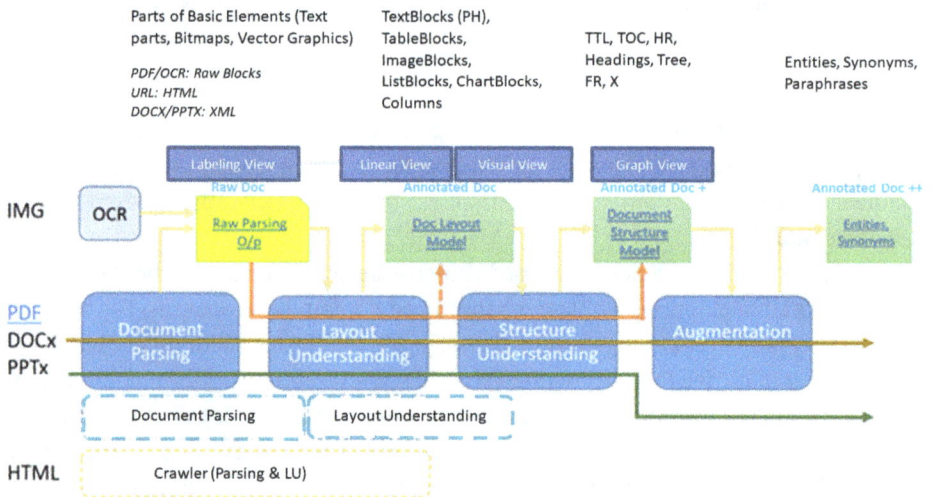

Figure 8.13: Knowledge extraction in 4 steps: 1. Parsing 2. Layout understanding 3. Structure understanding 4. Augmentation (Entity extraction with metadata).

In the first step, the documents are fetched by a **crawler**. Regardless of format, each document is transformed into an HTML file, typically containing both text and hypertext links. Also, documents from deep links are retrieved as the crawler scans the documents for hyperlinks and adds them to a job queue.

In a subsequent step, **layout understanding** plays a crucial role in grasping the document's layout elements. The documents are parsed by extracting basic building blocks, such as columns, tables, lists, graphics, and paragraphs from the source documents. Utilizing the top-down **page segmentation** technique known as the *recursive X-Y cut*, documents are cut into distinct blocks while accounting for nested elements. Noteworthy for its language-agnostic nature and low computational costs, this preprocessing technique for document analysis operates efficiently in linear time. This means that the time it takes to segment a document is proportional to the document's size.

In a subsequent stage, named **structure understanding**, each element undergoes tagging through a rule-based approach. This approach employs explicit rules to label elements, such as headers, footers, table of content, index, watermark, table, image, table caption, image caption, heading, heading level, and answers. *Agglomerative clustering* is then used to group similar data points, or content, based on parameters such as font and style. As a result, one can identify similar sections and subsections within the document, constructing an **intent**[9] **tree**, a structured representation of different intents, organized in a tree-like structure whereby the tree reflects the **relationships** between different intents, with more general intents at the higher levels and more specific intents as branches or leaves. For example, **leaf nodes** are identified as **QA pairs**. Despite its high computational demands resulting from its hierarchical nature and the necessity to compute pairwise similarities at each step, this technique is preferred for its efficacy in managing the hierarchical structure of input documents, ultimately delivering commendable results.

Finally, the intent tree is further **augmented** and annotated with **extracted actions or entities**,[10] such as "USB Port" (entity), "switch off" (action), or "engine oil" using **conditional random field (CRF)**-based sequence labeling (see Chapter 6), a statistical modeling technique for labeling sequential data. **Intents** that are repeated in and across documents are further augmented with their parent intent, adding more context to resolve potential ambiguities (see "Laptop" in Figure 8.14). The top-level intents are extracted

```
TITLE 1 = tag1: USB-A
          Subtitle = tag1: USB-A tag2: Laptop
                   text
TITLE 2 = tag2: Mini-USB
          Subtitle = tag1: X PORT tag2: Laptop
                   text
```

Figure 8.14: Example of Extracted Entities.

---

9 In the context of NLP and conversational AI, intent represents the user's purpose or goal in their communication with a system.

10 Words that consistently refer to the same object, classified into categories (person name, location, company name, date, etc.)

for every document and every top-level intent is connected to child-nodes. The number of levels is not fixed and depends on the structure of the document.

Logical Block 2: Retrieval

Once the raw data has been extracted and valuable **metadata** has been added, QnA-Maker uses a **search engine** as its retrieval layer to filter out the most relevant results (Level L1) and to then implement a reranking mechanism (Level L2) on top (see Figure 8.15). This two-layer retrieval and ranking system ensures both computational efficiency and improved relevancy of the results.

More concretely, the search service enables the creation of an index—a data structure that enables efficient searching over textual content—and utilizes **inverted indexing** to map each unique term in the corpus to the list of documents containing that term. The retrieval of top results is based on **term frequency–inverse document frequency (TF-IDF)** scores, a technique which evaluates the importance of a word within a document relative to the entire corpus. To consider possible spelling mistakes, **fuzzy matching** based on edit-distance is supported. Furthermore, it also incorporates **lemmatization** and **tokenization** as a pre-processing step (see Section 8.3.5.1). An index

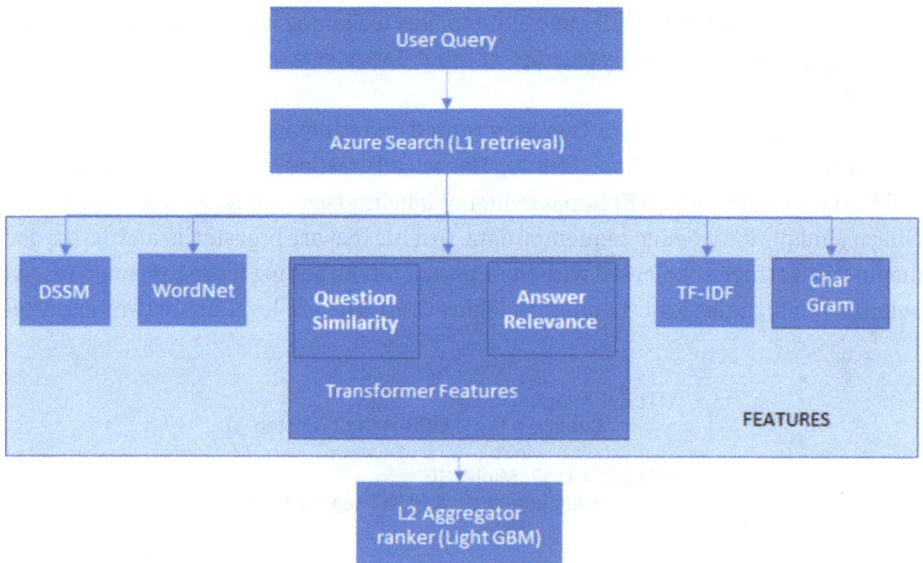

Figure 8.15: Retrieval and ranking.[11]

---

11 The expanded form of the terms "DSSM" and "GBM" is provided for clarity and context. However, it is not essential for the reader to fully grasp these concepts to understand the main text. DSSM = Deep Semantic Structured Modeling; Light GBM = Gradient Boosting Model.

created in a search service can scale up to millions of documents, lowering the burden on the QnAMaker application, receiving fewer than 100 results for **reranking**, the next step.

Logical Block 3: Ranking

Once the initial set of top 100 results is returned, the next step involves enhancing accuracy through advanced **deep learning** models. Given the diverse nature of content domains, variations in the number and length of QAs, and the presence of alternate questions per QA, the QnAMaker's ranker model employs generic features applicable across various use cases. It deliberately avoids any domain-specific features and relies on generic **similarity** and relevance-based signals. However, users can use the QnA-Maker for certain domain specific data because the features are generic enough to be able to fit a wide range of domains.

Transformer features (Figure 8.16) support the **query question similarity** and **query-answer relevance** modules and play a crucial role in determining the ranking of

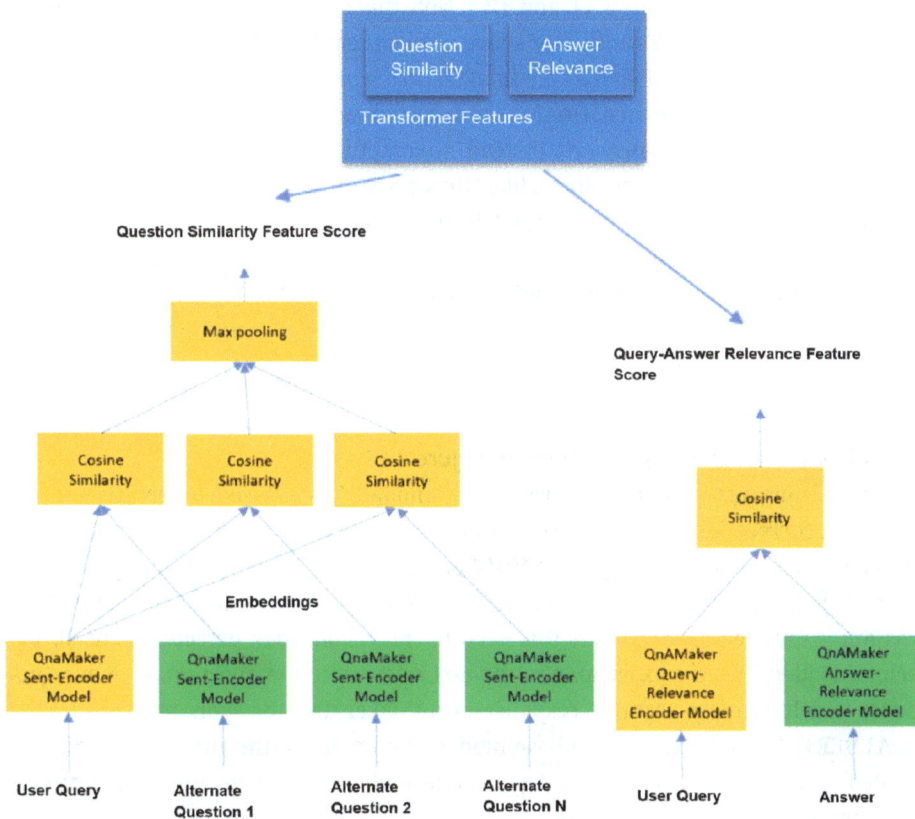

Figure 8.16: Transformer features. The user query is sent to 2 different modules: Query question similarity (left- hand side) and query-answer relevance (right-hand side).

results. The Transformer architecture, detailed in Section 8.3.3. of this chapter, is used to encode and represent questions, answers, and other relevant information. During the ranking phase, these encoded representations are used to compute **similarity scores** and measure the **relevance** of potential answers to a given query.

As stated above, the ranking layer uses transformers for two modules, namely **query-question similarity** and **query-answer relevance**. Whenever a bot developer creates a **knowledge base** in QnAMaker, all the QA pairs are encoded into **embeddings** (Section 8.3.4) and cached. Embeddings are vector-based representations of entities in a vector space where the distance between the vectors capture semantic relationships. For instance, the mathematical representations of "apple" and "orange" have a smaller distance between them than the distance between "apple" and "house."

- **Query-question similarity.** When a user initiates a query, the QnAMaker-Sent Encoder Model converts the query into an **embedding**. This embedding is then compared to precomputed and cached embeddings stored in the system, using a **distance metric**. More information on static embeddings and contextual embeddings can be found in Sections 8.3.1. and 8.3.4. Note that a single QA can contain alternate questions. Hence, the **similarity score** of the question with which the user query has the best match is taken as the final feature score value.

- **Query-answer relevance.** QnAMaker doesn't assess the similarity between a user query and potential answers; instead, it computes a **relevance score**. This approach recognizes that questions like "How are you," "How do you do," and "What's up" share similarity, yet "I am fine" is only a relevant answer to "How are you." To achieve query-answer matching, a separate transformer-based model, trained on question-answer relevance tasks, comes into play. This is achieved by training different encoders for queries (query-relevance encoder) and answers (answer-relevance encoder).

Logical Block 4: Answer span detection (Figure 8.17)

Sometimes, just picking a correct answer isn't enough. What the end-user might need is not only a correct, but also a concise and to the point answer. For instance, the answer "Apple was founded by Steve Jobs in 1976" potentially answers 3 different question intents: "Who founded Apple?", "What was founded by Steve Jobs in 1976?", and "When was Apple founded?". To address this need for precision, span detection is used. This technique identifies the exact span in the text relevant to the question, allowing QnAMaker to deliver more accurate responses. QnAMaker uses the span detection trained on **ALBERT**, A Lite BERT v2, as a base model. The model is **fine-tuned** using datasets created by inhouse crowdsourcing. An example applying span detection is illustrated in the figure below.

Finally, a few words about **active learning**. QnAMaker's ranking system attempts to reply to user questions with the most suitable answer, but sometimes there are several

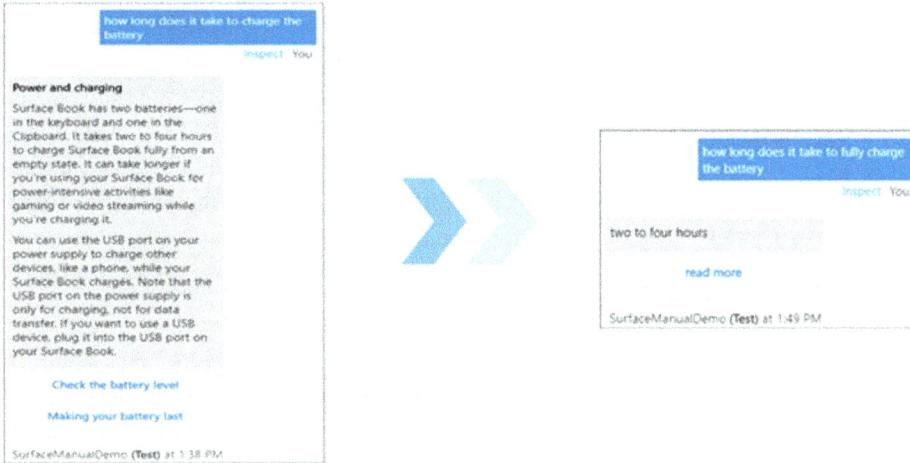

Figure 8.17: Before span detection (left) and after span detection (right).

candidates for the answer. To deal with such uncertainty, QnAMaker relies on an **ambiguity detection** mechanism to identify such user queries and generate suggestions for bot developers, giving them the capability to disambiguate between answers for the suggested user query by clicking on approve or reject (Figure 8.18). **Clustering** is used to group all similar suggestions into one representative suggestion (cluster head), which pops up as a recommendation. This is done to ensure that the suggestions displayed are diverse enough and similar suggestions are not shown. If the bot developer approves the recommendation, it gets added as an alternate question to the same answer in the knowledge base.

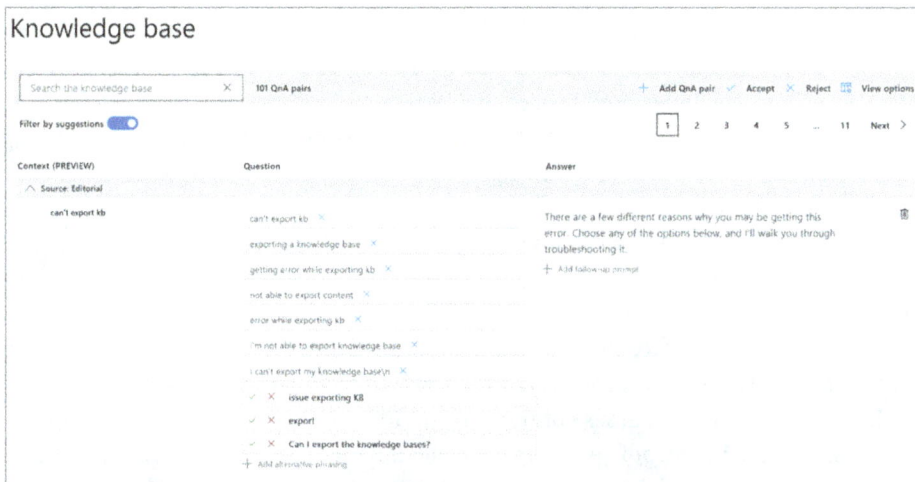

Figure 8.18: Disambiguation by the developer. Source: https://learn.microsoft.com/en-us/azure/ai-services/qnamaker/how-to/improve-knowledge-base (third picture).

In conclusion, with a **transformer** based ranking system, QnAMaker is able to further reduce the effort of content editors by understanding semantic matches to a great extent. The gains observed vary a lot based on the end-user content and requirements. While certain end users have experienced a substantial enhancement of over 40 % in query-answer quality, others less reliant on semantic understanding, perceived the quality to be comparable to non-deep learning models. Enhancements were brought since then and a newer version with question answering capability is available as part of **Azure AI language service**. QnAMaker, as a standalone service, will be retired in March 2025.

8.5.2 Extract information from contracts using NLP, the case of Daimler

Sunu Engineer, Anneleen Artois

Facing a multitude of more than 400,000 suppliers in its supply chain, Daimler, a producer of premium cars and manufacturer of commercial vehicles, used to manage its sourcing processes in old and fragmented systems across multiple platforms. These processes were supported by manual inspections of the contractual documents. This was a labor-intensive process as contracts not only come in different shapes and formats (e. g., paper, digital), but also have become longer over the last decade. Knowing that companies process an average of 10,000 to 15,000 contracts a year with an average size of 8 pages, it comes as no surprise that contractual knowledge often lies mired in pages and pages of complex legalese.

On-boarding new suppliers (i. e., issuing RFPs, selecting possible suppliers, tracking their performance as well as their contract terms) required Daimler's **procurement team** to go through multiple disparate systems. This naturally promoted errors, complicated thorough analysis, and limited the scope of contractual compliance controls, thereby raising legal and compliance challenges.

Therefore, Daimler decided to overhaul its **procurement process** by streamlining and automating its supplier contracting processes. Information had to be automatically extracted and classified from contracts. Leveraging **Icertis'** enterprise contract management (ECM) system, Daimler has now moved to a solution where all contracts, at all levels in the supply chain, are automatically subject to compliance analysis and checks with the required contractual conditions, irrespective of geography or language. As this solution reduces cycle-time, eliminates errors, and reduces risks by automating key aspects of the contract review process, it proves to be a highly profitable activity.

Icertis automatically parses contracts from different languages, sources, and formats (scanned document, pdf, jpg, word) into meaningful building blocks, identifies contractual clauses, classifies them, and then extracts relevant metadata and its meaning or semantics (e. g., identifying that "Mars" means Mars Inc., the global food manufacturing company, not the planet). It then makes the processed information available for structured, rich semantics-based searches and reports. Privacy of the information is strictly maintained as the NLP engines are designed to identify only the required clauses and process them accordingly.

Daimler's procurement teams are now assisted by a solution that semiautomatically (as contracts still must be validated by the legal teams) processes more than 25,000 contracts of 15–20 pages in average length per year, reducing the time spent on manual checks to the bare minimum. To address the need for trust between the different involved parties in the supply chain, the system also uses a highly secured **distributed ledger** based on **blockchain technologies** to ensure an immutable audit trail with strong role-based access control models.[12]

For the sake of this use case, a **contract** is defined as an agreement between several parties, recorded in a document. The language that is used is decided upon by Daimler's legal department. In this case, the legal department has a set of legal clauses, which are categorized in 72 different **clause types**. When a contract (NDA, MSA, etc.) is drawn up, it is then assembled out of these clause types. Inside the clauses, we find **parameters** or **attributes** such as the name of a person and the amount to be paid for the service, and these attributes are what determines the structure of the contract. In other words, the business aspect of the contract is encoded into these parameters. For example: Party A, Party B, Address of Party A, Address of Party B, Start Date, End Date, Duration, Execution Date, Place of Jurisdiction (optional), Signatory A (optional), Signatory B (optional), Penalty Terms (optional), etc.

The representational structure or form that exists in relation to a contract is provided by the **document**. The document is usually divided into **sections**, which consist of **paragraphs**, **sentences**, and **words**. The legal view of the contract, as discussed earlier, is hierarchically divided into sections, clauses, and attributes. The legal contract's hierarchy may be mapped to the document's hierarchy in a variety of ways (one paragraph per legal clause/multiple paragraphs per legal clause/multiple legal clauses per paragraph). Embedded in the paragraphs are "attributes," which are treated as **named entities**. Named entities of interest are structures, which identify the parties to the contract, the objects that constitute the state, and a variety of other objects that provide the context and environment in which the contract is executed. The relationships as described below are illustrated with excerpts from a sample Master Services Contract (MSC), which can be found in the Addendum of this chapter. The mapping can be visualized as follows:

Legal Contract
– Sections
— Clauses

— Attributes

Document
– Sections
— Paragraphs
— Sentences
— Words

---

12 However, this is not in scope of this article.

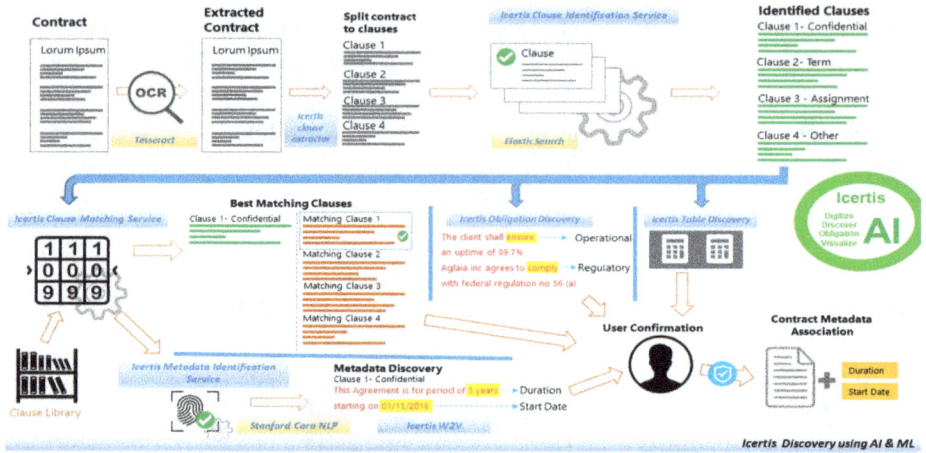

Figure 8.19: End-to-end pipeline. Logical flow of Icertis solution.

The logical flow as depicted in Figure 8.19 will be followed so the reader can witness the journey of a real document. The aim is to show the logical steps to introduce document processing and document modeling techniques. The main problem, as discussed above, is that Daimler deals with a large number of contracts, written in the past, present, and future tense, which need to be converted into a flow of information that is structured to improve business processes. The conversion of a document to structured data and actionable information goes through the following sequence of steps:

Logical Block 1: Scanning
Daimler has different types of source documents such as digital documents in a computer readable format and physical documents in paper form (mainly legacy contracts). The first step in the process is therefore to scan physical contracts and to transform them into high resolution images. Meanwhile, additional error correction processes and noise removal techniques (e. g., image rotation, shot noise removal, margin noise removal) are applied.

Logical Block 2: Document cracking
The second step is to convert these images into machine encoded text. This is called **document cracking**. Legal documents contain text, images, tables as well as other structures in various layouts and formats and are usually written in the legal language of a corresponding natural language such as English. Extracting information from these documents involves dealing with both form and content. As such, Icertis extracts both data and metadata from individual contracts and stores it into a database. To extract both printed and handwritten text from high resolution images, **optical character recogni-**

tion $(OCR)^{13}$ is used. The efficacy of the OCR and the correctness of the output depend on many factors, such as the deterioration of the physical document (through age or storage), the scanning quality (resolution and contrast), possible noise introduced while scanning, orientation errors, etc. Some of these errors can be reduced by using other components such as spell checkers and dictionary lookup engines. When all content is extracted, it is mapped into a JSON[14] data structure to facilitate storage and to pass it around the network by using application programming interface (API) calls.

Logical Block 3: Clause identification

Next, the machine encoded text, which emerges from the OCR pipeline, is passed to an NLP module which identifies the legal clauses embedded in the text. The contract is separated into sections and legal clauses. This step is called **clause identification**. Legal contracts traditionally are made out of "**legal clauses.**"[15] Clause delineation involves breaking up the document into different clauses. Different techniques such as whitespace-based delineation and section-title identification to discover the start and end of paragraphs and sentences are used. As discussed earlier, the clauses may be mapped to one or more paragraphs, or one paragraph can contain multiple legal clauses. The process of breaking up the contract into disjoint pieces involves finding the logical beginning and ending of each clause. Breaking paragraphs into sentences and subsequently employing a **rule-based engine** to reassemble them into clauses proves to be a beneficial technique during the clause delineation process. This is especially valuable when dealing with clauses that may span across multiple segments. On a technical note, the clauses are vectorized using a TF-IDF vectorizer and trained against an annotated dataset. Additional vectorizations such as GLOVE and BERT are used to check the results. Note that there is a recursive identification problem at hand: The clauses must be identified correctly to find the contract category or contract type belonging to the contract as a whole. Knowing the contract type allows one to disambiguate some clause categories, which may be mistakenly identified in return. Therefore, additional training data is sometimes added, or manual annotation and correction are needed.

---

13 Discovering tables and embedded images within documents requires special processing of the images of the pages to detect lines and regions, partitioning the space into smaller cells, and doing OCR on the contents of the cells to get the text within.

14 A lightweight, human-readable data interchange format that uses key-value pairs to represent data.

15 The document is broken up into clauses via a multistage algorithm. The initial document is broken into paragraphs based on the bounding box information and the paragraphs are classified as belonging to different legal clauses based on the keyword content. In addition, headings and section titles are distinguished and used to separate clauses. The paragraphs are recombined to form legal clauses depending on the classification and adjacency of the paragraphs.

Logical Block 4: Clause classification

To classify clauses, Daimler relies on the knowledge of legal coworkers which define a priori the types and names of the clause types. There is also a need for a domain expert to define **extraction masks** (e. g., a mask for the social security number). The actual clauses may be based on predefined structures called **templates**: structures with place-holders embedded into the text, which can be replaced by suitable values. Such a collection of templates is referred to as a **clause library**.[16] A contract may be templated as well, being made up of an ordered sequence of clauses from the clause library. There are several ways in which classification algorithms can be trained. There exist both supervised and unsupervised learning for data classification and both binary and multiclass classification algorithms. Given that legalese is a much more constrained environment (more constrained than natural language), a TF-IDF based vectorization algorithm trained on a large set of contracts (700,000 approximately) has led to a **clause identifier** routine, which is capable of classifying the clauses and deriving the contract type. The accuracy of the trained system, a linear SVM (see Chapter 7) is above 85 % across the different contract types seen in business contexts today.

Example. This clause is identified as a Preamble.

IMPORTANT-READ CAREFULLY: THIS MASTER SERVICES AGREEMENT (THIS "MSA") IS BINDING AND ENFORCEABLE
BETWEEN YOU ("CUSTOMER") AND XXXXXXXXXX., A DELAWARE CORPORATION HAVING ITS PRINCIPAL PLACE OF
BUSINESS AT 2000 NORTH AMERICAN STREET, SUITE 2, MINNEAPOLIS, MINNESOTA, FOR ITSELF AND ITS AFFILIATES
("XXXXXXXXX"). "YOU" REFERS TO THE ENTITY OR ORGANIZATION USING THE PLATFORM, PRODUCTS, AND/OR SERVICES
DESCRIBED IN THIS MSA. BY SIGNING AN ORDER FORM TO USE THE PLATFORM AND/OR PRODUCTS AND/OR TO RECEIVE
SERVICES, YOU ARE ACCEPTING AND AGREEING TO BE BOUND BY THIS MSA. YOU SHALL INFORM ALL USERS OF THE
PRODUCTS OF THE TERMS AND CONDITIONS OF THIS MSA.

Logical Block 5: Entity extraction

Once the clauses are identified, the required attributes are extracted. This involves parsing the clauses into sentences, applying **named entity recognition (NER)** algorithms, extracting the attributes, and classifying them using various models and rule engines in terms of the required known categories. The NER system operates on the obtained clauses to extract metadata with their semantics, which the Contract Management Platform refers to as "attributes." This system is based on a conventional codebase using a

---

16 Note: A given clause type can have multiple template forms such as multiple ways of formulating a "termination clause."

linear **conditional random field (CRF)**, (Chapter 6) algorithm, which tags the attributes as Organization, Place, Date, Number, etc. The attributes that are tagged by the NER system are presented to the user if requested for manual validation and learning in the pipeline. A separate NER system utilizing a bidirectional **LSTM**[17] is used to accept the user feedback and retrain the classifier. This framework is an option to make the system more accurate and adaptive. For example, in the above text, Minneapolis and Minnesota are labeled as "location."

Logical Block 6: Common model for contracts
The contract with the identified type, the clauses and clause categories, the attributes and their values, and the sentences (along with the layout information when relevant) are sent to a **rule-based engine**, which applies a set of predefined rules. For example: The Effective Date should precede the Expiry Date, the difference between Expiry Date and Effective Date must match the Duration. The "understanding" of the contract that emerges from this step allows the system to create **a common data model** (see Figure 8.20) encompassing multiple categories of contracts and connecting them together.

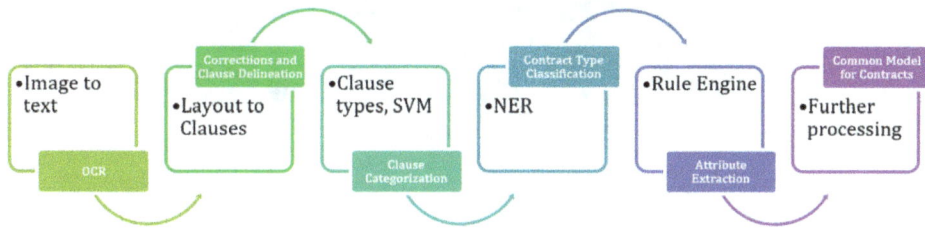

Figure 8.20: Steps to create a common data model.

Logical Block 7: Obligations and entitlements in code
In parallel, the clauses are analyzed for **obligations** and **entitlements** such as deliveries and payments, which are embedded within. Obligations and entitlements are logical actions to be performed as part of the transactions identified in the contract. This logical sequence of actions and associated conditions and constraints form the active core of the contract. By parsing the clauses identified as potentially containing obligations and entitlements, one can identify the implied modalities in the clauses. *One must, One should, One has to, One may*, etc. are mapped to a second-order deontic logic model, which allows these obligations to be modeled in code. This, in conjunction with the entities and other attributes identified in the previous stage, allows the contract to be mod-

---

17 One can describe this in detail. An example reference: https://towardsdatascience.com/named-entity-recognition-ner-meeting-industrys-requirement-by-applying-state-of-the-art-deep-698d2b3b4ede

eled as a dynamic object. This is used in the smart autonomous contract part of Icertis' platform.

Logical Block 8: What If simulations
Once the above elements are extracted from the contract document, the contract is "understood". At this point, both the static structure, representing the informational content, and the dynamic structure, depicting the flow of activities associated with the contract, are extracted and encoded into a structured schema within a database. This schema follows a standardized format across all contracts and an object oriented or procedural code formulation whose execution simulates that of the contract. These computable representations allow the contracts to be simulated, *What If* scenarios to be examined, risks to be computed, and more (see Figure 8.21). From Daimler's point of view, these techniques allow them to understand their contracts more in depth.

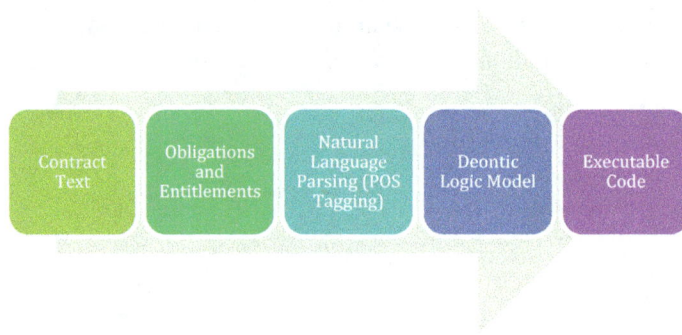

Figure 8.21: Contract flow.

In addition to getting a better understanding of the contracts, it is also necessary to ensure compliance (see Figure 8.22) to a class of certifications and sustainability criteria such as ethical practices, proper hazardous waste management, environmental sensitive operations, and so on during the creation of the contracts. To comply, Daimler turned once more to the use of the AI engine in conjunction with a consortium **blockchain** component that is built into the contract management platform. The consortium of suppliers who participate in the supplier blockchain is able to submit complete contracts or parts of their contracts into the blockchain and to have it verified automatically by an NLP program that matches the clauses in terms of structure and intent.

The engine is capable of **translating** clauses from other languages, comparing them with the required clause set via distance metrics such as *fuzzy matching* or *earth mover's distance* and of producing a report regarding the presence or absence of required clauses and estimated degree of conformity with the required clause text. This is very useful when the contract is framed in a foreign language, and we have to grade a contract in terms of its **degree of compliance** with the sustainability clauses.

The resulting information is modeled and put into a connected graph, which models the physical supply chain. A powerful visualization immediately indicates to Daimler the degree of compliance of the complete supply chain network to their sustainability clauses. Contracts which are red (meaning that they lack the required clauses), yellow (meaning that most of the clauses are present or that all are present, but modifications are required to become compliant), and green (fully compliant) are shown in the connected graphical network model (Figure 8.22). The reporting structure underlying it allows for messaging to the contract parties and subsequent corrections of the contracts to make them fully compliant before the supply chains are kicked off.

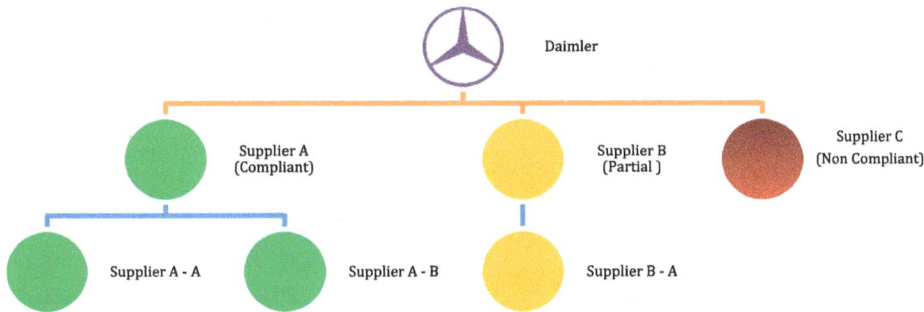

Figure 8.22: Compliance check.

Conclusion

The use of an intelligent contract repository analysis system allows for the extraction of useful knowledge. A variety of statistical measures permit Daimler to immediately see the different types of contracts and the connections between them. Analyzing specific contract types such as master service agreements, scope of works, human resources contracts, etc. in bulk also allows for the construction of large-scale business flow models. Post analysis, the static and dynamic structure of the contracts that emerge can be visualized in many ways. These make the contracts easy to decipher and understand. Moreover, it helps Daimler to detect errors, anomalies, and inconsistencies efficiently. On top of that, the large-scale statistical models of contract pricing, penalty clauses, force majeure clauses and alike allow for several risks to be modeled and estimated. These analyses also aid in the contract creation process by identifying the commonly negotiated clauses and acceptable parameter ranges. The analyses of obligations and entitlements allow for details related to payments, eligible discounts, and penalties to be detected automatically and operated upon without failure, thus minimizing losses and recovering money "left on the table." All the above strategies pave the way toward a high level of automation leading to "autonomous contracting" as the foundation of global commerce.

8.5.3 Analyze legal content for lawyers at AXA's legal protection unit

Pierre-Yves Thomas, Anneleen Artois

Legal Village NV., AXA's online legal assistance platform, is a claims settlement office specialized in handling legal aid files and in offering innovative legal services. **AXA Belgium** entrusts Legal Village NV. with the management of claims for all contracts in its insurance portfolio related to the legal assistance branch.

One of the most important tasks in the day-to-day work of lawyers is the retrieval of legal content from the internet and internal knowledge bases to identify and develop legal arguments for ongoing law cases. Legal data comes in different shapes and formats and is to be consulted in a wide spectrum of data sources. As a result, one of the biggest challenges for lawyers is to efficiently find the right content at the right time, as different sources typically have different, non-consistent descriptions of meta-metadata and no indicator for the relevance and completeness of documents. Crucial information might not be found.

At the same time, law and jurisprudence are a living matter, as they naturally result from people living together and interacting with each other. As a result, legal data is published at a fast-paced rhythm and knowledge mining has become indispensable in the field of case law.

To address the major challenges in the area of knowledge mining, Legal Village uses a **solution called LexagorIA**, developed by **PythagorIA**, a company based in Luxembourg, that offers professional solutions for knowledge management and text mining. LexagorIA not only retrieves and presents relevant content (concepts, legal references, named entities, etc.) from internal and external sources in a language independent manner, but also creates and extracts a homogeneous set of metadata to describe the meaning of the content and to build logical bridges between the different documents. This ensures that one can, among others, track the evolution of law.

The solution also anonymizes documents, and thus preserves the integrity of the people named in the legal documents by detecting and classifying named entities and their relations (persons, organizations, lawyers, judges, dates, etc.). This is especially important to protect confidentiality.

Note that the language model at the basis of LexagorIA's **suggestion engine**[18] has been trained on data previously published by LexagorIA. The main languages that were used are English, French, Dutch, and German. Approximately 4,500,000 legal documents classified as law and 3,200,000 documents classified as case law covering the territories

---

[18] The suggestions are filtered based on the user's role. The main idea is that one will never get a suggestion that doesn't point to a result. Hence, the suggestions are not limited to a « begin with … » suggestion. For example, as one types "concur," it can result into "clause de nonconcurrence," https://ksg.lexagoria.ai/#/search-results?u=law&search= as a suggestion.

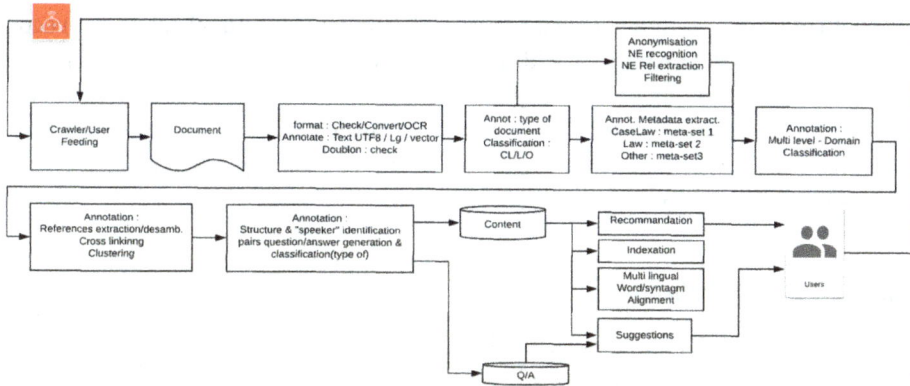

Figure 8.23: End-to-end pipeline of how LexagorIA processes the documents for lawyers.

of Belgium, France, the Netherlands, Luxembourg, Switzerland, the USA, and some European institutions (Court of Justice of the European Union, European Court of Human Rights and Official Journal of the European Union) have been used to this end.

Below we will take you step-by-step through the end-to-end pipeline of LexagorIA depicted on Figure 8.23.

Logical Block 1: Crawl and avoid duplicates

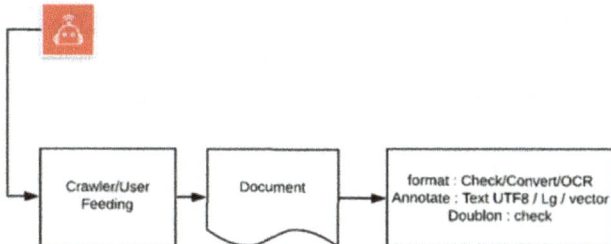

It's Monday evening. A divorce attorney finalizes writing a legal document and uploads it on the company's file server. LexagorIA, which **crawls** the internal storage systems (e.g., Google Suite, Microsoft SharePoint) on a daily basis, picks up the document and processes it so its content becomes searchable and gets linked to other legal data. Once the document is picked up by the crawler, LexagorIA verifies whether it has been processed before. Since the same document can be stored in different **formats** (e. g., pdf and word), the vectorized version of documents are compared to identify **doubles**. Vectorization is achieved based on the extraction of words and legal phrases stored in a dictionary. Each extracted item gets a vector value and an ID, which are stored in a database. Note that the same document can be stored multiple times in a different **language**. When this happens, both versions are retained, but some clustering is done later in the schema.

When the input document is a scan or a handwritten file,[19] **optical character recognition (OCR)** is used. This adds an additional complexity as some broader problems can be distinguished. As explained by Kukich (1992) we need to deal with complexities such as *non-word error detection* (e. g., recognizing lawsut for lawsuit), *isolated-word error correction* (e. g., correcting lawsut into lawsuit) and *context-dependent error detection and correction* (e. g., I suit you for I sue you).

Logical Block 2: Classification

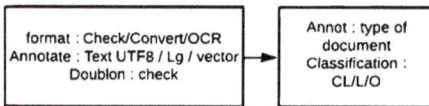

Once we know for sure that the crawled document is a unique entry in LexagorIA, we determine what type of legal document it is. Whereas all public content gets classified as **case law (CL), law (L), or other (O)**, internal client systems use more detailed classification patterns based on their professional needs. A One-vs.-All SVM[20] model has been trained on millions of legal documents to classify every new entry. Initially, there were around 6 million documents for the first step classification and 600,000 to build the learning model related to the domain classification. Hence, a multiclass problem has been solved by mapping it out onto several supervised biclass models.

Part of speech (PoS) Tagging has been used in one of the earlier versions of LexagorIA as well because in legal documents the tense of a verb (future, past, conditional, etc.) is indicative of the type of document. For example, documents labeled as law (L) are only written in the present or future tense. Different tag sets are used depending on the language and the corpus available to train with. Since PoS tags are also applied to punctuation (e. g., exclamation mark, parenthesis, etc.), it is important to perform **tokenization** beforehand.

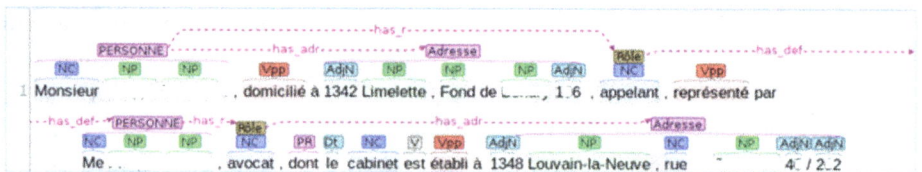

Figure 8.24: PoS tags, dependence analysis, and NER in IOB format.

---

19 LexagorIA is very careful with pdf documents containing a scan or picture of a physical copy of the document. Sometimes the pictures have been made with a smartphone and are of very poor quality. Therefore, the pdf version with the image is always published for the users to have a look at.

20 N different binary classifies are trained. For each classifier, one class is treated as positives and all other classes are treated as negative. For SVM see Chapter 7.

Logical Block 3a: Anonymization, entity, and relation extraction

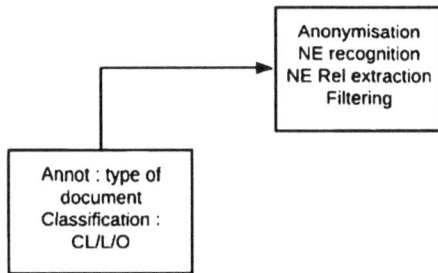

Anonymisation
NE recognition
NE Rel extraction
Filtering

Annot : type of
document
Classification :
CL/L/O

The next step consists of hashing entities to **anonymize** and filter out predefined entities. To this end, the extracted content of the document is used and not the vectors. LexagorIA's anonymization rules depend on the type of document and the role of the person who requests access. For example, if the text is classified as case law (CL) and the person who wishes to consult it is external to the company, the (full) name of the lawyers and judges as well as the location of the court can be kept AS IS, while all other names and locations have to be masked and replaced by a Dummy Token (e. g., Person1, Organization1, Address1, etc.).

First, a model for **entity extraction** must be built to find patterns in data. A very widespread and easy to use technique is **REGEX** (regular expressions). While REGEX is very useful when the underlying structure of the entity we want to extract (e. g., ID, e-mail, telephone number, etc.) is known, this technique won't help to detect names, job titles, and locations.[21] What is needed is a **probabilistic sequence classifier** which, "given a sequence of units (words, letters, morphemes, sentences, whatever), (...) compute[s] a probability distribution over possible labels and choose[s] the best label sequence" (Jurafsky and Martin, 2009).

A well-known example of such a sequence classifier is a (fully-connected) **hidden Markov model** (HMM, Chapter 6), which computes probabilities about both the observed events (the words, entities we try to classify) and hidden events (PoS tags). Apart from probabilistic sequence classifiers, we also have exponential sequence classifiers such as the Maximum Entropy Markov Model (MeMM) and the **conditional random field** (CRF) model. The latter model is chosen to classify entities in LexagorIA as it proves to be a good method for entity recognition.

Once the required entities are extracted, their **relationships** must be found (see Figure 8.25). This process is backed up by a cascade of tools and processes. The focus lies on two types of relationships. First, we want to know which person/location has what title (e. g., defender). Second, it is important that Ann Art, Ann, and Miss Ann are all recognized as the same entity. All the different versions of this name must be clustered as the same entity when anonymizing the document. To this end, a **similar strings identification** (SSI) is used based on the *n*-gram extraction of strings with the Jaccard algorithm as a similarity metric.

---

21 As goes for locations, there are lots of open data resources that contain all the addresses and cities in Belgium and Luxembourg.

The goal is to end up with a graph, which explains the **relationships** between all the named entities, for example, Jan is a defender of CompanyXYZ located in Brussels, Belgium.

Jugement Civil (IIe chambre)
(Jugement sur requête)
No 25/2016

Audience publique du vendredi, vingt-neuf janvier deux mille dix

Numéro du rôle : 999.999

Composition :

PERSON_1 juge présidente.
PERSON_2 juge,
PERSON_3 juge,

PERSON_4 greffier.

ENTRE :

PERSON_5 demeurant à **ADDRESS_1**,

demanderesse aux termes d'une requête en relevé de déchéance déposée le 15 décembre 2009,

comparant pas Maitre **PERSON_6** avocat à la cour, demeurant à **LOCATION_1**.

Figure 8.25: Example of a legal text with entity and relation extraction.

Logical Block 3b: Extraction of additional meta-data

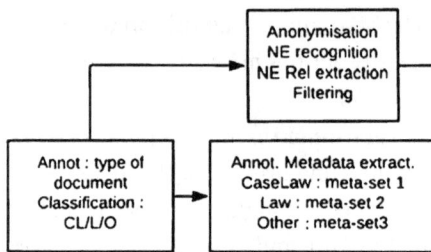

At the same time, some metadata related to the content is extracted. The set of metadata that is extracted depends on the kind of document that is processed. For example, if the document has been classified as case law (CL), we look for the date of delivery, the legal concept (e. g., labor court, cassation court, constitutional court), the name of the parties involved, the author, the title, and the publisher. Figure 8.26 demonstrates the full metadataset that can be extracted. Some metadata comes from the NER (e. g., the legal concept is related to the analysis of PoS tagging and the syntactic analysis of the content), other metadata comes from a different set of algorithms.[22]

---

22 Various algorithms are employed for specific metadata tags. For jurisdiction, resort, and dates, combinations of TensorFlow with BERT, alongside post-tagging or dependency parsing with rules, are utilized.

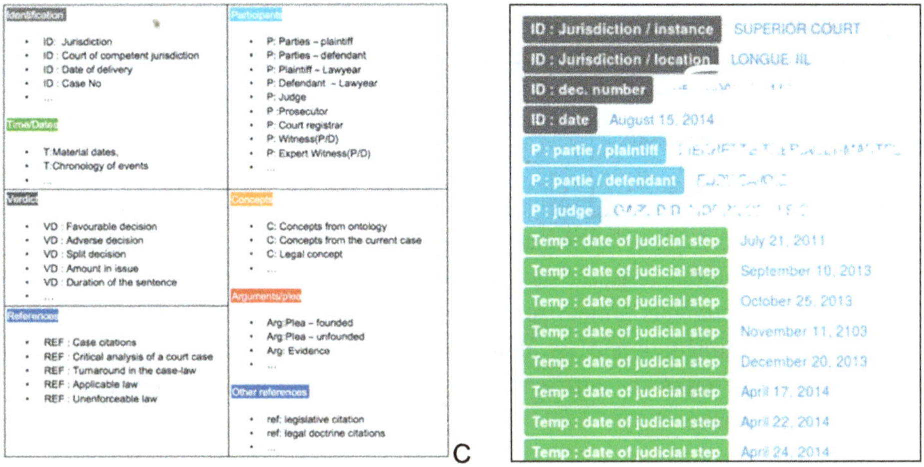

Identification	Participants
• ID : Jurisdiction • ID : Court of competent jurisdiction • ID : Date of delivery • ID : Case No • ...	• P: Parties – plaintiff • P: Parties – defendant • P: Plaintiff – Lawyear • P: Defendant – Lawyear • P: Judge • P :Prosecutor • P: Court registrar • P: Witness(P/D) • P: Expert Witness(P/D) • ...

Time/Dates	
• T:Material dates, • T:Chronology of events •	

Verdict	Concepts
• VD : Favourable decision • VD : Adverse decision • VD : Split decision • VD : Amount in issue • VD : Duration of the sentence • ...	• C: Concepts from ontology • C: Concepts from the current case • C: Legal concept • ...

References	Arguments/Plea
• REF : Case citations • REF : Critical analysis of a court case • REF : Turnaround in the case-law • REF : Applicable law • REF : Unenforceable law	• Arg:Plea – founded • Arg:Plea – unfounded • Arg: Evidence • ...

Other references
• ref: legislative citation • ref: legal doctrine citations • ...

Figure 8.26: The full metadataset that LexagorIA tries to capture.

Additionally, anonymization also applies to the schemas and drawings. When OCR is used, it is important to remove signatures, stamps, etc., but to retain pictures or drawings which, for example, describe the position of cars at the time an accident happened.

Logical Block 4: Multilevel domain classification

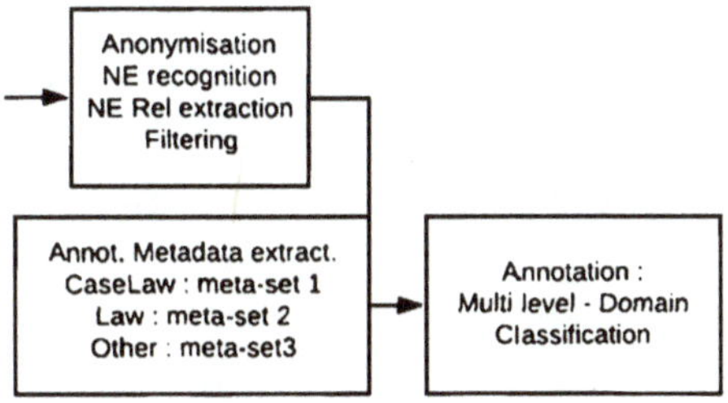

To label[23] document entries, **multilevel domain classification** is applied. LexagorIA has two-level classifications. A document can be classified as social right (level 1) and belong to the subclass of international social right (level 2). While for the first level, there is a single classifier, for the second level there is one classifier per class (for about 800 classes). Hence, one document can be classified into several subclasses of a first level class. The main technique used to this end is (semi) supervised text categorization using **long short-term memory** (LSTM) for region **embeddings** (see Section 8.3.2. of this chapter). Note that multilevel classification is independent of the language used in the search query.

Legal and domain metadata, on the other hand, utilize SVM or CNN in conjunction with regions and word embedding.

23 This is based on the UTU classification, a universal tree structure that allows for the organization and incorporation of the different sources of law (https://bartoc.org/en/node/369).

Logical Block 5: The legal references

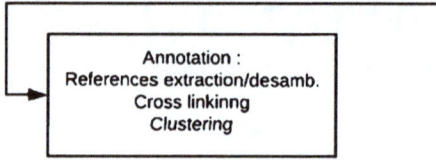

```
┌─────────────────────────────┐
│        Annotation :         │
│ References extraction/desamb.│
│       Cross linkinng        │
│         Clustering          │
└─────────────────────────────┘
```

The next step after classifying the texts by legal domain and subdomain is **to identify and disambiguate** the **legal references** in documents (see Figure 8.27, orange text). Occasionally, the reference to a particular law is very detailed, such as in the example below, but often there is only a single piece of information available (e. g., the date) about the referred law. In this case, LexagorIA depends on the documents context, for example, the existence of an extracted legal concept, to do a disambiguation. After identifying the law that the document refers to, a hyperlink is added to the law with a citation of all the other case laws that are found in the original document.

II. Procédure

Pour une ordonnance de 4 juillet 2018, l'affaire a été fixée à l'audience du 26 juillet 2018 à 10 heures.

La note d'observation et le dossier administratif ont été déposés.

La contribution et les droits visés respectivement aux articles **66** 6° et l'*article 70* de l'arrêté du Régent du 23 août 1948 déterminant la procédure devant la section du contentieux administratif du Conseil d'Etat ont été acquittés.

(…)

Il est fait application des dispositions relatives à l'emploi des langues, inscrites au titre VI, chapitre II, des *lois sur le Conseil d'Etat, coordonnées le 12 janvier 1973*.

III. Faits

1. Le 16 novembre 2017, le conseil d'administration de la partie adverse décida de retirer sa délibération du 28 septembre 2017 visant à attribuer les différents lots du marché « Traitement des déchets verts » précédent, de renoncer à l'attribution des 23 lots de ce marché et d'approuver le lancement d'une nouvelle procédure dont les conditions seront définies par le directeur général.

(…)

Figure 8.27: Legal references in a legal text (French).

As mentioned before, annotation is independent of the language used in the search query. If there are doubles in a different language, all versions are retained and shared metadata at analysis time is added to all the related documents in order to make a cluster in the system and to collapse all versions into one label. Hence, when searching for *"contrat de travail,"* LexagorIA automatically translates the search term from French to Dutch and English so all the documents in the system related to work law can be found, independently of the language in which they are published. This is called **cross-language search.**[24] Note that virtual clusters are also used to relate historical versions

---

24 Another reason to cluster the documents is because one gets different results for "contrat de travail" (French) as for "werkcontract" (Dutch) because the French term is composed of three words versus one.

of a document. When a law modifies, adds, or repeals an article of another law, a historical version of the modified law is created. LexagorIA manages these different versions and explains the differences between the old and new documents (Figure 8.28).

Figure 8.28: The historical and new version of a document.

In the previous steps, LexagorIA has annotated each document with generated metadata (metaset, extraction, of legal concepts, NER, etc.) and has linked semantically related documents in a semantic cluster. This results in a **semantic map** of every document. Imagine once again a divorce attorney trying to find information about the case of a colleague. The attorney searches for the name of colleague and the date of the court order, finds a hyperlink to the correct document, and clicks on it. At this moment, LexagorIA forgets about the search query itself and only semantically related content, based on the semantic map of this document, is suggested to the end user.

Logical Block 6: Question and Answer (QA) pairs

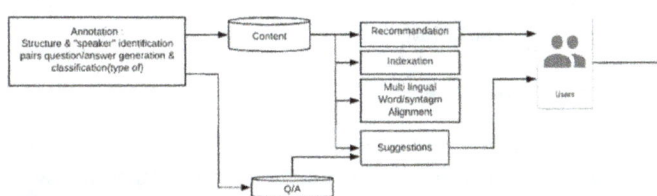

For this final feature, it is crucial to perform speaker identification. When searching for answers through LexagorIA, it is crucial that the response originates from a judge rather than a party's lawyer or an individual lacking legal authority. Without input from a legally authoritative source, the answer lacks any legal standing and remains a mere opinion. Currently, there are two approaches to manage speaker identification and legal authority. (1) The legacy approach for which the model must be adapted every time a new language is added. (2) The usage of a multilingual BERT model, which allows the coverage of different languages with the same learning data. The latter is easier to adopt new documents and to add additional languages as there is only one sample of learning data for all languages needed.

Imagine once again a divorce attorney with access to an interface to ask questions. Upon typing, a scroll-down list appears with generic questions to choose from. When clicking on a question, an answer immediately appears because the questions from the scroll-down list are stored as QA pairs in the **Q&A database**. If the attorney doesn't click on a suggested question, LexagorIA will try to find the answer in the content database, containing the whole corpus from the website. In this case a different QA process is used whereby some abstract candidates in full text that contain the answer are selected. Note that LexagorIA makes a clear distinction between **suggestions** and **recommendations**. Whereas the former are created at search time and are user agnostic, the latter are related to a user profile and match the cases the user works on with a new document in the system.

8.6 Addendum

Addendum 1.

PREAMBLE

THIS MASTER SERVICES AGREEMENT (THIS "MSA") IS BINDING AND
ENFORCEABLE
BETWEEN YOU ("CUSTOMER") AND ABCDE, INC., A DELAWARE CORPORATION
HAVING ITS PRINCIPAL PLACE OF
BUSINESS AT 200 NORTH WIND STREET, SUITE 2, INDIANAPOLIS, INDIANA 46207,
FOR ITSELF AND ITS AFFILIATES
("ABCDE"). "YOU" REFERS TO THE ENTITY OR ORGANIZATION USING THE
PLATFORM, PRODUCTS, AND/OR SERVICES
DESCRIBED IN THIS MSA. BY SIGNING AN ORDER FORM TO USE THE PLATFORM
AND/OR PRODUCTS AND/OR TO RECEIVE
SERVICES, YOU ARE ACCEPTING AND AGREEING TO BE BOUND BY THIS MSA.
YOU SHALL INFORM ALL USERS OF THE
PRODUCTS OF THE TERMS AND CONDITIONS OF THIS MSA.

This MSA includes the General Terms and Conditions set forth on the following pages and all terms and conditions set forth in all Product Addenda specific to the Products purchased as part of Your subscription. Product Addenda, as well as the most current version of this MSA, are available for review at www.abcde.com/legal/. The parties' complete agreement with respect to the subject matter set forth in the Order Forms executed by the parties during the Term includes this MSA (including all applicable Product Addenda) and all such Order Forms, all of which shall be hereinafter referenced as the "Agreement." You expressly agree that the terms and conditions of this MSA shall govern all Products and Services provided to You during the Term and are a material part of ABCDE's agreement to provide such Products and Services, whether or not the same is made express at the time of provision.

ABCDE hereby agrees to make the Products and/or Services described in each Addendum hereto available to You, and You agree to purchase such Products and/or Services from ABCDE, subject to the terms and conditions of the Agreement.

This MSA was last updated on March 4, 2013. It is effective between You and ABCDE as of the date of You accepting this MSA.

GENERAL TERMS AND CONDITIONS
1. DEFINITIONS
"Affiliate" shall mean, with respect to a party, any entity that directly or indirectly controls, is controlled by, or is under common control with such party, where "control" (or variants of it) shall mean the ability (whether directly or indirectly) to direct the affairs of another by means of ownership, contract, or otherwise.
"Applicable Law" shall mean any international, federal, state, or local statute, regulation, or ordinance, ex-pressly, including without limitation those relating to individual privacy or the distribution of email and other one-to-one digital messages.
"Confidential Information" shall have the meaning set forth in Section 6.
"Data" shall mean all data and other information uploaded by Customer to the Platform or to a Product.
"Malicious Code" shall mean viruses, worms, time bombs, Trojan horses, and other harmful or destructive code, files, scripts, agents, or programs.
"Order Form" shall mean the ordering documents for Customer's purchases of Products or Services from ABCD that are executed by the parties from time to time, which shall be governed by the terms of this MSA.

Bibliography

Bird S., Klein E., and Loper E. Natural language processing with Python: analyzing text with the natural language toolkit. O'Reilly Media, Inc., 2009.
Brown Tom B., Mann Benjamin, Ryder Nick, Subbiah Melanie, Kaplan Jared, Dhariwal Prafulla, Neelakantan Arvind, Shyam Pranav, Sastry Girish, Askell Amanda, Agarwal Sandhini, Herbert-Voss Ariel, Krueger Gretchen, Henighan Tom, Child Rewon, Ramesh Aditya, Ziegler Daniel M., Wu Jeffrey, Winter Clemens, Hesse Christopher, Chen Mark, Sigler Eric, Litwin Mateusz, Gray Scott, Chess Benjamin, Clark Jack, Berner Christopher, McCandlish Sam, Radford Alec, Sutskever Ilya, and Amodei Dario. Language Models are Few-Shot Learners, 2020, arXiv:2005.14165, [cs.CL].
Devlin Jacob, Chang Ming-Wei, Lee Kenton, and Toutanova Kristina. Bert: Pre-training of deep bidirectional transformers for language understanding, arXiv preprint arXiv:1810.04805, 2018.
Hochreiter S. and Schmidhuber J. Long short-term memory. Neural computation, 9(8):1735–1780, 1997.
Howard Jeremy and Ruder Sebastian. Universal Language Model Fine-tuning for Text Classification. 2018, arXiv:1801.06146.
Jurafsky Daniel and Martin James H. Speech and language processing: an introduction to natural language processing, computational linguistics, and speech recognition. Upper Saddle River, N.J.: Pearson Prentice Hall, 2009.
Kukich K. Techniques for automatically correcting words in text. ACM Computing Surveys, 24(4):377–439, 1992.
Locke W. N. and Booth D. A., eds. Machine Translation of Languages, pages 15–23. Cambridge, Massachusetts: MIT Press, 1955.
McTear M. Conversational AI: Dialogue Systems, Conversational Agents, and Chatbots. Synthesis Lectures on Human Language Technologies, 13(3):1–251, 2020.

Mikolov Tomas, Chen Kai, Corrado Greg, and Dean Jeffrey. Efficient Estimation of Word Representations in Vector Space, 2013, arXiv:1301.3781 [cs.CL].

Peters, Matthew E. et al. Deep contextualized word representations. 2018, arXiv:1802.05365.

Ramesh Aditya, Pavlov Mikhail, Goh Gabriel, Gray Scott, Voss Chelsea, Radford Alec, Chen Mark, and Sutskever Ilya. Zero-shot text-to-image generation. In International conference on machine learning, 8821–8831, 2021.

Stanovsky Gabriel, Smith Noah A., and Zettlemoyer Luke. Evaluating Gender Bias in Machine Translation. In Korhonen Anna, Traum David, and Màrquez Lluís, Proceedings of the 57th Annual Meeting of the Association for Computational Linguistics, Florence, Italy, pages 1679–1684. Association for Computational Linguistics, https://aclanthology.org/P19-1164 https://doi.org/10.18653/v1/P19-1164, jul, 2019.

Turing Alan Mathison. Mind, Mind, 59(236):433–460, 1950.

Vaswani A., Shazeer N., Parmar N., Uszkoreit J., Jones L., Gomez A. N., ... and Polosukhin I. Attention is all you need. In Advances in neural information processing systems, pages 5998–6008, 2017.

Erik Mannens

9 Some words about ethics. The angles of fairness and transparency

9.1 The challenges of ethical AI

While considering the fundamental problem-solving capabilities of any artificial intelligence (AI) technique, practitioners should foremost and upfront consider the ethical implications of any machine learning (ML) project they envision. As such, *ethical AI* is a nuanced, complex, and still emerging discipline with many angles to look at, which means there are few concrete guidelines already to follow today. But the definition of *ethical AI* definitely encompasses: (1) the safeguarding of a person's individual and fundamental rights; (2) respecting a person's end-to-end privacy; (3) being nondiscriminatory in the broadest sense; and (4) being nonmanipulative in any way.

Concretely, this first means AI systems need to be trained with the *right data* to take well-informed ethical decisions. An early example from 2016 being a bot trained with obviously biased and discriminatory Twitter feeds, hence the bot of course also answering normal questions with hostile and racist remarks, being a classic example of an AI system fed with the wrong training dataset, that is, the engineers not checking enough if the data set was balanced and without any ethical issues, for example, unbiased and inclusive towards gender, race, age, religion, political orientation, etc. Only if one is certain that the right data is used to train an AI system, one can truly say that AI system will take *fair* decisions.

Second, any AI solution in production needs to be *trustworthy*, that is, an AI system cannot start to degrade in quality or accuracy without flagging an alarm. This then would result in making false claims, and thus again possibly treating people differently over time. An AI model should therefore be as *robust* as an aircraft, with the same meticulous continuous monitoring and maintenance, for all people to always trust using it.

Third, any AI system should be *transparent*. We should strive to use AI systems that can always tell us why a certain decision was taken, that is, via the use of interpretable *AI* and more specifically *white box* algorithms. As such, by using interpretable AI, one enables the trust and transparency needed to also make the claim an AI system is *accountable* for its decisions.

Lastly, AI systems (as any other software system connected to the World Wide Web) should be end-to-end *secure*, so that malicious people or bots cannot exploit the AI system by changing the model or knowing how to trick the model from within and should guarantee the end-to-end *privacy* of the individual also when multiple data sets (that by merging them might tamper the individual privacy) are taken into account.

As such, there's globally a consensus the pillars of *ethical AI* can be summarized as follows:

https://doi.org/10.1515/9783111426143-009

- **Fairness**: AI systems should treat all people *fairly* and *equally* aiming to empower everyone *inclusively*. Therefore, multiple possible biases are to be considered. The translation system as discussed in Chapter 8 can be taken as an example, that is, when making a general translation system from Dutch to English, it is important that the vocabulary, speaking style, etc. of different Dutch geographical regions as well as different age groups are included.
- **Trustworthiness**: AI systems should be *reliable, robust*, and *safe* under any dynamically changing circumstances, that is, if the world is changing the AI model needs to dynamically adapt to represent the world at hand, relearn and retake the right decision. Only then does one deal with AI systems one can fully *trust*.
- **Transparency**: AI systems should be *understandable* and be able to *explain* why a certain decision was made by the algorithm. As such, completely fair and transparent AI systems are *accountable*.
- **Security**: AI systems should have built in *privacy-by-design*. Data driven solutions should always treat data in a *secure* manner with, among others, encryption at rest and in transit, and respect the *individual's data privacy*. Take the example of LexagorIA in Chapter 8 where, depending on the role of the data requester, some entities (such as name and address) are hashed as not everyone has the same level of data access.

Within this chapter, we will primarily focus on *Fairness* (more specifically bias) and *Transparency* (more specifically *interpretability*). The theory section will focus on the five different kinds of *biases* out there. Some of the biases are embedded in the data and some in the AI models, and we will describe techniques (preprocessing, in-processing, and post-processing) to handle all kinds of biases. Furthermore, interpretability can be described explicitly via two different umbrella techniques, that is, explainability by intrinsic global design and by post-hoc local interpretation.

Following the theory section, two concrete case studies are chosen to exemplify the use of the techniques and tooling described in the theory section. The first case study is from EY that built an AI solution for financial institutions to embed *Fairness* in their loan division. Here, the focus lies in the use of the **Fairlearn** tooling with embedded fairness techniques.

The second case study addresses *Interpretable AI*, with a model that detects fraudulent behaviors including the description of a feedback loop architecture that guarantees trust and transparency for the decision support system. To make the system accountable, the exemplified techniques from **interpretML** tooling were used.

9.2 Initial framing of fairness

To further fully explain and understand *algorithmic bias*, an initial question needs to be addressed: "What does *fairness* mean?" And this is already a difficult one, as there are tons of definitions for fairness. So, how to pick the right one? (spoiler alert: the "right" one cannot just be picked, as everything is context dependent as will be shown later on).

Let's start with the not-so-mathematical basic idea. A common paradigm for thinking about *fairness* in universal law is *disparate treatment* and *disparate impact*. Both terms refer to practices where a group of people sharing **protected characteristics** (e. g., age, disability, gender, marital status, pregnancy, race, religion, and sexual orientation) are **disproportionately disadvantaged**. These "protected characteristics" are traits such as race, gender, age, physical or mental disabilities, *where differences due to such traits cannot be reasonably justified*. Ideally, one should have a set of sensitive traits that can be checked against. But in reality, what constitutes "protected characteristics" *varies by context, culture, and country*. Next, the phrase "disproportionately disadvantaged" dismisses differences in treatment due to statistical randomness. The difference between disparate treatment and disparate impact can be summarized as **explicit intent**. Disparate treatment is explicitly intentional, while disparate impact is implicit or unintentional.

Let's use Amazon's Free Same-Day delivery service as an example. Since it is in the beta stage, Amazon wants to trial the service before rolling it out to everyone. Suppose Amazon implements a model that picks lucky neighborhoods to get a first glimpse on the Free Same-Day delivery service. Using race to decide who should get this service is certainly unjustified. So, if Amazon had explicitly used racial composition of neighborhoods as an input feature for the model, that would be **disparate treatment**. In other words, disparate treatment occurs when protected characteristics are used as input features. Disparate treatment is relatively easy to spot and resolve once the set of protected characteristics is determined, as one just must make sure none of the protected characteristics are explicitly used as input features. On the other hand, Amazon might have been cautious about racial bias and deliberately excluded racial features from their model and, for example, used ZIP codes instead. Focusing on ZIP codes with high density of Amazon Prime members makes perfect business sense. But what if the density of Prime members correlates with racial features?

The image (Figure 9.1) below from a 2016 Bloomberg article by David Ingold shows that there is indeed a large racial bias in the selected neighborhoods, nevertheless. Despite not using any racial features, the resulting model appears to make recommendations that disproportionately exclude predominantly black ZIP codes. This *unintentional bias* can be seen as **disparate impact**. In general, disparate impact occurs when protected characteristics are not used as input features but the resulting outcome still exhibits *disproportional disadvantages*. Disparate impact is more difficult to fix since it can come from multiple sources, such as: a nonrepresentative dataset, a dataset that already encodes unfair decisions, or input features that are proxies for protected characteristics in the first place.

So, how does one know how much disparity is *unfair*? What is unfair in one case might be justified in another, depending on the specific contextual circumstances. One can easily find and calculate some basic *fair* metrics (Verma et al., 2018), among others,

The northern half of Atlanta, home to 96%
of the city's white residents, has same-day
delivery. The southern half, where 90%
of the residents are black, is excluded.

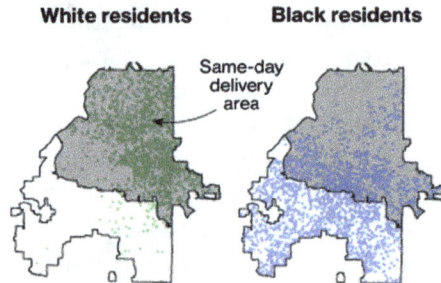

White residents   **Black residents**

Same-day
delivery
area

Figure 9.1: Racial bias in Amazon's Prime Members "same day delivery" regions.

Predictive Parity, Counterfactual Fairness, Demographic Parity, Group Fairness, Equalized Odds, Conditional Use Accuracy Equality, Overall Accuracy Equality, Treatment Equality, Calibration, or Fairness through Awareness. The awesome thing about these metrics is that they can be put into a loss function. Then a model can be trained to optimize the function to act fair with respect to that specific fairness metric. Except, it doesn't necessarily work like that. A major issue with these metrics (besides the question of how to pick the right and best ones) is that they neglect the larger context. The fairness metrics can be a systematic way to check for bias, but they are only one piece of the puzzle. A complete assessment for fairness needs one to get down and dirty with the complete problem at hand. Most of the fairness metrics focus on equality in the rates of true positives, true negatives, false positives, false negatives, or some combination of these. But remember that these metrics are insufficient when they exclude the larger context of the AI system and neglect contextual justifications. After all, notions of fairness are heavily based on context and culture. To get this context into the equation, try asking the following sociotechnical questions to a group of *real* end-users of your AI application to be:

- What is the ultimate aim of the application?
- What are the main privacy concerns for those who are impacted by the application?
- Are there discrepancies between the application's privacy policy and the application's workings?
- What are the pros and cons of an AI system versus other solutions?
- What are the AI systems (un)intended effects?
- What is the current system that the AI system will be replacing?
- How can the AI system be misused by unknowing or malicious actors?
- What examples of fair and unfair predictions can one find and why are they fair/unfair?
- What are the relevant protected traits in this problem?
- Which fairness metrics should we prioritize?

– When we detect some disparities among protected groups, is this disparity justified or is it considered unfairness?

9.2.1 Generic definition on bias

As AI is already becoming mainstream as part of the solution in a myriad of end-user applications, among others, any mobile phone's voice-controlled personal assistant, it is of the uttermost importance that one has a transparent and unambiguous understanding of why and how unwanted side effects and consequences arise in any kind of AI-driven services. Whatever harmful decisions are taken to particular individuals and/or groups of people are often attributed to **"biased"** data, but by default merely blaming imperfect AI-driven applications using those two words is too coarse-grained and often just faulty. As one will find out hereafter, there are at least five distinct categories of "biased" data that downstream harm the end-to-end machine learning pipeline from data generation up to the evaluation of an AI model. Hereafter will be described how these different flavors of bias arise, how they are relevant to specific AI-driven applications, and how they push toward different kinds of solutions. As such, one gets a clearer picture of specific reasonable claims, rather than generically relying on what may or may not be called "fair" in the eye of the beholder.

9.2.2 What is the problem to solve?

There are a "trillion" types of biases as stated in (Mehrabi et al., 2019), but to not make things overly complicated, the focus will be on five major types of bias that creep into the different stages of AI systems, that is, Historical Bias, Representation Bias, Measurement Bias, Aggregation Bias, and Evaluation Bias.

Hereafter, one can identify issues that commonly pop up in AI applications that lead to undesired and occasionally societally malignant outcomes. As such, decomposing the consequences of a particular algorithm should begin with the complete comprehension of the data generation pipeline followed by the machine learning processing pipeline that led to its output in the first place. As one will see, the origins of bias investigated originate at different stages in that very pipeline (see Figure 9.2a and 9.2b, Suresh and Guttag).

1. **Historical bias** is a normative concern in the world as one knows it. It is a structural and fundamental issue within the first step of the data generation process and can arise even when perfect sampling and proper feature selection has been done. For instance, even if one has access to the perfectly measured feature "crime," it might still echo historical factors that have led to more "crime" in pauperis neighborhoods, because in some systems "crime" is used as a proxy for "arrested." Such systems, even if they precisely reflect the world "as is," can still damage certain parts

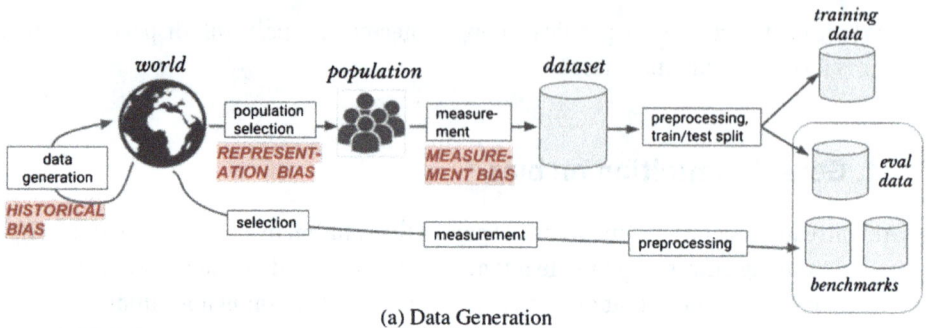

(a) Data Generation

Figure 9.2a: Data generation pipeline.

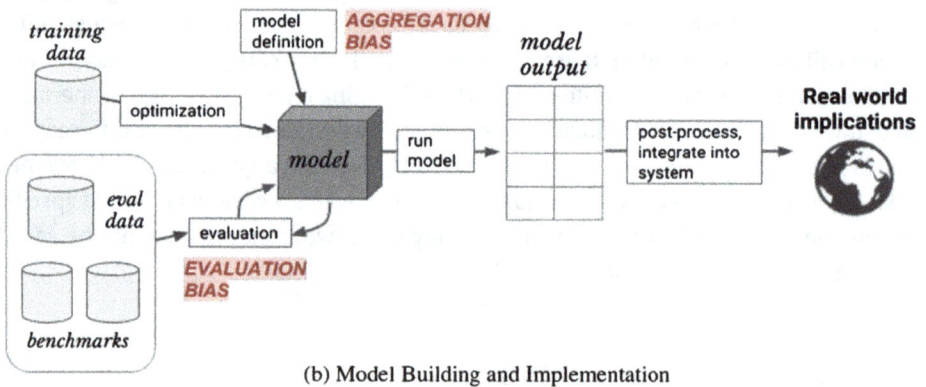

(b) Model Building and Implementation

Figure 9.2b: Model building and implementation pipeline.

of a population. Contemplating on historical bias implicates evaluating the representational harm (such as intensifying stereotypes) to particular identity groups. As such, recognizing historical bias requires a retrospective notion of both the machine learning application and the data generation process over time.

Example. Back in 2018, only 5 % of the Fortune 500 CEOs were female. Should image search results for "CEO" echo that number? Eventually, a variety of stakeholders, including affected representatives of the use case, should evaluate the particular disadvantages that this result could cause in this particular and adjacent use cases and voice a final opinion. Their decision may be at odds with the current available data even if that data is a perfect reflection of the world as one knows it. Indeed, Google has recently altered their Image Search results for "CEO" to display a higher ratio of women. For example, today's search term for "CEO" on Google resulted in 10 % percent of CEO pictures to depict women.

2. **Representation bias** heavily depends on the definition of the population and the sampling thereof, that is, when certain parts of the input space are underrepre-

sented. As such, representation bias arises if the probability distribution over the input space samples too few examples from a particular part of that input space. When one lacks data about some part of the input space, the learned mapping will be more uncertain for new data pairs in that area. It is worth noting that even if some group is a minority that only makes up 5 % of the true distribution, then even perfect sampling, that is, with no representation bias, from the true data distribution will likely lead to a significantly less robust model for this group anyway. Representation bias can occur due to at least these two reasons:

- The sampling techniques only reach some part of the population. For example, datasets harvested via smartphone applications can under-represent lower-income or older groups, who are less likely to own or heavily use smartphones. Likewise, medical data for a particular disease is maybe only collected from the population of patients who were considered serious enough to bring in for further in-depth screening.
- The population of interest has altered or is different from the population used during model training. Data that is representative for the population of Brussels, for example, may not be representative if used to analyze the population of London. Likewise, data representative of Brussels 50 years ago will most certainly not reflect today's population.

Example. ImageNet is a broadly utilized image dataset consisting of 1.2 million labeled images. About 45 % of the images in ImageNet were photographed in the US, and the bulk of the remaining images are taken in Canada or Western Europe. Respectively, 1 % and 2.1 % of the images come from China or India. As such, it can come to no surprise that the performance of a classifier trained on ImageNet is significantly worse for various categories (e. g., "bride" or "groom") on pictures that are crowdsourced from vastly underrepresented countries such as Pakistan, India, or Vietnam versus pictures taken from North America and Western Europe.

3. **Measurement bias** originates from subsequently choosing and measuring only some particular features of interest. Valid measured data are regularly a proxy for some ideal features and labels, that is, arrest rates are frequently used as a proxy for crime rates. If the measurement method just appends random noise, the model parameters will converge to those which would eventually anticipate upon with the correctly measured features (given abundant data). On the other hand, measurement bias often arises because proxies are generated differently across populations (also known as differential measurement error). Measurement bias can occur due to at least these three reasons:
   - The granularity of data varies across populations, that is, if a group of factory workers is more rigorously or regularly monitored, more errors will be noticed in that group. This can also lead to a feedback loop wherein the group is subject to even additional monitoring because of the obviously higher rate of mistakes.

- The quality of data varies across populations, that is, structural discrimination can lead to systematically higher error rates in a certain population. For example, women are more likely to be misdiagnosed or not diagnosed at all for conditions where self-reported pain is a symptom (in this case "diagnosed with condition X" is a biased proxy for "has condition X"), as they have a higher pain threshold than men, and thus are underrepresented in the self-reported pain feature.
- The defined classification job is an oversimplification. In order to build a supervised machine learning model, some label to predict must be picked. Diminishing a decision to a single attribute can create a biased proxy label as it only captures a specific aspect of what one genuinely wants to measure. Contemplate on the prediction task of deciding whether a student will be successful or not in a university admission context. Completely capturing the outcome of "successful student" in terms of one single measurable attribute is almost impossible because of its inherent complexity. In cases such as these, algorithm modelers just depend on some available label such as "GPA score," which ignores different talents and abilities demonstrated by parts of the population as other indicators of success.

Example. As mentioned earlier, in predictive policing applications, the proxy variable "arrest" is often used to measure "crime" or some underlying notion of "risk to commit crimes." As minority communities are often more heavily patrolled and, as a consequence, have higher arrest rates, there is a different mapping from crime to arrest for people from these communities. Former arrests and friend/family arrests were two of many wrong measured proxy variables used in the recidivism risk prediction tool COMPAS. These were unmistakably factors that in the end led to higher false positive rates for black versus white defendants. It is worth noting that in the same realm even an evaluation by the proxy label "rearrest" used to measure "recidivism" is doubtful to be correct, too.

4. **Aggregation bias** originates from flawed assumptions about the population affect model definition, that is, when a one-size-fits-all model is used for populations with different conditional distributions. Underlying aggregation bias is an assumption that the mapping from inputs to labels is consistent across populations. In reality, this is frequently not the case. Population membership can be indicative of different backgrounds, cultures, or norms, and a given variable can signify something entirely different for a person in a different population. Aggregation bias can lead to a model that is suboptimal for any population, or a model that is just fit to the dominant population (if combined with representation bias). If there is a nonlinear relation between population membership and outcome, for example, any single linear classifier will have to offer performance on one or both populations. In some cases, embodying information about population differences into the design

of a model can lead to simpler learned functions that improve performance across these populations.

Example. Diabetes patients have known odds in associated complications across ethnicities. Studies have also suggested that HbA1c levels (widely used to diagnose and monitor diabetes) differ in complex ways across ethnicities and genders. Because these factors have different meanings and magnitudes within different subpopulations, a single model is unlikely to be best suited for any group in the population even if they are equally represented in the training data.

5. **Evaluation bias** emerges during model iteration and evaluation, that is, when the evaluation and/or benchmark data for an algorithm is not representative for the target population. A model is optimized on its training data, but its quality is often measured on benchmarks (MovieLens, FaceScrub, ImageNet, etc.). As such, a misrepresentative benchmark fosters the development of models that only perform well on a subset of the population. Evaluation bias originates because of the necessity to objectively compare models against one another. Applying different models to some set of external datasets tries to serve this purpose, but it is often further used to make generic statements about how good a model is. These generalizations are frequently not statistically valid, and thus can lead to overfitting to particular sets of benchmarks. This is particularly problematic if the benchmark is furthermore not that representative. This is a self-fulfilling process, as the more successful a benchmark is if more and more people use it for comparative assessments, the more serious this problem of overfitting will become. Evaluation bias can be further sharpened by specific metrics that are used to report performance, that is, by only looking at the single metric "accuracy" varied disparities in other types of errors like "false positives" are likely to be hidden as well.

Example. It is known that current commercial facial analysis algorithms, executing tasks such as gender or smile detection on dark-skinned females perform badly. Just by looking at some common facial analysis benchmark datasets, it becomes obvious why such algorithms should be considered inappropriate for use as just 5 % of the images in these benchmark datasets are of dark-skinned female faces. Algorithms that underperform on this slice of the population therefore suffer quite little in their evaluation performance on these benchmarks, as the algorithms' underperformance is likely due to representation bias in the training data, but the benchmarks (alas suffering from the same representation bias) failed to discover and penalize this. Since this malfunctioning was discovered, other algorithms have been benchmarked on more balanced face datasets, changing the overall development process to encourage models that perform well across different diverse populations.

9.2.3 How do we mitigate it?

Currently, the two most elaborated (and still constantly evolving) tool kits are *IBM's initiated AIF 360 tool kit* and *Microsoft's initiated FairnLearn* tool kit. Hereafter, the bias mitigation algorithms of both frameworks are discussed as they rely on a myriad of +75 (jointly overlapping) metrics. The bias mitigation algorithm categories are based on the location where these algorithms can intervene in a complete machine learning pipeline. If the algorithm is allowed to modify the training data, then preprocessing can be used. If it is allowed to change the learning procedure for a machine learning model, then in-processing can be used. If the algorithm can only treat the learned model as a black box without any ability to modify the training data or learning algorithm, then only post-processing can be used. This is illustrated in Figure 9.3 below.

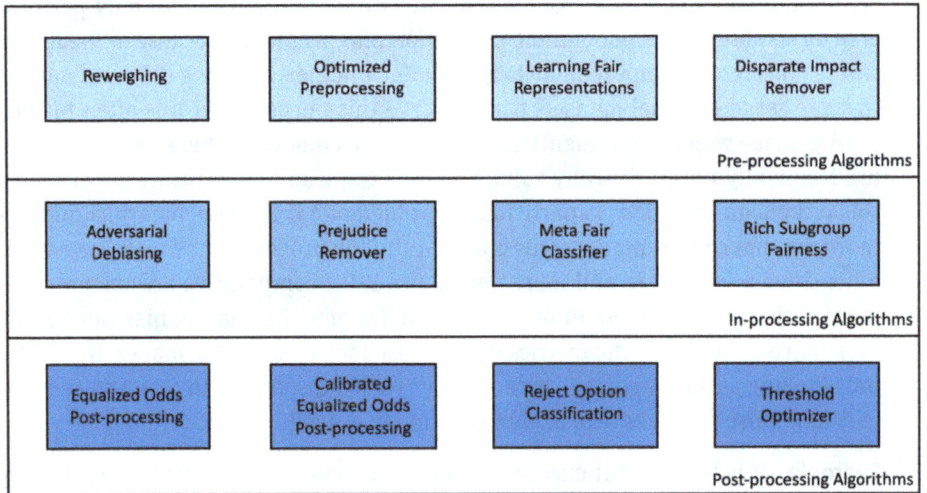

Figure 9.3: Bias mitigation strategies for ML models.

Pre-processing algorithms involve modifying the training data, with the aim of preventing the algorithmic model from learning discriminatory decision-making rules in the training stage. This can be accomplished by, for example, modifying the training data itself, for instance by changing the values of specific attributes for individual records or even removing attributes entirely. *Reweighing* (Kamiran et al., 2012) generates different weights for the training examples in each (group, label) combination to ensure fairness before actual classification. The idea is to apply appropriate weights to different tuples in the training dataset to make the training dataset discrimination-free with respect to the sensitive attributes. Instead of reweighing, one could also apply techniques (nondiscrimination constraints) such as suppression (remove sensitive attributes) or massaging the dataset—modify the labels (change the labels appropriately to remove discrimination from the training data). However, the reweighing technique is more effective than the other two mentioned earlier. *Optimized preprocessing* (Calmon et al., 2017), on the other hand, learns a probabilistic transformation that edits the features and labels in the data with group fairness, individual distortion, and data fidelity constraints and objectives. Furthermore, *learning fair representations* (Zemel et al., 2013) finds a latent representation that encodes the data well but obfuscates information about the protected attributes. *Disparate impact remover* (Feldman et al., 2015) edits feature values to increase group fairness while preserving rank-ordering within the groups. The last two algorithms are "reduction" algorithms that cause the reweighting and relabeling of the input data. The key idea is to reduce fair classification to a sequence of cost-sensitive classification problems, whose solutions yield a randomized classifier with the lowest (empirical) error subject to the desired constraints. This reduces the problem back to standard machine learning training and is thus more generic than the other preprocessing algorithms. *Exponentiated gradient* (Agarwal et al., 2018) allows for any definition of fairness that can be formalized via linear inequalities on conditional moments, such as demographic parity or equalized odds metrics. It shows how binary classification subject to these constraints can be reduced to a sequence of cost-sensitive classification problems. Only black-box access to a cost-sensitive classification algorithm is required, which does not need to have any knowledge of the desired definition of fairness or protected attribute. The solutions to this sequence of cost-sensitive classification problems yield a randomized classifier with the lowest (empirical) error subject to the desired fairness constraints. In some situations, though, it is preferable to select a deterministic classifier, even if that means a lower accuracy or a modest violation of the fairness constraints, that is, when the protected attribute is binary then *Grid search* (Agarwal et al., 2019) can in fact be conducted in a single dimension. Indeed, when the number of constraints is very small, as is the case for demographic parity or equalized odds with a binary protected attribute, it is also reasonable to consider a grid of constraint values, calculate the best response for each value, and then select the value with the desired trade-off between accuracy and fairness.

In-processing algorithms methods involve modifying the algorithmic model itself. One approach is to train separate models for each protected group in isolation of one another, and then only use the relevant model for decisions concerning that group as this would, however, be difficult in many situations where certain individuals may belong to multiple categories (Asian and female, for example). Another is to change the criteria that result in "branches" in a decision tree to ignore or correct the influence of protected characteristics. However, many in-processing methods require personal data regarding protected characteristics to be available, which cannot be taken for granted due to the legal sensitivity of this data. The legal status and necessity of monitoring for bias means well-intentioned data scientists wishing to detect discrimination in their systems can face barriers to obtaining the necessary data or information about protected characteristics, especially if they did not seek consent to gather and process the data for these purposes from the beginning. *Adversarial debiasing* (Zhang et al., 2018) learns a classifier to maximize prediction accuracy and simultaneously reduce an adversary's ability to determine the protected attribute from the predictions. This approach leads to a fair classifier as the predictions cannot carry any group discrimination information that the adversary can exploit. *Prejudice remover regularizer* (Kamishima et al., 2012), on the other hand, adds a discrimination-aware regularization term to the learning objective. Furthermore, *Meta fair classifier* (Celis et al., 2018) takes the fairness metric as part of the input and returns a classifier optimized regarding the fairness metric. Finally, *Rich subgroup fairness* (Kearns et al., 2018) is an algorithm for learning classifiers that are fair in respect to rich subgroups, which are defined by (linear) functions over the sensitive attributes and the statistical fairness notions (false positive, false negative, and statistical parity rates). The implementation uses a maximum of two regressions as a cost-sensitive classification oracle, and supports linear regression, support vector machines, decision trees, and kernel regression.

Post-processing algorithms methods involve removing discriminatory rules or otherwise modifying a model (e. g., confidence intervals, weights, probabilities, predicted classes, or labels) after it has been trained. This might mean, for example, modifying a model so that it places less significance on particular postcodes, which could be closely correlated with one specific ethnic group. Outcomes or decisions can also be artificially adjusted to ensure equitable treatment across groups within the affected population. For example, if it is known that a probation risk assessment algorithm consistently ranks one ethnic group as a higher risk than others, any risk assessment relating to an individual from that group might be downgraded by a human probation officer to ensure an equitable outcome. *Equalized odds* post-processing (Hardt et al., 2016) solves a linear program to find probabilities with which to change output labels to optimize equalized odds. *Calibrated equalized odds* post-processing (Pleiss et al., 2017), on the other hand, optimizes over calibrated classifier score outputs to find probabilities with which to change output labels with an equalized odds objective. Furthermore, *Reject option classification* (Kamiran et al., 2012) gives favorable outcomes to unprivileged groups and unfavorable outcomes to privileged groups in a confidence band around the decision boundary with the highest uncertainty. Finally, *Threshold Optimizer* (Hardt et al., 2016) takes as input an existing classifier and the sensitive feature and derives a monotone transformation of the classifier's prediction to enforce the specified parity constraints. The idea is that the classifier is obtained by applying group-specific thresholds to the provided estimator. These thresholds are chosen to optimize the provided performance objective subject to the provided fairness constraints.

9.2.4 When do we apply these techniques?

With the increasing popularity of AI and machine learning over the past decade, and their already exponential spread in different applications, safety and fairness constraints should become a huge issue for researchers and engineers. Machine learning is used in courts to assess the probability that a defendant recommits a crime. It is used in different medical fields, in childhood welfare systems (Chouldechova et al., 2018), and autonomous vehicles. All of these applications have a direct effect in our lives and can harm our society if not designed and engineered correctly, with consideration to fairness. Current AI systems (Osoba et al., 2017) (Howard et al., 2018) already affect our daily lives with their inherent biases, such as the existence of bias in AI chatbots, face recognition, search engines, employment matching, flight routing, and automated legal aid for immigration algorithms, and search and advertising placement algorithms. Therefore, it is important for researchers and engineers to be concerned about the downstream applications and their potential harmful effects when modeling an algorithm or a system. Hence, the question is not "When" as the answer is "Always," but all involved in developing these new AI systems should be made aware to the fullest of possible bias creeping into the data and models used for any envisioned (AI) application. As such, "fairness" should be one of the key requirements that should constantly be on a checklist when software applications are being architected, developed, tested, and (re)deployed.

9.2.5 What are the limitations?

Unknown unknowns. The introduction of bias is not always crystal clear during a model's construction because one may not realize the downstream impacts of one's data and choices there upon until much later. Once one does, it is hard to retroactively identify where that bias came from and then figure out how to get rid of it. In Amazon's AI recruiting case, when the engineers initially discovered that its tool was penalizing female candidates, they reprogrammed it to ignore explicitly gendered words like "women." They soon discovered that the revised system was still picking up on implicitly-gendered words—verbs that were highly correlated with men over women, such as "executed" and "captured"—and using that to make its (biased) decisions.

Imperfect processes. First, many of the standard practices in machine learning are not designed with bias detection in mind, among others, deep-learning models are tested for performance before they are deployed, creating what would seem to be a perfect opportunity for catching bias. However, in practice, testing usually looks like this: (male) computer scientists randomly split their data before training into one group that is actually used for training and another that is reserved for validation once training is done. That means the data one uses to test the performance of one's model has the same biases as the data one used to train it. Thus, it will fail to flag skewed or prejudiced results.

Lack of social context. Similarly, the way in which computer scientists are taught to frame problems often is not compatible with the best way to think about social problems. Within computer science, it is considered good practice to design a system that can be used for different tasks in different contexts. But what that does is ignore a lot of the social context. One cannot have a system designed in Brussels and then apply it directly in Bangalore because chances are that different communities have different versions of fairness. Likewise, one cannot have a system that one applies for "fair" criminal justice results and then immediately apply it to employment. How one thinks about fairness in those contexts is just totally different, hence "the portability trap" is encountered (Selbst et al., 2019).

The definitions of fairness. It is also not clear what the absence of bias should look like. This isn't true just in computer science—this question has a long history of debate in philosophy, social science, and law. What is different about computer science is that the concept of fairness must be defined in mathematical terms, like balancing the false positive and false negative rates of a prediction system. But as researchers have discovered, there are many different mathematical definitions of fairness that are also mutually exclusive. Does fairness mean, for example, that the same proportion of black and white individuals should get high risk assessment scores? Or that the same level of risk should result in the same score regardless of race? It is impossible to fulfill both definitions at the same time, so at some point one must pick one. But whereas in other fields this decision is understood to be something that can change over time, the computer science field has a notion that it should be fixed. "By fixing the answer, one solves a problem that looks very different than how society tends to think about these issues," says Selbst.

Identifying and mitigating bias in AI systems is essential to building trust between humans and machines that learn. As AI systems find, understand, and point out human inconsistencies in decision making, they could also reveal ways in which we are partial, parochial, and cognitively biased, leading us to adopt more impartial or egalitarian views. In the process of recognizing our bias and teaching machines about our common values, one may improve more than AI. We might just improve ourselves along the way. As such, it is not merely technological solutions that will get the bias out of our data and AI systems. Further socio-economic and sociocultural (read: all but technological) recommendations can also make the much-needed awareness' difference, that is:

- Identify critical services and subsystems that require "human-in-the-loop" decision making. Selection criteria may include high-risk systems or systems that require special accountability. Limit the role of artificial agents in these systems to a strictly advisory capacity. Emphasize the need for the ability to audit the results of these advisory artificial agents.
- Establish best practices for auditing algorithmic decision-making aids designed for use in government services and policy domains (e. g., the criminal justice system and social services administration). This should include specific guidance discouraging the use of unaccredited third-party black-box algorithmic solutions. Audit procedures should also address questions of disparate impact.
- Adopt standardized disclosure practices to inform stakeholders when decisions affecting them are algorithmically generated. Institute standard procedures for appealing or reviewing such decisions.

- Invest science research funds in research on algorithmic disparate impact. Engage with the commercial AI community to share best practices.
- Address diversity issues in the science, technology, engineering, and math educational pipeline. Update accreditation guidelines for engineering schools to include more training on the effects of technology on society and sociotechnical systems more generally.

9.3 Initial framing of Interpretability

Some machine learning models are simple and easy to understand. We know how changing the inputs will affect the predicted outcome and can make justification for each prediction. However, with the recent advances in machine learning and AI, models have become very complex, including complex deep neural networks and ensembles of different models. These complex models are referred to as black box models.

Unfortunately, the complexity that gives extraordinary predictive abilities to black box models also make them very difficult to understand and trust. The algorithms inside the black box models do not expose their secrets. They don't, in general, provide a clear explanation of why they made a certain prediction. They just give us probability, and they are opaque and hard to interpret. Sometimes there are thousands (even millions) of model parameters, there is no one-to-one relationship between input features and parameters, and often combinations of multiple models using many parameters affect the prediction. Some of them are also data hungry. They need enormous amounts of data to achieve high accuracy. It is hard to figure out what they learned from those datasets and which of those data points have more influence on the outcome than the others.

Due to all those reasons, it is very difficult to understand the process and the outcomes from those techniques. It is also difficult to figure out whether one can trust the models and whether one can make fair decisions when using them. What happens if they learn the wrong thing? What happens if they are not ready for deployment? There is a risk of misrepresentation, oversimplification, or overfitting. Thus, one needs to be careful when using them, and one should better understand how those models work. As such, **interpretability** means giving explanations to the end-users for a particular decision or process. More specifically, it entails:
- Understanding the main tasks that affect the outcomes.
- Explaining the decisions that are made by an algorithm.
- Finding out the patterns/rules/features that are learned by an algorithm.
- Being critical about the results.
- Exploring the unknown unknowns for your algorithm.

It is not merely about understanding every detail about how a model works for each data point in the training data.

9.3.1 What is the problem to solve?

Why is accuracy not enough? In machine learning, accuracy is measured by comparing the output of a machine learning model to the known actual values from the input dataset. A model can achieve high accuracy by memorizing the unimportant features or patterns in a dataset. If there is a bias in an input dataset, this can also affect its model afterwards. In addition, the data in the training environment may not be a good representation of the data in the production environment in which the model is deployed. Even if it is sufficiently representative initially, considering that the data in the production environment is not stationary as it can become outdated very quickly. Thus, one cannot rely only on the prediction accuracy achieved for a specific dataset. One needs to know more to further demystify the black box machine learning models and improve transparency and interpretability to make them more trustworthy and reliable.

Why is interpretability needed? Well, interpretability is important to different people for different reasons:

– Data scientists want to build models with high accuracy. They want to understand the details to find out how they can pick the best model and improve that model. They also want to get insights from the model so that they can communicate their findings to their target audience.
– End-users want to know why a model gives a certain prediction. They want to know how they will be affected by those decisions. They want to know whether they are being treated fairly and whether they need to object to any decision. They want to have a certain measure of trust when they are shopping online or clicking ads on the web.
– Regulators and lawmakers want to make the system fair and transparent. They want to protect consumers. With the inevitable rise of machine learning algorithms, they are becoming more concerned about the decisions made by models.

All those users want similar things from the black box models. They want them to be transparent, trustworthy, and explainable.

1. Transparent: The system can explain how it works and/or why it gives certain predictions
2. Trustworthy: The system can handle different scenarios in the real world without continuous control.
3. Explainable: The system can convey useful information about its inner workings, for the patterns that it learns and for the results that it gives.

In a typical machine learning pipeline, one has control over the dataset used to train the model, one has control over the model being used, and one has control over how those models are being assessed and deployed.

Two types of pipeline and interpretability are instantiated—during training and during inference. On the one hand, in the pipeline during training, it is important to have transparency and see the feature importance as a tool to verify data leakage and remove noise, to design the model and optimize the model for bias and performance. Here, *global explanations* are useful, looking at the aggregated view of the feature importance.

During inference, on the other hand, it is key to see each individual prediction—how the model came to its conclusion, for example, for a classification—fraud or not fraud—give transparency and trustworthiness for each case, where continuous monitoring can detect changes of patterns and data drift. Here, *local explanation* is useful, looking at the individual predictions and its feature importance.

9.3.2 How do we mitigate it?

Explainable AI is only beginning to get the attention it really deserves, both in academia and in industry. As this is new ground to cover, hereafter some initial frameworks are discussed that make hints toward solving the explainability issue by extending current AI modeling techniques.

9.3.2.1 Explainability by intrinsic global design

One can *by design* choose to avoid certain machine learning algorithms of a black box nature, such as neural network-based algorithms and instead select tree-based algorithms, which by nature have a white box design, that is, do not require Mimic explainers and surrogate models afterwards to make the black, white again.

GIRP (*global interpretation via recursive partitioning*) (Yang et al., 2018) builds a global interpretation tree for a wide range of machine learning models based on their local explanations. That is, one recursively partitions the input variable space by maximizing the difference in the contribution of input variables averaged from local explanations between the divided spaces. By doing so, one ends up with a binary tree that is called the interpretation tree describing a set of decision rules that is an approximation of the original machine learning model. *NBDT* (*neural-backed decision trees*) (Wan et al., 2020), on the other hand, build modified hierarchical classifiers that use trees constructed in weight space. As such, NBDT achieves both interpretability and neural network accuracy. It preserves interpretable properties, for example, leaf purity and a nonensembled model, and demonstrates interpretability of model predictions both qualitatively and quantitatively. In short, an NBDT is a hierarchical classifier that uses a hierarchy derived from model parameters to avoid overfitting, and that can be created from any existing classification neural network without architectural modifications Furthermore, it retains interpretability by using a single model, sequential discrete decisions, and pure leaves. NBDT is built in 2 steps, that is, first it constructs a tree structure using the weights of a trained network, dubbed an induced hierarchy; and second, it retrains or fine-tunes that classification network with an extra hierarchy-based loss term, called a tree supervision loss. For the forward pass, one needs to run the fully connected layer as embedded decision rules, which are variants of oblique decision rules for arbitrary branching factors. Finally, *RFEX* (*random forest explainer*) (Petkovic et al., 2019) goes far beyond simple ranking of features and provides many other measures to enhance random forest explainability, for example, trade-offs vs. accuracy, ranking of feature combinations, and feature interactions via feature cliques. It further implements the permutation feature importance (PFI) technique used to explain classification and regression models. At a high level, the way it works is by randomly shuffling data one feature at a time for the entire dataset and calculating how much the performance metric of interest changes. The larger the change, the more important that feature is. As such, PFI can explain the overall global behavior of any underlying model but does not explain individual predictions.

These kinds of model-specific interpretability methods are limited to specific model classes, as intrinsic global methods are by definition model-specific. The drawback of this practice is that when one requires a particular type of interpretation, one is limited in terms of choice to models that provide it, potentially at the expense of using a more predictive and representative model. Therefore, there has been a recent surge in interest in model-agnostic local interpretability methods as they are model-free. As such, even though a multitude of techniques is used in literature to enable global interpretability, arguably, global model interpretability is hard to achieve in practice, especially for models that exceed a handful of parameters. Analogically to humans, who focus effort on only part of the model in order to comprehend the whole of it, local interpretability as mentioned hereafter is more readily applicable.

9.3.2.2 Explainability by post-hoc local Interpretation

The current "hot" kid in town is *SHAP* (*SHapley additive exPlanations*) (Lundberg et al., 2017), which is a game theoretic approach to explain the output of any machine learning model. The goal of *SHAP* is to explain the prediction of an instance x by computing the contribution of each feature to the prediction. The *SHAP* explanation method computes Shapley values from coalitional game theory. The feature values of a data instance act as players in a coalition. Shapley values tell us how to fairly distribute the "payout" (= the prediction) among the features. A player can be an individual feature value, for example, for tabular data. A player can also be a group of feature values. For example, to explain an image, pixels can be grouped to super pixels and the prediction distributed among them. One innovation that *SHAP* brings to the table is that the Shapley value explanation is represented as an additive feature attribution method, a linear model. As such, it connects optimal credit allocation with local explanations using the classic Shapley values from game theory and their related extensions. In a way, these *SHAP* values act as a unified measure of feature importance. By now, there are already a few

SHAP explainers around for different specific machine learning models, that is, *SHAP LinearExplainer* computes SHAP values for a linear model, optionally accounting for interfeature correlations. *SHAP TreeExplainer* focuses on a polynomial time fast SHAP value estimation algorithm specific to trees and ensembles of trees. *SHAP DeepExplainer* is a high-speed approximation algorithm for SHAP values in deep learning models that builds on DeepLIFT (Shrikumar et al., 2017), and finally *SHAP KernelExplainer* uses a specially weighted local linear regression to estimate SHAP values for any model (and thus makes it model agnostic).

LIME (*local interpretable model-agnostic explanation*) (Ribeiro et al., 2016) is another good algorithm to provide a technique for explaining a predictive model in an interpretable and faithful manner. Local surrogate models are interpretable models that are used to explain individual predictions of black box machine learning models. *LIME* is a concrete implementation of such local surrogate models. Surrogate models are trained to approximate the predictions of the underlying black box model. Instead of training a global surrogate model, *LIME* focuses on training local surrogate models to explain individual predictions. The idea is quite intuitive. First, forget about the training data and imagine one only has the black box model where one inputs data points and gets the predictions of the model. One can probe the box as often as wanted. The goal is to understand why the machine learning model made a certain prediction. *LIME* tests what happens to the predictions when giving variations of the data into the machine learning model. *LIME* generates a new dataset consisting of permuted samples and the corresponding predictions of the black box model. On this new dataset, *LIME* then trains an interpretable model, which is weighted by the proximity of the sampled instances to the instance of interest. This newly learned model should be a good approximation of the machine learning model predictions locally, but it does not have to be a good global approximation. This kind of accuracy is called local fidelity.

Other older, but nevertheless interesting, *feature selection methods* to look at and to further learn from are *L2X*, which seeks a variational approximation of the mutual information and makes use of a Gumbel-softmax relaxation of discrete subset sampling during training (Chen et al., 2018), *Saliency*, which computes the gradient of the selected class with respect to the input feature and uses the absolute values as importance scores (Simonyan et al., 2013), and *DeepLift*, which decomposes the output prediction of a neural network on a specific input by backpropagating the contributions of all neurons in the network to every feature of the input. As such, it compares the activation of each neuron to its "reference activation" and assigns contribution scores according to the difference (Shrikumar et al., 2017). Older, more *visual interactive data mining methods* are also worth studying and having a look at, that is, *partial dependence plots* (PDP), which show the marginal effect one or two features have on the predicted outcome of a machine learning model (Friedman, 2001), *individual conditional expectation* (ICE) displays one line per instance that shows how the instance's prediction changes when a feature changes (Goldstein et al., 2017), and *accumulated local effects* (ALE), which describe how features influence the prediction of a machine learning model on average. *ALE plots* are a faster and unbiased alternative to *PDP plots* (Apley et al., 2019).

One can also choose the best of both worlds, that is, mix global with local interpretability, to expand the bucket of choice for machine algorithms. As such, since the white box model may not score the best, here explainability needs to be designed by oneself. A common approach is to use *mimic explainer with surrogate models* to mimic the black box model as a white box model. InterpretML's Mimic explainer is based on the idea of training Global Surrogate Models to mimic black box models. A global surrogate model is an intrinsically interpretable model that is trained to approximate the predictions of any black box model as accurately as possible. Data scientists can then interpret the surrogate model to draw conclusions about the black box model. Within Mimic Explainer, one can use one of the following interpretable models as your Surrogate Model: LightGBM (LGBMExplainableModel), Linear Regression (LinearExplainableModel), Stochastic Gradient Descent explainable model (SGDExplainableModel), and Decision Tree (DecisionTreeExplainableModel). With these techniques, one gets both *Global* and *Local* relative feature importance. On top of that, one will also get a *Global* and *Local* feature prediction relationship, which gives us the same information as a white box model as, for example, by Design in a *SHAP TreeExplainer*. The benefit of a Mimic explainer is that it is model agnostic, that is, it can both handle tree-based models, linear models, or even deep learning models. This is in comparison to most SHAP explainers, which are mostly model specific, the exception being the *SHAP KernelExplainer*, which also is model agnostic.

9.3.3 When do we apply these techniques?

If interpretability is needed, first one needs to ask why it is needed and in which stage of this process interpretability is needed? It may not be necessary to understand how a model makes its predictions for every application. However, one might need to know it if those predictions are used for high-stakes decisions. After the purpose is defined, one should focus on what techniques are needed in which stage of the process, that is, we have the following:

Interpretability in premodeling (interpretability of model inputs) Understanding the dataset is very important before one starts building models. One can use different exploratory data analysis and visualization techniques to have a better understanding of the dataset. This can include summarizing the main characteristics of the dataset, finding representative or critical points in the dataset, and finding the relevant features from that dataset. After one has an overall understanding of the dataset, one needs to think about which features are going to be used in modeling. If one wants to explain the input–output relationship after the modeling, one needs to start with meaningful features. While highly engineered features (such as those obtained from t-distributed stochastic neighbor embedding, random projections, etc.) can boost the accuracy of your model, they will not be interpretable when one puts the model to use.

Interpretability in modeling Models can be categorized as white box (transparent) and black box (opaque) models based on their simplicity, transparency, and explainability, that is:

1. **White box (transparent) models**: Decision trees, rule-lists, and regression algorithms are usually considered in this category. These models are easy to understand when used with few predictors. They use interpretable transformations and give one more intuition about how things work, which helps one understand what is going on in the model. One can explain them to a technical audience. However, if one has hundreds of features and one builds a very deep, large decision tree, things can still become complicated and uninterpretable.

2. **Grey box (semitransparent) models**: There is also a solution in between, where one can check to some extent why an algorithm made a certain decision. This is called a "grey box" algorithm, for example, consider a mix of linear regression (white box) with a neural network (black box). The ultimate goal is to make as many types of algorithms as "white" as possible (Adadi et al., 2018). A trace of provenance metadata can also make a substantial difference in turning "black box" models into "grey" and / or "white" box models (Mannens et al., 2012). As with predictability, it is another way to provide proof why an algorithm made that decision, which makes the model transparent.

3. **Black box (opaque) models**: Deep neural networks, random forests, and gradient boosting machines can be considered in this category. They usually use many predictors and complex transformations. Some of them have many parameters. It is usually hard to visualize and understand what is going on inside these models. They're harder to communicate with a target audience. However, their prediction accuracy can be much better than other models. Recent research in this area hopes to make these models more transparent. Some of that research includes techniques that are part of the training process. Generating explanations in addition to the predictions is one way to improve transparency in these models. Another improvement is to include visualization of features after the training process.

Interpretability in post-modeling (post hoc interpretability) Interpretability in the model predictions helps to inspect the dynamics between input features and output predictions. Some post-modeling activities are model-specific, while the others are model-agnostic. Adding interpretability at this phase can help to understand the most important features for a model, how those features affect the predictions, how each feature contributes to the prediction, and how sensitive the model is to certain features. As said, there are local model-agnostic techniques such as SHAP, LIME, PDP, and ICE, in addition to the global model-specific techniques, such as variable importance output from random forest.

9.3.4 What are the limitations?

Taking some limitations into account from the above-mentioned solutions, one can already state that the *SHAP KernelExplainer* is slow. This makes it impractical to use when you want to compute Shapley values for many instances, as all global SHAP methods, such as SHAP feature importance, require computing Shapley values for all instances. Furthermore, *SHAP KernelExplainer* also ignores feature dependence. Most other permutation-based interpretation methods have this problem, too. By replacing feature values with values from random instances, it is usually easier to randomly sample from the marginal distribution. However, if features are dependent, for example, correlated, this leads to putting too much weight on unlikely data points. *SHAP TreeExplainer* solves this problem by explicitly modeling the conditional expected prediction. On the other hand, *SHAP TreeExplainer* can produce unintuitive feature attributions. While *SHAP TreeExplainer* solves the problem of extrapolating to unlikely data points, it introduces a new problem. *SHAP TreeExplainer* changes the value function by relying on the conditional expected prediction. With the change in the value function, features that have no influence on the prediction can get a *SHAP TreeExplainer* value different from zero. Lastly, the disadvantages of ordinary Shapley values also apply to *SHAP*, that is, Shapley values can be misinterpreted and access to data is needed to compute them for new data.

As for pinpointing some limitations from that other currently used interpretability solution framework, within *LIME* the correct definition of the neighborhood is a very hard problem, which even remains unsolved when using *LIME* with tabular data. For each application, one must try different kernel settings and see whether the explanations make any sense. Also, sampling could be improved in the current implementation of *LIME*. Data points are sampled from a Gaussian distribution, ignoring the correlation between features. This can lead to unlikely data points, which can then be used to learn local explanation models. Furthermore, the complexity of the explanation model has to be defined in advance. In all, a minor remark, as in the end the user always has to define the compromise between fidelity and sparsity. Another really big problem though is the instability of the explanations, that is, if you repeat the sampling process, then the explanations that come out can be different. Instability means that it is difficult to trust the explanations, and one should be very critical about that if it comes to interpretability.

The following scenarios illustrate *when one does not need or even does not want interpretability* of machine learning models. First, **interpretability is not required if the model has no significant impact.** Let's say Sofie is an engineering student working on a machine learning side-project to predict where her friends will go on holiday based on their Facebook data. Sofie just likes to make educated guesses about where her friends will spend their next summer. In fact, it is not a problem if her model is wrong. It is also not a problem if Sofie cannot explain the output of her model. As such, it is perfectly fine not to have interpretability in this particular scenario. The situation would change, though, if Sofie started building a business around these holiday destination predictions. If her model is wrong, the business could go bankrupt, or her model may work worse for some kinds of people as of learned racial bias. As soon as the model has a significant impact, be it financial or social, interpretability becomes relevant. Second, **interpretability is also not required when the problem at hand is well studied.** A good example here is a machine learning model for optical character recognition that processes images from postal envelopes and extracts the addresses. There already are decades of experience with these systems and they clearly work. As such, one is not really interested in gaining additional insights about this task at hand. Third, **interpretability might enable people or programs to manipulate the system.** Problems with users who deceive a system result from a mismatch between the goals of the creator and the user of a model. Credit scoring is, for example, such a system because banks want to ensure that loans are only given to applicants who are likely to return them, and some applicants aim to get the loan even if they already know the bank will not give one to them. This mismatch between the goals introduces incentives for applicants to trick the system to increase their chances of getting a loan. If an applicant knows that having more than two credit cards negatively affects his score, he simply returns his third credit card to improve his score and organizes a new card after the loan has been approved. While his score improved, the actual probability of repaying the loan remained unchanged. A system can only be tricked if the inputs are proxies for a causal feature, but do not actually cause the outcome. Whenever possible, proxy features should be avoided as they make models prone to fraud. As such, models should ideally only use real causal features because these are not gameable.

9.4 Industry examples

9.4.1 Fair loan adjudication models with Fairlearn at EY

Joakim Åström, Yanyun Hu, Mario Schlener, Jason Tuo, Yara Elias

One of the biggest barriers to current adoption of AI is a lack of trust. Professional services firm EY is therefore committed to providing the frameworks and tools that organizations need to support and monitor the responsibility application on top of their AI systems. This helps these organizations, for example, to better understand their customers, identify fraud, and security breaches sooner, and make fair loan decisions faster. The EY Trusted AI Platform primarily uses the open-source machine learning fairness toolkit Fairlearn to assess and mitigate unfairness in machine learning models, to further help their customers—and their regulators—to develop confidence in their machine learning applications.

When, for instance, a bank grants or denies a loan, the reasons must be appropriate, fair, and defensible for every application. The US Equal Credit Opportunity Act therefore prohibits banks from discriminating against credit applicants based on things like race, religion, and sex. But what if an algorithmic system denies a loan? How do you know it has done so for the right unbiased reasons? As financial services organizations begin to use AI and machine learning to optimize their operations, this has become an important question. With the ability to analyze vast amounts of data, AI holds great potential to help finance companies, for example, to better understand their customers, identify fraud and security breaches sooner, and make loan decisions faster and more efficiently. At the same time, finance companies have concerns about adding AI to their daily business practices. Exactly what factors do machine learning models take into account? How does an organization know whether its AI system is behaving unfairly? Without solid answers to these questions, many companies won't fully embrace AI. In fact, only 4 % of respondents use AI across multiple processes to perform advanced tasks today, even though 71 % consider AI an important topic for executive management, according to a recent European joint study by EY—a global leader in assurance, tax, transaction, and advisory services—and Microsoft.

"AI represents such a broad spectrum of technologies that organizations struggle to gain the skills, capabilities, and frameworks to fully assess the risks and feel comfortable that they've got them all under control," says Cathy Cobey, EY Global Trusted AI Advisory Leader. As mentioned earlier in the chapter, AI systems can behave unfairly for many reasons, including societal biases reflected in the datasets used to train them. To help customers determine and improve the trustworthiness of their AI systems, EY built their EY Trusted AI Platform. The platform identifies areas of risks and suggests ways to mitigate them. Using it also helps organizations develop a robust AI risk management system. EY's AI developers use Fairlearn to assess a model's fairness by looking at its performance across different demographics. Then they use one of the Fairlearn algorithms to mitigate any observed unfairness by retraining the model as will be thoroughly explained hereafter.

9.4.1.1 Introduction to Fairlearn usage and capability at a high level

Fairlearn, an open-source Python package released by Microsoft, aims to help data scientists and developers of AI systems to assess and improve the fairness of their systems. The design of Fairlearn reflects the understanding that there is no single definition of fairness and that prioritizing fairness in AI often means making trade-offs based on competing priorities. Fairlearn therefore enables data scientists and developers to select a fairness metric that is appropriate for their setting, to navigate trade-offs between fairness and model performance, and to select an unfairness mitigation algorithm that best fits their needs. Fairlearn focuses on negative impacts for groups of people, such as those defined in terms of race, sex, age, or disability status. Fairlearn supports a wide range of fairness metrics for assessing a model's impacts on different groups of people, covering both classification and regression tasks. The fairness metrics can be evaluated using an interactive visualization dashboard, which also helps with navigating trade-offs between fairness and performance. Besides the assessment component, Fairlearn also provides a range of unfairness mitigation algorithms appropriate for a wide range of contexts.

9.4.1.2 Fairness metrics

Fairlearn provides a wide range of fairness metrics that quantify the extent to which a model satisfies a given notion of fairness. Fairlearn covers several standard notions of fairness for binary classification as well as some additional notions appropriate for regression. These notions either require a parity in performance (e. g., accuracy rate, error rate, precision, recall) or a parity in selection rate (e. g., loan approval rate) between different groups defined in terms of a sensitive feature like "sex" or "age." One should note that the sensitive feature need not be used as an input feature to the model, as it is only required to evaluate the fairness metrics. For example, in classification settings where a more accurate prediction corresponds to a better user experience (e. g., spam detection or fraud detection), the following notions might be appropriate:

- **Bounded group loss:** The accuracy rate within each group should be above some threshold corresponding to an acceptable level of service. The corresponding fairness metric is the worst-case accuracy rate (the lowest accuracy rate across all groups).
- **Accuracy-rate parity:** The accuracy rates across all groups should be equal. The corresponding fairness metric is the difference between the largest and smallest group-level accuracy rate.

On the other hand, in classification settings where being classified as "positive" results into an allocation of resource (e. g., loan approval) and having a positive label in the data means the individual is "qualified," the following notions might be appropriate:

- **Demographic parity:** All groups should receive the positive outcome at equal rates. Equivalently, selection rates should be equal across all groups.
- **True-positive-rate parity:** The qualified individuals in each group should receive the positive outcome at equal rates. Equivalently, true-positive rates should be equal across all groups.
- **Equalized odds:** The qualified individuals in each group should receive the positive outcome at equal rates, and the unqualified individuals in each group should receive the positive outcome at equal rates. Equivalently, true-positive rates should be equal across all groups, and false-positive rates should be equal across all groups.

Note that in the loan adjudication case study that EY performed, the meaning of the "positive" label is to withhold the resource (loan), and so the meaning of positive and negative label is flipped. This has no effect on demographic parity and equalized odds (since they treat positive and negative labels symmetrically), but the interpretation of true-positive-rate parity is changed, so a symmetric notion of true-negative-rate parity might be more appropriate.

9.4.1.3 Unfairness mitigation algorithms

Fairlearn includes two types of unfairness mitigation algorithms—post-processing algorithms and reduction algorithms—that are intended to help users improve the fairness of their AI systems. Both types operate as "wrappers" around any standard classification or regression algorithm. All the constraints (which can be specified by the selected fairness metric) currently supported by reduction algorithms are group-fairness constraints. Note that the choice of a fairness metric and fairness constraints is a crucial step in the AI development and deployment, and that choosing an unsuitable constraint can lead to harms instead of desired unfairness mitigation.

9.4.1.4 Post-processing threshold optimizer algorithms

Fairlearn's post-processing threshold optimizer algorithms take an already-trained model and transform its predictions so that they satisfy the constraints implied by the selected fairness metric (e. g., demographic parity) while maximizing model performance (e. g., accuracy rate); there is no need to retrain the model. For example, given a model that predicts the probability of defaulting on a loan, a post-processing algorithm will try to find a threshold above which an applicant should be rejected. This threshold typically needs to be different for each group of people (defined in terms of the selected sensitive feature). The post-processing algorithm is based on a specific technique (Hardt et al., 2016), which takes as input an existing classifier and the sensitive feature and derives a monotone transformation of the classifier's prediction to enforce the specified parity constraints.

One can emphasize that this limits the scope of post-processing algorithms, because sensitive features may not be available to use at deployment time, or may be inappropriate to use, or (in some domains) prohibited by law, such as the ECOA. Overall, the post-processing mitigation techniques have the following common advantages and disadvantages:

Advantages:
–   The technique can be applied on any classifiers' result.
–   They have a good performance in fairness measures.
–   They do not need to modify the classifier.

Disadvantages:
–   The techniques need to access the protected attribute in test time.
–   The techniques cannot be applied to regression models.
–   There is a lack of flexibility for picking any accuracy and fairness trade-off.
–   They require to set a different threshold for each protected group to achieve fairness which may be inappropriate to use by regulation.

9.4.1.5 Reduction algorithms

At a high level, the reduction algorithms within Fairlearn enable unfairness mitigation for an arbitrary machine learning model with respect to user-provided fairness constraints. Fairlearn's reduction algorithms wrap around standard classification or regression algorithm, and iteratively (a) reweight the training data points and (b) retrain the model after each re-weighting. After many iterations, this process results in a model that satisfies the constraints implied by the selected fairness metric while maximizing model performance. Note that reduction algorithms do not need access to sensitive features at deployment time, and work with many different fairness metrics. These algorithms also allow for training multiple models that make different trade-offs between fairness and model performance, which users can compare using Fairlearn's interactive visualization dashboard.

Exponentiated Gradient and Grid Search, which are explained briefly below, are two optimization approaches under the reduction algorithms within Fairlearn. Both approaches are backed up by a mathematical theory (Agarwal et al., 2018).

Exponentiated gradient. The idea is to incorporate fairness into the training algorithm itself and framing the problem as a constrained optimization problem solvable by the Lagrange multipliers method. The Lagrange multipliers is a technique for constrained optimization, and the base problem is maximization of accuracy denoted by the predictor f over nonsensitive attributes x and sensitive attributes y. By adjusting λ, the Lagrange equation (i. e., $L(x, y, \lambda) = f(x, y) + \lambda g(x, y)$) can result in multiple solutions but all solutions will satisfy the fairness constraint imposed by the new fairness constraint g. One can refer to a published work (Agarwal et al., 2018) for details about the Exponentiated Gradient reduction algorithm, which has been implicitly implemented within Fairlearn.

Grid search. The same idea as the exponentiated gradient approach is desired which is incorporating fairness into the training algorithm itself. However, framing the problem as a constrained optimization problem in the grid search approach simplifies the problem as a deterministic searching problem by sacrificing accuracy. The grid search predefines a grid of λ values and calculates the model prediction $f(x, y)$ for each λ value, and then selects the model with the desired trade-off between accuracy and fairness. Grid search is a deterministic approach, which involves moderately violation of fairness constraints allowed by the user. For regression, the grid-search variant of the algorithm is used (Agarwal et al., 2019).

The reduction algorithms are types of in-training mitigation techniques, which have some advantages and disadvantages as listed below:
Advantages:
– Flexibility to choose the trade-off between accuracy and fairness measures based on the user's favors.

- High accuracy (exponentiated gradient).
- Computationally simplified (grid search).

Disadvantages:
- Computationally expensive (exponentiated gradient).
- Grid Search has lower accuracy compared to exponentiated gradient.

Probability of default models

In the case study, the goal was to develop probability of default (PD) models for automatic loan rejection—the setting which can be viewed through the lens of binary classification. PD is formulated here as the probability that the applicant falls behind on payments by more than 90 days during the coming year. The study was constructed with the following steps:

- *First*, a PD model (as the initial model) is trained on the historical loan application data with a standard machine learning algorithm (specifically, LightGBM) which shows unfairness across groups defined in terms of the sensitive feature "sex" even though "sex" is not used as an input feature to the model
- *Second*, Fairlearn is introduced to assess and mitigate this unfairness. In this step, two types of unfairness mitigation algorithms from Fairlearn have been explored: Post-processing threshold optimizer and reduction grid search.

From the perspective of the financial services organization, there are two kinds of adverse events caused by a classifier: false positives and false negatives. False positives are rejections of applicants that would not default, which reduces the organization's profits. False negatives are approvals of applicants that default, which increases the organization's loan-default risk. The costs of these two kinds of events are not equal.

The performance metrics of the PD model include: (1) false positive rate (FPR) and false negative rate (FNR) to measure frequency of the two adverse events, and (2) cost rate and weighted error rate to measure the cost impact on the business.

For fairness assessment and unfairness mitigation, the negative impacts on "male" and "female" groups defined in terms of the "sex" feature have been evaluated by assessing between-group differences in the occurrence of the two adverse events:

- **FPR difference**: the absolute difference between false positive rates for the "male" group and the "female" group, defined as |FPR("male")-FPR("female")|.
- **FNR difference**: the absolute difference between false negative rates for the "male" group and the "female" group, defined as |FNR("male")-FNR("female")|.
- **Equalized odds difference**: the maximum of the FPR difference and the FNR difference.

When the equalized odds difference equals zero, the two groups (i. e., "male" and "female") have equal false positive rates and equal false negative rates. This property corresponds to the standard quantitative fairness definition, which we introduced earlier, called equalized odds, hence the name. These fairness metrics, alongside many other metrics, are part of the Fairlearn module fairlearn.metrics.

With both Fairlearn's post-processing threshold optimizer and reduction grid search algorithms, the observed unfairness (in this case, the equalized odds difference) in the initial model can be mitigated without much impact on the performance metric (in this case, weighted error rate).

In addition, reduction algorithms are also used to navigate trade-offs between fairness and performance. The plot below (Figure 9.4) shows the fairness and overall performance of the initial model (at the performance-optimizing single threshold), the model obtained using the post-processing algorithm (ThresholdOptimizer), as well as several models obtained using the reduction grid search algorithm (GridSearch).

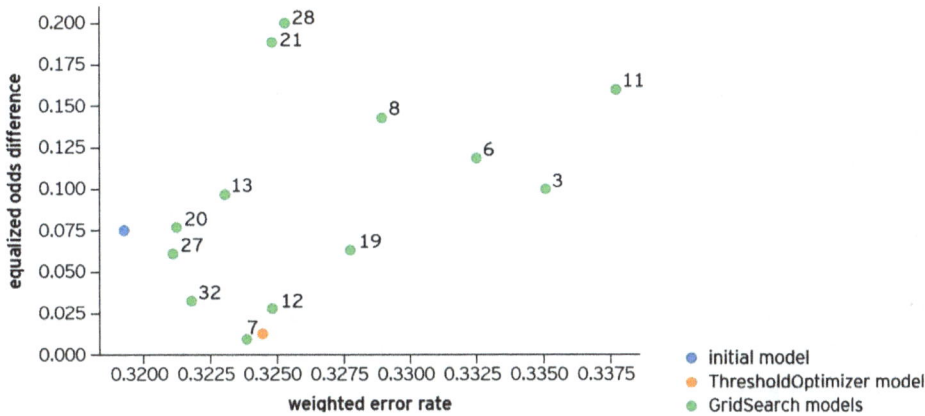

Figure 9.4: Trade-offs between fairness and performance using Fairlearn.

When EY put the Fairlearn framework to the test with real mortgage adjudication data—including transactions, payment histories, and other unstructured and semistructured data—it improved the fairness of loan decisions. Before mitigation, the models had a disparity of 7 % between men and women in the accuracy of approving or denying loans. After mitigation, the disparity was less than 0.5 %. In constructing and implementing the Trusted AI Platform, the EY team supports fairness considerations and streamlines development throughout the full end-to-end machine learning life cycle. EY data scientists and IT teams use automation, monitoring, validation, and governance capabilities as part of their machine learning operations to accelerate their work while maintaining fairness levels. "Due to scrutiny from regulators, one wants the transparency and repeatability fully integrated into the machine learning operations," says Alex Mohelsky, Partner and Advisory Data, Analytic, and AI Leader at EY Canada. "It is now understood at every stage in the process how the system is sourcing and selecting data and guiding models to a particular result, revealing any fairness issues."

EY customers using the Trusted AI Platform can choose several ways to mitigate unfairness through the Fairlearn package. That is important, because unfairness can look different depending on the context and situation. "Customers often ask, 'What's the best approach to ensuring fairness? What are the different ways of looking at fairness?'" says Mario Schlener, Partner, Financial Services Risk Management and AI Risk/Validation Lead at EY. By incorporating Fairlearn into the interactive visualizations in its Trusted AI Platform, EY gives customers the ability to navigate any trade-offs between fairness and performance and to select the mitigation strategy that best fits their needs.

EY believes that fairness and interpretability go hand-in-hand. By combining these two areas, any company can bring a level of transparency in its models that ensures greater trust in their results. "Every time a model changes with new data, its fairness and interpretability will change as well," explains Jason Tuo, Senior Manager, Quantitative Risk Advisory at EY. "For that reason, one monitors them at every step of the process." Not just customers but regulators too appreciate that ability to understand how machine learning models work. EY hopes that clarity about the inner workings of machine learning models will lead to greater trust in AI and give organizations more confidence to use it to improve their businesses and their service to all customers.

We emphasize that fairness in AI is a sociotechnical challenge, and so no software or analytical tool will "solve" fairness in all AI systems. That is not to say that software tools cannot play a role in developing fair AI systems—simply that they need to be precise and targeted, embedded in a holistic risk management framework that considers the sociocultural context of the systems being developed, and supplemented with additional resources. Fairlearn is one such tool and here it has been shown how EY and other companies can use Fairlearn to prioritize fairness in AI systems.

9.4.2 Detect and reduce fraud for loyalty services using InterpretML to respect ethical AI principles
Joakim Åström

Fraud in loyalty programs is a common problem for, for example, hotel chains, gas companies, airlines, or the retail industry. These loyalty programs are a target for fraud because loyalty points are valuable monetary assets, where scammers try to gain points in multiple ways to redeem them for their own usage. Using machine learning as an efficient tool to detect fraud is less exceptional today, but here you might find a responsible AI framework attractive for multiple reasons. When accusing someone of fraud via AI systems, it has been a top priority to also have full transparency of how the model comes to such conclusions:

> "How did the AI model come to its conclusion of fraud—as I'm actually accusing someone of fraud, I need full transparency on its decision."

One key takeaway when talking to this Microsoft customer was that they consider it extremely important to fight fraud to ensure the end-to-end integrity of their loyalty program. Deploying a transparent AI model to solve the fraud problem was a major boost for the validity and the trustworthiness of their loyalty program. For AI systems to be fully transparent, interpretability techniques come into play. As such, the customer is using the AI interpretability library InterpretML and its integration within Azure's machine learning package. By using these two packages, the customer together with Microsoft was able to debug the model, get full transparency, and perform the right features selection.

Another takeaway is that interpretability was needed for two different scenarios at different points in time. There was both a need to detect real-time fraudulent behavior when illegal points are being redeemed by scammers, but also to proactively look for historic fraudulent behaviors. The customer successfully used these AI techniques for both the historical and real-time inference scenarios.

Historical and real-time inference are jointly supported by the open source InterpretML package within Microsoft's Azure reference architecture (partly seen in Figure 9.5). Together, they cover the full chain of ethical aspects, that is, via the automated retraining feedback loop in the architecture both reliability and historical biases are taken care of. The final AI model is made interpretable via two ways: the InterpretML toolkit is used to boost transparency, accountability, enhance the privacy of the model, whereas on top different techniques of interpretability are used to support the overall process flow via Azure's machine learning package and retraining feedback loop.

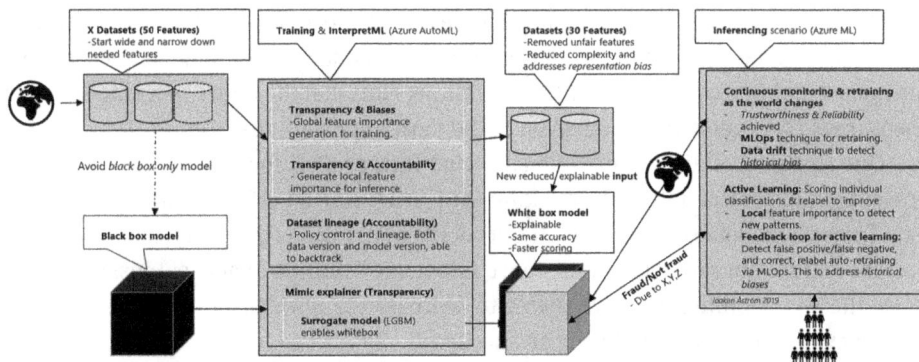

Figure 9.5: Architecture: To get a *trustworthy*, *reliable*, *transparent* fraud detection system that actively learns and adapts to real world changes.

The fraud detection model can both be served and deployed as an online web service at inference time, or as a scheduled batch inference pipeline to score the historical data. In both cases, the customer can use the scored data with further local feature importance techniques to be presented in a UI for their manual fraud team to take the final decisions based on the AI fraud alerting. By keeping *the-human-in-the-loop*, human corrective actions can be taken if AI errors would occur by flagging nonfraudulent transactions, thus the machine learning model can be retrained with the corrective data for better future fraud prognosis.

Naturally, one must ensure that by using the fraud detection model no false accusation of fraud against an innocent customer would be made. Hence, always keep a human in the equation to make the final decision. As such, it is of the uttermost important that the machine learning model is solely based on explainable factors and is not the output of a "black box."

InterpretML provides both *global feature importance* at the time of model training, and *local feature importance* at the time of real-time inference, for example, when the AI webservice is called, as such getting case-by-case importance. Indeed, both techniques are valuable but are used for different purposes, that is, global features (*"Explainability by global design"*) for training purposes and global transparency, and local features (*"Explainability by local interpretation"*) for individual transparency, bias detection, and reliability.

9.4.2.1 Applicable for airline and hotel industry

Let's look at a real-world customer example that needed both *global* and *local* interpretability and the process for this. Building a machine learning model is an interactive process, that is, one starts with a large feature set and iteratively one can use different interpretability features (using both *global feature importance* for global explanation, and *individual local feature importance* for local explanation) in the machine learning life cycle to debug and compare different training runs that the Azure machine learning keeps track of, to come to a smaller set of the most important features.

One customer in the airline industry used InterpretML for exactly these two purposes: both historical and real-time inference. They started off with a combination of data from seven data sources, and after a couple of iterations they detected the less important features to remove and the more important features to get real insights from. Using these capabilities, they were able to decrease the number of initial features immensely with ~40 % without losing any accuracy.

Reducing the number of features to fuel a model is good for many reasons: (1) autoremoval of sensitive features for privacy reasons; (2) increased performance during training and inference; and (3) decreased complexity in the model to debug, with full transparency. By using InterpretML, they boosted feature importance both during training and during inference. As such, their initial goal to both open the "black box" and have a more performant feature selection (by reducing the number of features) was achieved.

This customer example meets the goals of responsible AI by jointly using the reference architecture for active learning and InterpretML, as can be seen in Figure 9.5

showing the process with embedded techniques for a fraud prevention solution. The following ethical AI goals are being addressed by this solution architecture:

1. **Full transparency.** Interpretability is fused into the "black box" by using a *Mimic explainer* in InterpretML. In this case, it utilized a surrogate model via a *light gradient boosting machine*. With this approach, we again get a "white box" solution, which enabled full visibility on how the model came to its conclusion, what is in fact key for any accountability.

2. **Accountability.** While removing less important data by using the feature importance analysis of InterpretML, it remains key to track data sources to get accountability of the AI system. Due to the overall Azure ML integration (which incorporates this needed tracking feature within AutoML), one knows what dataset version the model was trained on. This lineage is achieved by using a technique built into AutoML of both saving the model and its training data, together with all accompanying training runs and results, all connected by a metadata versioning number.

3. **Trustworthiness and reliability.** This is case-by-case achieved by InterpretML's technique of continuous monitoring and improvement at prediction time, together with its accompanying feature importance visualization. Customers are thus able to detect both feature pattern changes and data drifting. An extra custom *feedback loop* technique (Figure 9.5) is further able to correct and retrain the model to always keep it reliable and trustworthy. This approach then addresses *historical bias*, too.

4. **Fairness.** This desired behavior is further strengthened by excluding protected features that have some risk to be unfair, yet are of low importance for the accuracy of the model.

5. **Biases.** By removing features via the feature importance technique of the surrogate model *light gradient boosting machine* to get a "white box" model, the customer also automatically removed possible *representation bias*. In the process of starting wide with 54 features and reducing it to just 31 features by iterative feature importance weighing and accompanying correlation analysis, *measurement bias* was also addressed. Within Azure's AutoML, the customer further also used the built-in cross-validation k-fold technique as an addition to a hold-out test set to further address possible *evaluation bias*, too. The default number of folds depends on the number of data rows, that is, if the dataset is less than 1,000 rows, ten folds are used. If the number of rows is between 1,000 and 20,000, three folds are used. Above 20,0000 rows, some custom folding can be chosen.

Besides addressing the main goals of deploying a responsible AI system, some further additional positive side effects were obtained, too:

- A **less complex model** was deployed, less maintenance needed, and easier to debug.
- **Ease of usage** at real-time inferences, as there were less features to consider, which in the end also gave a **better overall performance**.
- **Less cost** since less data needed to be processed **and further easier processing**. Since the customer was able to remove high cardinality features, it saved on com-

puting power due to them high cardinality features being performance heavy for the machine learning algorithms. These chucked features were autodetected by using *InterpretML* within Azure's *AutoML* machine learning platform.

By using InterpretML techniques and a Mimic explainer with a surrogate model, one gets the benefits of explaining the entire global model behavior or just individual local predictions for engineered features. Another benefit with the Mimic explainer is that it is model agnostic. The client gathered this information and used an extra visualization dashboard to interact with these model explanations. They also deployed a scoring explainer alongside the model to observe explanations during further inferencing.

During the training phase, the mimic explainer was also used. As such, the client has the option of choosing different machine learning algorithms, all to turn the "black box" model to a "white box" model. While scoring a specific scoring tree explainer was used during inference. It was used as a feedback loop technique to keep the model reliable and trustworthy, and further correct the model during retraining.

One other learning from the field of fraud detection was that by using these interpretability techniques, not only did the client gain full understanding of the models, but they also got deeper insights into the levels of fraud that was earlier hard to find. If someone tried to commit fraud on a large scale, it was relatively easy to find because they stood out. Now by using machine learning with embedded interpretability, they are much better equipped to identify fraud both on a small scale and at a large scale.

The overall benefits can thus be summarized as follows:

- Keeping the loyalty points as a TRUE premier service (hotel room, seats on an airplane, etc.) for their TRUE loyal customers without being hassled by scammers trying to downgrade their service.
- Surfacing more fraudulent transactions than earlier detection software on the same data and this by combining more real-time data sources with continuously retuning and optimizing their model.
- Detecting both the small and big frauds.
- Decreasing the number of features without losing any accuracy.
- Detecting and dropping features with high cardinality, which further lead to saving computing costs.

9.5 Final conclusions on AI ethics

As seen, AI systems can behave unfairly for a variety of reasons. Sometimes, it is because of societal biases reflected in the training data and in the decisions made during the development and deployment of these systems. In other cases, AI systems behave unfairly not because of societal biases, but because of characteristics of the data (e. g., too few data points about some group of people) or characteristics of the systems themselves. It can be hard to distinguish between these reasons, especially since they are not mutually exclusive and often exacerbate one another. Therefore, one defines whether an AI system is behaving unfairly in terms of its impact on people—that is, in terms of *harms*—and not in terms of specific causes, such as societal biases, or in terms of intent, such as prejudice. As such, theory or practice, in the end it boils down to what *harm* can be done to an individual or a group of individuals. To that extent, two types of harm can be distinguished: *allocation harm* and *representation harm*.

Allocation harm can occur when AI systems extend or withhold opportunities, resources, or information. Some of the key applications are in hiring, school admissions, and lending. *Representation harm* can occur when a system does not work as well for one person as it does for another, even if no opportunities, resources, or information are extended or withheld. Examples here include varying accuracy in face recognition, document search, or product recommendation.

Therefore, the way forward to get technical solutions that are as *fair* as possible, we need to adopt an interdisciplinary way of working. As such, these ethical AI issues are at least as much *societal* as they are *technical*. Hence, one needs to create maximum awareness of the possible flaws at hand in the current solutions and put a maximum effort on developing the right software tooling, among others, robust and trustworthy AI algorithms via interdisciplinary teams, that is, social scientists together with law and business researchers, data engineers, and computer science researchers. Let's hope that this chapter pointed you toward possible solutions in the *bias and interpretability* realm of AI and above all created the awareness of the current pitfalls thereof. In the end, when it comes to the fairness of machine learning, one always has to ask the following two questions and act accordingly during all phases of any AI & software development project: (1) "Who is going to benefit from the systems one is building?" and (2) "Who might be harmed?"

Bibliography

Adadi A. and Berrada M. Peeking Inside the Black-Box: A Survey on Explainable AI (XAI). IEEE Access, 6:52138–52160, 2018. doi: https://doi.org/10.1109/ACCESS.2018.2870052.

Agarwal, Beygelzimer, Dudik, Langford, and Wallach. A Reductions Approach to Fair Classification, In ICML, 2018, https://arxiv.org/pdf/1803.02453.pdf.

Agarwal, Dudik, and Wu. Fair Regression: Quantitative Definitions and Reduction-based Algorithms, In ICML, 2019, https://arxiv.org/pdf/1905.12843.pdf.

AI, https://www.sas.com/en_us/insights/analytics/what-is-artificial-intelligence.html.

Apley et al. 2019, https://arxiv.org/pdf/1612.08468.pdf.

Azure's machine learning, https://review.docs.microsoft.com/en-us/azure/machine-learning/how-to-machine-learning-interpretability?branch=pr-en-us-119762.

Azure's machine learning package, https://github.com/Azure/MachineLearningNotebooks.

Calmon et al. 2017, http://papers.nips.cc/paper/6988-optimized-pre-processing-for-discrimination-prevention.

Celis et al. 2018, https://arxiv.org/abs/1806.06055.

Chen et al. 2018, https://arxiv.org/pdf/1802.07814.pdf.

Chouldechova et al. 2018, http://proceedings.mlr.press/v81/chouldechova18a.html.

Fairlearn, https://fairlearn.org/.

Feldman et al. 2015, https://dl.acm.org/doi/10.1145/2783258.2783311.

Friedman. 2001, https://statweb.stanford.edu/~jhf/ftp/trebst.pdf.

Global Surrogate Models, https://christophm.github.io/interpretable-ml-book/global.html.

Goldstein et al. 2017, https://cran.r-project.org/web/packages/ICEbox/index.html.

Hardt Price, and Srebro. Equality of Opportunity in Supervised Learning, In NeurIPS, 2016, https://papers.
    nips.cc/paper/6374-equality-of-opportunity-in-supervised-learning.pdf.

Howard et al. 2018, https://link.springer.com/article/10.1007/s11948-017-9975-2.

IBM's initiated AIF 360 tool kit, https://github.com/Trusted-AI/AIF360.

Impact, https://christophm.github.io/interpretable-ml-book/interpretability-importance.html.

Implicitly-gendered words, https://www.reuters.com/article/us-amazon-com-jobs-automation-insight/
    amazon-scraps-secret-ai-recruiting-tool-that-showed-bias-against-women-idUSKCN1MK08G.

In the case study, EY & Microsoft Whitepaper. Assessing and Mitigating Unfairness in Credit Models with
    the Fairlearn Toolkit, 2020.

Kamiran et al. 2012, https://link.springer.com/article/10.1007/s10115-011-0463-8.

Kamiran et al. 2012, https://ieeexplore.ieee.org/document/6413831.

Kamishima et al. 2012, https://rd.springer.com/chapter/10.1007/978-3-642-33486-3_3.

Kearns et al. 2018, https://arxiv.org/abs/1711.05144.

Lundberg et al. 2017, https://arxiv.org/pdf/1705.07874.pdf.

Mannens E., Coppens S., Verborgh R., Hauttekeete L., Van Deursen D. and Van de Walle R. Automated Trust
    Estimation in Developing Open News Stories: Combining Memento & Provenance. In 2012 IEEE 36th
    Annual Computer Software and Applications Conference Workshops, Izmir, 2012, pages 122–127, doi:
    https://doi.org/10.1109/COMPSACW.2012.32.

Mehrabi et al. 2019, https://arxiv.org/pdf/1908.09635.pdf.

Microsoft's initiated FairnLearn, https://fairlearn.github.io/.

Osoba et al. 2017, https://www.rand.org/content/dam/rand/pubs/research_reports/RR1700/RR1744/RAND_
    RR1744.pdf.

Petkovic et al. 2019, https://www.biorxiv.org/content/10.1101/819078v1.full.pdf.

Pleiss et al. 2017, https://papers.nips.cc/paper/7151-on-fairness-and-calibration.

Ribeiro et al. 2016, http://sameersingh.org/files/papers/lime-kdd16.pdf.

Same level of risk, https://www.washingtonpost.com/news/monkey-cage/wp/2016/10/17/can-an-
    algorithm-be-racist-our-analysis-is-more-cautious-than-propublicas/?utm_term=.2276d78de3c1.

Same proportion, https://www.propublica.org/article/machine-bias-risk-assessments-in-criminal-
    sentencing.

Selbst et al. 2019, https://dl.acm.org/doi/10.1145/3287560.3287598.

SHAP explainers, https://github.com/slundberg/shap.

Shrikumar et al. 2017, https://arxiv.org/pdf/1704.02685.pdf.

Simonyan et al. 2013, https://arxiv.org/pdf/1312.6034.pdf.

Study by EY, https://pulse.microsoft.com/en/business-leadership-en/na/fa1-articial-intelligence-report-at-a-
    glance/.

Suresh Harini and Guttag John, https://arxiv.org/pdf/1901.10002.pdf.

Verma et al. 2018, http://fairware.cs.umass.edu/papers/Verma.pdf.

Wan et al. 2020, https://arxiv.org/pdf/2004.00221.pdf.

Yang et al. 2018, https://arxiv.org/pdf/1802.04253.pdf.

Zemel et al. 2013, http://proceedings.mlr.press/v28/zemel13.html.

Zhang et al. 2018, https://arxiv.org/abs/1801.07593.

Oussama Chelly and Hendrik Blockeel

10 Industry examples where different AI techniques are combined

In this chapter, we demonstrate how applications can combine different artificial intelligence (AI) techniques from the previous chapters for more advanced applications and how they can cooperate.

The first example shows how an AI based chatbot helps **KBC Group**, a Belgian bank-insurance group, to automate 50 % of the processing of the insurance claims using a chatbot. The second example from the **manufacturing industry** uses a combination of **machine learning** methods and **symbolic AI** techniques to offer a digital engineering assistant that can automatically extract relevant information from engineering drawings and assist the engineers with their choice.

10.1 An AI-enabled chatbot for the Casco[1] insurance industry, an example from KBC Group

Oussama Chelly, Michaël Mariën

Motor-vehicle insurance has been the largest nonlife insurance market over the past decade. In Europe, it accounted for 36 % of the global Property and Casualty (P&C) market with the total motor premium income amounting to €149 bn in 2020.

With over three thousand insurers fiercely competing in the European market, improving the insurance offering for the B2C segment became crucial for every insurer to gain a competitive edge. Consequently, competition was no longer limited to offering better price-to-risk ratios but extended from competition in pricing to competition in the **quality of services** being offered to the end-customer. In this context, **digitally enabled vehicle insurance services** have been gaining more prominence and were estimated to cover as high as 12 % of the vehicle insurance market in Europe as of 2020.

Statistics reflect a **shift in consumer preferences** from the traditional B2C transactions usually carried out in person, over the phone, or by mail to a more digital communication. The change prompted more competition in the race to the digital transformation of B2C insurance policies.

From the consumer point of view, the **insurance claim process** has been—and still is—widely considered to be long and tedious. While the process length and complexity may vary from country to country based on the local regulations, and from insurer to

---

1 Casco stands for CASualty and Collision (automobile insurance).

insurer based on the service license agreements, the driver is rarely satisfied with the quality of service. It should be highlighted that a driver involved in an accident is typically not in the best psychological condition to objectively evaluate the quality of the insurance service. To add more context to the situation in which such claims typically take place, one should rewind back to the moment when the vehicle is involved in a traffic accident. From that moment, the driver is usually under a high amount of stress. Besides any physical injury, post-traumatic effects, or legal consequences, the driver could be faced with the financial implications of healthcare assistance, repairing his vehicle, as well as being deprived of it for an unknown amount of time. That is without mentioning the impact of the accident on his personal and professional plans. Consequently, the claim handling process becomes very sensitive.

In the traditional process displayed in Figure 10.1, the customer notifies the insurance company about the accident by contacting an agent from the company over the phone. In addition to the notification, the customer indirectly expresses his insecurity by inquiring about his current insurance policy. The type of policy, the deductible, and the financial limits are among the most frequently asked questions. Then the customer registers the data about the accident. This data includes the circumstances leading to the accident, information about any other vehicles or persons that were involved, and damage to the parties involved. Once the data is registered, the customer waits while the claim is being processed by the insurance. During this time, the agent has to manually log the data in the company's system and communicate the case to the claim handler. The latter checks for potential fraud, verifies the coverage and liability of the driver, and determines the value of the damage, before issuing the payment to the customer. This process lasts anywhere between a few days and a few weeks. Customer-obsessed insurance companies typically perform this process within 2 to 3 days. While this waiting time is reasonable and competitive, it leaves the customer in a situation of uncertainty, insecurity, and anxiety. In most cases if not all, he is only relieved when the claim is paid out.

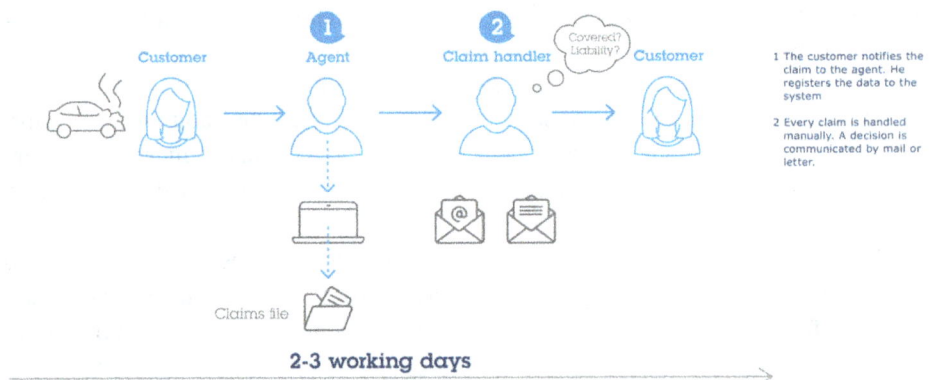

Figure 10.1: The traditional process of claim handling in car insurance.

From the insurers' point of view, the claim processing is a time-consuming and costly endeavor. Even in the most straightforward claim cases, the decision is manually made by a claim handler. Any partial automation of the process would significantly enhance the claim handling capacity of the company, increase consistency of the claim decisions, decrease the costs of handling, and reduce the processing time for both the company, and more importantly, the customer.

Many insurers have invested significantly in the modernization of their services to automate several parts of the process. **KBC Group**, an integrated bank-insurance group from Belgium (see chapter 7), offers a **AI-enabled service** to its customers since November 2018.

In KBC's service, the claim process has been drastically shortened in particular for the straightforward cases (*cf.* Figure 10.2). When a vehicle has an accident, all the driver needs to do is to connect to the insurance app on his phone to start the process. Additional digital channels including the company's website are also an option that the customer can opt for. The process that follows can be divided into three phases: categorizing the claim, assessing the damage, and making a decision.

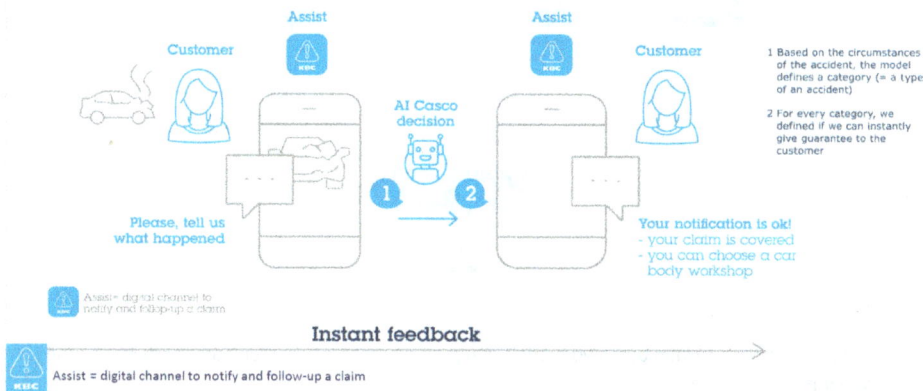

Figure 10.2: The modern claim handling process for car insurance in the KBC Group application.

In the first phase, the customer interacts with a **chatbot**, which plays the role of the company's agent. The chatbot is implemented using **Rasa**, which is an open-source framework for building intelligent chatbots. The chatbot can be decomposed into an **input module**, which uses **natural language understanding** (NLU) algorithms to understand user inputs, and an **output module**, which performs **natural language generation** (NLG) to produce human-like text. The NLU module starts with vectorization to convert the text into a **vector**, then a classification to associate the vector with an **intent**. In parallel, each sentence is tokenized, then chunked, before **named entity recognition** is performed and, therefore, entities are identified. With Intents and Entities identified from each step in the discussion, business logic is used to allow the chatbot to react to every user input through the output module. This output module relies on a **long-short-term memory** (LSTM) network. As explained in Chapter 7, this type of **recurrent neural network** (RNN) has the ability to maintain a neural representation of the dialog history. Consequently, the context of each sentence in the dialog is inherently maintained.

Besides answering the user's questions on the process and insurance policies, this chatbot captures the story of the accident in text format. The text is then forwarded to two natural language processing (NLP) **classification models** in order to categorize the accident in one of many categories such as "collision with an animal" or "collision with a vehicle." The first of the two models is a **k-nearest neighbors (KNN)** model, which is a well-established classification technique, while the second is a more recent **recurrent neural network** (RNN). The use of two models is motivated by the "four-eye principle" as explained in Figure 10.3.

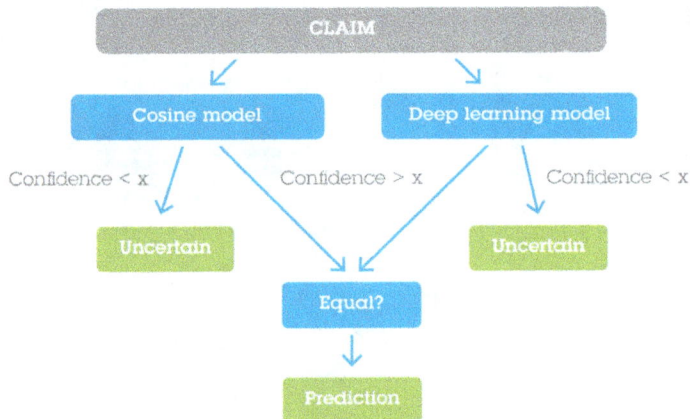

Figure 10.3: Handling the claim uses two models. A decision is only made when both models yield the same category.

In order to use the k-NN model, the customer description of an accident is first transformed into a binary vector indicating the presence of certain predefined key words. Let's assume that our keywords are the following: (car, deer, highway, wall). If the customer says "I was driving on the highway when a deer crossed the road, and I could not avoid it," then the corresponding vector would be $(0, 1, 1, 0)$. The resulting vector is then compared to a set of known vectors extracted from a historical database. The distance metric used to assess similarity is the **cosine similarity**. In other words, the similarity between two vectors is measured by the cosine of the angle separating them. After comparing the vector to those in the historical database, the k **most similar vectors** are extracted. The vector is labeled with most represented category in those k vectors.

To use the RNN model, the phrase provided by the driver is first tokenized. After this process, we obtain a sequence of numbers representing the position of each word in the tokens' dictionary. The vector is then fed into a series of **long short-term memory** (LSTM) networks, which can transform the representation of the vector based on the words interdependencies. Therefore, the order of the words plays an important role in the classification. Finally, a dense network is used to label the transformed vector and, therefore, the original corresponding sentence.

Based on the test data, both models agree on a category with a high confidence level in 65.2 % of the cases. On these cases, the **accuracy of the classification** is 99.4 %. In this scenario, the claim is labeled with the detected category and **processed instantly**. In the other scenario where the models yield a low confidence score or disagree on the category, the claim is then **processed manually**.

In the second phase of the claim, the customer submits photos of the accident. The photos are processed using a **convolutional neural network** to identify the type and extent of the damage. This data is then crossed with information on the car to assess the cost of repairs.

In the third and final phase, the insurance app is able to retrieve the customer data and instantly confirm if his insurance policy covers the claimed accident. The chatbot also offers the customer the chance to ask his questions which are then categorized in the same way using an LSTM model to reply with the appropriate answer.

The whole process takes less than a minute before the claim is categorized and resolved. This alleviates the uncertainty on the customer side. It is worthy to mention that during the process the claim data is processed in the background to **detect potential fraud** which was covered in Chapter 7.

In order to improve the quality of its models, KBC implemented a **retraining loop** for the claims where the two models were uncertain or did not agree (Figure 10.4) . In addition to those claims, 10 % of all claims labeled by the two models are evaluated by

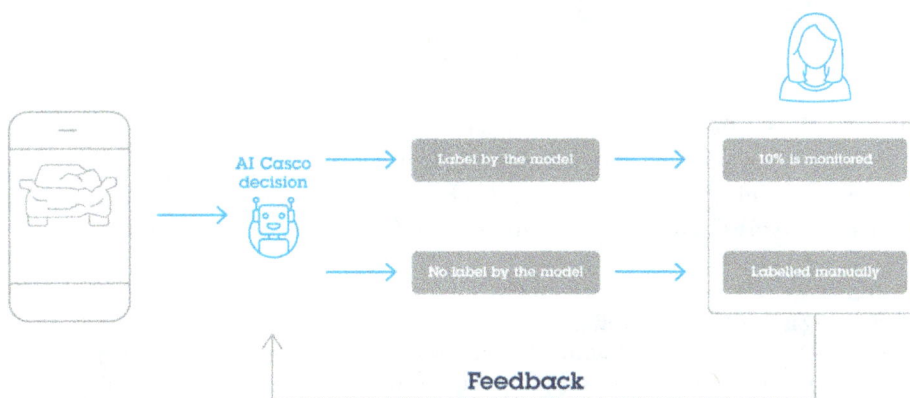

Figure 10.4: The retraining loop in the AI models used for claim categorization.

Figure 10.5: The distribution of claims based on their complexity in the traditional and new claim handling process at KBC Group.

human agents to **monitor the quality** of the model. All manual labels are then fed back into the training of the models to improve classification quality.

The outcomes of the new product are overwhelming for KBC. In fact, **50.4 % of the claims** are currently fully automated (*cf.* Figure 10.5). Moreover, by replacing human intervention with AI in the core process of claim handling, the organization has become more cost-efficient and focused on complex claims. With AI processing half of the claims, the complex claim can get more time and attention, which results in better service quality for the insurance customers.

10.2 An automated engineering assistant that uses a mix of learning and reasoning techniques in manufacturing[2]

Hendrik Blockeel, Wannes Meert, Joost Vennekens

This example is taken from a collaboration between **KU Leuven University** and a multinational company active in design and manufacturing. The company produces parts for all kinds of machines. Given a specification of the functionality of some required part, this part needs to be designed and manufactured. The goal of the project was to create an

---

2 Section 10.2 is based on two scientific papers: (1) Van Daele et al., 2021; (2) Aerts et al., 2022. The research was supported by Flanders Innovation & Entrepreneurship (VLAIO O&O project 'Digital Engineer'), the Flemish Government ("Onderzoeksprogramma Artificiële Intelligentie Vlaanderen"), and the European Research Council (ERC) (Horizon 2020 research and innovation programme, grant agreement No. 694980, SYNTH: Synthesising Inductive Data Models).

automated engineering assistant using AI technology to help with this. The assistant should allow design engineers to work more efficiently by better **disclosing expertise** already available in the company. Below, we provide more context and give details on the solution that was developed during the project.

10.2.1 The problem setting

The main type of document used by design engineers is the **technical drawing**. Such a drawing typically consists of 2D and 3D visual descriptions of machines or parts, together with annotations such as measurements, a bill of materials, etc. When designing a new object, engineers produce such a technical drawing.

A customer will typically contact the company with a requirements specification for the product they need. This includes its functionality, the conditions under which it will operate (e. g., extreme temperatures), and so on.

Sometimes, a standard solution is already available for what the customer needs, and the sales department can immediately handle the order. When that is not the case, an engineer is faced with the task of **designing a new product**. Often, they can start from a basic type of design that they know was already deployed and evaluated in the field and adapt it to the needs of the customer; more rarely, they need to design something from scratch.

Engineers obviously use their own expertise when designing a product, but they do not have direct access to their (former) colleagues' expertise. The company therefore keeps a database in which earlier designs are stored, so that engineers can tap into it. Being able to find relevant earlier designs that are close to what is needed in a new use case can boost the **engineers' productivity**. Moreover, such a database helps retain to some extent the expertise of retired engineers.

At the start of this project, a large **database with product designs** was available. This database is linked to databases on sales and after-sales that provide additional relevant information (e. g., what kind of unexpected problems were frequently encountered with a given design). Partially because the database spans many years of expertise, different companies, and multiple regions, it is very **heterogenous**: recent technical drawings are typically stored in a digital format, but older ones are simply scans of drawings on paper. This database can be **searched based on keywords**. There is considerable variance in what keywords are used to describe a design: terminology may differ between different company locations and even among engineers at one location; new types of materials become available over the years; insights on what are the most relevant keywords evolve; typing mistakes, etc. Thus, there is **heterogeneity** not only in the designs themselves but also in the **metainformation** about them. This makes it hard to search the database effectively. Looking up information can take a substantial amount of time and effort from the engineer: it may take many attempts before a sufficiently good combination of keywords is found (one that yields a relatively small set of previous designs that are sufficiently relevant to be useful). Even then, there is no guarantee of completeness: perhaps the best design to start from is not even in this set.

A better way of disclosing the database could significantly increase the efficiency with which engineers can do their job. Hence, **one goal** of the project described here was to build an AI-based system that helps the engineer to find **relevant information** in this database.

By themselves, however, relevant past designs provide only a limited amount of information. For instance, the engineer has no way of knowing whether the past design was actually successful. In addition, the solutions that used to be optimal, perhaps are no longer optimal today (e.g., because new, superior materials have been invented). Perhaps most importantly, a design drawing tells the engineer which design choices were made, but not *why* these choices were made.

The company therefore also had a **second goal**, which was to extract the knowledge of key senior engineers and to explicitly store it in a formal **knowledge base**, such that it will remain available for future generations of engineers. In addition, this knowledge can then be used to provide **flexible and explainable decision support** to the engineers.

10.2.2 How are those problems solved ? A mix of techniques from Chapters 4, 5, and 7

The developed software uses multiple AI technologies to assist the design engineers: it combines **computer vision, inductive logic programming, pattern mining, knowledge representation, logic reasoning, and constraint reasoning** (Chapters 4, 5, and 7).

10.2.2.1 Reading the drawings

Technical drawings consist of a 2D and 3D drawing (the "CAD" drawing) together with annotations including measurements, a list of parts and/or materials, and so on. A lot of relevant information is in the 2D drawing itself. A vision component was developed that can analyze a drawing and extract relevant information from it. This vision component reads the drawing as a bitmap image (so it works as well with scans of designs on paper as with digital drawings). The image is first segmented using standard **computer vision** methods, and segments are then **classified** as "table," "two-dimensional CAD," or "irrelevant." This classification determines the next processing step: table segments are handled differently from CAD segments, whereas irrelevant segments are ignored. The segmentation and classification were found to be 100 % accurate in the available data (which is in line with the fact that line drawings are generally not very hard to segment).

10.2.2.2 Reading the tables

Data in tables are organized partly through annotation (e. g., column or row titles) and partly through positioning (e. g., all cells below some column title belong to that column). Different tables may have a different organization, however, and tables do not always have a simple matrix form (m rows, n columns): cells may span multiple columns or rows, a cell may contain a subtable, etc. The system therefore needs to learn how to parse tables.

An **inductive logic programming** (ILP) approach was used for this. The ILP system takes as input, descriptions of cells (cell text, cell location), relational information derived from this (relative cell positions, neighboring cells, the order in which cells occur), and labels of the cells. It produces as output "mini programs" that state how to derive the label of a cell from the other information. An example of a rule that the system finds is:

$$\text{author(A):- cell\_contains(B, "drawn"), above(B, A),}$$

which states: when the cell above this one contains the text "drawn," this cell contains the name of the author. Figure 10.6 shows another example of a program and how it interprets a table.

Figure 10.6: Example of a table and a mini-program defining the concepts "bill of materials" and "header." The highlighting shows what the program defines; it is not part of the original drawing. (Figure by Van Daele et al., 2021.)

Rules similar to these were introduced for all labels. Some cells can be classified quite well using such basic rules, others are harder to classify accurately. The overall quality was substantially improved by introducing a novel element into this ILP approach, called **bootstrapping**. The basic idea is to let classification rules for difficult labels exploit the results of other classification rules for easier labels. A dependency graph is constructed, where labels are ranked according to the accuracy with which they can be predicted and the size of the program predicting it. After a first learning run, simple rules with high accuracy are added to the background knowledge that the ILP system can exploit, then a second learning run is made that can exploit the additional knowledge that has become available in this way. This process is repeated for each consecutive run. One task that benefits from this procedure is to recognize the bill of materials, where first the concept of a header is detected, after which recognizing rows of materials is easy.

10.2.2.3 Reading the CAD drawings

To perform searches based on CAD drawings, a ***meaningful* similarity** measure for such drawings needs to be available. Specifically, two drawings should be considered maximally similar if they represent the same design (making abstraction of rotations, translations, mirror symmetries, etc.) To learn a suitable similarity measure, **self-supervised learning** is used: for each image, 10 more images are constructed with irrelevant variations of the original (e. g., rotating the image); then a "**siamese network**" is trained that for any pair of images should output whether they represent the same design or not (using so-called ***contrastive learning***, where pairs of images derived from the same original are positive pairs, and random samples of image pairs are negatives). The siamese network processes each image in a given pair using the same network (a convolutional neural network with ResNet architecture), then combines the outputs of these networks using a few fully connected layers. In this way, a neural network is trained that can assess to what extent two drawings represent the same design.

10.2.2.4 Identifying relevant designs

With the functionality described in previous sections, it becomes possible to define a measure for the **similarity** between two designs. Each design is first represented using two feature vectors:
- The first relates to the **tables**. As the description of the table resulting from 2.2 uses a logical format, an ILP system called Warmr is used to find **frequent patterns** in the logical description that are likely relevant for determining similarity. A feature is introduced for each relevant pattern. These features form the "tabular" feature vector. For tabular feature vectors, which have binary values, similarity is defined as the proportion of features that have the same value.

– The second feature vector, which relates to the **CAD part**, contains the nodes in the penultimate layer of the **siamese network** that determines whether two drawings represent the same design (these are obviously relevant for the similarity between designs). For these CAD feature vectors, the cosine similarity measure is used.

The overall similarity between two designs is finally defined as the geometric mean of both similarities.

10.2.2.5 The knowledge base

To complement the database of designs, a **knowledge base** was built that captures the knowledge of key domain experts. To construct this knowledge base, several interactive workshops were held, in which experts from different sites worldwide participated, guided by a knowledge engineer. The knowledge base is written in the **FO(.) language**, which is a rich extension of classical **first-order logic** (Chapter 4).

During such workshops, it is important to represent the knowledge in a **formal language** that not only the knowledge engineers but also the domain experts can understand. In this way, the domain experts can immediately check whether the knowledge engineer has correctly understood what they are saying, which greatly reduced the number of mistakes that end up in the knowledge base.

While the FO(.) language is powerful and easy to use for trained experts, it can be challenging for people who first encounter it. The workshops therefore made use of the decision model and notation (DMN) standard, and its extension cDMN. This offers an intuitive table-based representation, which has been specifically developed to be usable by domain experts. An example of a cDMN table is shown in Figure 10.7.

Component Materials				
E*	**Component**	**Component is Used**	**Design Type**	**Material of Component**
1	Body	True	-	M1, M2, M3
2	Spring , Spacer	True	-	M1, M3, M5
3	-	False	-	null
4	Body	True	Closed	Not(M2)

Figure 10.7: A cDMN table that defines which materials can be used for which components in which types of design. For instance, in a closed design type, any material apart from M2 can be used to manufacture the body of the component.

10.2.2.6 Interactive decision support

To build a usable **decision support** tool, it is key that the tool can adapt to the way of working of the engineers, rather than forcing the engineers to adapt their way of working to the tool. A decision support tool was therefore developed using KU Leuven's

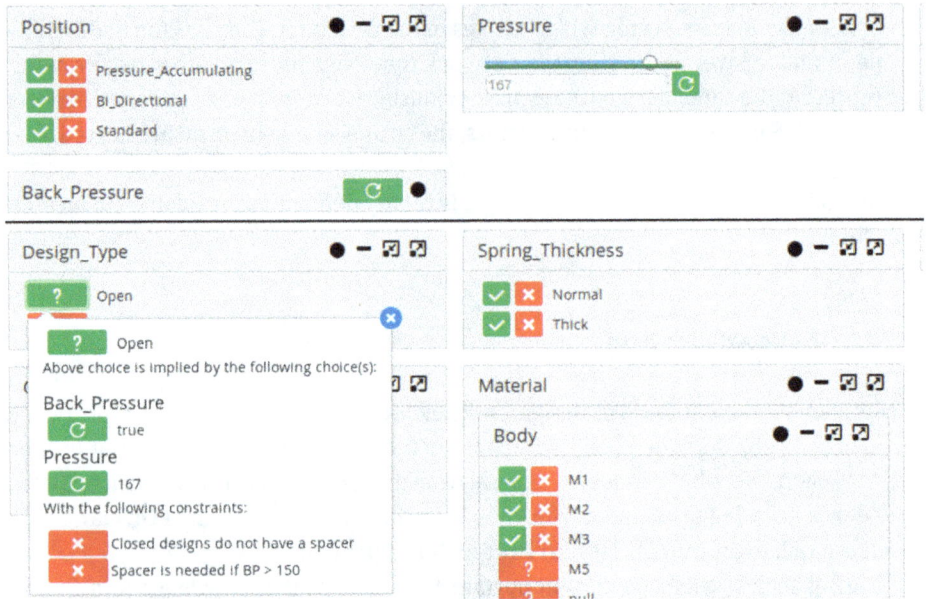

Figure 10.8: The Interactive Consultant providing an explanation for why an open design must be used. This is the case because the engineer has entered the requirements that the design should be able to release back pressure and to cope with pressures up to 167. Other information that the system has already derived is that material M5 cannot be used to manufacture the body of the component.

generic Interactive Consultant interface (idp-z3.be), which is powered by the **IDP-Z3 reasoning system** for FO(.). Figure 10.8 shows a screenshot of this interface.

An important property of this system is that it adheres to the ***knowledge base paradigm***: the knowledge base itself is purely a declarative representation of knowledge (i. e., by itself, it does not *do* anything), to which different logical inference algorithms can then be applied to derive different kinds of conclusions from different input. In this way, the system can give the engineer the freedom to work in whichever way they choose. They can start from the requirements, they can start from a specific design, they can start by choosing materials, etc. Whatever information the engineer chooses to enter, the system will use the knowledge base to derive further conclusions from this. In this way, the engineer and the AI system cooperate to gradually reduce the number of options that remain, until finally a single, complete design remains. Because all the information that the system provides is derived by means of **logic reasoning** from the knowledge base that has been constructed and verified by the domain experts, information coming from this system is at least as reliable as information that an expert would provide themselves. Moreover, all the output is also **explainable**, in the sense that the system can always point to a precise combination of choices made by the engineer and parts of its knowledge base that suffice to reach this output.

At any point during the design process, the engineer can use the same user interface to inspect the database of previous designs. In this case, the current state of the design process is used to return only designs that match this specific context. Moreover, the properties of each design are shown using the same concepts that are used in the configuration interface. This makes it easy for the engineer to spot the key differences between different designs and to copy relevant parts over to their design.

10.2.3 How well does it work?

The time that engineers need to come up with a good design given specifications was substantially reduced. The time spent searching the database to see what already exists, what were the problems with some earlier designs, which designs were successful, and which were not (and under what circumstances) is reduced significantly, by 15–30 minutes per use. This comes on top of the fact that engineers rarely spend more than an hour on this search: if they cannot find anything fast enough, they start designing from scratch, which may take dozens of hours. The AI provides much smarter access to the database, and the interaction allows the engineer to quickly zoom in on the most relevant cases. The integration into a configurator allows the engineer to quickly come up with new variations that meet specific requirements.

Bibliography

Aerts B., Deryck M., and Vennekens J. Knowledge-based decision support for machine component design: A case study, Expert Systems with Applications, 187, 2022.

Van Daele D., Decleyre N., Dubois H., and Meert W. An Automated Engineering Assistant: Learning Parsers for Technical Drawings, In Proceedings of the 33rd Annual Conference on Innovative Applications of Artificial Intelligence, 2021.

Emmanuel Gillain

11 Conclusion – Moving forward

Academic review by Professor Hendrik Blockeel

11.1 So far...

By this stage, readers are expected to have gained a fundamental understanding of various artificial intelligence (AI) techniques and their applicability in addressing common industry challenges. This book provided a comprehensive, albeit non exhaustive, overview of prevalent AI techniques that can help humans to make decisions more effectively and efficiently by

- **searching** and **planning** (Chapter 3),
- **reasoning** with symbolic AI algorithms, which explicitly represent knowledge and reason over that knowledge using different syntaxes and logics such first-order logic (Chapter 4) or descriptive logics (Chapter 5),
- **reasoning in nondeterministic environment**, with probabilistic graphical models that can capture uncertain knowledge and infer conclusions (Chapter 6), and
- **learning**, from raw data, or from rewards in interacting with the environment (Chapter 7).

In Chapter 8, we have learned how **natural language processing (NLP)** methods enable human-machine communication. These methods allow machines to either produce or understand human language, often using many of the methods of the other chapters, while also possessing their own unique features. By elucidating the limitations of various AI methods, we hope that the reader has also gained some insight into the challenges that the current AI-based systems encounter. Furthermore, we discussed how the shortcomings of certain techniques can sometimes be mitigated by others. Chapter 9 particularly focused on addressing some of the **ethical challenges** that AI systems face. Chapter 10 illustrated how different AI techniques can also be combined together for even richer applications.

11.2 Moving forward

The field of AI is evolving extremely rapidly to address the limitations of the current state of the art. This book does not intend to advocate for any specific direction in AI research. Most likely, no one has a definitive answer as to which directions will prove most promising. We suggest however to conclude the book by delving into some promising prospects: the concepts of **hybrid AI**, seen by a large community as one possible promising approach that reconciles the two fundamental AI paradigms covered in this book, as well as the principles of **causal AI**, also known as causal inference or reasoning, as it represents a significant and relatively recent paradigm shift. Causal inference theory fills indeed an important theoretical gap in the AI techniques toolkit, paving the way towards stronger or more general AI. This theory focuses on studying cause-and-effect relationships to address questions about the effect of one variable on another. Finally, we'll end with a few additional words about **generative AI (GenAI) trends**, as generative AI is revolutionizing the way people work and learn and will have significant effects on the business and society overall.

11.2.1 Hybrid AI systems to merge the best of both worlds

Symbolic AI and **statistical data-driven AI** have unfortunately been somehow disconnected disciplines in practice whereas they offer obvious complementarity: machine learning (ML) and deep neural networks are very powerful at certain types of learning, modeling, and action[1] but have currently limited capability for abstraction, symbolic reasoning, and for the inclusion of prior structured knowledge. Traditional machine learning techniques also have challenges to **generalize** the learning if the environment differs, sometimes even in small ways, from the context on which they are trained. And it is still uncertain and controversial among researchers how well **large language models** in the field of NLP can do abstraction and reasoning, and how far they can go.[2]

Symbolic AI, on the other hand, is powerful at modeling and reasoning over abstractions, can easily integrate prior structured knowledge, is compositional (i.e., can combine different concepts), but deals poorly with empirical data to establish links and correlations to make new hypotheses from raw data. Accordingly, hybrid AI systems, which bring the strengths of both statistical and symbolic approaches together, are seen by some researchers in the AI community as the way to move toward stronger, more general AI. It seems to be one of the promising domains of research: the capacity to learn from large scale data sets associated with the capacity of symbolic AI to represent abstract representations[3] and reason about those.

So-called **neuro-symbolic AI** systems, for example, pursue such a concept. Deep neural networks are used for their pattern recognition capabilities, which include feature learning and feature engineering. These networks learn domain knowledge from raw data. This knowledge is then represented in a structured symbolic form using a logical representation. Following this, symbolic AI is employed to apply logical reasoning to this structured data. This process uncovers new facts and relations And vice versa:

---

1 when applied in the context of reinforcement learning.

2 This is also an active area of research.

3 especially when there are only small datasets.

knowledge about an application domain can help guide the learning process by inject-ing prior knowledge. The main concept of neuro-symbolic AI systems is well expressed by Sebastian Bader and Pascal Hitzler: *"A symbolic system at the front-end is used to provide symbolic (partial) expert knowledge to a neural or connectionist system that can be trained on raw data, possibly considering the symbolic knowledge that is represented internally. Knowledge acquired through the learning process can then be extracted back to the symbolic system (which now also acts as a back end), and made available for fur-ther processing in symbolic form"* (Sebastian Bader and Pascal Hitzler, *"Dimensions of Neural-symbolic Integration—A Structured Survey,*[4]*"* November 2005).

Interested readers will find a recent survey about the advancements in neural-symbolic or neuro-symbolic systems from distinct perspectives in the research paper titled "A Survey on Neural-symbolic Learning Systems" (Dongran Yu, Bo Yang, Dayou Liu, Hui Wang, Shirui Pan, June 2023).

11.2.2 Cause-effect and causal AI

« *Correlation isn't causation* » is probably a quote familiar to the reader. Classical statis-tics and ML techniques don't capture cause-effect relationships. For a long time, they have focused much on **correlation** without uncovering the real **causal relationships**. They uncover patterns, find correlations in the data, make predictions but do not pro-vide a **language for causality** with a causal model. Analyzing the $P(Y|X)$ by observ-ing the impact of the X values on Y doesn't guarantee to find causality. To estimate the causality of a variable X (like a drug) on the variable Y (like the recovery), the effects of a **confounding variable** Z that can influence either the taking of the drug or the recov-ery (like the age or a healthy life) must be removed. Even if the designer of a Bayesian network models his network with causal links between variables, the fundamental as-sumption of his model remains (e. g., independence, conditional independence, etc.).

Models that can capture causal relationships are more generalizable and allow us to understand what would happen if some of the assumptions in the model change, like predicting the effect of an action without actually taking that action or answer causal question like *"Does that drug cause recovery?"* This is particularly useful for examples where we can't apply experiments and **randomized controlled trial (RCT)**,[5] a classical approach to **causal inference**. In addition to this, causal inference models can predict facts that didn't happen. They can compare the observed world to a counterfactual one, something that experiments and data alone can not achieve. This capability is essential

---

[4] https://arxiv.org/abs/cs/0511042

[5] RCT typically randomly allocate subjects to 2 groups, treat them differently (like a treatment group vs placebo), and then compare them with respect to a measured outcome like recovery. The reason of randomization is to remove possible effects from confounders, dissociates the variable of interest (drug, no drug) from other variables that would otherwise affect them both, like the age or a healthy life.

to answer the *"Why"* questions: *"What is factor X that caused Y?"* Imagine that Sarah started a new fitness routine and lost 10 pounds a month later. We are interested in determining whether her new routine might have caused her weight loss. To answer this question, we need to envision a scenario where Sarah was about to start the exercise but changed her mind. Would she have still lost 10 pounds? Classical statistics do not provide a framework for posing this counterfactual question. Causal inference not only provides a formal notation for such questions but also offers a method for finding a solution.

Judea Pearl[6] is a prominent figure in the field of AI, particularly known in the field of **probabilistic reasoning and causal theory**. He created and developed a general theory of causation with a mathematical language to formally express causal questions and handle causes and effects. His work on causality has significantly advanced the understanding of cause-and-effect relationships in various fields, including philosophy, psychology, and computer science. In recognition of his groundbreaking work on the theory of probability and causal inference, he was awarded the Turing Award, one of the most prestigious awards in computer science, in 2011.

Without going into the details, here are some fundamental ideas. Simply observing $P(y|x)$, as in classical statistics, is what Judea Pearl sees as the **first level** in his causal hierarchy, the level of **association**, because it purely invokes statistical relations (Chapter 6). Leveraging graph models and **causal diagrams** to make it operational, his first key idea is about an **intervention** formalized by introducing an explicit operator "do" in the form of $P(y|do(x), z)$ for predicting the effect of a deliberate action X[7] set to x, observe event $Z = z$, and learn the effect y of that intervention. Such expressions can be estimated by RCT or analytically by using causal Bayesian network. Judea Pearl calls it the **second level** in his causal hierarchy. Building on that second intervention layer, the next key concept is to handle **counterfactuals**, a world that cannot be seen in the data because it contradicts what is seen in the data. This counterfactual layer sits at the top of Judea Pearl's hierarchy to address counterfactual questions such as *"What if Sarah didn't exercise?"* to continue with the above example. The formalization is then expressed as follows with a subscript: *"*$P(Y = y_{X=x}|X = x', Y = y')$, which stands for *"The probability that event y would be observed had X been x, given that we actually observed X to be x' and Y to be y'."* For example, the probability that Sarah wouldn't have lost 10 pounds had she not taken that fitness, given that she lost 10 pounds and took that exercise, all the rest being the same.

Judea Pearl's fundamental ideas are well summarized as follows: *"The main point is this: While probabilities encode our beliefs about a (uncertain) static world, causality tells us whether and how probabilities change when the world changes, be it by intervention or by act of imagination. The Causal Revolution has been to explain how to predict the effects of an intervention without actually enacting it and predict facts that didn't happen, both*

---

6 Judea Pearl – A. M. Turing Award Laureate (acm.org).

7 X being the control variable.

things that ML can't do". Excerpt from *"The Book of Why: The New Science of Cause and Effect"*, Judea Pearl and Dana Mackenzie, 2018.

11.2.3 Generative AI

Generative AI (GenAI) techniques are based on the techniques covered in Chapter 8. They are poised to significantly impact the way people work across all industries and learn. To explain GenAI and illustrate its power in practice, the paragraph that follows has been written based on questions entirely answered by using ChatGPT, a conversational agent launched by OpenAI in November 2022. At the time of testing it in February 2023, ChatGPT used GPT-3.5, a **fine-tuned version** of GPT-3[8] to be a general purpose chatbot. GTP-4 version was launched the month after, in March 2023.

Question	Answers captured in February 2023 from the public ChatGPT tool, which was based on GPT-3.5 at the time.
Who are you ?	I am ChatGPT, a language model developed by OpenAI. My purpose is to engage in natural language conversations with humans and provide helpful responses to their questions and prompts. I use a combination of machine learning algorithms and a large dataset of text to generate responses that are coherent and contextually relevant to the input I receive. I do not have a physical form or consciousness; I am a program that operates in a computer system. My creators designed me to be a helpful tool for people seeking information or assistance with various topics.
What is generative AI?	Generative AI refers to a type of artificial intelligence that is designed to generate new, original content rather than simply responding to inputs or executing predefined actions. Generative AI models are trained on large datasets and can generate text, images, music, or other types of content that resemble human-created content. These models are often based on deep learning techniques such as neural networks, which enable them to learn patterns in data and use those patterns to generate new content. Some examples of generative AI models include language models like GPT-3, which can generate text in a wide range of styles and genres, and generative adversarial networks (GANs), which can generate realistic images, videos, and sound.
How old is the data you've been trained on?	The data I've been trained on comes from a variety of sources and covers a wide range of topics. The specific datasets used to train me were collected between 2015 and 2021, with the most recent data being used to train me being up to the knowledge cutoff date of September 2021. It's worth noting that the training data used to create me is constantly updated and refreshed by my creators at OpenAI to ensure that I stay up-to-date with the latest developments in language and knowledge.

---

[8] Up to 175 billion parameters.

Question	Answers captured in February 2023 from the public ChatGPT tool, which was based on GPT-3.5 at the time.
What can generative AI do in practice?	Generative AI can be used to create a wide range of outputs, depending on the type of model and the training data used. Here are some practical applications of generative AI: – Text generation: Generative language models like GPT-3 can be used to generate text in a variety of styles and genres. This can be useful for content creation, chatbots, language translation, and more. – Image generation: Generative adversarial networks (GANs) can be used to generate realistic images, which can be useful in fields like fashion design, architecture, and art. – Music generation: Generative AI models can be trained to generate music in various styles and genres, which can be useful in music composition, sound design, and other creative applications. – Video generation: GANs can also be used to generate realistic videos, which can be used in fields like film and animation. – Data augmentation: Generative AI can be used to generate synthetic data that can be used to augment existing datasets for training machine learning models. – Drug discovery: Generative AI can be used to generate new molecules and predict their properties, which can be useful in drug discovery. Overall, generative AI has the potential to revolutionize many fields and enable new applications that were previously impossible. However, as with any powerful technology, there are also concerns about its potential misuse and unintended consequences.
What's the impact of generative AI on the industry?	Generative AI has the potential to impact various industries in significant ways. Here are some potential impacts: – Creative industries: Generative AI can help automate and augment creative processes, such as music composition, art, and writing. For example, AI-generated music and artwork have already gained popularity, and AI-assisted content creation could become more common in the future. – Content creation: Generative AI can help generate content at scale, such as product descriptions or news articles, which could reduce the need for human writers in some cases. – Personalization: Generative AI can help personalize content for individuals, such as personalized news feeds, advertisements, and music playlists. – Customer service: Generative AI can improve customer service by automating responses to common queries and enabling more efficient communication. – Healthcare: Generative AI can help accelerate drug discovery by generating and testing new molecules, which could lead to faster and more cost-effective drug development. – Financial services: Generative AI can help automate financial analysis and risk management, leading to more accurate predictions and better-informed decision-making. However, the widespread adoption of generative AI could also lead to job displacement and ethical concerns, such as the potential misuse of the technology for propaganda, disinformation, and surveillance. As with any disruptive technology, it is important to consider its potential impacts and to develop strategies to mitigate negative consequences.

NLP language models can produce high-quality text and images and show impressive results. Recent advancements in **multimodality**, have recently been substantial, particularly in the development of multimodal large language models, supporting multimodal inputs or outputs (processing text, images, audio, video). Multimodality represents a significant step forward in the evolution of NLP, enabling models to better understand and interact with a more complex, multi-faceted world. For instance, a multimodal model can analyze an image along with its accompanying text to get a better context, something that a text-only model would miss. This makes multimodal models more versatile and capable of handling real-world applications more effectively.

With their fundamentally **stochastic** approach, there is also much research ongoing to better understand and address the phenomenon of **hallucinations** and how far **large language models** can go for **reasoning** (see chapter 2, limitations of natural language processing). An interesting debate within the scientific community revolves around the question of whether large language models are merely **stochastic parrots** that mimic statistical patterns in their training data, or if they're capable of more complex understanding and original thought. See for example the article of VentureBeat titled *"With GPT-4, dangers of 'Stochastic Parrots' remain, say researchers. No wonder OpenAI chief executive officer (CEO) is a 'bit scared'."*[9] Finally, **small language models** are another promising direction for the future of language models. These models are significantly smaller (typically a few billion parameters) than large language models like GPT-3 or GPT-4, which have hundreds of billions to trillions of parameters, but aim to achieve performance on par with the larger models. Despite their smaller size, some models, like the **Phi-2 model**[10] released by Microsoft Research end 2023, already demonstrate remarkable performance on a variety of benchmarks. Even though it shouldn't be forgotten that those models reflect the **bias** of their training data, all the aspects mentioned above are important areas of continuous research that keep advancing the AI abilities at an extremely fast rate.

11.3 Final word

We have seen that there are many AI-related techniques that can help humans in various tasks. These can be roughly classified into 2 main and complementary AI paradigms: the symbolic approaches and the statistical, data-driven, approaches. Both approaches have their strengths and weaknesses but they also have synergy, meaning that they can sometimes be combined for more complex applications.

---

9 https://venturebeat.com/ai/with-gpt-4-dangers-of-stochastic-parrots-remain-say-researchers-no-wonder-openai-ceo-is-a-bit-scared-the-ai-beat/

10 https://www.microsoft.com/en-us/research/blog/phi-2-the-surprising-power-of-small-language-models/

Due to the rapidly evolving nature of developments and the book writing process, the information provided may not be the most up to date. Nevertheless, the knowledge acquired should empower the reader with concepts that facilitate an understanding of the broad domain of AI, its different fields and the more recent advancements.

The aim of this introductory book was to clarify some of the main concepts that underpin AI applications today, by combining basic theoretical foundations with industry examples. The authors and reviewers of this book accepted the challenge to provide a rather comprehensive view of the AI world, while trying to balance the depth of the formal academic methods and a higher-level language that can be grasped by business practitioners, all in a limited number of pages. A real challenge. We'll let the reader judge whether that challenge has been achieved!

Bibliography

Bader Sebastian and Hitzler Pascal. Dimensions of Neural-symbolic Integration—A Structured Survey, November 2005, https://arxiv.org/abs/cs/0511042.

ChatGPT, https://chat.openai.com/chat.

Mackenzie Dana. https://www.bing.com/search?q=Dana+Mackenzie&filters=ufn%3a%22Dana+Mackenzie%22+sid%3a%2289a7576b-ea0d-2dd1-b1ef-44f33474aac4%22+catguid%3a%228fd6dca4-f0de-2334-b43f-3fb86b14c729_21672fa4%22+segment%3a%22generic.carousel%22+gsexp%3a%228fd6dca4-f0de-2334-b43f-3fb86b14c729_bXNvL2Jvb2sud3JpdHRlbl93b3JrLmF1dGhvcnxUcnVl%22&FORM=SNAPST.

Pearl Judea. https://www.bing.com/search?q=Judea+Pearl&filters=ufn%3a%22Judea+Pearl%22+sid%3a%220818342c-daf7-0539-2386-2d777800a5cd%22+catguid%3a%228fd6dca4-f0de-2334-b43f-3fb86b14c729_21672fa4%22+segment%3a%22generic.carousel%22+gsexp%3a%228fd6dca4-f0de-2334-b43f-3fb86b14c729_bXNvL2Jvb2sud3JpdHRlbl93b3JrLmF1dGhvcnxUcnVl%22&FORM=SNAPST.

Pearl Judea – A. M. Turing Award Laureate (acm.org), https://amturing.acm.org/award_winners/pearl_2658896.cfm.

Pearl Judea and Mackenzie Dana. The Book of Why: The New Science of Cause and Effect, 2018.

Yu Dongran, Yang Bo, Liu Dayou, Wang Hui, and Pan Shirui. A Survey on Neural-symbolic Learning Systems, June 2023, https://arxiv.org/pdf/2111.08164.pdf.

Trademarks

The citation of registered names, trade names, trademarks, etc. in this work does not imply, even in the absence of a specific statement, that such names are exempt from laws and regulations protecting trademarks, etc. and, therefore, free for general use. Any liability to the content of websites cited within this book is disclaimed. Product names and registered trademarks used in this book:

The following product or service names	are a registered trademark of
AlphaGo	DeepMind Technologies, Limited
Amazon Alexa	Amazon Technologies, Inc.
Apple Siri	Apple Inc.
ChatGPT	OpenAI
DISQOVER	ONTOFORCE N.V.
Google Assistant	Google LLC
Google Now	Google LLC
Microsoft Azure	Microsoft Corporation
Microsoft Azure AI Document Intelligence	Microsoft Corporation
Microsoft Azure Form Recognizer	Microsoft Corporation
Microsoft Bing Maps	Microsoft Corporation
Microsoft Copilot	Microsoft Corporation
Microsoft Cortana	Microsoft Corporation
Microsoft Dynamics 365	Microsoft Corporation
Microsoft Excel	Microsoft Corporation
Microsoft Flash Fill	Microsoft Corporation
Microsoft MSN	Microsoft Corporation
Microsoft News	Microsoft Corporation
Microsoft Office 365	Microsoft Corporation
Microsoft QnA Maker	Microsoft Corporation
Microsoft Sentinel	Microsoft Corporation
Microsoft TrueSkill	Microsoft Corporation
Microsoft Xbox, Microsoft Xbox Live	Microsoft Corporation
Wikipedia	Wikimedia Foundation, Inc.

The following company brands	are a registered trademark of
Accenture	Accenture Global Services Limited, part of Accenture plc
Alibaba	Alibaba Group Holding Limited
Amazon	Amazon Technologies, Inc.
AMD	Advanced Micro Devices, Inc.
Arcelor	ArcelorMittal S.A.
ASML	ASML Netherlands B.V., part of ASML Holding N.V.
Axa Insurance	Axa S.A.
Daimler	Mercedes-Benz Group AG
Deloitte	Deloitte Touche Tohmatsu Limited
EY	Ernst & Young Global Limited
Google	Google LLC
Icertis	Icertis, Inc.
IDC	International Data Group, Inc.
Imandra	Aesthetic Integration Limited
KBC	KBC Group N.V.
Legal Village	Axa Belgium
Maersk Container Industry A/S	A.P. Møller - Mærsk A/S
McKinsey	McKinsey & Company
Meta	Meta Platforms, formerly Facebook, Inc.
Microsoft	Microsoft Corporation
Netflix	Netflix, Inc.
O'Reilly	O'Reilly Media, Inc.
ONTOFORCE	ONTOFORCE N.V.
OpenAI	OpenAI, Inc.
Pepsi	PepsiCo Inc.
PricewaterhouseCoopers, PwC	PricewaterhouseCoopers International Limited
Pythagoria	PythAgoria S.A.R.L.
Robovision	Robovision N.V.
Sioen Textiles	Sioen Industries N.V.
Spotify	Spotify AB, Spotify Technology S.A.
Statista	Statista GmbH
Veranneman Technical Textiles	Sioen Industries N.V.
Viu More	Viu More N.V.

Index